AF352974

Induced Impairment of Neurogenesis and Brain Diseases

Induced Impairment of Neurogenesis and Brain Diseases

Editor

FengRu Tang

MDPI • Basel • Beijing • Wuhan • Barcelona • Belgrade • Manchester • Tokyo • Cluj • Tianjin

Editor
FengRu Tang
Radiation Physiology Lab
Singapore Nuclear Research
and Safety Initiative
National University
of Singapore
Singapore

Editorial Office
MDPI
St. Alban-Anlage 66
4052 Basel, Switzerland

This is a reprint of articles from the Special Issue published online in the open access journal *Cells* (ISSN 2073-4409) (available at: www.mdpi.com/journal/cells/special_issues/ Impairment-Neurogenesis).

For citation purposes, cite each article independently as indicated on the article page online and as indicated below:

LastName, A.A.; LastName, B.B.; LastName, C.C. Article Title. *Journal Name* **Year**, *Volume Number*, Page Range.

ISBN 978-3-0365-6214-8 (Hbk)
ISBN 978-3-0365-6213-1 (PDF)

Contents

Preface to "Induced Impairment of Neurogenesis and Brain Diseases"

The impairment of neurogenesis may be induced after pre- and post-natal chemical and biological toxin, alcohol, or radiation exposure, drug treatment, hormone imbalances, stress, pain, hypoxia, brain trauma, malnutrition, and aging. It also occurs as a result of genetic disorders such as Down syndrome (DS), autism, fragile X syndrome (FXS), neurological disorders including Alzheimer's disease (AD), Parkinson's disease (PD), epilepsy, and Huntington's disease (HD), and neuropsychiatric disorders such as depression and schizophrenia. However, the causal relationship between the impairment of neurogenesis and neurological and neuropsychiatric disorders remains unknown. In this Special Issue entitled "Induced Impairment of Neurogenesis and Brain Diseases", original animal or cell experimental research and review papers were combined to discuss different causes of the impairment of neurogenesis, relevant neurobehavioral changes, molecular mechanisms, and therapeutic approaches. The aim is to update researchers and clinicians about the complexity of the development of the impairment of neurogenesis, the importance of the involvement of the impairment of neurogenesis in neurological and neuropsychiatric disorders, and to provide some clues for designing novel therapeutic approaches by targeting the impairment of neurogenesis to effectively prevent or treat different genetic, neurological, and neuropsychological disorders.

In a mouse radiation exposure model, Wang et al. revealed that early-life (postnatal day 3) irradiation induced hypoplasia and the impairment of neurogenesis in the dentate gyrus and adult depression, which were mediated by the microRNA-34a-5p/T-cell intracytoplasmic antigen-1 pathway [1]. In their review, Wang et al. described the effects of irradiation on the aging of different types of brain cells, including neurons, microglia, astrocytes, and cerebral endothelial cells, and the relevant molecular mechanisms, and highlighted how radiation-induced senescence in different cell types might lead to the brain aging and the subsequent development of various neurological and neuropsychological disorders [2]. Boyd et al. reviewed the control of neuroinflammation through radiation-induced microglial changes, and summarized the impacts of ionizing radiation on healthy brains by altering microglial function states, and low-dose ionizing radiation on neurodegenerative diseases [2].

In mutant animal models, Wang et al. investigated the biological functions of MCPH1's central domain, by constructing a mouse model which lacked the central domain of MCPH1 by deleting its exon 8 (designated as Mcph1-Δe8). Mcph1-Δe8 mice exhibited a reduced brain size and thinner cortex, likely caused by a compromised self-renewal capacity and the premature differentiation of Mcph1-Δe8 neuroprogenitors during corticogenesis [4]. Chithanathan et al. demonstrated the cell-specific function of miR-146b in the neuronal and astroglial organization of the mouse brain. In miR146b-/- mice, there was a higher density of neurons in the frontal cortex and enhanced hippocampal neurogenesis. No microglial activation or neuroinflammation was observed in miR146b-/- mice. At molecular level, they demonstrated that miR-146b deficiency was associated with the elevated expression of glial-cell-line-derived neurotrophic factor (Gdnf) mRNA, which might be related to hippocampal neurogenesis [5]. Sun et al. found that TRIM32 deficiency impaired the generation of pyramidal neurons in the developing cerebral cortex, leading to a smaller brain size. Reduced Notch signaling may be involved in the TRIM32-deficiency-induced impairment of pyramidal neurogenesis [6].

Neurogenesis is also regulated epigenetically. Blanco-Luquin et al. reported that Aβ-treated human neural progenitor cells (NPCs) exhibited transient decreases in mRNA expression for SEPT5-GP1BB and NXN and an increase in DNA methylation for NXN, suggesting that NXN gene epigenetic changes may be involved in the impairment of hippocampal neurogenesis in Alzheimer's disease [7]. Maria Guerra et al. reported that SMG6 regulated cell fate in a cell-type-specific manner and was more important for neuroprogenitors originating from the ganglionic eminence (GE) than for progenitors from the cortex. N6-methyladenosine (m6A), the most abundant modification in messenger RNAs (mRNAs), was deposited by methyltransferases ("writers") Mettl3 and Mettl14 and erased by demethylases ("erasers") Fto and Alkbh5. m6A can be recognized by m6A-binding proteins ("readers"), such as Yth domain family proteins (Ythdfs) and Yth domain-containing protein 1 (Ythdc1) [8]. Shu et al. summarized the current knowledge about the roles of m6A machinery, including writers, erasers, and readers, in regulating gene expression, and the function of m6A in neurodevelopment and neurodegeneration; perspectives for studying m6A methylation were also provided [9].

Ghrelin [10] and TGF-β/Smad signaling [11] are also involved in neurogenesis. In cultured cerebral cortex neurons, cerebellar granule neurons, and organotypic cerebral cortex slices from rat brains to hypoxia, Stoyanova et al. found that ghrelin stimulates neurogenic factors for the protection of neurons in a GHSR1-dependent manner in non-neurogenic brain areas such as the cerebral cortex after exposure to hypoxia [10]. Hiew et al. reviewed TGF-β/Smad signaling in neurogenesis and its implications for neuropsychiatric diseases, and suggested that TGF-β/Smad signaling was an important regulator of stress response and was implicated in the behavioral outcomes of mood disorders [11].

The influence of the gut microbiota on neurogenesis and the possible underlying mechanisms were reviewed by Sarubbo et al., and the potential implications of the emerging knowledge for the fight against neurodegeneration and brain aging through the gut microbiota were also provided [12]. Yow et al. systematically reviewed the therapeutic potential of complementary and alternative medicines in peripheral nerve regeneration [13].

Funding: This research was supported by the National Research Foundation of Singapore.

Acknowledgments: I would like to thank all the authors for their contribution to this Special Issue. I am grateful for all of the support that I have received from the *Cells* editorial staff.

Conflicts of Interest: The authors declare no conflicts of interest.

References

1. Wang, H., Ma, Z.W., Shen, H.Y., Wu, Z.J., Liu, L., Ren, B., Wong, P.Y., Sethi, G., Tang, F. Early life irradiation-induced hypoplasia and impairment of neurogenesis in the dentate gyrus and adult depression are mediated by microrna-34a-5p/t-cell intracytoplasmic antigen-1 pathway. Cells 2021, 10(9). doi:10.3390/cells10092476.

2. Wang, Q. Q., Yin, G., Huang, J.R., Xi, S. J., Qian, F., Lee, R. X., Peng, X.C., Tang, F. R. Ionizing radiation-induced brain cell aging and the potential underlying molecular mechanisms. Cells 2021, 10(12), doi:10.3390/cells10123570.

3. Boyd, A., Byrne, S., Middleton, R.J., Banati, R.B., Liu G.J. Control of Neuroinflammation through Radiation-Induced Microglial Changes. Cells 2021, 10(9), 2381; doi:10.3390/cells10092381.

4. Wang, Y.R., Zong, W., Sun, W.L., Chen, C.Y., Wang, Z.Q., Li, T.L., The Central Domain of MCPH1 Controls Development of the Cerebral Cortex and Gonads in Mice. Cells 2022, 11(17), 2715; doi:10.3390/cells11172715.

5.   Chithanathan, K., Somelar, K., Jürgenson, M., Žarkovskaja, T., Periyasamy, K., Yan, L., Nathaniel Magilnick, N., Boldin, M.P., Rebane, A., Tian, L., and Zharkovsky, A. Enhanced Cognition and Neurogenesis in miR-146b Deficient Mice. Cells 2022, 11(13), 2002; doi:10.3390/cells11132002.

6.   Sun, Y.Y., Chen, W.J., Huang, Z.P., Yang, G., Wu, M.L., Xu, D.E., Yang, W.L., Luo, Y.C., Xiao, Z.C., Xu, R.X., Ma, Q.H. TRIM32 Deficiency Impairs the Generation of Pyramidal Neurons in Developing Cerebral Cortex. Cells 2022, 11(3), 449; doi:10.3390/cells11030449.

7.   Blanco-Luquin, I., Acha, B., Urdánoz-Casado, A., Gómez-Orte, E., Roldan, M., Pérez-Rodríguez, D.R., Juan Cabello, J., Mendioroz, M. NXN Gene Epigenetic Changes in an Adult Neurogenesis Model of Alzheimer's Disease. Cells 2022, 11(7), 1069; doi:10.3390/cells11071069.

8.   Maria Guerra, G., May, D., Kroll, T., Koch, P., Groth, M., Zhao-Qi Wang, Z.Q., Li T.L., Grigaravičius, P. Cell Type-Specific Role of RNA Nuclease SMG6 in Neurogenesis. Cells 2021, 10(12), 3365; doi:10.3390/cells10123365.

9.   Shu, L.Q., Huang, X.L., Cheng, X.J., Li, X.K. Emerging Roles of N6-Methyladenosine Modification in Neurodevelopment and Neurodegeneration.   Cells 2021, 10(10), 2694; doi:10.3390/cells10102694.

10. Stoyanova, I.I., Klymenko, A., Willms, J., Doeppner, T.R., Anton B. Tonchev, A.B., Lutz, D. Ghrelin Regulates Expression of the Transcription Factor Pax6 in Hypoxic Brain Progenitor Cells and Neurons. Cells 2022, 11(5), 782; doi:10.3390/cells11050782.

11. Hiew, L.F., Poon, C.H., You H.Z., Lim, L.W. TGF-β/Smad Signalling in Neurogenesis: Implications for Neuropsychiatric Diseases. Cells 2021, 10(6), 1382; doi:10.3390/cells10061382.

12. Sarubbo, F., Cavallucci, V., and Pani, G. The Influence of Gut Microbiota on Neurogenesis: Evidence and Hopes. Cells 2022, 11(3), 382; doi:10.3390/cells11030382.

13. Yow, Y.Y., Goh, T.K., Nyiew, K.Y., Lim, L.W., Phang, S.M., Lim, S.H., Ratnayeke, S., Wong, K.H. Therapeutic Potential of Complementary and Alternative Medicines in Peripheral Nerve Regeneration: A Systematic Review. Cells 2021, 10(9), 2194; doi:10.3390/cells10092194.

FengRu Tang

Editor

Article

Early Life Irradiation-Induced Hypoplasia and Impairment of Neurogenesis in the Dentate Gyrus and Adult Depression Are Mediated by MicroRNA-34a-5p/T-Cell Intracytoplasmic Antigen-1 Pathway

Hong Wang [1,†], Zhaowu Ma [2,†], Hongyuan Shen [1], Zijun Wu [3], Lian Liu [2], Boxu Ren [2], Peiyan Wong [4], Gautam Sethi [5,*] and Feng Ru Tang [1,*]

1   Radiation Physiology Lab, Singapore Nuclear Research and Safety Initiative, National University of Singapore, Singapore 138602, Singapore; snrwh@nus.edu.sg (H.W.); snrsh@nus.edu.sg (H.S.)
2   The School of Basic Medicine, Health Science Center, Yangtze University, 1 Nanhuan Road, Jingzhou 434023, China; zhaowu823@126.com (Z.M.); zifanqie_00@126.com (L.L.); boxuren188@163.com (B.R.)
3   Huaxi MR Research Center (HMRRC), Functional and Molecular Imaging Key Laboratory of Sichuan Province, Department of Radiology, West China Hospital, Sichuan University, Chengdu 610017, China; 2019324020159@stu.scu.edu.cn
4   Neuroscience Phenotyping Core, Department of Pharmacology, Yong Loo Lin School of Medicine, National University of Singapore, Singapore 117456, Singapore; wong_peiyan@nus.edu.sg
5   Department of Pharmacology, Yong Loo Lin School of Medicine, National University of Singapore, Singapore 117600, Singapore
*   Correspondence: phcgs@nus.edu.sg (G.S.); tangfr@gmail.com (F.T.)
†   These authors contributed equally to this work.

Citation: Wang, H.; Ma, Z.; Shen, H.; Wu, Z.; Liu, L.; Ren, B.; Wong, P.; Sethi, G.; Tang, F.R. Early Life Irradiation-Induced Hypoplasia and Impairment of Neurogenesis in the Dentate Gyrus and Adult Depression Are Mediated by MicroRNA-34a-5p/T-Cell Intracytoplasmic Antigen-1 Pathway. *Cells* **2021**, *10*, 2476. https://doi.org/10.3390/cells10092476

Academic Editor: Thorsten R. Doeppner

Received: 24 August 2021
Accepted: 15 September 2021
Published: 18 September 2021

Publisher's Note: MDPI stays neutral with regard to jurisdictional claims in published maps and institutional affiliations.

Abstract: Early life radiation exposure causes abnormal brain development, leading to adult depression. However, few studies have been conducted to explore pre- or post-natal irradiation-induced depression-related neuropathological changes. Relevant molecular mechanisms are also poorly understood. We induced adult depression by irradiation of mice at postnatal day 3 (P3) to reveal hippocampal neuropathological changes and investigate their molecular mechanism, focusing on MicroRNA (miR) and its target mRNA and protein. P3 mice were irradiated by γ-rays with 5Gy, and euthanized at 1, 7 and 120 days after irradiation. A behavioral test was conducted before the animals were euthanized at 120 days after irradiation. The animal brains were used for different studies including immunohistochemistry, CAP-miRSeq, Real-Time Quantitative Reverse Transcription PCR (qRT-PCR) and western blotting. The interaction of miR-34a-5p and its target T-cell intracytoplasmic antigen-1 (Tia1) was confirmed by luciferase reporter assay. Overexpression of Tia1 in a neural stem cell (NSC) model was used to further validate findings from the mouse model. Irradiation with 5 Gy at P3 induced depression in adult mice. Animal hippocampal pathological changes included hypoplasia of the infrapyramidal blade of the stratum granulosum, aberrant and impaired cell division, and neurogenesis in the dentate gyrus. At the molecular level, upregulation of miR-34a-5p and downregulation of Tia1 mRNA were observed in both animal and neural stem cell models. The luciferase reporter assay and gene transfection studies further confirmed a direct interaction between miR-34a-5p and Tia1. Our results indicate that the early life γ-radiation-activated miR-34a-5p/Tia1 pathway is involved in the pathogenesis of adult depression. This novel finding may provide a new therapeutic target by inhibiting the miR-34a-5p/Tia1 pathway to prevent radiation-induced pathogenesis of depression.

Keywords: γ-irradiation; depression; neurogenesis; miR-34a-5p; Tia1

1. Introduction

Depression is one of the most significant and long-term consequence of radiation exposure among the survivors of nuclear accidents or wars [1–4] and is associated with brain-structural changes including reduced dentate gyrus size and altered hippocampal volume [5–12]. The small dentate gyrus may be caused by the impairment of neurogenesis, and it supports the "neurogenesis hypothesis of depression" [13] and the "cellular plasticity hypothesis of depression" [14]. The therapeutic effect achieved by promoting neurogenesis to improve depression symptoms further supports the "neurogenesis hypothesis of depression" [11,15,16]. These neuropathological changes may be induced by pre- or post-natal radiation exposure [17–23]. While it has been well documented that oxidative stress and neuroinflammation are involved in radiation-induced brain damage, whether brain microRNA (miR) and its target gene are involved in radiation-induced hippocampal structural changes and subsequent depression remains unknown.

Recent studies suggest that brain miR changes may be involved in depression-like behavior or depressive symptoms [24–28]. For instance, miR-15b is upregulated in the medial prefrontal cortex of mice with depression-like behavior and inhibits neuronal progenitor proliferation [29]. miR-124 serves as a putative therapeutic target and a biomarker for depression [30], and the inhibition of miR-124 could reduce depression-like behavior in animals [31–33]. miR-34a was shown to be significantly up-regulated in a mouse model [34] and in patients [35] with depression. In the latter, the serum level of miR-34a-5p was positively correlated with the severity of depression [35]. It suggests that further study of brain miR-34a-5p and its target gene may be needed in order to correlate serum miR-34a-5p changes to brain changes. Establishment of the relationship between brain miR-34a-5p, its target gene and the impairment of neurogenesis or small dentate gyrus may provide new clues for understanding the mechanism of the development of depression and for novel therapeutic approaches to prevent the genesis of depression.

The formation of stress granules (SGs) is a key event in cells after exposure to physiological or environmental stressors. The suppression of SG generation may underlie the neuronal cell death observed in neurodegenerative diseases [36]. T-cell intracytoplasmic antigen-1 (Tia1), as an RNA-binding protein in brain tissues, functions as a posttranscriptional regulator of gene expression. It aggregates to form SG following cellular damage, which is strongly linked to the pathophysiology of neurodegeneration [37,38]. Evidence has demonstrated that Tia1 is a potential biomarker in the brain of a mouse model for Alzheimer's disease [39]. One study predicts that miR-34a may target Tia1 in the inhibition of myeloid-derived suppressor cell apoptosis [40]. However, the function of Tia1 on gamma irradiation-induced cellular damage has been poorly investigated.

This study aimed to examine if the acute irradiation with 5Gy at postnatal day 3 (P3) induced adult depression. The progressive neuropathological changes in the dentate gyrus, in particular, neurogenesis at 1, 7, and 120 day(s) after radiation exposure, were also investigated. We also elucidated the molecular mechanisms involved in radiation-induced neuropathological and neuropsychological changes, focusing on the miR and its targeted gene. We found that the γ-irradiation-activated miR-34a-5p/T-cell intracytoplasmic antigen-1 (Tia1) pathway in P3 mice was involved in the hypoplasia of the infrapyramidal blade of the stratum granulosum and the impairment of neurogenesis in the dentate gyrus and, therefore, participated in the pathogenesis of adult depression.

2. Materials and Methods

2.1. Radiation Exposure

Balb/c mice were purchased from InVivos Pte. Ltd. (Singapore) and housed with free access to water and food in the Comparative Medicine Facility, National University of Singapore. The experimental protocols were approved by the Institutional Animal Care and Use Committee (IACUC), National University of Singapore (IACUC number: R15-1576). Mice were exposed to 5 Gy gamma radiation (dose rate: 3.33 Gy/m) in a γ-Irradiator (BIOBEAM GM 8000, The Gamma-Service Medical GmbH, Leipzig, Germany) at postnatal day 3 (P3). Animals were euthanized at 1, 7 and 120 day(s) after irradiation and brain samples were collected for different experimental studies. For animals kept for 120 days, different neurobehavioral tests were performed before the animals were euthanized.

2.2. Behavioral Tests

2.2.1. Open Field (Locomotor) Test

Mice were allowed to explore freely for 1 h in a square open field (40 cm $\times$ 40 cm) cage. Locomotor activity in terms of total distance travelled was recorded using the TopScan Behavioural Analysing system (Cleversys, Reston, VA, USA).

2.2.2. Tail Suspension Test

Mice were suspended by their tails for 6 min [41]. Immobility time was detected by a strain gauge in the Tail Suspension Chamber (Med Associates Inc., St. Albans, VT, USA). A duration of time in which the force of the mouse's movement did not exceed a set threshold was counted as immobility time.

2.2.3. Forced Swim Test

The mice were put inside a cylinder filled with water for 6 min. The movement of the animal was recorded and analyzed using ForcedSwimScan (Cleversys, Reston, VA, USA). Floating time (during which the animal remained almost immobile and with its head above water) was used as a parameter to indicate depression-like behavior.

2.3. Immunohistochemical Staining

Animals were anaesthetized at 1, 7 and 120 day(s) post-irradiation. After perfusion with 4% paraformaldehyde, brain tissues were dissected, postfixed and then kept in 30% sucrose in 0.1 M phosphate buffer (pH: 7.4). Sagittal brain sections were cut at 40 μm and processed for Ki67, NeuN and doublecortin (DCX) immunohistochemistry in free floating sections. After treatment with 3% H_2O_2 and blocking with serum, free-floating sections were incubated with antibodies against DCX (1:500; Santa Cruz Biotechnology Inc., Santa Cruz, CA, USA), NeuN (1:500; Gene Tex, Hsinchu City, Taiwan), and Ki 67 (1:400; Gene Tex, Hsinchu City, Taiwan) overnight. The sections were then washed and incubated with respective secondary antibodies followed by avidin–biotin complex (ABC) reagent (Vector Laboratories Inc., Burlingame, CA, USA). After reaction in DAB Peroxidase Substrate (Vector Laboratories Inc., Burlingame, CA, USA), the sections were washed, mounted, counterstained and covered. The slides were examined and photographed under microscopy (Leica Microsystems GmbH, Wetzlar, Germany). The Stereologer System (Stereology Resource Center, Biosciences Inc. Tampa, FL, USA) was used to unbiasedly analyze the number of DCX and Ki67 immunopositive cells in the subgranular zone, and indicated the number/volume of the hilus of the dentate gyrus (mm^3).

2.4. RNA Extraction from the Mouse Brain

RNA extraction from brain samples was performed using the miRNeasy Mini Kit (Qiagen, Hilden, Germany). The cerebrum was homogenized, and RNA extraction was performed according to the manufacturer's instructions. RNA concentration and integrity were checked using the Nanodrop and Bioanalyzer system (Agilent Technologies, Santa Clara, CA, USA) before being subjected to miR sequencing and qPCR analysis.

2.5. Systematic miR Sequencing (miRSeq) Analysis

miRSeq was carried out using CAP-miRSeq (Molecular Genomics Pte Ltd., Singapore). The detected miRs were further compared between the control and irradiated animals and summarized using a heatmap.

2.6. Real-Time Quantitative Reverse Transcription PCR (qRT-PCR) Analysis of miR

RNA was first reversely transcribed using the miScript II RT kit (Qiagen, Hilden, Germany). The 20 µL reaction mixture, including 4 µL 5× HiSpec buffer, 2 µL 10× nucleotide mix, 2 µL reverse transcripts mix, 5 µL template RNA and 7 µL nuclease-free water, was incubated at 37 °C for 1 h followed by 95 °C for 5 min. The resulting cDNA was then diluted by adding 80 µL of nuclease-free water and stored at −80 °C.

Twenty microliters of master mix, for real time PCR, was prepared as follows: 2 µL diluted cDNA, 10 µL 2× miScript SYBR green PCR master mix, 2 µL 10× miScript universal primer and 2 µL primer for target miR, and 4 µL nuclease-free water. Samples were denatured at 95 °C for 15 min, followed by 40 cycles of denaturation at 94 °C for 15 sec, annealing at 55 °C for 30 sec, and extension at 70 °C for 30 s PCR amplification was carried out in QuantStudio 6 Real-Time PCR Systems (Thermo Fisher Scientific, Waltham, MA, USA). Fluorescence data were then collected. The average expression of miR-68 and miR-64 was used as internal control.

2.7. Predication of miR-34a-5p Targets and Luciferase Reporter Assay

Several online databases (TargetScan, miRanda, TarBase, miR2Disease, miRTarBase, miRecords, miRWalk) were used to analyze and predict the potential target genes of miR-34a-5p, one of the miRs that shows significant changes after irradiation. Tia1, one of the miR-34a-5p targets, was selected to further validate their direct interaction because down-regulation of Tia1 increases inflammatory response and chronic stress, which may prevent neurogenesis [42–45]. The luciferase reporter assay was based on the previous description [46]. The fragments of 3′UTR of mouse Tia1 containing the binding sequence of miR34a-5p were amplified by RT-PCR. The primers used were 5′-CAC GAT GGT GGA TGT TTG CC-3′ and 5′-GAT GCG GCG AGG ACT TAT CA-3′. The amplified fragments were directionally cloned into the PmeI and XhoI unique restriction enzyme sites of psiCHECK-2 plasmid (C8021, Promega Corporation, Madison, WI, USA), which are downstream of the Renilla luciferase gene. Transfection efficiency was normalized using firefly luciferase. The miR-34a-5p seed region of Tia1 was mutated using the Phusion Site-Directed Mutagenesis Kit (Thermo Fisher Scientific, Waltham, MA, USA) with primer sequences of 5′-phosphate ATT CCT TTT TTA AAA ATA AGA GGC-3′ and 5′-phosphate GTC AAT CCC TGC ATT TGT CTT TG. HEK293T cells were co-transfected with 0.2 µg psiCHECK-2 constructed with 3′UTR binding sites of miR-34a-5p and 100 nM miR-34a-5p mimic (Dharmacon, Lafayette, CO, USA) or scrambled mimic control (Dharmacon, Lafayette, CO, USA), using X-tremeGENE siRNA Transfection Reagent (Roche, Basel, Switzerland) according to the manufacturer's protocol. The Dual-Luciferase® Reporter Assay System (Promega Corporation, Madison, WI, USA) was used to measure luciferase and renilla signals 48 h after transfection.

2.8. qRT-PCR Analysis of Tia1 mRNA

The RNA was reversely transcribed using Maxima first strand cDNA synthesis kits (Thermo Fisher Scientific, Waltham, MA, USA). One microgram of RNA was added to 4 µL 5× Reaction Mix and 2 µL Maxima Enzyme Mix, and topped up to 20 µL with nuclease-free water. Tubes were incubated at 25 °C for 10 min, followed by 50 °C for 45 min and 85 °C for 5 min. The resulting cDNA was then diluted by adding 100 µL of nuclease-free water and stored in aliquots at −20 °C.

Twenty microliters of master mix, for real time PCR, were prepared as follows: 2 µL diluted cDNA, 10 µL 2 x Maxima SYBR Green qPCR Master Mix, 2 µL 10 x forward and reverse primers for target genes, and 4 µL nuclease-free water. The primers were used for Tia1: 5′-GAGAAGGGCTATTCGTTT-3′ and 5′-CCATACTGTTGTGGGTTT-3′; GAPDH: 5′-GCACCGTCAAGGCTGAGAAC-3′ and 5′-TGGTGAAGACGCCAGTGGA-3′. Samples were denatured at 95 °C for 10 min, followed by 40 cycles of: denaturation at 95 °C for 15 sec, annealing at 60 °C for 30 sec, extension at 72 °C for 30 s PCR amplification was carried out in QuantStudio 6 Real-Time PCR Systems (Thermo Fisher Scientific, Waltham,

MA, USA). Fluorescence data were then collected. The expression of GAPDH was used as the internal control.

2.9. Western Blot

The mouse cerebrum was homogenized in CelLytic MT Mammalian Tissue Lysis/Extraction Reagent (Sigma-Aldrich Corporation, St. Louis, MO, USA) containing 100 X Halt™ Protease and Phosphatase Inhibitor Cocktail (Thermo Fisher Scientific, Waltham, MA, USA). The lysate was centrifuged at $15,000 \times g$ for 15 min, and the supernatant was transferred into a clear Eppendorf tube. Pierce™ BCA Protein Assay Kit (Thermo Fisher Scientific, Waltham, MA, USA) was used to measure protein concentration.

Protein samples were separated by 10% SDS-PAGE, and then transferred to a nitrocellulose membrane. The membrane was blocked by Blotting One (Nacalai Tesque Inc., Kyoto, Japan), incubated with the respective primary antibodies (β-actin, 1:1000, Cell Signalling Technology, Beverly, MA, USA; Tia1, 1:1000, Invitrogen, Thermo Fisher Scientific, Waltham, MA, USA) overnight at 4 °C followed by HRP-conjugated secondary antibodies (1:10,000 dilution) at room temperature for 1h. The WesternBright Sirius Chemiluminescent Detection Kit (Advansta Inc, San Jose, CA, USA) was used to detect immunoreactive proteins. Membranes were then visualized and quantified using the Bio-Rad Gel Doc system. Band densities were measured by ImageJ and normalized by the respective loading control β-actin. The fold change relative to the control was calculated.

2.10. Culture of Neural Stem Cells

Mouse cortical neural stem cells (NSCs) were purchased from R&D Systems, Inc. (Minneapolis, MN, USA). Cell culture flasks or plates were pre-coated with matrigel (Gibco, Thermo Fisher Scientific, Waltham, MA, USA). Cells were grown in NeuroCult™ basal medium (STEMCELL Technologies Singapore Pte Ltd., Singapore), and supplemented with NeuroCult™ proliferation supplement (STEMCELL Technologies Singapore Pte Ltd., Singapore), epidermal growth factor (EGF) and basic fibroblast growth factor (bFGF) (Invitrogen, Thermo Fisher Scientific, Waltham, MA, USA) in a humidified incubator at 37 °C with 5% CO_2. Cells were passaged using accutase (Gibco, Thermo Fisher Scientific, Waltham, MA, USA).

2.11. RNA Isolation from NSCs and qRT-PCR Analysis for miR and mRNA

Cells were seeded in a matrigel-coated 6-well plate. RNA isolation and qRT-PCR analysis for miR and mRNA were performed as mentioned above.

2.12. Western Blot for NSCs

Cells were harvested in CelLytic MT Mammalian Tissue Lysis/Extraction Reagent (Sigma-Aldrich Corporation, St. Louis, MO, USA) containing $100 \times$ Halt™ Protease and Phosphatase Inhibitor Cocktail (Thermo Fisher Scientific, Waltham, MA, USA), and lysed by violent vortex several times before incubating on ice for 20 min on a shaker. After centrifugation at $15,000 \times g$ for 15 min, the supernatant was transferred into a clear Eppendorf tube. The protein concentration was measured and samples were separated as described above.

2.13. Overexpression of Tia1 in NSCs

Mammalian vector pCMV6-AC-GFP, containing Tia1 (NM_009383) Mouse Tagged ORF Clone (No: MG226372, OriGene Technologies, Inc., Rockville, MD, USA) or blank control pCMV6-AC-GFP with C-terminal tGFP tag (No: PS100010, OriGene Technologies, Inc., Rockville, MD, USA), was transfected into NSCs using Lipofectamine 3000 reagent (Thermo Fisher Scientific, Waltham, MA, USA) according to the manufacturer's instructions.

2.14. Statistical Analysis

Data were expressed as mean $\pm$ SEM, and $p < 0.05$ was considered as statistically significant. The Student's t-test was used to compare two sets of quantitative data. For the comparison among three or more groups, one-way analysis of variance (ANOVA) was used followed by Bonferroni-corrected pairwise post-hoc tests.

3. Results

3.1. γ-Irradiation at P3 Induced Depression-Like Behavior in Adult Mice

The open-field test showed no difference in distance travelled between non-irradiated and irradiated mice (Figure 1A). Irradiated mice displayed a significantly increased time of immobilization in the tail suspension test (Figure 1B) and the forced swim test (Figure 1C) compared to the control, suggesting that γ –irradiation-induced depression-like behavior in adult mice.

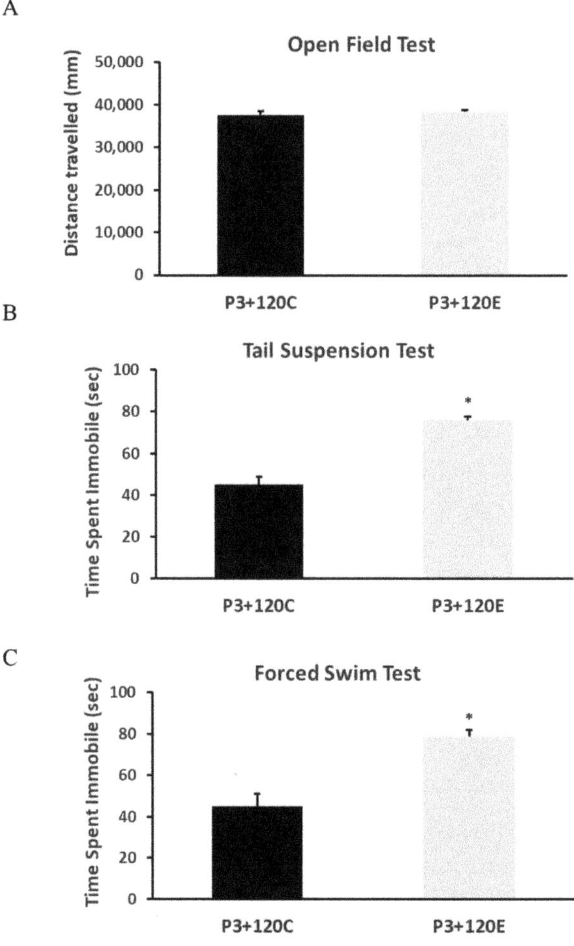

Figure 1. γ-irradiation with 5Gy at postnatal day 3 (P3) induced depression-like behavior in adult mice without locomotor activity change. (**A**) The total distance travelled in the open field (locomotor) test; (**B**) the immobile time in the tail suspension test; (**C**) the immobile time in the forced swim test. Data are expressed as mean $\pm$ SEM, and analyzed by Student's *t*-test ($n = 12$). * $p < 0.05$ vs. P3+120C.

3.2. γ-Irradiation with 5Gy Induced Hypoplasia of The Infrapyramidal Blade of The Stratum Granulosum, and Aberrant and Impaired Neurogenesis in the Subgranular Zone of the Dentate Gyrus

NeuN immunohistochemistry revealed hypoplasia of the infrapyramidal blade of the stratum granulosum of the dentate gyrus, which appeared at 1 day (Figure 2A,B) after irradiation at P3, and persisted from 7 (Figure 2C,D) to 120 days (Figure 2E,F) after irradiation. In the suprapyramidal blade of the stratum granulosum of the dentate gyrus, the loss of NeuN immunopositive neurons also occurred (Figure 2 B,D,F), particularly at 120 days after irradiation (Figure 2F), compared to the age-matched control (Figure 2A,C,E).

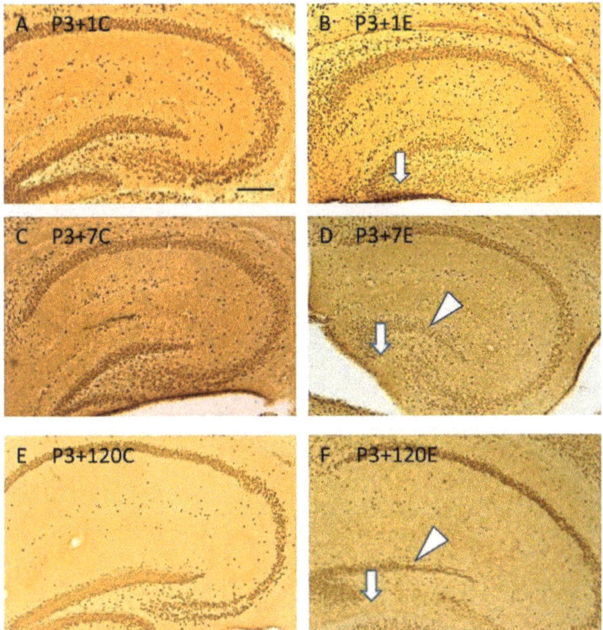

Figure 2. NeuN immunostaining shows that γ-irradiation with 5Gy at postnatal day 3 induced hypoplasia of the infrapyramidal blade of the stratum granulosum of the dentate gyrus (arrows) at 1 ((**B**) compared to (**A**) in the control), 7 ((**D**) compared to (**C**) in the control), and 120 ((**F**) compared to € in the control) days after radiation exposure. Loss of NeuN immunopositive neurons also occurred in the suprapyramidal blade of the stratum granulosum of the dentate gyrus, particularly at 7 (arrowhead in (**D**)) and 120 (arrowhead in (**D**)) days after irradiation when compared to the age-matched control mice (**C,E**) ($n = 5$). Scale bar = 200 µm in A applies to (**B–F**).

Ki67 immunohistochemistry indicated a significant reduction in Ki67 immunopositive cells in the hilus, including the subgranular zone of the dentate gyrus, which started at 1 day (Figure 3A–D,I) after irradiation of P3 mice, and persisted from 7 days (Figure 3E,F and I) to 120 days (Figure 3G–I) after irradiation exposure. Counterstaining with hematoxyin showed many apoptotic bodies in the hilus (including the subgranular zone) of the dentate gyrus at 1 day after irradiation (Figure 3D). At 120 days after irradiation, aberrant Ki67 immunopositive dividing cells were observed in the molecular layer of the dentate gyrus (Figure 3H).

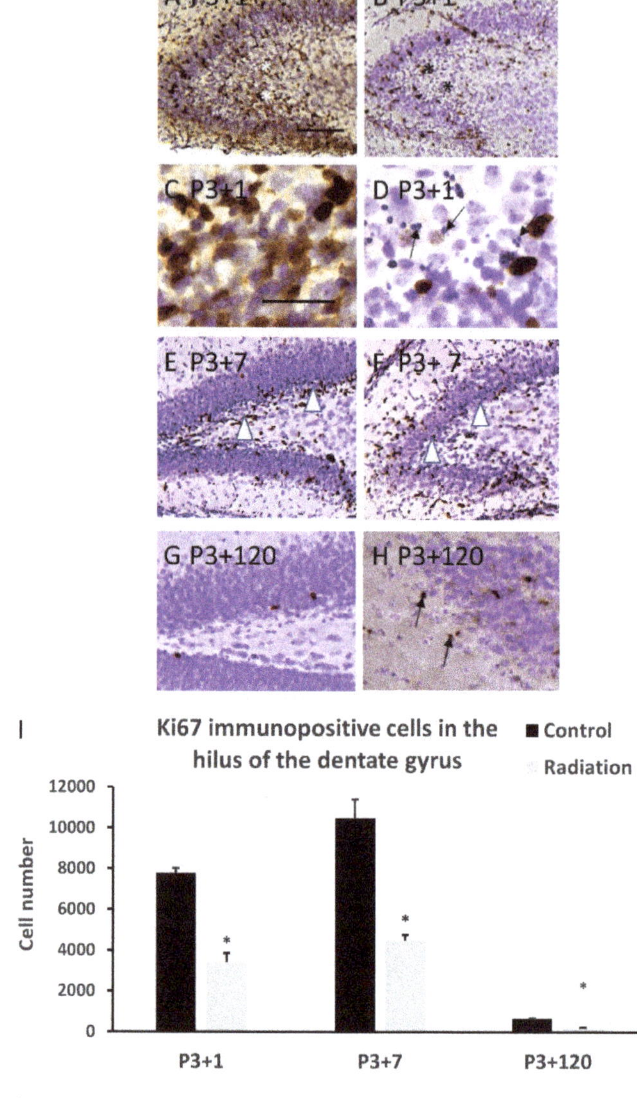

Figure 3. Ki67 immunostaining shows that γ-irradiation with 5Gy at postnatal day 3 induced a significant reduction in Ki67 immunopositive cells in the hilus of the dentate gyrus, including the subgranular zone (asterisks), at 1 (**B,D** compared to (**A**), (**C**) in the control, (**C,D**) are magnified figures from (**A,B**), respectively, (**I**)) (arrows in (**D**) indicate apoptotic bodies), 7 (**F** compared to (**E**) in the control, white arrowheads indicate Ki67 immunopositive cells, (**I**)), and 120 ((**H**) compared to (**G**) in the control, **I**) days after radiation exposure. Aberrant Ki67 immunopositive cells appeared in the molecular layer of the dentate gyrus at 120 days after radiation exposure (arrow in (**H**)) ($n = 5$). * $p < 0.05$ vs. respective control. Scale bar =100 μm in (**A**) applies to (**B,E–H**). Scale bar = 50 μm in C applies to (**D**).

DCX immunohistochemistry showed a homogenous distribution of DCX in the dentate gyrus at postnatal day 4 (P3+1) (Figure 4A) and day 10 (P3+7) (Figure 4C) in the control (Figure 4A,C) and experimental (Figure 4B,D) mice. In the experimental mice (Figure 4B,D),

DCX immunostaining in the entire dentate gyrus was weaker than in the control mice (Figure 4A,C), particularly in the subgranular zone at 7 days after irradiation (Figure 4D compared to Figure 4C). At 120 days after irradiation, the number of DCX immunopositive cells in the subgranular zone of the dentate gyrus was significantly reduced in the irradiated mice (Figure 4F,H,I) compared to the control (Figure 4E,G). Furthermore, aberrant newly generated DCX immunopositive cells were observed in the molecular layer of the dentate gyrus (Figure 4F,H).

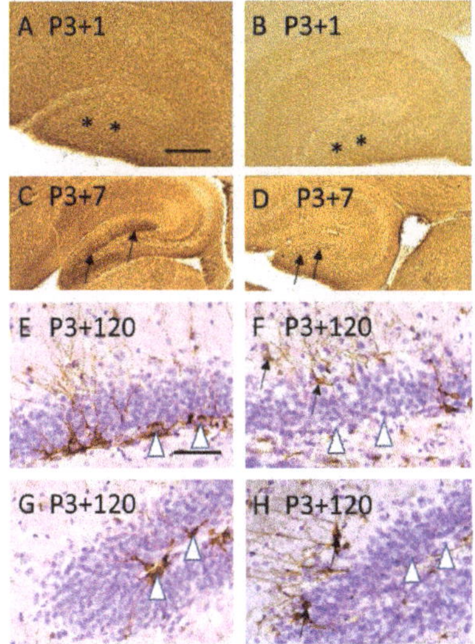

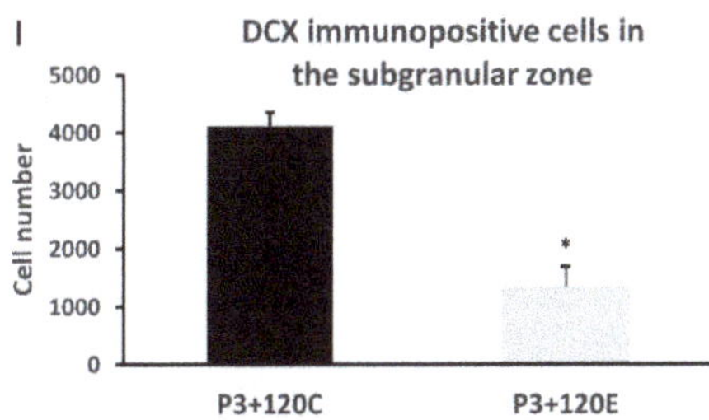

Figure 4. DCX immunostaining shows that γ-irradiation with 5Gy at postnatal day 3 induced an obvious reduction in DCX immunopositive product in the dentate gyrus at 1 day ((**B**) compared to (**A**) in the control, asterisks) and 7 days ((**D**) compared to (**C**) in the control, in particular in the subgranular zone, arrows indicate DCX immunopositive cells). Quantitative study indicates a significant reduction in DCX immunopositive neurons in the subgranular zone of the dentate gyrus at 120 days after radiation exposure (white arrowheads in (**F**), (**H**) indicate subgranular zone with fewer DCX immunopositive cells compared to those in (**E,G**) which indicate more DCX immunopositive cells. (**I**)). Aberrant DCX immunopositive neurons also appeared in the molecular layer of the dentate gyrus at 120 days after radiation exposure (arrows in (**F,H**)) ($n = 5$). * $p < 0.05$ vs P3+120C. Scale bar =200 μm in (**A**) applies to (**B–D**). Scale bar = 100 μm in E applies to (**F–H**).

3.3. Systematic miRSeq and Real Time RT-PCR

A total of 771 miRs was detected and analyzed. Seven miRs with significant changes between control and irradiated mice at P3+1 and P3+7 groups were summarized by the heatmap (Figure 5A). Statistical analysis indicated upregulation of miR-34a-5p at 1 and 7 day(s), but not 120 days after irradiation (Figure 5B). qRT-PCR further validated miRSeq and showed a significantly increased expression of miR-34a-5p in the mouse brain at 1 and 7 day(s) after irradiation (Figure 5C).

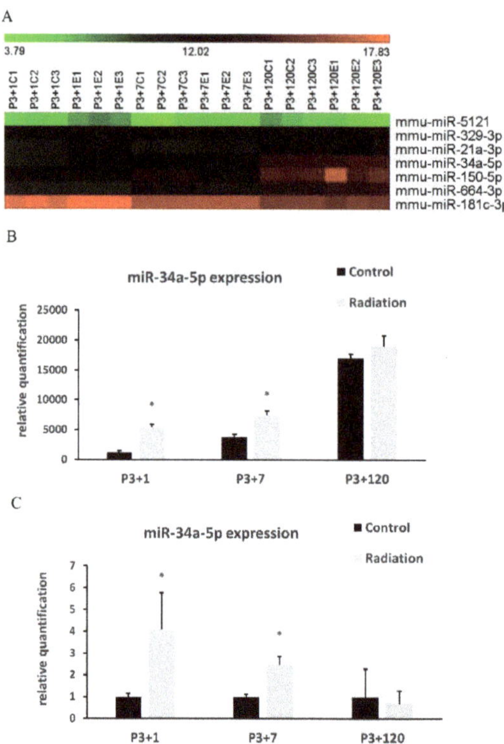

Figure 5. A heatmap displays seven differentially expressed miRNAs identified by systematic deep microRNA sequencing (miRSeq) among mice at 1, 7, 120 day(s) after irradiation at P3 and the respective control (**A**). The color bar from left to right represents the expression levels from low to high, and the number indicates the base-2 logarithm of miR expression. Among these miRNAs, the expression of miR-34a-5p is summarized in (**B**). Real-Time Quantitative Reverse Transcription PCR (qRT-PCR) validates the increased expression of miR-34a-5p at 1 and 7 day(s) after radiation exposure (**C**). Data are expressed as mean ± SEM ($n = 3$). * $p < 0.05$ vs respective control.

3.4. miR-34a-5p Targeted on mRNA of Tia1

The binding site of mouse miR-34a-5p exists in the position 269 to 275 of Tia1 3′ UTR (Figure 6A), and their direct interaction, was confirmed by luciferase reporter assay (Figure 6B).

No change in luciferase activity was observed when the miR-negative control (miR-NC) or miR-34a-5p mimic was transfected into the blank plasmid psiCHECK-2 (Figure 6B). However, when miR-34a-5p mimic was co-transfected with psiCHECK-2 containing mouse Tia1 3′ -UTR binding sequence into HEK 293 cells, the luciferase intensity was reduced significantly compared to the negative control (Figure 6B), indicating the binding of the miR-34a-5p and 3′ -UTR regions of Tia1. miR-34a-5p did not bind to the mutant Tia1 3′-UTR sites (Figure 6B), suggesting that Tia1 may function as the direct target of miR-34a-5p.

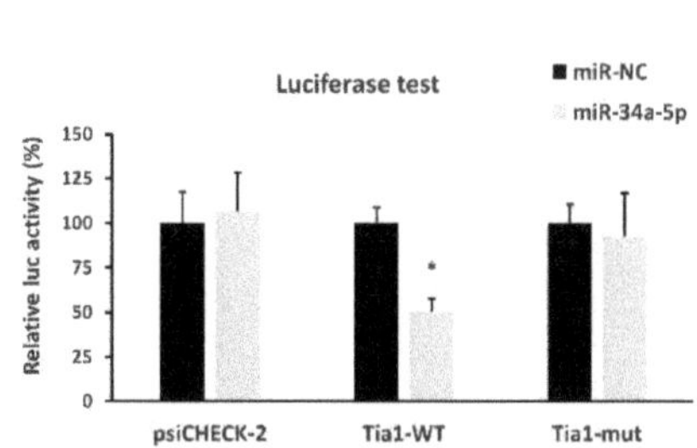

Figure 6. Direction interaction of miR-34a-5p with Tia1 examined by luciferase reporter assay. (**A**) Sequence alignment of putative miR-34a-5p binding sites in Tia1 3′ UTRs, and the mutation sequences. (**B**) Activity of luciferase gene linked to the 3′ UTR of Tia1 mRNA. HEK293T cells were co-transfected with psiCHECK-2 constructed with 3′ UTR binding sites of miR-34a-5p and miR-34a-5p mimic or scrambled mimic control. Luciferase and renilla signals were measured 48 h after transfection. Data are expressed as mean ± SEM. * $p < 0.05$.

3.5. mRNA and Protein Expression of Tia1 in Mice Brain after Γ-Irradiation with 5Gy

Tia1 mRNA decreased significantly at 1 and 120 day(s), but not at 7 days, after γ-irradiation with 5Gy compared to the respective controls (Figure 7A). A significant decrease in Tia1 protein occurred at 1 day (Figure 7B,C), but not 7 or 120 days (Figure 7B,C), after γ-irradiation.

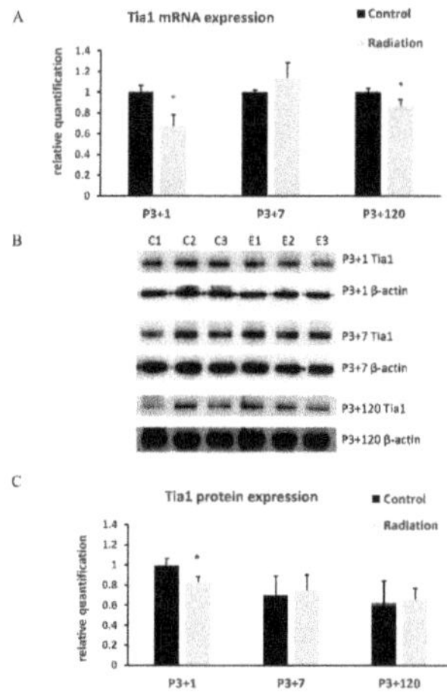

Figure 7. The expression of Tia1 in animal brains. Balb/c mice were γ-irradiated with 5Gy at P3, and brain samples were collected at 1, 7, and 120 day(s) after irradiation. (**A**) mRNA expression of Tia1 by qRT-PCR; (**B**) western blot images of Tia1; (**C**) statistical results of Tia1 protein levels. Data are expressed as mean ± SEM (n = 3). * $p < 0.05$ vs respective control.

3.6. Γ-Irradiation Increased the Expression of miR-34a-5p in NSCs and Decreased the mRNA Expression of Tia1

qRT-PCR indicated the increased expression of miR-34a-5p (Figure 8A) and decreased Tia1 miRNA (Figure 8B) in NSCs from 4 to 8h, but not at 1h after γ-irradiation with 5Gy. This negative correlation between the expression of miR and its target mRNA further confirms that Tia1 is a direct target of miR-34a-5p.

3.7. Γ-Irradiation Dose- and Time-Dependently Decreased Protein Expression of Tia1 in NSCs

NSCs were γ-irradiated with different doses from 0.2 to 5Gy. Western blot results showed that the γ-irradiation dose-dependently decreased the protein levels of Tia1 in NSCs (Figure 8C). When NSCs were γ-irradiated with 5 Gy, the reduction in Tia1 protein levels was time-dependent (Figure 8D).

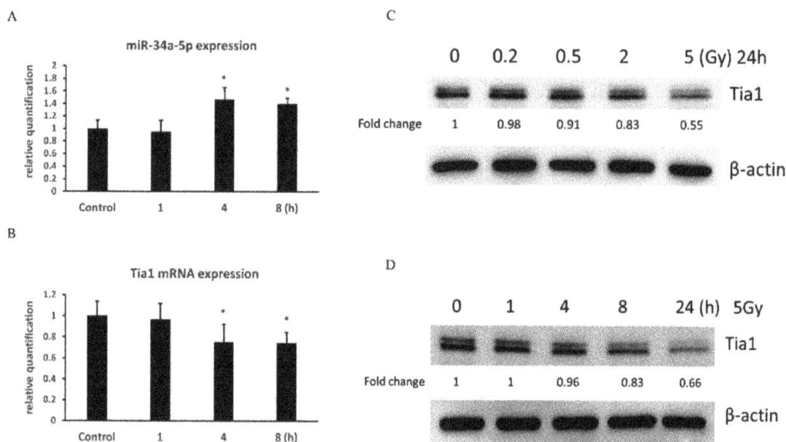

Figure 8. The expression of miR-34a-5p and Tia1 in NSCs after γ-irradiation. Time-dependent change of miR-34a-5p (**A**) and Tia1 (**B**) mRNA expression in NSCs after γ-irradiation with 5Gy. Data are expressed as mean ± SEM (n = 3). * $p < 0.05$ vs control. The protein expression of Tia1 in NSCs examined by western blot is dose- (**C**) and time-dependent (**D**) after γ-irradiation.

3.8. Overexpression of Tia1 Partially Blocked Γ-Irradiation-Induced Impairment of Neurogenesis

Γ-irradiation significantly decreased cell proliferation in NSCs transfected with either blank or Tia1 plasmid when compared with the respective non-irradiated groups (Figure 9A). The transfection of Tia1 plasmid partially blocked the loss of proliferating cells induced by γ-irradiation in NSCs (Figure 9A). The overexpression of Tia1 also enhanced cell proliferation in the non-irradiated groups as compared to NSCs transfected with blank plasmid (Figure 9A).

Western blotting showed that the transfection of Tia1 into NSCs enhanced the Tia1 protein levels compared to the blank plasmid group (Figure 9B). γ-irradiation with 5Gy reduced the Tia1 protein levels in NSCs transfected with blank plasmid, while this decrease was reversed by the transfection of Tia1 plasmid (Figure 9B), suggesting that promoting Tia1 expression in brain neurogenesis niches may be a novel therapeutic approach to prevent the radiation-induced impairment of neurogenesis.

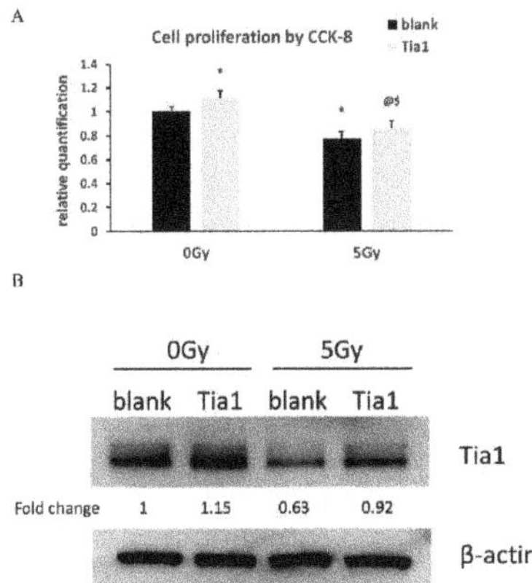

Figure 9. Effects of overexpression of Tia1 in NSCs after γ-irradiation. pCMV6-AC-GFP containing Tia1 or blank control was transfected into NSCs, which were then γ-irradiated with 5Gy or not irradiated as a control. (**A**) Cell proliferation in transfected NSCs. Data are expressed as mean ± SEM (n = 5). * $p < 0.05$ vs blank 0 Gy; @ $p < 0.05$ vs. Tia1 0 Gy; $ $p < 0.05$ vs. blank 5 Gy. (**B**) Protein expression of Tia1 in transfected NSCs examined by western blot.

4. Discussion

4.1. Radiation, Hippocampal Neuropathology and Depression

Survivors of major nuclear accidents or war may experience brain damage and depression [1,2,47,48]. In Chernobyl clean-up workers and liquidators with depression, the radiation-induced dysfunction of the cortico-limbic system in the left dominant hemisphere of the brain, with a specific involvement of the hippocampus, is considered to be the key cerebral basis of post-radiation brain damage [1,2,47,48]. Patients with depressive symptoms have smaller dentate gyrus [5]. This was supported by a recent study showing that early life adversity increased major depressive disorder (MDD) and suicide risk and could potentially affect the dentate gyrus, leading to a smaller dentate gyrus and fewer granular neurons in MDD [12]. In the present study, irradiation at P3 induced depression in adult animals. A neuropathological study showed that irradiation-induced hypoplasia of the infrapyramidal blade of the stratum granulosum of the dentate gyrus and neuronal loss occurred as early as one day after irradiation. At 120 days after irradiation, hypoplasia of the infrapyramidal blade of the stratum granulosum of the dentate gyrus persisted; meanwhile, aberrant cell division and neurogenesis in the molecular layer of the dentate gyrus were found. Our animal study strongly supports the conclusion from human data [12] and suggests that early life radiation-induced hypoplasia of the infrapyramidal blade of the stratum granulosum, as well as impairment and aberrant neurogenesis of the dentate gyrus, may be related to smaller dentate gyrus sizes and the development of depression. These pathological changes also support the neurogenesis and cellular plasticity hypotheses of depression. Radiation was shown to induce depression in mouse models [17–20]. However, previous studies on adult animals did not show hypoplasia of the infrapyramidal blade of the stratum granulosum of the dentate gyrus and aberrant cell division and neurogenesis. The pathological changes at 1 day after irradiation of P3 mice strongly suggest that therapeutic approaches to prevent early life adversity may impede the development of different neuropsychological disorders at the late stages of human life.

4.2. miR-34a-5p in Neurogenesis and Depression

Increased miR-34a was detected in the brain, blood and cerebrospinal fluids of patients or animals with depression [28]. Of the differentially expressed miRs in the present study, we chose to validate miR-34a-5p and investigate its target due to the fact that it negatively regulates neural stem cell proliferation, differentiation, neuronal migration and maturation [49–51]. Brain miR-34a-5p changes were also found to be involved in other brain insults [52]. Radiation-induced miR-34a-5p changes in the blood may be involved in the development of Alzheimer' s disease, depression and schizophrenia, suggesting that miR-34a-5p may be involved in the common brain neuropathological changes in these diseases, such as impairment of neurogenesis [28]. Altered miR-34a-5p in the cerebrospinal fluid and blood was considered as an early biomarker of major depression disorder [53] or Alzheimer's disease [54]. Plasma miR-34a-5p expression may be used to distinguish radiation exposure levels in mice [55]. Furthermore, miR-34a-5p contributes to synaptic plasticity via dis-inhibition of the translation of key plasticity-related molecules [56]. The upregulation of miR-34a-5p was revealed at 1 and 7 day(s) after irradiation at P3 in the present study. Radiation-induced upregulation of miR-34a-5p and reduced cell proliferation were further confirmed in a neural stem cell model. miR-34a is a tumor-suppressor gene and overexpression of miR-34a suppressed the expression of 136 neuronal progenitor genes [57]. The radiation-induced upregulation of brain miR-34a-5p in both animal and stem cell models may prevent brain neurogenesis, leading to hypoplasia of the infrapyramidal blade of the stratum granulosum, impairment of neurogenesis, aberrant cell division and neurogenesis in the molecular layer of the dentate gyrus, and subsequent depression.

4.3. miR-34a-5p Targets Tia1 to Prevent Neurogenesis

miR-34a targets E2F3, Numbl, platelet-derived growth factor receptor A (PDGFRA), synaptotagmin 1 (Syt1), autophagy-related 9a (Atg9a), CD44, cyclic AMP response-element binding protein (CREB), brain-derived neurotrophic factor (BDNF) to inhibit cell proliferation, migration, invasion and adhesion [58–65]. It regulates extracellular signal-regulated kinase 1/2 (ERK) and p38, and increases nuclear factor kappa light chain enhancer of activated B cells (NF-κB), Sirtuin 1 (SIRT1) and Bcl2, to promote or reduce apoptosis [66–68]. The miR-34a- A central nucleotide-binding oligomerization domain (NOD)-, C-terminal leucine rich repeat domain (*LRR*)– and caspase activation and recruitment domains (CARDs) -containing 5 (NLRC5)

-NFκB signaling pathway may be involved in HIV-1 Tat-mediated microglial inflammation [69]. miR-34a-5p suppresses tumorigenesis by targeting the Notch signalling pathway [70] or high mobility group AT-hook 2 (HMGA2) [71]. It inhibited proliferation, migration, and cell invasion, accompanied by targeting of matrix metalloproteinase 9 (MMP9) activity and the microtubule-associated protein 2 (MAP2) protein, to reduce their expression [63]. miR-34a-5p also inhibits N-methyl D-aspartate(NMDA) receptors, leading to neuroplasticity changes and Alzheimer's disease development [72]. Radiation-induced up-regulation of miR-34a-5p in the small intestine and peripheral blood down-regulated hippocampal BDNF, leading to the cognitive impairment [55]. In the present study, we chose Tia1 from many miR-34a-5p targets to confirm its gene and protein expression and validate its direct interaction with miR-34a-5p. This is because Tia1 reduction increases detrimental inflammatory responses in different types of cells and tissues such as bone [44], endometrium [42], peritoneal macrophages [45], and the central nervous system during chronic stress [43]. It controls a large network of immune system genes with modulatory roles in synaptic plasticity and long-term memory, which may be involved in stress-related psychiatric conditions [73]. Tia1 mRNA is detectable in the mouse brain during embryogenesis [74], and is linked to the pathophysiology of neurodegeneration, e.g., Alzheimer' s disease [39]. Previous studies suggested that Tia1 acts as the target of several miRs, e.g., miR-19a, miR-487a and miR-599 [75–77]. In both animal and neural stem cell models, we confirmed that irradiation reduced Tia1 gene and protein expression. In the neural stem cell models, γ-irradiation dose- and time-dependently decreased cell proliferation in NSCs, accompanied by the increase in miR-34a-5p expression as well as the reduction in Tia1 mRNA and protein expression. The luciferase reporter assay indicated a direct interaction between miR-34a-5p and Tia1. The overexpression of Tia1 in NSCs partially reversed the decrease in cell proliferation induced by γ-irradiation, strongly suggesting the involvement of Tia1 in neurogenesis. Our results, therefore, suggest that the radiation-induced interaction of miR-34a-5p and Tia1 may increase inflammatory responses of neural stem cells, leading to the impairment of neurogenesis and the subsequent hypoplasia of the infrapyramidal blade of the stratum granulosum, and aberrant cell division and neurogenesis in the molecular layer of the dentate gyrus.

5. Conclusions

The present study showed that irradiation in P3 mice induced hypoplasia of the infrapyramidal blade of the stratum granulosum, and aberrant and impaired cell division and neurogenesis in the dentate gyrus. Hypoplasia of the infrapyramidal blade of the stratum granulosum and impairment of cell division and neurogenesis in the dentate gyrus may be related to the radiation-induced upregulation of miR-34a-5p and downregulation of its target gene Tia1. Our results may provide new clues for understanding the mechanism of the development of depression and for the development of novel therapeutic approaches to the prevention or treatment of depression.

Author Contributions: Conceptualization, F.R.T.; methodology, F.R.T., H.W., Z.M., H.S.; software, F.R.T., Z.M., P.W.; validation, H.W., Z.M., H.S., Z.W. and L.L.; formal analysis, F.R.T., H.W., Z.M., P.W.; investigation, H.W., Z.M., H.S., Z.W. and L.L.; resources, F.R.T.; data curation, H.W., F.R.T.; writing—original draft preparation, H.W., F.R.T.; writing—review and editing, F.R.T., G.S., B.R.; visualization, F.R.T., H.W., Z.M.; supervision, F.R.T.; project administration, F.R.T.; funding acquisition, F.R.T., G.S. All authors have read and agreed to the published version of the manuscript.

Funding: This research was funded by National Research Foundation of Singapore, grant number Block grant.

Institutional Review Board Statement: The study was conducted according to the guidelines of the Declaration of Helsinki, and approved by the Institutional Animal Care and Use Committee (IACUC), National University of Singapore (IACUC number: R15-1576, 22/04/2016).

Informed Consent Statement: Not applicable.

Data Availability Statement: The data presented in this study are available on request from the corresponding author. The data are not publicly available due to ethical restrictions.

Acknowledgments: This study was supported by grants from the National Research Foundation of Singapore to the Singapore Nuclear Research and Safety Initiative (F.T., G.S.) and the behavioral experiments were carried out at the Neuroscience Phenotyping Core Facility, which is supported by the NMRC NUHS Centre Grant—Neuroscience Phenotyping Core (NMRC/CG/M009/2017_NUH/NUHS).

Conflicts of Interest: The authors declare no conflict of interest.

Abbreviations

ANOVA	Analysis of variance
Atg9a	Autophagy-related 9a
BDNF	Brain-derived neurotrophic factor
bFGF	Basic fibroblast growth factor
CREB	Cyclic AMP response-element binding protein
DCX	Doublecortin
EGF	Epidermal growth factor
ERK	Extracellular signal-regulated kinase $\frac{1}{2}$
HEK	Human embryonic kidney
HMGA2	High Mobility Group AT-Hook 2)
MAP2	Activity and microtubule-associated protein 2
MDD	Major depressive disorder
miR	MicroRNA

miRSeq	MicroRNA sequencing
MMP9	Matrix metalloproteinase 9
NF-κB	Nuclear factor kappa light chain enhancer of activated B
NLRC5	A central nucleotide-binding oligomerization domain (NOD)-, C-terminal leucine rich repeat domain (LRR)- and caspase activation and recruitment domains (CARDs) -containing 5
NMDA	N-methyl D-aspartate
NUMBL	NUMB Like Endocytic Adaptor Protein
NSC	Neural stem cell
PDGFRA	Platelet-derived growth factor receptor A
qRT-PCR	Real-time quantitative reverse transcription polymerase chain reaction
SEM	Standard error of the mean
SIRT1	Sirtuin 1
Syt1	Synaptotagmin 1
Tia1	T-cell intracytoplasmic antigen-1
UTR	Untranslated region

References

1. Loganovsky, K.N.; Vasilenko, Z.L. Depression and ionizing radiation. *Probl. Radiatsiinoi Medytsyny Ta Radiobiolohii* **2013**, *18*, 200–219.
2. Remennick, L.I. Immigrants from Chernobyl-affected areas in Israel: The link between health and social adjustment. *Soc. Sci. Med.* **2002**, *54*, 309–317. [CrossRef]
3. Loganovsky, K.N.; Loganovskaja, T.K.; Nechayev, S.Y.; Antipchuk, Y.Y.; Bomko, M.A. Disrupted development of the dominant hemisphere following prenatal irradiation. *J. Neuropsychiatry. Clin. Neurosci.* **2008**, *20*, 274–291. [CrossRef] [PubMed]
4. Contis, G.; Foley, T.P., Jr. Depression, suicide ideation, and thyroid tumors among ukrainian adolescents exposed as children to chernobyl radiation. *J. Clin. Med. Res.* **2015**, *7*, 332–338. [CrossRef] [PubMed]
5. Gold, S.M.; Kern, K.C.; O'Connor, M.F.; Montag, M.J.; Kim, A.; Yoo, Y.S.; Giesser, B.S.; Sicotte, N.L. Smaller cornu ammonis 2-3/dentate gyrus volumes and elevated cortisol in multiple sclerosis patients with depressive symptoms. *Biol. Psychiatry* **2010**, *68*, 553–559. [CrossRef] [PubMed]
6. Sapolsky, R.M. Why stress is bad for your brain. *Science* **1996**, *273*, 749–750. [CrossRef]
7. Satterthwaite, T.D.; Cook, P.A.; Bruce, S.E.; Conway, C.; Mikkelsen, E.; Satchell, E.; Vandekar, S.N.; Durbin, T.; Shinohara, R.T.; Sheline, Y.I. Dimensional depression severity in women with major depression and post-traumatic stress disorder correlates with fronto-amygdalar hypoconnectivty. *Mol. Psychiatry.* **2016**, *21*, 894–902. [CrossRef] [PubMed]
8. Sheline, Y.I.; Wang, P.W.; Gado, M.H.; Csernansky, J.G.; Vannier, M.W. Hippocampal atrophy in recurrent major depression. *Proc. Natl. Acad. Sci. USA* **1996**, *93*, 3908–3913. [CrossRef] [PubMed]
9. Sheline, Y.I. 3D MRI studies of neuroanatomic changes in unipolar major depression: The role of stress and medical comorbidity. *Biol. Psychiatry* **2000**, *48*, 791–800. [CrossRef]
10. Zheng, L.S.; Hitoshi, S.; Kaneko, N.; Takao, K.; Miyakawa, T.; Tanaka, Y.; Xia, H.; Kalinke, U.; Kudo, K.; Kanba, S.; et al. Mechanisms for interferon-α-induced depression and neural stem cell dysfunction. *Stem Cell Rep.* **2014**, *3*, 73–84. [CrossRef]
11. Santarelli, L.; Saxe, M.; Gross, C.; Surget, A.; Battaglia, F.; Dulawa, S.; Weisstaub, N.; Lee, J.; Duman, R.; Arancio, O.; et al. Requirement of hippocampal neurogenesis for the behavioral effects of antidepressants. *Science* **2003**, *301*, 805–809. [CrossRef]
12. Boldrini, M.; Galfalvy, H.; Dwork, A.J.; Rosoklija, G.B.; Trencevska-Ivanovska, I.; Pavlovski, G.; Hen, R.; Arango, V.; Mann, J.J. Resilience Is Associated With Larger Dentate Gyrus, While Suicide Decedents With Major Depressive Disorder Have Fewer Granule Neurons. *Biol. Psychiatry* **2019**, *85*, 850–862. [CrossRef] [PubMed]
13. Jacobs, B.L.; van Praag, H.; Gage, F.H. Adult brain neurogenesis and psychiatry: A novel theory of depression. *Mol. Psychiatry* **2000**, *5*, 262–269. [CrossRef] [PubMed]
14. Kempermann, G.; Kronenberg, G. Depressed new neurons—Adult hippocampal neurogenesis and a cellular plasticity hypothesis of major depression. *Biol. Psychiatry* **2003**, *54*, 499–503. [CrossRef]
15. Vogel, G. Neuroscience. Depression drugs' powers may rest on new neurons. *Science* **2003**, *301*, 757. [CrossRef]
16. Warner-Schmidt, J.L.; Duman, R.S. Hippocampal neurogenesis: Opposing effects of stress and antidepressant treatment. *Hippocampus* **2006**, *16*, 239–249. [CrossRef]
17. Kang, J.; Kim, W.; Seo, H.; Kim, E.; Son, B.; Lee, S.; Park, G.; Jo, S.; Moon, C.; Youn, H.; et al. Radiation-induced overexpression of transthyretin inhibits retinol-mediated hippocampal neurogenesis. *Sci. Rep.* **2018**, *8*, 8394. [CrossRef]
18. Snyder, J.S.; Soumier, A.; Brewer, M.; Pickel, J.; Cameron, H.A. Adult hippocampal neurogenesis buffers stress responses and depressive behaviour. *Nature* **2011**, *476*, 458–461. [CrossRef]
19. Son, Y.; Yang, M.; Kim, J.S.; Kim, J.; Kim, S.H.; Kim, J.C.; Shin, T.; Wang, H.; Jo, S.K.; Jung, U.; et al. Hippocampal dysfunction during the chronic phase following a single exposure to cranial irradiation. *Exp. Neurol.* **2014**, *254*, 134–144. [CrossRef] [PubMed]
20. Son, Y.; Yang, M.; Wang, H.; Moon, C. Hippocampal dysfunctions caused by cranial irradiation: A review of the experimental evidence. *Brain Behav. Immun.* **2015**, *45*, 287–296. [CrossRef] [PubMed]

21. Tang, F.R.; Loke, W.K.; Wong, P.; Khoo, B.C. Radioprotective effect of ursolic acid in radiation-induced impairment of neurogenesis, learning and memory in adolescent BALB/c mouse. *Physiol. Behav.* **2017**, *175*, 37–46. [CrossRef]
22. Guo, Y.R.; Liu, Z.W.; Peng, S.; Duan, M.Y.; Feng, J.W.; Wang, W.F.; Xu, Y.H.; Tang, X.; Zhang, X.Z.; Ren, B.X.; et al. The Neuroprotective Effect of Amitriptyline on Radiation-Induced Impairment of Hippocampal Neurogenesis. *Dose-Response A Publ. Int. Hormesis Soc.* **2019**, *17*, 1559325819895912. [CrossRef]
23. Wang, S.W.; Ren, B.X.; Qian, F.; Luo, X.Z.; Tang, X.; Peng, X.C.; Huang, J.R.; Tang, F.R. Radioprotective effect of epimedium on neurogenesis and cognition after acute radiation exposure. *Neurosci. Res.* **2019**, *145*, 46–53. [CrossRef] [PubMed]
24. Prabu, P.; Poongothai, S.; Shanthirani, C.S.; Anjana, R.M.; Mohan, V.; Balasubramanyam, M. Altered circulatory levels of miR-128, BDNF, cortisol and shortened telomeres in patients with type 2 diabetes and depression. *Acta Diabetol.* **2020**, *57*, 799–807. [CrossRef] [PubMed]
25. Wingo, T.S.; Yang, J.; Fan, W.; Min Canon, S.; Gerasimov, E.S.; Lori, A.; Logsdon, B.; Yao, B.; Seyfried, N.T.; Lah, J.J.; et al. Brain microRNAs associated with late-life depressive symptoms are also associated with cognitive trajectory and dementia. *NPJ Genom. Med.* **2020**, *5*, 6. [CrossRef] [PubMed]
26. Zhang, H.P.; Liu, X.L.; Chen, J.J.; Cheng, K.; Bai, S.J.; Zheng, P.; Zhou, C.J.; Wang, W.; Wang, H.Y.; Zhong, L.M.; et al. Circulating microRNA 134 sheds light on the diagnosis of major depressive disorder. *Transl. Psychiatry* **2020**, *10*, 95. [CrossRef]
27. Maffioletti, E.; Cattaneo, A.; Rosso, G.; Maina, G.; Maj, C.; Gennarelli, M.; Tardito, D.; Bocchio-Chiavetto, L. Peripheral whole blood microRNA alterations in major depression and bipolar disorder. *J. Affect. Disord.* **2016**, *200*, 250–258. [CrossRef]
28. Segaran, R.C.; Chan, L.Y.; Wang, H.; Sethi, G.; Tang, F.R. Neuronal Development-Related miRNAs as Biomarkers for Alzheimer's disease, Depression, Schizophrenia and Ionizing Radiation Exposure. *Curr. Med. Chem.* **2020**. [CrossRef]
29. Guo, L.; Zhu, Z.; Wang, G.; Cui, S.; Shen, M.; Song, Z.; Wang, J.H. microRNA-15b contributes to depression-like behavior in mice by affecting synaptic protein levels and function in the nucleus accumbens. *J. Biol. Chem.* **2020**. [CrossRef] [PubMed]
30. Dwivedi, Y. microRNA-124: A putative therapeutic target and biomarker for major depression. *Expert Opin. Ther. Targets* **2017**, *21*, 653–656. [CrossRef]
31. Wang, S.S.; Mu, R.H.; Li, C.F.; Dong, S.Q.; Geng, D.; Liu, Q.; Yi, L.T. microRNA-124 targets glucocorticoid receptor and is involved in depression-like behaviors. *Prog. Neuro-Psychopharmacol. Biol. Psychiatry* **2017**, *79*, 417–425. [CrossRef]
32. Gu, Z.; Pan, J.; Chen, L. MiR-124 suppression in the prefrontal cortex reduces depression-like behavior in mice. *Biosci. Rep.* **2019**, *39*. [CrossRef]
33. Yang, W.; Liu, M.; Zhang, Q.; Zhang, J.; Chen, J.; Suo, L.; Chen, Q. Knockdown of miR-124 reduces depression-like behavior in rats by targeting CREB1 and BDNF. *Curr. Neurovascular Res.* **2020**, *17*, 196–203. [CrossRef] [PubMed]
34. Lo Iacono, L.; Ielpo, D.; Accoto, A.; Di Segni, M.; Babicola, L.; D'Addario, S.L.; Ferlazzo, F.; Pascucci, T.; Ventura, R.; Andolina, D. MicroRNA-34a Regulates the Depression-like Behavior in Mice by Modulating the Expression of Target Genes in the Dorsal Raphe. *Mol. Neurobiol.* **2020**, *57*, 823–836. [CrossRef]
35. Kuang, W.H.; Dong, Z.Q.; Tian, L.T.; Li, J. MicroRNA-451a, microRNA-34a-5p, and microRNA-221-3p as predictors of response to antidepressant treatment. *Braz. J. Med. Biol. Res. Rev. Bras. Pesqui. Med. Biol.* **2018**, *51*, e7212. [CrossRef] [PubMed]
36. Arimoto-Matsuzaki, K.; Saito, H.; Takekawa, M. TIA1 oxidation inhibits stress granule assembly and sensitizes cells to stress-induced apoptosis. *Nat. Commun.* **2016**, *7*, 10252. [CrossRef] [PubMed]
37. Apicco, D.J.; Ash, P.E.A.; Maziuk, B.; LeBlang, C.; Medalla, M.; Al Abdullatif, A.; Ferragud, A.; Botelho, E.; Ballance, H.I.; Dhawan, U.; et al. Reducing the RNA binding protein TIA1 protects against tau-mediated neurodegeneration in vivo. *Nat. Neurosci.* **2018**, *21*, 72–80. [CrossRef] [PubMed]
38. Vanderweyde, T.; Apicco, D.J.; Youmans-Kidder, K.; Ash, P.E.A.; Cook, C.; Lummertz da Rocha, E.; Jansen-West, K.; Frame, A.A.; Citro, A.; Leszyk, J.D.; et al. Interaction of tau with the RNA-Binding Protein TIA1 Regulates tau Pathophysiology and Toxicity. *Cell Rep.* **2016**, *15*, 1455–1466. [CrossRef] [PubMed]
39. Arisi, I.; D'Onofrio, M.; Brandi, R.; Felsani, A.; Capsoni, S.; Drovandi, G.; Felici, G.; Weitschek, E.; Bertolazzi, P.; Cattaneo, A. Gene expression biomarkers in the brain of a mouse model for Alzheimer's disease: Mining of microarray data by logic classification and feature selection. *J. Alzheimer's Dis. JAD* **2011**, *24*, 721–738. [CrossRef] [PubMed]
40. Huang, A.; Zhang, H.; Chen, S.; Xia, F.; Yang, Y.; Dong, F.; Sun, D.; Xiong, S.; Zhang, J. miR-34a expands myeloid-derived suppressor cells via apoptosis inhibition. *Exp. Cell Res.* **2014**, *326*, 259–266. [CrossRef]
41. Crowley, J.J.; Jones, M.D.; O'Leary, O.F.; Lucki, I. Automated tests for measuring the effects of antidepressants in mice. *Pharmacol. Biochem. Behav.* **2004**, *78*, 269–274. [CrossRef]
42. Karalok, H.M.; Aydin, E.; Saglam, O.; Torun, A.; Guzeloglu-Kayisli, O.; Lalioti, M.D.; Kristiansson, H.; Duke, C.M.; Choe, G.; Flannery, C.; et al. mRNA-binding protein TIA-1 reduces cytokine expression in human endometrial stromal cells and is down-regulated in ectopic endometrium. *J. Clin. Endocrinol. Metab.* **2014**, *99*, E2610–E2619. [CrossRef] [PubMed]
43. LeBlang, C.J.; Medalla, M.; Nicoletti, N.W.; Hays, E.C.; Zhao, J.; Shattuck, J.; Cruz, A.L.; Wolozin, B.; Luebke, J.I. Reduction of the RNA Binding Protein TIA1 Exacerbates Neuroinflammation in Tauopathy. *Front. Neurosci.* **2020**, *14*, 285. [CrossRef] [PubMed]
44. Phillips, K.; Kedersha, N.; Shen, L.; Blackshear, P.J.; Anderson, P. Arthritis suppressor genes TIA-1 and TTP dampen the expression of tumor necrosis factor alpha, cyclooxygenase 2, and inflammatory arthritis. *Proc. Natl. Acad. Sci. USA.* **2004**, *101*, 2011–2016. [CrossRef] [PubMed]
45. Piecyk, M.; Wax, S.; Beck, A.R.; Kedersha, N.; Gupta, M.; Maritim, B.; Chen, S.; Gueydan, C.; Kruys, V.; Streuli, M.; et al. TIA-1 is a translational silencer that selectively regulates the expression of TNF-alpha. *EMBO J.* **2000**, *19*, 4154–4163. [CrossRef] [PubMed]

46. Ma, Z.; Liu, T.; Huang, W.; Liu, H.; Zhang, H.M.; Li, Q.; Chen, Z.; Guo, A.Y. MicroRNA regulatory pathway analysis identifies miR-142-5p as a negative regulator of TGF-beta pathway via targeting SMAD3. *Oncotarget* **2016**, *7*, 71504–71513. [CrossRef]
47. Zonenberg, A.; Leoniak, M.; Zarzycki, W. The effect of Chernobyl accident on the development of non malignant diseases. *Endokrynol. Pol.* **2006**, *57*, 38–44.
48. Loganovsky, K.N.; Bomko, M.O.; Abramenko, I.V.; Kuts, K.V.; Belous, N.I.; Masiuk, S.V.; Gresko, M.V.; Loganovska, T.K.; Antypchuk, K.Y.; Perchuk, I.V.; et al. Neuropsychobiological mechanisms of affective and cognitive disorders in the chornobyl clean-up workers taking into account the specific gene polymorphisms. *Probl. Radiatsiinoi Medytsyny Ta radiobiolohii* **2018**, *23*, 373–409. [CrossRef]
49. Aranha, M.M.; Santos, D.M.; Sola, S.; Steer, C.J.; Rodrigues, C.M. miR-34a regulates mouse neural stem cell differentiation. *PLoS ONE* **2011**, *6*, e21396. [CrossRef]
50. Jauhari, A.; Singh, T.; Singh, P.; Parmar, D.; Yadav, S. Regulation of miR-34 Family in Neuronal Development. *Mol. Neurobiol.* **2018**, *55*, 936–945. [CrossRef]
51. Jauhari, A.; Yadav, S. MiR-34 and MiR-200: Regulator of Cell Fate Plasticity and Neural Development. *Neuromolecular Med.* **2019**, *21*, 97–109. [CrossRef]
52. Liang, T.Y.; Lou, J.Y. Increased Expression of mir-34a-5p and Clinical Association in Acute Ischemic Stroke Patients and in a Rat Model. *Med. Sci. Monit. Int. Med. J. Exp. Clin. Res.* **2016**, *22*, 2950–2955. [CrossRef]
53. Wan, Y.; Liu, Y.; Wang, X.; Wu, J.; Liu, K.; Zhou, J.; Liu, L.; Zhang, C. Identification of differential microRNAs in cerebrospinal fluid and serum of patients with major depressive disorder. *PLoS ONE* **2015**, *10*, e0121975. [CrossRef]
54. Cosín-Tomás, M.; Antonell, A.; Lladó, A.; Alcolea, D.; Fortea, J.; Ezquerra, M.; Lleó, A.; Martí, M.J.; Pallàs, M.; Sanchez-Valle, R.; et al. Plasma miR-34a-5p and miR-545-3p as Early Biomarkers of Alzheimer's Disease: Potential and Limitations. *Mol. Neurobiol.* **2017**, *54*, 5550–5562. [CrossRef] [PubMed]
55. Cui, M.; Xiao, H.; Li, Y.; Dong, J.; Luo, D.; Li, H.; Feng, G.; Wang, H.; Fan, S. Total abdominal irradiation exposure impairs cognitive function involving miR-34a-5p/BDNF axis. *Biochim. Biophys. Acta Mol. Basis. Dis.* **2017**, *1863*, 2333–2341. [CrossRef]
56. Joilin, G.; Guévremont, D.; Ryan, B.; Claudianos, C.; Cristino, A.S.; Abraham, W.C.; Williams, J.M. Rapid regulation of microRNA following induction of long-term potentiation in vivo. *Front. Mol. Neurosci.* **2014**, *7*, 98. [CrossRef] [PubMed]
57. Chang, S.J.; Weng, S.L.; Hsieh, J.Y.; Wang, T.Y.; Chang, M.D.; Wang, H.W. MicroRNA-34a modulates genes involved in cellular motility and oxidative phosphorylation in neural precursors derived from human umbilical cord mesenchymal stem cells. *BMC Med. Genom.* **2011**, *4*, 65. [CrossRef] [PubMed]
58. Fineberg, S.K.; Datta, P.; Stein, C.S.; Davidson, B.L. MiR-34a represses Numbl in murine neural progenitor cells and antagonizes neuronal differentiation. *PLoS ONE* **2012**, *7*, e38562. [CrossRef]
59. Morgado, A.L.; Xavier, J.M.; Dionísio, P.A.; Ribeiro, M.F.; Dias, R.B.; Sebastião, A.M.; Solá, S.; Rodrigues, C.M. MicroRNA-34a Modulates Neural Stem Cell Differentiation by Regulating Expression of Synaptic and Autophagic Proteins. *Mol. Neurobiol.* **2015**, *51*, 1168–1183. [CrossRef] [PubMed]
60. Namgyal, D.; Chandan, K.; Sultan, A.; Aftab, M.; Ali, S.; Mehta, R.; El-Serehy, H.A.; Al-Misned, F.A.; Sarwat, M. Dim Light at Night Induced Neurodegeneration and Ameliorative Effect of Curcumin. *Cells* **2020**, *9*, 2093. [CrossRef] [PubMed]
61. Welch, C.; Chen, Y.; Stallings, R.L. MicroRNA-34a functions as a potential tumor suppressor by inducing apoptosis in neuroblastoma cells. *Oncogene* **2007**, *26*, 5017–5022. [CrossRef] [PubMed]
62. Xu, L.; Zheng, Y.L.; Yin, X.; Xu, S.J.; Tian, D.; Zhang, C.Y.; Wang, S.; Ma, J.Z. Excessive Treadmill Training Enhances Brain-Specific MicroRNA-34a in the Mouse Hippocampus. *Front. Mol. Neurosci.* **2020**, *13*, 7. [CrossRef] [PubMed]
63. Córdova-Rivas, S.; Fraire-Soto, I.; Mercado-Casas Torres, A.; Servín-González, L.S.; Granados-López, A.J.; López-Hernández, Y.; Reyes-Estrada, C.A.; Gutiérrez-Hernández, R.; Castañeda-Delgado, J.E.; Ramírez-Hernández, L.; et al. 5p and 3p Strands of miR-34 Family Members Have Differential Effects in Cell Proliferation, Migration, and Invasion in Cervical Cancer Cells. *Int. J. Mol. Sci.* **2019**, *20*, 545. [CrossRef]
64. Lu, H.; Hao, L.; Yang, H.; Chen, J.; Liu, J. miRNA-34a suppresses colon carcinoma proliferation and induces cell apoptosis by targeting SYT1. *Int. J. Clin. Exp. Pathol.* **2019**, *12*, 2887–2897. [PubMed]
65. Wang, B.; He, G.; Xu, G.; Wen, J.; Yu, X. miRNA-34a inhibits cell adhesion by targeting CD44 in human renal epithelial cells: Implications for renal stone disease. *Urolithiasis* **2020**, *48*, 109–116. [CrossRef] [PubMed]
66. Kiang, J.G.; Smith, J.T.; Anderson, M.N.; Elliott, T.B.; Gupta, P.; Balakathiresan, N.S.; Maheshwari, R.K.; Knollmann-Ritschel, B. Hemorrhage enhances cytokine, complement component 3, and caspase-3, and regulates microRNAs associated with intestinal damage after whole-body gamma-irradiation in combined injury. *PLoS ONE* **2017**, *12*, e0184393. [CrossRef]
67. Ji, Q.; Han, J.; Wang, L.; Liu, J.; Dong, Y.; Zhu, K.; Shi, L. MicroRNA-34a promotes apoptosis of retinal vascular endothelial cells by targeting SIRT1 in rats with diabetic retinopathy. *Cell Cycle.* **2020**, *19*, 1–11. [CrossRef]
68. Khanna, A.; Muthusamy, S.; Liang, R.; Sarojini, H.; Wang, E. Gain of survival signaling by down-regulation of three key miRNAs in brain of calorie-restricted mice. *Aging* **2011**, *3*, 223–236. [CrossRef]
69. Periyasamy, P.; Thangaraj, A.; Bendi, V.S.; Buch, S. HIV-1 Tat-mediated microglial inflammation involves a novel miRNA-34a-NLRC5-NFκB signaling axis. *Brain Behav. Immun.* **2019**, *80*, 227–237. [CrossRef]
70. Xu, H.; Zhang, Y.; Qi, L.; Ding, L.; Jiang, H.; Yu, H. NFIX Circular RNA Promotes Glioma Progression by Regulating miR-34a-5p via Notch Signaling Pathway. *Front. Mol. Neurosci.* **2018**, *11*, 225. [CrossRef]

71. Ma, S.; Fu, T.; Zhao, S.; Gao, M. MicroRNA-34a-5p suppresses tumorigenesis and progression of glioma and potentiates Temozolomide-induced cytotoxicity for glioma cells by targeting HMGA2. *Eur. J. Pharmacol.* **2019**, *852*, 42–50. [CrossRef] [PubMed]
72. Xu, Y.; Chen, P.; Wang, X.; Yao, J.; Zhuang, S. miR-34a deficiency in APP/PS1 mice promotes cognitive function by increasing synaptic plasticity via AMPA and NMDA receptors. *Neurosci. Lett.* **2018**, *670*, 94–104. [CrossRef] [PubMed]
73. Rayman, J.B.; Hijazi, J.; Li, X.; Kedersha, N.; Anderson, P.J.; Kandel, E.R. Genetic Perturbation of TIA1 Reveals a Physiological Role in Fear Memory. *Cell Rep.* **2019**, *26*, 2970–2983.e2974. [CrossRef] [PubMed]
74. Lowin, B.; French, L.; Martinou, J.C.; Tschopp, J. Expression of the CTL-associated protein TIA-1 during murine embryogenesis. *J. Immunol.* **1996**, *157*, 1448–1454. [PubMed]
75. Bi, J.W.; Zou, Y.L.; Qian, J.T.; Chen, W.B. MiR-599 serves a suppressive role in anaplastic thyroid cancer by activating the T-cell intracellular antigen. *Exp. Ther. Med.* **2019**, *18*, 2413–2420. [CrossRef]
76. Liu, Y.; Liu, R.; Yang, F.; Cheng, R.; Chen, X.; Cui, S.; Gu, Y.; Sun, W.; You, C.; Liu, Z.; et al. miR-19a promotes colorectal cancer proliferation and migration by targeting TIA1. *Mol. Cancer* **2017**, *16*, 53. [CrossRef] [PubMed]
77. Yang, X.; Wang, M.; Lin, B.; Yao, D.; Li, J.; Tang, X.; Li, S.; Liu, Y.; Xie, R.; Yu, S. miR-487a promotes progression of gastric cancer by targeting TIA1. *Biochime* **2018**, *154*, 119–126. [CrossRef] [PubMed]

Review

Ionizing Radiation-Induced Brain Cell Aging and the Potential Underlying Molecular Mechanisms

Qin-Qi Wang [1,2,†], Gang Yin [3,†], Jiang-Rong Huang [4,†], Shi-Jun Xi [1,2], Feng Qian [5], Rui-Xue Lee [6], Xiao-Chun Peng [1,2,*] and Feng-Ru Tang [6,*]

1   Laboratory of Oncology, Center for Molecular Medicine, Health Science Center, School of Basic Medicine, Yangtze University, Jingzhou 434023, China; 201971481@yangtzeu.edu.cn (Q.-Q.W.); 201971484@yangtzeu.edu.cn (S.-J.X.)
2   Health Science Center, Department of Pathophysiology, School of Basic Medicine, Yangtze University, Jingzhou 434023, China
3   Department of Neurology, Jingzhou Central Hospital, Jingzhou 434023, China; 201971487@yangtzeu.edu.cn
4   Health Science Center, Department of Integrative Medicine, School of Health Sciences, Yangtze University, Jingzhou 434023, China; Hjr@yangtzeu.edu.cn
5   Health Science Center, Department of Physiology, School of Basic Medicine, Yangtze University, Jingzhou 434023, China; qianfeng@yangtzeu.edu.cn
6   Radiation Physiology Laboratory, Singapore Nuclear Research and Safety Initiative, National University of Singapore, Singapore 138602, Singapore; snrlrx@nus.edu.sg
*   Correspondence: pengxiaochun@yangtzeu.edu.cn (X.-C.P.); tangfr@gmail.com (F.-R.T.)
†   These authors contributed equally to this work and should be considered co-first authors.

Citation: Wang, Q.-Q.; Yin, G.; Huang, J.-R.; Xi, S.-J.; Qian, F.; Lee, R.-X.; Peng, X.-C.; Tang, F.-R. Ionizing Radiation-Induced Brain Cell Aging and the Potential Underlying Molecular Mechanisms. *Cells* **2021**, *10*, 3570. https://doi.org/10.3390/cells10123570

Academic Editors: Jean Marie Billard and Cord Brakebusch

Received: 22 October 2021
Accepted: 16 December 2021
Published: 17 December 2021

Publisher's Note: MDPI stays neutral with regard to jurisdictional claims in published maps and institutional affiliations.

Abstract: Population aging is occurring rapidly worldwide, challenging the global economy and healthcare services. Brain aging is a significant contributor to various age-related neurological and neuropsychological disorders, including Alzheimer's disease and Parkinson's disease. Several extrinsic factors, such as exposure to ionizing radiation, can accelerate senescence. Multiple human and animal studies have reported that exposure to ionizing radiation can have varied effects on organ aging and lead to the prolongation or shortening of life span depending on the radiation dose or dose rate. This paper reviews the effects of radiation on the aging of different types of brain cells, including neurons, microglia, astrocytes, and cerebral endothelial cells. Further, the relevant molecular mechanisms are discussed. Overall, this review highlights how radiation-induced senescence in different cell types may lead to brain aging, which could result in the development of various neurological and neuropsychological disorders. Therefore, treatment targeting radiation-induced oxidative stress and neuroinflammation may prevent radiation-induced brain aging and the neurological and neuropsychological disorders it may cause.

Keywords: ionizing radiation; aging; brain; oxidative stress; mitochondrial dysfunction; DNA damage

1. Introduction

Global population aging is currently occurring at an unprecedented rate. There has been a demographic shift toward an older population, and this may have far-reaching consequences. Population aging is considered a crisis from a global economy and healthcare perspective [1]. In most species, the geriatric stage of life involves the impairment of adaptation and self-balancing mechanisms, leading to increased susceptibility to environmental or internal pressure, disease, and mortality [2]. In humans, aging is associated with progressive cognitive and physical impairment, as well as an increasing risk of diseases such as neurodegenerative diseases. Age-related disability and morbidity negatively affect the quality of human life, ultimately increasing the risk of mortality, leading to problems at the individual, family, and community levels [3].

Brain aging, which involves complex cellular and molecular mechanisms that ultimately lead to cognitive decline, is the primary contributor to neurodegeneration [4].

Aging causes a gradual deterioration in the brain's functional capacity, which leads to impaired learning and memory, attention deficits, reduced decision-making speed, and impaired sensory and motor incoordination [5]. The age-related deterioration of brain function occurs almost parallel to the functional deterioration of other organ systems, and the decline in performance is significantly accelerated after the age of 50 years [6]. Nonetheless, aging-related alterations in cellular integrity and molecular pathways are shared across tissues, including the brain [7]. These alterations include mitochondrial dysfunction; intracellular accumulation of oxidative damage to macromolecules; dysregulation of energy metabolism; impairments in cellular waste disposal (autophagy-lysosome and proteasome functions), adaptive stress response signals, and DNA repair; and inflammation. Further, abnormal neuronal network activity, altered Ca^{2+} processing in neurons, and reduced neurogenesis are also observed in the aging brain [8,9].

All living organisms undergo aging and are exposed to ionizing radiation (IR) throughout their lifespan. Several studies have linked IR to accelerated aging [10]. Kuzmic et al. used glp-1 sterile *Caenorhabditis elegans* to evaluate the impact of chronic gamma radiation on lifespan and confirmed that IR can accelerate aging [11]. Exposure to IR is known to cause a wide array of physiological changes. IR can lead to DNA double-strand breaks (DSBs), which cause genetic instability DNA damage and oxidative stress, leading to brain endothelial cell senescence and cell death [12,13]. Cellular senescence, an irreversible state of growth stagnation, can help us understand the relevance of aging to several other biological processes, from embryonic development to tissue repair and aging-related diseases [14,15]. High-dose exposure may cause acute radiation sickness, whereas prolonged exposure to low-dose radiation often results in chronic disorders such as neurodegenerative diseases. While the harmful effects of high-dose/dose-rate IR on human health are well-established, the effect of low-dose/dose-rate exposure is often overlooked despite its ubiquitous nature. The use of IR in medical diagnosis and cancer treatment has increased significantly. Consequently, nuclear waste from hospitals accounts for approximately 14% of the world's total annual radiation exposure [16]. Several studies have shown that long-term exposure to low-dose IR in catheterization laboratories increases the risk of cardiovascular diseases, indicating that it causes enhanced vascular aging and early atherosclerosis [17]. More than 50% of cancer patients will be treated with radiotherapy. Radiotherapy will kill the tumor tissue while also damaging the surrounding normal tissues, leading to radiation toxicity. [18]. Radiotherapy exposes both tumor tissue and surrounding healthy IR, causing DNA damage, which triggers the DNA damage response (DDR). In this reaction, ionizing radiation will cause the cell cycle to stop and cause cell damage, and then these damaged cells will automatically repair. If the DNA is fully repaired, these cells can recover as before. However, when internal and external factors affect the ability of DNA repair, senescence (i.e., permanent cell cycle arrest) or cell death (such as apoptosis or mitotic disaster) will occur. [19,20]. Furthermore, as IR is often required for obtaining high-resolution images during neuroimaging, the contribution of low-dose/dose-rate IR toward neurodegenerative diseases must be examined [21].

Radiation exposure, particularly natural radiation exposure, occurs in daily human life. Some radioactive elements in the earth's crust such as uranium (238U), potassium (40K), thorium (232T), and their radioactive decay products, e.g., radon (222Rn) and radium (226Ra) act as natural sources of radiation exposure [22]. Areas with high levels of background radiation are considered ideal for investigating the long-term effects of chronic low-dose radiation exposure in humans [23]. Some studies on high natural background radiation have been performed in Brazil, China, India, and Iran [24–26]. High-dose radiation exposure can cause cancer. In addition, it is worth noting that the high natural background radiation observed in Yangjiang, China also increases the incidence of some non-cancer diseases, such as tuberculosis, digestive diseases and cerebrovascular diseases. [27]. High natural background radiation in the environment can be considered as a type of natural pollution. It can reach the human body through both internal and external sources, and it can damage human DNA. In addition, natural background radiation also enters the ecosystem through human activities, affecting the health and quality of life of individuals residing in areas with high natural background radiation.

A long-term follow-up study of the 1986 Chernobyl disaster revealed an increased incidence of an extensive array of diseases in exposed individuals across all contaminated regions assessed. In particular, alterations to the central nervous system (CNS), resulting in radiation-induced neurocognitive dysfunction, were observed in many individuals [28]. Further, surviving Chernobyl liquidators showed signs of inflammation that could be associated with premature aging [29]. Radiation-induced immune system impairments are important contributors to the physiological changes that occur shortly after radiation exposure and have been implicated in delayed effects of radiation such as tumor development and early aging [30]. Chronic low-dose IR exposure can accelerate the aging of blood vessels, including cerebral vessels. This has been shown to correlate with age-related encephalopathy in individuals over 40 years of age, as well as with systemic atherosclerosis [31,32]. Of the 306 workers exposed to the Chernobyl nuclear accident examined in a previous study, 81% and 77% of men and women, respectively, exhibited signs of accelerated aging. In addition, those younger than 45 years of age appeared to be more susceptible to radiation-induced accelerated aging [33]. In humans, sensitivity to radiation decreases with age until an individual matures. However, this sensitivity increases in old age.

Experimental data from animals also supports the theory that IR induces aging. Brizzee observed that with increasing age, some changes occur in the cerebral cortices of Rhesus monkeys and albino rats [34]. Analyses of transcriptomic profiles from murine brains revealed that the molecular responses observed hours after full-body low-dose irradiation (100 mGy) were similar to those associated with premature cognitive decline, Alzheimer's disease, and various neuropsychiatric disorders [35,36]. The transcriptomic profiles of microglia obtained one day and one month post-irradiation were also similar to those observed during aging, pointing to the aging-enhancement effects of radiation [37]. In vitro high-dose (2–8 Gy) irradiation of primary cerebrovascular endothelial cells in rats promotes a secretory phenotype associated with aging, characterized by the upregulation of pro-inflammatory cytokines and chemokines, including IL-6, IL-1α, and MCP-1 [38]. It has been reported that IR increases cellular senescence, and senescence-associated β-galactosidase (SA-β-Gal) and senescence specific genes (p16, p12, and Bcl-2) are highly expressed in irradiated bone marrow derived macrophages [39]. These findings corroborate the in vivo evidence pointing to the potential senescence-inducing effects of radiation on the endothelial cells of cerebral blood vessels.

Altogether, it is clear that the biological effects of IR exposure, not limited to oxidative stress, chromosomal damage, apoptosis, stem-cell failure, and inflammation, all contribute to accelerated aging [40]. Furthermore, the contribution of IR exposure to the development of non-malignant conditions such as neurodegenerative diseases is also becoming evident through epidemiological studies [21], and many medical conditions have been found to be related to exposure to different types of low-dose/dose rate radiation (Table 1).

Table 1. Medical conditions caused by different sources of low-dose/dose rate IR.

Radiation Source	Effects	References
Medical radiation (radiographs, computed tomography scans)	Cardiovascular disease, premature aging, inflammation, and neurodegenerative diseases	[16,17,21]
Natural background radiation	Inflammation, immunosenescence, thyroid cancer, and childhood leukemia	[41,42] [43,44]
Nuclear disasters	"Chernobyl AIDS," CNS damage, premature aging, atherosclerosis, and senile encephalopathy	[28–30] [31,32]

In this review, we will focus on the impact of IR on brain aging, including the aging of various CNS cell types (microglia, astrocytes, cerebral endothelial cells, and neurons). Further, the relevant molecular mechanisms will be discussed, and future research directions aimed at elucidating the true impact of radiation-induced brain aging will be proposed.

2. Radiation-Induced Senescence of Different Types of Brain Cells

The understanding of how IR affects brain aging begins with an elucidation of its influence on individual CNS cell types. CNS cells are broadly classified into two categories, glial cells—including microglia, astrocytes, and oligodendrocytes—and neurons [45]. The endothelial cells of cerebral blood vessels are also closely associated with the brain; they lead to the formation of the blood–brain barrier (BBB) and are important for maintaining CNS integrity [46]. IR has been proven to cause to happen aging in all these cell types, especially microglia, astrocytes, cerebral vascular endothelial cells, and neurons (Table 2). The cumulative effects of senescence in these cells may lead to brain aging, related neurological and neuropsychological disorders, and a shortened lifespan [40].

Table 2. Radiation-induced senescence in different cell types.

Cell Types	Models	Radiation Type & Dose/Dose-Rate	Radiation-Induced Changes	Reference
Microglia	Murine microglial cells BV2 and neuronal cells HT22	3 Gy/min (Clinac iX) (X-ray) 2 Gy/min (X-ray irradiator)	SA-β-Gal, p16^{INK4a}, MMP3↑	[47]
	Primary microglia from adult male C57BI6/J mice	Single dose of 10/20 Gy at a dose rate of 3 Gy/min (Clinac iX) (X-ray)	SA-β-Gal, p16^{INK4a}↑	[47]
Astrocytes	Non-cancerous tissue from cancer patients having received cranial radiation	IR (X-Rad 320 biologic irradiator) (X-ray)	p16^{INK4a}, Hp1γ↑	[48]
	Primary human astrocytes	0.5–20 Gy (X-ray)	SA-β-Gal, p16^{INK4a}, p21, IL-1, IL-6, IL-8↑IGF-1, GFAP↓ DNA damage	[48]
Brain endothelial cells	ATCC-derived murine brain endothelial cells, bEnd.3	X-ray (20 Gy)	SA-β-Gal, p21, p16^{INK4a}, ICAM-1, PAI-1↑	[49]
Neurons	Male rats aged 8, 18 or 28 months	Whole-brain radiation with a single dose of 10 Gy (X-ray)	Greater inflammatory response; decrease in newborn neurons	[50]

2.1. Microglia

Microglia are yolk sac-derived phagocytes located in the CNS [51]. These cells are involved in immune responses and the maintenance of brain homeostasis. Microglia also respond to changes in the tissue environment by upregulating different cell surface receptors and producing a multitude of secreted factors [52]. As such, this class of glial cells has been affected in a glut of neurological diseases. Under normal physiological conditions, there is a balance between pro- and anti-inflammatory mediators in the brain [53]. However, a shift towards the pro-inflammatory state is observed during brain aging. Interestingly, this shift is also observed after radiation exposure [54]. Thus, radiation-induced neuroinflammation could be a potential contributor to the development of brain aging and cognitive impairment [55].

Although microglial activation is necessary for protection against foreign substances, beyond a certain threshold, such activation can be damaging. Activated microglia exhibit a neurotoxic phenotype and can cause neuronal damage and death. This phenotype has been implicated in various neurodegenerative diseases, radiation-induced brain injury, and brain aging [47,56]. Several features of this neurotoxic phenotype resemble those of aging. In microglia from aged mice, pro-inflammatory cytokines such as IL-6, IL-1β, and tumor necrosis factor-α (TNF-α) are upregulated [57], and telomeres appear to be shortened [58]. These factors are also upregulated in murine models of accelerated senescence [59]. In addition, microglia abnormally activated by radiation continuously produce neurotoxic cytokines, including IL-1β, TNFα, and IL-6 [60]. Moreover, these cells show elevated reactive oxygen species (ROS) levels, inducing oxidative stress and consequent DNA damage [61]. Furthermore, in vitro, chronically activated microglia exhibit multiple features of aging, leading to SA-β-Gal activity, metachromatic focus formation, and growth arrest [62].

Chronic aging is considered to be the driving force for age-related tissue dysfunction. Studies show that SA-β-Gal and p16^{INK4a}, key markers of senescence, are up-regulated in microglia treated with irradiation. Additionally, these markers continue to be expressed even one month post-irradiation. Senescent cells often secrete large amounts of cytokines and matrix metalloproteinases (MMPs) and show sustained oxidative and genotoxic damage, eventually leading to tissue impairments and aging [63].

IR leads to progressive DNA damage; moreover, the consequent induction of oxidative stress can trigger the aging of normal cells [64]. Studies have demonstrated that IR can induce senescence in microglia [65], characterized by inflammation, DDR, and metabolic changes [66]. Irradiated microglia are involved in the pathology of radiation-induced brain damage [56] and aging-related diseases [48].

2.2. Astrocytes

Astrocytes, one of the most common brain cells, were previously considered non-functional cells providing packing for brain networks. Nevertheless, recent research has demonstrated their functional roles in several processes. Previous studies have reported that microglia and astrocytes interact with each other, and astrocytes are known to participate in immune activity [67]. Astrocytes provide osmotic balance and therefore contribute to the maintenance of CNS homeostasis [68]. They also provide metabolic support to neurons [69] and help in the establishment and maintenance of the BBB [70]. Astrocytes promote neuronal communication and are involved in neurotransmitter recovery. They also help protect the brain against trauma, infections, and neurodegeneration, thus maintaining its health and function [71].

With age, the number of astrocytes expressing p16^{INK4a} and MMP3 (protease closely associated with the senescence-associated secretion phenotype) increases [72]. Primary astrocytes isolated from human brain tissues acquire senescence-related characteristics after multiple passages [48]. Incidentally, exposure to IR has also been found to induce the senescence-associated secretory phenotype (SASP) and aging in irradiated human astrocytes, likely due to excessive DNA damage accumulation [73]. In a mouse model of radiation-induced brain injury, changes in the expression of TNF-α and IL-1β mRNA and related signaling pathways have been observed in the hippocampus. It has also been observed that the release of pro-inflammatory cytokines and inhibition of hippocampal neurogenesis may be related to the activation of microglia and may play a critical role in radiation-induced brain injury [74]. Additionally, irradiated astrocytes showed an increase in the expression of the senescence-related markers p16^{INK4a} and p21, cell size, and number of multinucleated cells [75]. In contrast, they showed a decrease in cell number and the expression of glial fibrillary acidic protein (GFAP), which is also observed during aging [76]. Altogether, these findings suggest that irradiated astrocytes not only promote neuroinflammation but also contribute to radiation-induced accelerations in brain aging.

2.3. Brain Endothelial Cells

Radiation-induced senescence is also observed in brain microvascular endothelial cells [49]. Cell surface proteins on brain endothelial cells communicate with both the blood and brain. Thus, these cells are involved in signal transmission and transduction across the BBB. Exposure to IR can cause premature degeneration of endothelial cells and thinning of the cerebral blood vessels, resulting in the temporary loss of contextual learning, interruption of working memory, gradual spatial learning impairments, and an increased risk of dementia [77].

Both cell culture and live animal studies have shown that radiation promotes stress-induced progeria-like phenotypes in endothelial cells [78–80]. Studies of brain microvascular endothelial cells from rats exposed to γ-irradiation have demonstrated that cerebral vascular endothelial cells are more radiosensitive than microglia and neurons. γ-irradiation has also been observed to destroy the clone formation and proliferative abilities of cerebral vascular endothelial cells, which are essential for intracranial angiogenesis. Cerebral microvascular damage caused by γ-irradiation has also been found to promote the accelerated aging of healthy tissues and result in progressive cognitive decline in <50% of tumor patients receiving radiotherapy [81].

In addition, acute γ-irradiation leads to increased production of ROS in cerebral vascular endothelial cells, causing premature aging [82]. Concomitantly, IR may also induce the expression of p16^{INK4a}, the main driver of cell cycle arrest during cerebral vascular endothelial cell senescence, leading to permanent cell cycle arrest [83,84]. Such radiation also increases the proportion of SA-β-Gal-positive cerebral vascular endothelial cells [3]. Similar to astrocytes, cerebral vascular endothelial cells also express the IR-induced SASP, including an enhance in the production of pro-inflammatory molecules, cytokines, chemokines, growth factors, and MMPs [85]. Therefore, radiation-induced SASP may promote neuroinflammation and cause neuronal damage by altering the microenvironment of endothelial cells [86]. Recently, Remes et al. examined the epidemiology of cerebrovascular disease in long-term childhood brain tumor survivors 20 years after the end of radiotherapy. They found that the incidence of ischemic infarction, microhemorrhage, and lacunar infarction in this population was similar to or higher than that observed in the general population over 70 years of age [87]. Such clinical data supports the hypothesis that radiation triggers the accelerated aging of the cerebrovascular system.

2.4. Neurons

Neurons are the most basic structural and functional unit of the nervous system. They interact with other functional cells in the brain and influence each other. Microglial senescence has a profound effect on neuronal activity and cognition during natural aging [88]. In a study on the effects of age on the response to radiation, it was observed that although the number of immature neurons in old rats did not decrease continuously after whole-brain irradiation, the inflammatory reaction was greater than that in younger rats. Thus, this reaction may have a greater contribution to the development of radiation-induced cognitive impairments in older adults [50].

Furthermore, neurons are one of the most highly oxygenated cells and experience oxidative genomic damage after long-term exposure to endogenous ROS, a by-product of cellular respiration [88]. These cells are extremely vulnerable to the DNA damage induced by genotoxic substances such as oxidative stress and IR. When DNA damage occurs, powerful DNA repair mechanisms are activated to limit the accumulation of oxidative damage [89]. However, the accumulation of unrepaired DNA may cause aging and several neurodegenerative diseases [90]. Chronic low-dose-rate γ-irradiation can induce brain aging and reduce neuronal density [34]. In mouse studies, high-dose rate γ-irradiation was shown to reduce the activity of superoxide dismutase and increase the amounts of free radicals, which may be related to aging [91].

In mice, several hours after whole-body irradiation (100 mGy), expression-level changes in molecules and networks involved in cognitive function, advanced aging, Alzheimer's disease, and neuropsychiatric diseases are observed [35,36]. Tang et al. studied the expression of γH2AX in the brains of mice exposed to radiation on different postnatal days. They suggested that persistent radiation-induced DNA damage at 120 days and 15 months after irradiation in the early life of mice may be associated with brain aging and shortened life expectancy [92]. Radiation-induced oxidative stress and inflammation prevent neurogenesis in the subgranular zone and induce aging of the granule cell assembly, leading to cognitive impairment [93–95].

3. Effect of Radiation-Induced Brain Aging

With an increase in life expectancy owing to improvements in medical interventions, it has become crucial to understand the advantages of radiation protection among older individuals. Increased inflammation, loss of the redox balance, continued telomere wearing, decreased efficiency of the DDR, mitochondrial dysfunction and autophagy are all changes that occur during aging and can negatively impact genome integrity (Figure 1). Further, as radiation can exacerbate these changes, it is important to understand the mechanisms involved in the effects of radiation exposure and brain aging. Radiation-induced ROS can directly cause mitochondrial respiratory chain breakage and water molecular decomposition inducing respiratory chain dysfunction and reduced antioxidant capacity. NADPH oxidase is a family of multi-subunit complex enzymes that activate the conversion of oxygen to superoxide anions (O_2^-) with NADPH as the electron source, and exists in vascular endothelial cells. In addition, cyclooxygenases-2 (COX-2) and 5-lipoxygenase (5-LPO) catalyze the production of prostaglandin H2 (PGH2) and the formation of ROS during arachidonic acid metabolism.

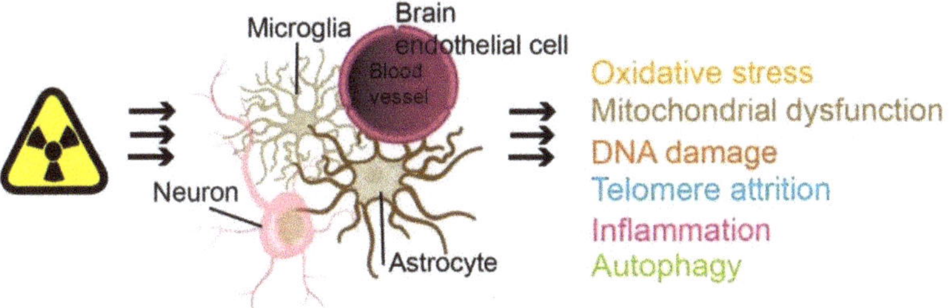

Figure 1. Radiation-induced brain aging includes oxidative stress, mitochondrial dysfunction, DNA damage, telomere attrition, inflammation, and autophagy.

Cells exposed to IR can be destroyed directly by secondary electrons and/or indirectly by ROS, resulting in DSBs [18] followed by the DDR [19,20,96–100]. The onset of this response generally does not exceed a few minutes after DNA damage. The damaged DNA is recruited by a complex reaction network that depends on damaged cells. The response produced by the cell depends on the type of DNA damage and environmental factors. For example, a response through transient activation of cell cycle checkpoints and DNA repair, namely cell survival. [101]. DSBs are usually repaired by error-prone non-homologous terminal junctions, which are coordinated by DNA-dependent protein kinases. Homologous recombination is operated only during the S or G2 phases of the cell cycle [101]. Unrepaired and/or incorrectly repaired DSBs may lead to genomic instability and cell death or cell senescence (an irreversible state of cell cycle arrest) [19,102]. Radiation exposure directly alters mitochondrial DNA, most notably the common deletion mutation. IR also indirectly alters mitochondrial dysfunction by producing ROS, resulting in disruption of the electron transport chain, and increases the production of antioxidant enzymes through nuclear factor E2-related factor 2 (Nrf2) [103].

Cells enter senescence with shortened telomeres, and, in fact, short telomeres increase the sensitivity of cells to radiation, with human cells that are sensitive to radiation having shorter telomeres than normal cells [104]. Individuals with short telomeres have a higher frequency of radiation damage than individuals with long telomeres [105]. The mechanism of telomere maintenance is directly or indirectly related to DNA damage [106]. Telomere shortening is an important sign of aging.

Under normal conditions, nuclear factor κB (NF-κB) associates with the inhibitory protein (IKB) to compose a protein complex and remains sleepy. However, when cells are exposed to IR, NADPH oxidase activity and other oxidative stress reactions are increased, mitochondrial electron transfer is impaired and occurs rapidly. ROS are over-generated in oxidative stress reactions, and IL-1 and TNF are produced by inflammatory cells to link to IL-1 receptor and TNF receptor, respectively, which in turn activates downstream NF-κB signaling and leads to transcription of inflammation-related genes. Activated NF-κB induces COX-2 and 5-LPO expression, leading to ROS production, which forms a positive feedback loop to increase inflammation and oxidative stress. Intracellular ROS directly stimulate NF-κB which may up-regulate the expression of cytokines including IL-1 and TNF. These cytokines increase inflammation by attracting white blood cells and activating NF-κB.

The way autophagy maintains protein production includes promoting misfolding and degradation of aggregated proteins. Mitochondrial quality control involves removing damaged mitochondria through autophagy. Autophagy cargo receptors recognize ubiquitin-modified mitochondria, facilitating their sequestration within autophagosomes and lysosome-mediated breakdown. The molecular pattern related to mitochondrial damage is also one of the reasons for triggering the production of inflammatory cytokines. The method of autophagy to regulate senescence is to promote the disintegration and degradation of the nuclear lamina. Free amino acids released from the lysosome during senescence support anabolic activities, including the production of inflammatory cytokines that make up the SASP (Figure 2).

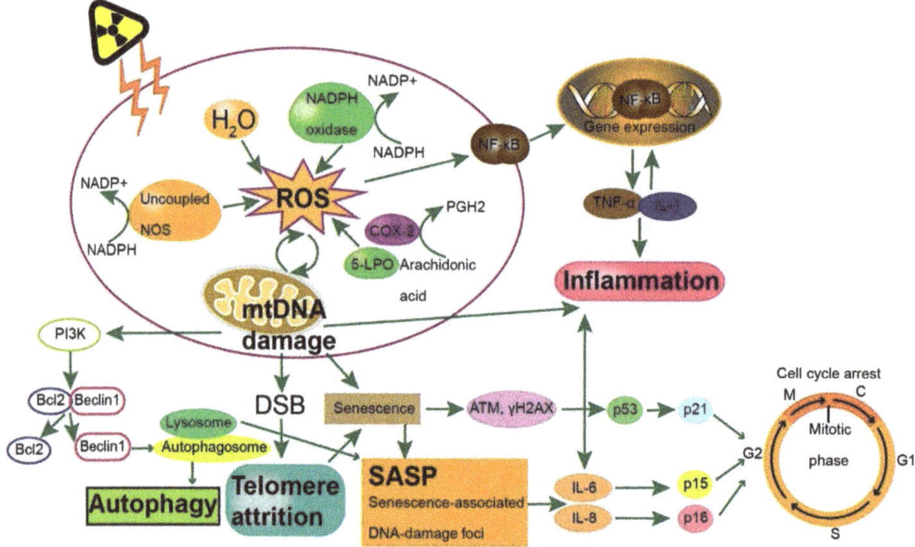

Figure 2. Interactions between various mechanisms of brain aging induced by ionizing radiation.

3.1. Oxidative Stress

Endogenous ROS production is a byproduct of regular cell metabolism, but exogenous ROS production may also occur due to radiation and chemical compounds [107]. ROS can be produced from a number of sources after irradiation [108]. Classical radiobiology shows that compared with most other oxidative stresses, the amount of ROS produced by the radiolysis of water is smaller and the maintenance time is shorter. Exposure to ionizing radiation will generate ROS in the 2 nm range of DNA and form complex DSBs, thereby triggering high cytotoxicity. The mitochondrial membrane-bound nicotinamide adenine dinucleotide phosphate oxidases (NOX) or other oxidases may have an indirect connection with the DDR pathway. These oxidases are the main source of cellular ROS caused by oxidative stress. [109–111]. Radiation can also cause ROS generation from these sources by damaging the mitochondria, stimulating NOX or other oxidases [110,111], leading to ATP release, ion channel activation [112] and purinergic signaling [113]. One example of a damage-associated molecular pattern (DAMP) observed after irradiation is expression of the high-mobility group box 1 (HMGB1) protein, a chromatin binding nuclear protein that acts via toll-like receptor 4 (TLR4) signaling to promote further ROS production [114]. High ROS activity can directly destroy macromolecules such as lipids, nucleic acids, and proteins. DNA damage, typically in the form of strand breaks and cross-links, leads to genomic mutations. Furthermore, ROS cause high oxidative stress in affected cells. In cells, oxidative damage depends on ROS concentration and on the balance between the relative levels of ROS and antioxidants. When the oxidant-antioxidant balance is lost, oxidative stress occurs, altering and destroying several intracellular macromolecules, as mentioned previously [115].

The free-radical theory of aging was put forth by Harman in 1956, and he subsequently demonstrated that mitochondrial respiration is the primary endogenous source of oxidative stress [116]. Aging is accompanied by increases in ROS levels and decreases in the activity and expression of antioxidant enzymes, including superoxide dismutase, catalase, and glutathione peroxidase [117,118]. Additionally, radiation-induced damage shows several characteristics often typical of cellular wear-and-tear, such as somatic mutations, which can trigger to the development of aging-related diseases [119]. ROS and reactive nitrogen species (RNS) attack macromolecules and cause oxidative stress, and this process has been implicated in several diseases. It has been also found that even low levels of ROS and RNS can lead to brain aging [120].

The inherent sensitivity of cells to radiation is thought to be dependent on the resultant ROS production. The irradiation of aging cells that already contain large amounts of active oxygen will undoubtedly overwhelm the antioxidant system responsible for removing excess amounts of oxygen metabolites [121]. The antioxidant protection against therapeutic and relieved doses of IR in human blood decreases with age [122]. These studies illustrate the role of oxidative stress regulation systems in determining the radiation sensitivity of senescent cells.

3.2. Mitochondrial Dysfunction

Mitochondria are present in neuronal dendrites and axons of neurons, and they produce the adenosine triphosphate (ATP) required for electrochemical neurotransmission and cell maintenance and repair [123]. As cells and organisms age, the efficiency of the electron transport chain (ETC) tends to decline, increasing electron leakage and reducing ATP production [124]. Most brain cells show the gradual accumulation of dysfunctional mitochondria, as observed in comparative studies of neurons and astrocytes in mice of different age groups [125,126]. With increasing age, mitochondrial dysfunction leads to increased ROS production, in turn causing further mitochondrial degradation and overall cell damage [127].

The main role of active oxygen is to activate the compensatory steady-state response. If ROS levels increase over a certain threshold, an imbalance in ROS homeostasis occurs, ultimately aggravating age-related damage [128]. Mitochondrial dysfunction can also cause aging via ROS-independent pathways. Mitochondrial defects may affect apoptosis signals by increasing the susceptibility of mitochondria to stress responses [129] and trigger the inflammatory response by promoting ROS-mediated and/or permeability-related inflammasome activation [124]. In addition, dysfunctional mitochondria could exert a direct influence on cell signaling and inter-organ crosstalk via negative effects on the outer mitochondrial membrane–endoplasmic reticulum interface [130].

Mitochondria isolated from animal brain tissue show many age-related changes, including increased mitochondrial DNA oxidative damage [131], mitochondrial enlargement or fragmentation [132], increased numbers of mitochondria with depolarizing membranes [133], and impaired ETC function [134]. Mutations and deletions of mitochondrial DNA (mtDNA) in older individuals may also lead to aging [135]. Due to the oxidative microenvironment in the mitochondria, mtDNA lacks protective histones. The efficiency of mtDNA repair mechanisms is also lower than that of nuclear DNA repair mechanisms. Hence, mtDNA is thought to be the main target of aging-related somatic mutations [136]. During brain development, abnormal mitochondrial breakage can lead to mitochondrial dysfunction and excessive ROS production, ultimately resulting in brain cell aging, cognitive impairment, and abnormal behavior [137].

IR exposure can induce mitochondrial dysfunction, which indirectly triggers aging. The time course of changes that occur after exposure to five Gy of γ-irradiation is as follows. First, cellular ROS levels increase significantly during the first few minutes, but reduce within 30 min. Subsequently, mitochondrial dysfunction is detected 12 h post-irradiation, as demonstrated by a decrease in the activity of nicotinamide adenine dinucleotide (NADH) dehydrogenase, the primary regulator of ROS release from the ETC [138].

Limoli et al. examined the mitochondrial membrane potential in unstable GM10115 cells after radiation exposure. They observed that the number of dysfunctional mitochondria was increased, and the mitochondrial membrane potential was decreased [139]. IR ionizes water molecules (H_2O), mainly resulting in the production of •OH, the ROS with the highest damage-causing ability [140]. •OH can oxidize biological molecules, such as proteins and lipids [121,141]. The inner mitochondrial membrane contains phospholipids, such as phosphatidylcholine, phosphatidylethanolamine, and cardiolipin, which are required for optimization and aid the functions of various enzymes of the mitochondrial ETC [142,143]. Any change in the lipid profile of the membrane, such as a decrease in the lipid content and peroxidation, may lead to or enhance the production of O_2^- via electron leakage from ETC enzymes. Therefore, the •OH produced due to irradiation induce mitochondrial oxidation via phospholipid peroxidation, thereby promoting O_2^- production.

3.3. Telomere Attrition

Telomeres are special nuclear protein complexes that protect the ends of linear chromosomes in eukaryotic cells. Telomeres are bound by a characteristic polyprotein complex termed shelterin [144] which prevents the entry of DNA repair into telomeres which would otherwise be "repaired" owing to the presence of apparent DNA breaks, leading to low capacity for repairing DNA damage in this region. Therefore, telomere damage often induces cellular senescence and/or apoptosis [145,146]. The loss-of-function of shelterin components induces the rapid weakening of tissue regeneration and accelerates aging, even when telomeres are of a normal length [147].

IR can induce cell proliferation, apoptosis and senescence, all of which are associated with telomeres, through oxidative damage and DNA disruption. A study on peripheral blood obtained from 83 Chernobyl cleaners showed that compared with those in healthy blood donors, the relative length of telomeres in Chernobyl cleaners was significantly shorter. This study suggested that low-dose irradiation led to telomere shortening, and the alterations were sustained even 20 years post-irradiation [148]. Microglial senescence may also be related to the shortening of telomeres. The reduction in telomere length in microglia may lead to the ability of these cells to respond appropriately to CNS damage or infection, leading to apoptosis [149]. It is important to note that in the brain of patients with Alzheimer's disease, malnutrition of microglia shows the strongest association with the degeneration of tau positive neurons, i.e., tau pathology is associated with microglial malnutrition. Thus, instead of brain inflammation, the lack of microglial support presents to be the essential cause of neurodegeneration. The age of microglia may be related to telomere shortening. Telomeres are the ends of eukaryotic chromosomes and shorten with age, leading to a sense of "replication" in microglia with self-renewal capabilities [150].

A study of 20 elderly patients with advanced head and neck cancer by Unryn et al. showed that radiation therapy resulted in severe shortening of telomere length in all patients [151]. Zhang et al. examined the effects of telomere dysfunction through the telomeric repeat binding factor 2 (TRF2)-mediated inhibition of neurons and mitotic nerve cells (astrocytes and neuroblastoma cells). They demonstrated that telomere dysfunction triggers DDRs and induces the activation of p53 and p21 and senescence [152]. Cellular responses to IR include cell cycle checkpoint arrest and programmed cell death. Since radiation outcomes double strand breaks in DNA leading to a reduction in telomere length, the radiation response appears to result from inappropriately induced cellular senescence [153].

The mechanism underlying accelerated aging due to IR is the same as that underlying ROS-mediated aging. Radiation damage to telomeres is also similar to the oxidative damage observed in these regions and leads to further shortening of telomeres and accelerated cell aging. A large dose of IR can cause tremendous cell death, resulting in compensatory cell division. Accelerated proliferation leads to telomere shortening, thereby accelerating aging in the entire organism, as observed in individuals exposed to IR [154]. Therefore, telomere shortening can be considered one of the mechanisms underlying radiation-induced aging [155].

3.4. DNA Damage

IR induces DNA damage through both direct and indirect pathways. The direct pathway refers to DNA ionization via radiation energy, and the indirect pathway refers to the generation of a large number of ROS after the radiolysis of water molecules. The latter pathway can induce DNA damage through a variety of mechanisms, including base damage and release, depolymerization, cross-linking, and chain breaking [156]. Such DNA damage, especially DSBs, triggers the complex and highly regulated DDR and repair pathways. DNA ionization directly causes damage to genetic macromolecules, whereas cytosol ionization leads to the generation of active substances such as •OH. Hydrated electrons and hydrogen atoms, which diffuse into the nanoscale area surrounding ionization events, inevitably react with the DNA components and indirectly damage them. Some of this damage occurs on the genetic molecular chain, leading to the generation of aggregated DNA damage or multiple damage sites. Owing to the endogenous nature of such damage, it is more likely to undergo error-prone repair than sparsely distributed damage, leading to irreversible cell damage [157].

At the molecular level, irradiated microglia show the upregulation of genes associated with DDRs, cellular stress, cell cycle arrest, and oxidative stress pathways [158,159]. Cells have a highly conserved and complex DNA damage recognition and repair network (DDR) that they use to respond to various types of DNA damage [160]. Studies using commercial cell lines and primary culture have shown that DNA damage can lead to permanent cell cycle arrest [161], resulting in an irreversible state in which damaged cells can survive but cannot proliferate, known as cellular senescence [162]. In mature neurons, homologous recombination and non-homologous terminal junctions are not sufficient to appropriately repair DSBs, likely because these cells do not divide. Therefore, it is generally believed that neurons increase unrepaired DNA damage over time, which may contribute to neurodegenerative diseases [163]. At acute stages after radiation exposure, radiation leads to p53-mediated speedy primary apoptosis and tardy secondary apoptosis (associated with mitotic mutations), thereby eliminating seriously damaged cells. Cells in the brain often respond to p53 activation through permanent cell cycle arrest instead of apoptosis [164].

With aging, DNA damage accumulates, inducing the loss of cellular function and the degeneration of cells and tissues. However, faulty repair can result in mutations and chromosomal aberrations. Unrepaired DNA damage often causes cell dysfunction or senescence, leading to multiple pathologies and cell death during aging [165,166]. The stimulation of DDR by IR triggers innate and adaptive immune regulation. Continued activation of the DDR promotes the production of inflammatory cytokines, involving IL-6 and IL-8 [167], initiating inflammatory responses that may damage surrounding tissues.

Notably, DNA damage can occur not only in the nucleus but also in the mitochondria, and mtDNA is more vulnerable to damage than nuclear DNA. ROS is the primary source of mtDNA mutations, which can accumulate with age and disease progression [168,169].

Furthermore, the DDR leads to cell senescence, and continued senescence induces the SASP, in which inflammatory cytokines are released. These mediators influence neighboring cells and trigger various pathologies [170].

Unrepaired DDR induces cell senescence via the p53 pathway, activates the SASP, causes the secretion of pro-inflammatory cytokines, and further activates the innate immune response, resulting in tissue senescence and age-related diseases [171–173].

Senescent cells are characterized by increased activity of the most common senescence marker, SA-β-gal and the enlarged expression of $p21^{WAF1/Cip1}$. While p21 can promote cellular senescence under exposure to IR, it can also promote G1 cell cycle arrest. However, this is not always correlated with p53 activity. Both p53 and p21 activation in senescent cells are temporary, since p53 and p21 expression decreases eventually and p16Ink4A maintains growth arrest in senescent cells [174]. In direct contrast to replicative senescence, stress-induced premature senescence (SIPS) caused by DNA damage is independent of telomere length or function [175,176].

3.5. Inflammation

Inflammation is a defensive response in the body. Chronic inflammation is linked to the onset and/or progression of a variety of diseases, such as age-related lesions and neurodegenerative diseases [177,178]. IR induces microglial activation and the release of inflammatory cytokines and chemokines. Inflammation is a common characteristic of microglial aging and plays a vital role in radiation-induced brain damage [179] and aging-related diseases [55]. As mentioned previously, radiation exposure often results in abnormal microglial activation, causing these cells to continuously produce neurotoxic cytokines, the levels of which increase with the dose of radiation [4]. DDR signaling can also be mediated by paracrine/systemic mechanisms that shape the systemic environment by regulating tissue repair and immune responses. Sustained DNA damage signals (telomere attrition) can cause DDR to send extracellular signals and induce SASPs [180–182]. The DDR/SASP signaling pathway regulates several bioactive pro-inflammatory mediators, such as interleukin-chemokine growth factor matrix degrading enzymes and ROS [183]. Moreover, the proinflammatory transcription of NF-kB and the inflammasome are the primary factors that set up the secretome, further highlighting the functional contribution of this pathway in the response to tissue injury [184–186]. NF-kB transcription triggers s to the production of several inflammatory features of SASP, such as IL-6, IL-1, and TNF-α, which are vital autonomic cellular modulators of aging [187,188]. Furthermore, a single dose of 10-Gy γ-irradiation can increase the levels of IL-6 and IL-8 in human endothelial cells in vitro [189]. In addition, an elevation in the inflammatory mediators TNF-α, IL-6, and IL-10 is observed with increasing radiation doses and age among survivors of atomic bombs [190].

Both IR and inflammation are related to an increase in ROS levels in tissues. In a mouse limb ischemia model, acute irradiation with two Gy was found to promote mast cell recruitment and tissue revascularization [191]. High-dose irradiation of the rat abdomen leads to neutrophil recruitment into the irradiated tissue [192]. Radiation-activated microglia express an inducible NO synthase and generate large amounts of NO, leading to neuronal oxidative damage. In addition, microglial toll-like receptors (TLRs) are involved in neuroinflammation, thereby contributing to age-related brain diseases [193]. Chronic inflammation may lead to excess ROS and RNS production, resulting in DNA damage and disease. The persistent presence of ROS and RNS in the microenvironment can lead to the further development of chronic inflammation, causing oxidative damage to DNA and DNA repair pathways, further leading to senescence and age-related diseases. In age-related neurodegenerative disease models, experimental activation of microglial TLRs can aggravate neuron degeneration, and pharmacological inhibition of microglial activation shows neuroprotective effects [194]. In addition to microglial activation, radiation-induced telomere shortening can also contribute to inflammation. Short telomeric ends induce DNA damage repair responses, leading to the production of NF-kB, a key regulator of inflammatory components such as the nod-like receptor 3 inflammasome and the secretion of inflammatory cytokines in the brain [195].

3.6. Autophagy

IR can cause macromolecular (mainly DNA) damage and endoplasmic reticulum (ER) stress induction, both of which can induce autophagy. [196]. Among the key molecules activated during radiation exposure, the inducible nitric oxide synthase (iNOS) gene and nitric oxide (NO) are involved in radiation induced autophagy and apoptosis [197,198]. The activation of the iNOS promoter will increase the production of NO, leading to the induction of autophagy mediated by protein nitration. The activation of iNOS promoter is related to its containing multiple transcription factor motifs such as NF-κB and kruppel like factor 6 (KLF6). [197]. Radiation-induced oxidative stress not only causes DNA damage, but also causes ER stress, impaired mitochondrial function, and protein misfolding. Most of these factors have been shown to induce autophagy [199,200].

Radiation-induced mitochondrial dysfunction and biogenesis are known to be related to mitochondrial autophagy [201]. Under conditions of extensive mitochondrial damage, the cell undergoes mitophagy so as to eliminate the damaged and dysfunctional mitochondria. Radiation induces a variety of responses, including autophagy and senescence. It is commonly thought that autophagy and senescence may promote cell survival. However, preclinical studies have demonstrated that autophagy can sometimes have opposite effects, such as cytotoxicity or other non-protective effects [202]. Free amino acids released by lysosomes during aging support the production of inflammatory cytokines that synthesize SASP [203].

Autophagy regulation is the core link of aging, age-related diseases and neurodegenerative diseases. In the process of aging and neurodegeneration, the regulation of autophagy will have step defects, leading to the accumulation of damaged organelles and protein aggregates, affecting cell metabolism and homeostasis, thereby exacerbating autophagy-related dysfunction and forming a vicious circle, which eventually leads to neuronal damage and cell death. [204]. Impaired autophagy in neurons contributes to the aggregation of toxic proteins and damaged organelles associated with neurodegenerative diseases [205]. Age is a vital risk factor for many neurodegenerative diseases, such as Alzheimer's disease, Parkinson's disease, and tauopathy [206]. Over time, the age-dependent decline in autophagy and the corresponding decline in protein metabolism and accumulation of protein toxicity together contribute to disease development and/or progression. Since post-mitotic neurons cannot eliminate protein-toxic damage in daughter cells during mitosis, they are more susceptible to age-related protein toxicity [207]. Impaired autophagy in glial cells, which have a critical homeostatic role in the central nervous system, may influence autophagic activities in neurons [208]. Altogether, these factors can interact and promote the degeneration of specific neurons in different neurodegenerative diseases, suggesting that neuronal population-specific therapeutic approaches may be warranted [204].

4. Conclusions and Future Research Directions

Current experimental studies on animal brains suggest that radiation induces aging in neural stem cells; mature and immature neurons; glial cells, including astrocytes, microglia, and oligodendrocytes; and endothelial cells of cerebral vessels. Cumulatively, these effects result in brain aging, leading to cognitive impairment and the development of aging-related brain disorders in individuals who are exposed to radiation, such as survivors from the Chernobyl nuclear power plant accident or individuals receiving radiotherapy. At the molecular level, radiation-induced oxidative stress and neuroinflammation may trigger different signal transduction pathways, resulting in the shortening of telomeres in brain cells, and eventually, brain aging.

Our current understanding of radiation-induced brain aging remains quite limited. With an increase in deep-space exploration, including space tourism, and the use of IR in medical diagnosis and treatment, extensive studies are required to obtain an in-depth understanding of how low-dose radiation affects brain aging and the molecular mechanisms that underlie this process. Furthermore, most radiation-related brain aging studies involve high doses of radiation, and few studies have examined the sensitivity of each CNS cell type and its progenitors to radiation and radiation-induced aging.

Senescent cells are considered effective treatment targets because they accumulate due to aging and other exogenous effects. Senotherapeutics drugs, a new class of drugs, can selectively kill senescent cells (senolytics) or suppress their disease-causing phenotypes (senomorphics/senostatics). Since 2015, several senolytics have been identified and examined via clinical trials. Preclinical data indicate that senolytics alleviate disease related effects in numerous organs, improve physical function and resilience, and suppress all causes of mortality, even among old patients [209]. In addition, new drugs can delay the patient's disease recurrence. Accurate assessment of radiation response may provide the possibility to increase the sensitivity of cancer cells to radiation therapy while reducing damage to normal tissues [202].

Many senolytics have already been shown to be effective as they mediate the activation or inactivation of redox-sensitive hubs. Consequently, ROS-dependent pathways that specifically mediate the apoptosis of senescent cells may represent novel preventive/therapeutic targets for increasing treatment efficacy. As cells divide, telomere shortening, a process linked to cellular senescence occurs. Therefore, although senolytics temporarily alleviate cellular senescence and its deleterious effects, they could potentially cause accelerated aging and related dysfunction [210]. Drugs targeting aging-related mitochondrial dysfunction or specifically targeting mitochondrial ROS, may also allow alterations in the SASP and downstream negative outcomes. However, further elucidation of the complex mechanisms via which redox-regulated signaling pathways or mitochondria affect the SASP are required. It is necessary to note that exacerbated antioxidation could also lead to severe adverse effects. Only a tight control of redox homeostasis can eventually allow effective senomorphic-based therapies [211].

Hence, we propose that future exploration should focus on the following areas: (1) effect of low-dose IR on the aging of different cell types; (2) radiosensitivity of different progenitors and differentiated cells in the brain to radiation-induced aging; (3) epidemiology of brain aging in patients exposed to frequent radiodiagnosis and radiotherapy for brain disorders; and (4) the application of different -omics approaches for understanding the molecular mechanisms underlying low dose radiation-induced brain aging. This information could significantly aid in the development of protective and therapeutic approaches against radiation-induced brain aging and other related neurological and neuropsychological disorders.

Author Contributions: All authors have read and agreed to the published version of the manuscript.

Funding: The present study was supported by the Nature Science Foundation of Hubei Province (grant no. 2017CFB786), the Hubei Province Health and Family Planning Scientific Research Project (grant no. WJ2016Y10), the Jingzhou Science and Technology Bureau Project (grant no. 2017-93), the Graduate innovation fund of the Health Science Center, the Yangtze University (200201), the National innovation and entrepreneurship training program for College Students (grant no. 202010489017) and the National Research Foundation of Singapore to Singapore Nuclear Research and Safety Initiative (TFR).

Data Availability Statement: Data sharing is not applicable to this article, as no datasets were generated or analyzed during the present study.

Conflicts of Interest: The authors declare no conflict of interest.

Abbreviations

ATP	Adenosine triphosphate
BBB	Blood-brain barrier
CNS	Central nervous system
COX2	Cyclooxygenases 2
DDR	DNA damage response
DSBs	DNA double-strand breaks
DAMP	Damage-associated molecular pattern
ER	Endoplasmic reticulum
ETC	Electron transport chain
GFAP	Glial fibrillary acidic protein
H_2O	Water molecules
HMGB1	High-mobility group box 1
IKB	Inhibitory protein
iNOS	Inducible nitric oxide synthase
IR	Ionizing radiation
KLF6	Kruppel like factor 6
5-LPO	5-lypoxygenase
MMPs	Matrix metalloproteinases
mtDNA	Mitochondrial DNA
NADH	Nicotinamide adenine dinucleotide
NO	Nitric oxide
NF-κB	Nuclear factor kappa B
Nrf2	Nuclear factor E2-related factor 2
$O_2{}^-$	Superoxide anion
PGH2	Prostaglandin H 2
RNS	Reactive nitrogen species
ROS	Reactive oxygen species
SASP	Senescence associated secretion phenotype
SA-β-Gal	Senescence-associated-β-galactosidase
SIPS	Stress-induced premature senescence
TLR4	Toll-like receptor 4
TLRs	Toll-like receptors
TNF-α	Tumor necrosis factor-α

References

1. Martin, L.G. Population aging policies in East Asia and the United States. *Science* **1991**, *251*, 527–531. [CrossRef] [PubMed]
2. McLean, A.J.; Le Couteur, D.G. Aging biology and geriatric clinical pharmacology. *Pharmacol. Rev.* **2004**, *56*, 163–184. [CrossRef] [PubMed]
3. Kritsilis, M.; Rizou, S.V.; Koutsoudaki, P.N.; Evangelou, K.; Gorgoulis, V.G.; Papadopoulos, D. Ageing, Cellular Senescence and Neurodegenerative Disease. *Int. J. Mol. Sci.* **2018**, *19*. [CrossRef]
4. Pluvinage, J.V.; Wyss-Coray, T. Systemic factors as mediators of brain homeostasis, ageing and neurodegeneration. *Nat. Rev. Neurosci.* **2020**, *21*, 93–102. [CrossRef] [PubMed]
5. Alexander, G.E.; Ryan, L.; Bowers, D.; Foster, T.C.; Bizon, J.L.; Geldmacher, D.S.; Glisky, E.L. Characterizing cognitive aging in humans with links to animal models. *Front. Aging Neurosci.* **2012**, *4*, 21. [CrossRef]
6. Mendonca, G.V.; Pezarat-Correia, P.; Vaz, J.R.; Silva, L.; Heffernan, K.S. Impact of Aging on Endurance and Neuromuscular Physical Performance: The Role of Vascular Senescence. *Sports Med.* **2017**, *47*, 583–598. [CrossRef]
7. Lopez-Otin, C.; Blasco, M.A.; Partridge, L.; Serrano, M.; Kroemer, G. The hallmarks of aging. *Cell* **2013**, *153*, 1194–1217. [CrossRef]
8. Bueler, H. Mitochondrial and Autophagic Regulation of Adult Neurogenesis in the Healthy and Diseased Brain. *Int. J. Mol. Sci.* **2021**, *22*. [CrossRef] [PubMed]
9. Jo, D.; Kim, B.C.; Cho, K.A.; Song, J. The Cerebral Effect of Ammonia in Brain Aging: Blood-Brain Barrier Breakdown, Mitochondrial Dysfunction, and Neuroinflammation. *J. Clin. Med.* **2021**, *10*. [CrossRef]
10. Aliper, A.M.; Bozdaganyan, M.E.; Orekhov, P.S.; Zhavoronkov, A.; Osipov, A.N. Replicative and radiation-induced aging: A comparison of gene expression profiles. *Aging* **2019**, *11*, 2378–2387. [CrossRef]
11. Kuzmic, M.; Galas, S.; Lecomte-Pradines, C.; Dubois, C.; Dubourg, N.; Frelon, S. Interplay between ionizing radiation effects and aging in C. elegans. *Free Radic. Biol. Med.* **2019**, *134*, 657–665. [CrossRef]

12. Kim, H.K.; Song, I.S.; Lee, S.Y.; Jeong, S.H.; Lee, S.R.; Heo, H.J.; Thu, V.T.; Kim, N.; Ko, K.S.; Rhee, B.D.; et al. B7-H4 downregulation induces mitochondrial dysfunction and enhances doxorubicin sensitivity via the cAMP/CREB/PGC1-alpha signaling pathway in HeLa cells. *Pflügers Arch.-Eur. J. Physiol.* **2014**, *466*, 2323–2338. [CrossRef]

13. Kim, S.B.; Heo, J.I.; Kim, H.; Kim, K.S. Acetylation of PGC1alpha by Histone Deacetylase 1 Downregulation Is Implicated in Radiation-Induced Senescence of Brain Endothelial Cells. *J. Gerontol. A Biol. Sci. Med. Sci.* **2019**, *74*, 787–793. [CrossRef]

14. van Deursen, J.M. The role of senescent cells in ageing. *Nature* **2014**, *509*, 439–446. [CrossRef]

15. Munoz-Espin, D.; Serrano, M. Cellular senescence: From physiology to pathology. *Nat. Rev. Mol. Cell Biol.* **2014**, *15*, 482–496. [CrossRef]

16. Tang, F.R.; Loganovsky, K. Low dose or low dose rate ionizing radiation-induced health effect in the human. *J. Environ. Radioact.* **2018**, *192*, 32–47. [CrossRef] [PubMed]

17. Andreassi, M.G.; Piccaluga, E.; Gargani, L.; Sabatino, L.; Borghini, A.; Faita, F.; Bruno, R.M.; Padovani, R.; Guagliumi, G.; Picano, E. Subclinical carotid atherosclerosis and early vascular aging from long-term low-dose ionizing radiation exposure: A genetic, telomere, and vascular ultrasound study in cardiac catheterization laboratory staff. *JACC Cardiovasc. Interv.* **2015**, *8*, 616–627. [CrossRef] [PubMed]

18. De Ruysscher, D.; Niedermann, G.; Burnet, N.G.; Siva, S.; Lee, A.W.M.; Hegi-Johnson, F. Radiotherapy toxicity. *Nat. Rev. Dis. Primers* **2019**, *5*, 13. [CrossRef] [PubMed]

19. Mahamud, O.; So, J.; Chua, M.L.K.; Bristow, R.G. Targeting DNA repair for precision radiotherapy: Balancing the therapeutic ratio. *Curr. Probl. Cancer* **2017**, *41*, 265–272. [CrossRef]

20. Morgan, M.A.; Lawrence, T.S. Molecular Pathways: Overcoming Radiation Resistance by Targeting DNA Damage Response Pathways. *Clin. Cancer Res.* **2015**, *21*, 2898–2904. [CrossRef]

21. Kempf, S.J.; Azimzadeh, O.; Atkinson, M.J.; Tapio, S. Long-term effects of ionising radiation on the brain: Cause for concern? *Radiat. Environ. Biophys* **2013**, *52*, 5–16. [CrossRef]

22. Hendry, J.H.; Simon, S.L.; Wojcik, A.; Sohrabi, M.; Burkart, W.; Cardis, E.; Laurier, D.; Tirmarche, M.; Hayata, I. Human exposure to high natural background radiation: What can it teach us about radiation risks? *J. Radiol. Prot.* **2009**, *29*, A29–A42. [CrossRef] [PubMed]

23. Sharma, N.K.; Sharma, R.; Mathur, D.; Sharad, S.; Minhas, G.; Bhatia, K.; Anand, A.; Ghosh, S.P. Role of Ionizing Radiation in Neurodegenerative Diseases. *Front. Aging Neurosci.* **2018**, *10*, 134. [CrossRef] [PubMed]

24. Hosoda, M.; Tokonami, S.; Omori, Y.; Sahoo, S.K.; Akiba, S.; Sorimachi, A.; Ishikawa, T.; Nair, R.R.; Jayalekshmi, P.A.; Sebastian, P.; et al. Estimation of external dose by car-borne survey in Kerala, India. *PLoS ONE* **2015**, *10*, e0124433. [CrossRef]

25. Kudo, H.; Tokonami, S.; Omori, Y.; Ishikawa, T.; Iwaoka, K.; Sahoo, S.K.; Akata, N.; Hosoda, M.; Wanabongse, P.; Pornnumpa, C.; et al. Comparative dosimetry for radon and thoron in high background radiation areas in China. *Radiat. Prot. Dosim.* **2015**, *167*, 155–159. [CrossRef]

26. Omori, Y.; Tokonami, S.; Sahoo, S.K.; Ishikawa, T.; Sorimachi, A.; Hosoda, M.; Kudo, H.; Pornnumpa, C.; Nair, R.R.; Jayalekshmi, P.A.; et al. Radiation dose due to radon and thoron progeny inhalation in high-level natural radiation areas of Kerala, India. *J. Radiol. Prot.* **2017**, *37*, 111–126. [CrossRef]

27. Nugraha, E.D.; Hosoda, M.; Mellawati, J.; Tamakuma, Y.; Ikram, A.; Syaifudin, M.; Yamada, R.; Akata, N.; Sasaki, M.; Furukawa, M.; et al. Comprehensive exposure assessments from the viewpoint of health in a unique high natural background radiation area, Mamuju, Indonesia. *Sci. Rep.* **2021**, *11*, 14578. [CrossRef] [PubMed]

28. Yablokov, A.V.; Nesterenko, V.B.; Nesterenko, A.V. Consequences of the Chernobyl catastrophe for public health and the environment 23 years later. *Ann. N. Y. Acad. Sci.* **2009**, *1181*, 318–326. [CrossRef] [PubMed]

29. Eglite, M.E.; Zvagule, T.J.; Rainsford, K.D.; Reste, J.D.; Curbakova, E.V.; Kurjane, N.N. Clinical aspects of the health disturbances in Chernobyl Nuclear Power Plant accident clean-up workers (liquidators) from Latvia. *Inflammopharmacology* **2009**, *17*, 163–169. [CrossRef]

30. Sasaki, H.; Kodama, K.; Yamada, M. A review of forty-five years study of Hiroshima and Nagasaki atomic bomb survivors. Aging. *J. Radiat. Res.* **1991**, *32*, 310–326. [CrossRef]

31. Yablokov, A.V. Accelerated aging as a consequence of the Chernobyl catastrophe. *Ann. N. Y. Acad. Sci.* **2009**, *1181*, 55–57. [CrossRef] [PubMed]

32. Holodova, N.B.; Zhavoronkova, L.A.; Ryzhov, B.N. [Complex pathogenetic treatment schemes of vascular dyscirculatory disorders in the remote period after exposure to low dose radiation]. *Radiat. Biol. Radioecol.* **2013**, *53*, 525–535.

33. Polyukhov, A.M.; Kobsar, I.V.; Grebelnik, V.I.; Voitenko, V.P. The accelerated occurrence of age-related changes of organism in Chernobyl workers: A radiation-induced progeroid syndrome? *Exp. Gerontol.* **2000**, *35*, 105–115. [CrossRef]

34. Brizzee, K.R. Quantitative histological studies on aging changes in cerebral cortex of Rhesus monkey and albino rat with notes on effects of prolonged low-dose ionizing irradiation in the rat. *Prog. Brain Res.* **1973**, *40*, 141–160. [CrossRef] [PubMed]

35. Lowe, X.R.; Bhattacharya, S.; Marchetti, F.; Wyrobek, A.J. Early brain response to low-dose radiation exposure involves molecular networks and pathways associated with cognitive functions, advanced aging and Alzheimer's disease. *Radiat. Res.* **2009**, *171*, 53–65. [CrossRef]

36. Lowe, X.; Wyrobek, A. Characterization of the early CNS stress biomarkers and profiles associated with neuropsychiatric diseases. *Curr. Genomics* **2012**, *13*, 489–497. [CrossRef]

37. Li, J.; Meng, Z.; Zhang, G.; Xing, Y.; Feng, L.; Fan, S.; Fan, F.; Buren, B.; Liu, Q. N-acetylcysteine relieves oxidative stress and protects hippocampus of rat from radiation-induced apoptosis by inhibiting caspase-3. *Biomed. Pharmacother.* **2015**, *70*, 1–6. [CrossRef]
38. Ungvari, Z.; Podlutsky, A.; Sosnowska, D.; Tucsek, Z.; Toth, P.; Deak, F.; Gautam, T.; Csiszar, A.; Sonntag, W.E. Ionizing radiation promotes the acquisition of a senescence-associated secretory phenotype and impairs angiogenic capacity in cerebromicrovascular endothelial cells: Role of increased DNA damage and decreased DNA repair capacity in microvascular radiosensitivity. *J. Gerontol. A Biol. Sci. Med. Sci.* **2013**, *68*, 1443–1457. [CrossRef]
39. Su, L.; Dong, Y.; Wang, Y.; Wang, Y.; Guan, B.; Lu, Y.; Wu, J.; Wang, X.; Li, D.; Meng, A.; et al. Potential role of senescent macrophages in radiation-induced pulmonary fibrosis. *Cell Death Dis.* **2021**, *12*, 527. [CrossRef] [PubMed]
40. Richardson, R.B. Ionizing radiation and aging: Rejuvenating an old idea. *Aging* **2009**, *1*, 887–902. [CrossRef]
41. Yarilin, A.A.; Belyakov, I.M.; Kusmenok, O.I.; Arshinov, V.Y.; Simonova, A.V.; Nadezhina, N.M.; Gnezditskaya, E.V. Late T cell deficiency in victims of the Chernobyl radiation accident: Possible mechanisms of induction. *Int. J. Radiat. Biol.* **1993**, *63*, 519–528. [CrossRef] [PubMed]
42. Li, K.; Li, W.; Jia, Y.; Liu, J.; Tan, G.; Zou, J.; Li, X.; Su, Y.; Lei, S.; Sun, Q. Long-term immune effects of high-level natural radiation on Yangjiang inhabitants in China. *Int. J. Radiat. Biol.* **2019**, *95*, 764–770. [CrossRef]
43. Little, M.P. The proportion of thyroid cancers in the Japanese atomic bomb survivors associated with natural background radiation. *J. Radiol. Prot.* **2002**, *22*, 279–291. [CrossRef]
44. Little, M.P.; Wakeford, R.; Lubin, J.H.; Kendall, G.M. The statistical power of epidemiological studies analyzing the relationship between exposure to ionizing radiation and cancer, with special reference to childhood leukemia and natural background radiation. *Radiat. Res.* **2010**, *174*, 387–402. [CrossRef]
45. Rasband, M.N. Glial Contributions to Neural Function and Disease. *Mol. Cell. Proteom.* **2016**, *15*, 355–361. [CrossRef] [PubMed]
46. Joo, F. Endothelial cells of the brain and other organ systems: Some similarities and differences. *Prog. Neurobiol.* **1996**, *48*, 255–273. [CrossRef]
47. Xu, A.; Li, R.; Ren, A.; Jian, H.; Huang, Z.; Zeng, Q.; Wang, B.; Zheng, J.; Chen, X.; Zheng, N.; et al. Regulatory coupling between long noncoding RNAs and senescence in irradiated microglia. *J. Neuroinflamm.* **2020**, *17*, 321. [CrossRef]
48. Turnquist, C.; Beck, J.A.; Horikawa, I.; Obiorah, I.E.; Von Muhlinen, N.; Vojtesek, B.; Lane, D.P.; Grunseich, C.; Chahine, J.J.; Ames, H.M.; et al. Radiation-induced astrocyte senescence is rescued by Delta133p53. *Neuro. Oncol.* **2019**, *21*, 474–485. [CrossRef]
49. McRobb, L.S.; McKay, M.J.; Gamble, J.R.; Grace, M.; Moutrie, V.; Santos, E.D.; Lee, V.S.; Zhao, Z.; Molloy, M.P.; Stoodley, M.A. Ionizing radiation reduces ADAM10 expression in brain microvascular endothelial cells undergoing stress-induced senescence. *Aging* **2017**, *9*, 1248–1268. [CrossRef]
50. Schindler, M.K.; Forbes, M.E.; Robbins, M.E.; Riddle, D.R. Aging-dependent changes in the radiation response of the adult rat brain. *Int. J. Radiat. Oncol. Biol. Phys.* **2008**, *70*, 826–834. [CrossRef] [PubMed]
51. Hickman, S.E.; Kingery, N.D.; Ohsumi, T.K.; Borowsky, M.L.; Wang, L.C.; Means, T.K.; El Khoury, J. The microglial sensome revealed by direct RNA sequencing. *Nat. Neurosci.* **2013**, *16*, 1896–1905. [CrossRef]
52. Harry, G.J.; Kraft, A.D. Neuroinflammation and microglia: Considerations and approaches for neurotoxicity assessment. *Expert Opin. Drug Metab. Toxicol.* **2008**, *4*, 1265–1277. [CrossRef] [PubMed]
53. Schmidt-Ullrich, R.K. Molecular targets in radiation oncology. *Oncogene* **2003**, *22*, 5730–5733. [CrossRef]
54. Lumniczky, K.; Szatmari, T.; Safrany, G. Ionizing Radiation-Induced Immune and Inflammatory Reactions in the Brain. *Front. Immunol.* **2017**, *8*, 517. [CrossRef]
55. Block, M.L.; Zecca, L.; Hong, J.S. Microglia-mediated neurotoxicity: Uncovering the molecular mechanisms. *Nat. Rev. Neurosci.* **2007**, *8*, 57–69. [CrossRef]
56. Xu, P.; Xu, Y.; Hu, B.; Wang, J.; Pan, R.; Murugan, M.; Wu, L.J.; Tang, Y. Extracellular ATP enhances radiation-induced brain injury through microglial activation and paracrine signaling via P2X7 receptor. *Brain Behav. Immun.* **2015**, *50*, 87–100. [CrossRef] [PubMed]
57. Soh, M.; Nguyen, T.; Silva, K.K.; Westerhout, R.; Antoszewski, J.; Keating, A.; Savvides, N.; Musca, C.; Dell, J.; Faraone, L. Short-wavelength infrared tuneable filters on HgCdTe photoconductors. *Opt. Express* **2005**, *13*, 9683–9694. [CrossRef] [PubMed]
58. Flanary, B.E.; Sammons, N.W.; Nguyen, C.; Walker, D.; Streit, W.J. Evidence that aging and amyloid promote microglial cell senescence. *Rejuvenation Res.* **2007**, *10*, 61–74. [CrossRef] [PubMed]
59. Tha, K.K.; Okuma, Y.; Miyazaki, H.; Murayama, T.; Uehara, T.; Hatakeyama, R.; Hayashi, Y.; Nomura, Y. Changes in expressions of proinflammatory cytokines IL-1beta, TNF-alpha and IL-6 in the brain of senescence accelerated mouse (SAM) P8. *Brain Res.* **2000**, *885*, 25–31. [CrossRef]
60. Osman, A.M.; Sun, Y.; Burns, T.C.; He, L.; Kee, N.; Oliva-Vilarnau, N.; Alevyzaki, A.; Zhou, K.; Louhivuori, L.; Uhlen, P.; et al. Radiation Triggers a Dynamic Sequence of Transient Microglial Alterations in Juvenile Brain. *Cell Rep.* **2020**, *31*, 107699. [CrossRef]
61. Betlazar, C.; Middleton, R.J.; Banati, R.B.; Liu, G.J. The impact of high and low dose ionising radiation on the central nervous system. *Redox Biol.* **2016**, *9*, 144–156. [CrossRef]
62. Yu, H.M.; Zhao, Y.M.; Luo, X.G.; Feng, Y.; Ren, Y.; Shang, H.; He, Z.Y.; Luo, X.M.; Chen, S.D.; Wang, X.Y. Repeated lipopolysaccharide stimulation induces cellular senescence in BV2 cells. *Neuroimmunomodulation* **2012**, *19*, 131–136. [CrossRef]
63. Childs, B.G.; Durik, M.; Baker, D.J.; van Deursen, J.M. Cellular senescence in aging and age-related disease: From mechanisms to therapy. *Nat. Med.* **2015**, *21*, 1424–1435. [CrossRef]

64. Le, O.; Palacio, L.; Bernier, G.; Batinic-Haberle, I.; Hickson, G.; Beausejour, C. INK4a/ARF Expression Impairs Neurogenesis in the Brain of Irradiated Mice. *Stem Cell Rep.* **2018**, *10*, 1721–1733. [CrossRef]
65. Schneider, L.; Pellegatta, S.; Favaro, R.; Pisati, F.; Roncaglia, P.; Testa, G.; Nicolis, S.K.; Finocchiaro, G.; d'Adda di Fagagna, F. DNA damage in mammalian neural stem cells leads to astrocytic differentiation mediated by BMP2 signaling through JAK-STAT. *Stem Cell Rep.* **2013**, *1*, 123–138. [CrossRef] [PubMed]
66. Bachstetter, A.D.; Xing, B.; de Almeida, L.; Dimayuga, E.R.; Watterson, D.M.; Van Eldik, L.J. Microglial p38alpha MAPK is a key regulator of proinflammatory cytokine up-regulation induced by toll-like receptor (TLR) ligands or beta-amyloid (Abeta). *J. Neuroinflamm.* **2011**, *8*, 79. [CrossRef]
67. Priego, N.; Valiente, M. The Potential of Astrocytes as Immune Modulators in Brain Tumors. *Front. Immunol.* **2019**, *10*, 1314. [CrossRef]
68. Abbott, N.J.; Ronnback, L.; Hansson, E. Astrocyte-endothelial interactions at the blood-brain barrier. *Nat. Rev. Neurosci.* **2006**, *7*, 41–53. [CrossRef] [PubMed]
69. Newington, J.T.; Harris, R.A.; Cumming, R.C. Reevaluating Metabolism in Alzheimer's Disease from the Perspective of the Astrocyte-Neuron Lactate Shuttle Model. *J. Neurodegener. Dis.* **2013**, *2013*, 234572. [CrossRef]
70. Lecuyer, M.A.; Kebir, H.; Prat, A. Glial influences on BBB functions and molecular players in immune cell trafficking. *Biochim. Biophys. Acta* **2016**, *1862*, 472–482. [CrossRef] [PubMed]
71. Baker, D.J.; Petersen, R.C. Cellular senescence in brain aging and neurodegenerative diseases: Evidence and perspectives. *J. Clin. Investig.* **2018**, *128*, 1208–1216. [CrossRef] [PubMed]
72. Bhat, R.; Crowe, E.P.; Bitto, A.; Moh, M.; Katsetos, C.D.; Garcia, F.U.; Johnson, F.B.; Trojanowski, J.Q.; Sell, C.; Torres, C. Astrocyte senescence as a component of Alzheimer's disease. *PLoS ONE* **2012**, *7*, e45069. [CrossRef]
73. Zou, Y.; Zhang, N.; Ellerby, L.M.; Davalos, A.R.; Zeng, X.; Campisi, J.; Desprez, P.Y. Responses of human embryonic stem cells and their differentiated progeny to ionizing radiation. *Biochem. Biophys. Res. Commun.* **2012**, *426*, 100–105. [CrossRef]
74. Monje, M.L.; Toda, H.; Palmer, T.D. Inflammatory blockade restores adult hippocampal neurogenesis. *Science* **2003**, *302*, 1760–1765. [CrossRef]
75. Eriksson, D.; Stigbrand, T. Radiation-induced cell death mechanisms. *Tumour Biol.* **2010**, *31*, 363–372. [CrossRef]
76. Crowe, E.P.; Tuzer, F.; Gregory, B.D.; Donahue, G.; Gosai, S.J.; Cohen, J.; Leung, Y.Y.; Yetkin, E.; Nativio, R.; Wang, L.S.; et al. Changes in the Transcriptome of Human Astrocytes Accompanying Oxidative Stress-Induced Senescence. *Front. Aging Neurosci.* **2016**, *8*, 208. [CrossRef] [PubMed]
77. Hladik, D.; Dalke, C.; von Toerne, C.; Hauck, S.M.; Azimzadeh, O.; Philipp, J.; Ung, M.C.; Schlattl, H.; Rossler, U.; Graw, J.; et al. CREB Signaling Mediates Dose-Dependent Radiation Response in the Murine Hippocampus Two Years after Total Body Exposure. *J. Proteome Res.* **2020**, *19*, 337–345. [CrossRef]
78. Seol, M.A.; Jung, U.; Eom, H.S.; Kim, S.H.; Park, H.R.; Jo, S.K. Prolonged expression of senescence markers in mice exposed to gamma-irradiation. *J. Vet. Sci.* **2012**, *13*, 331–338. [CrossRef] [PubMed]
79. Azimzadeh, O.; Sievert, W.; Sarioglu, H.; Merl-Pham, J.; Yentrapalli, R.; Bakshi, M.V.; Janik, D.; Ueffing, M.; Atkinson, M.J.; Multhoff, G.; et al. Integrative proteomics and targeted transcriptomics analyses in cardiac endothelial cells unravel mechanisms of long-term radiation-induced vascular dysfunction. *J. Proteome Res.* **2015**, *14*, 1203–1219. [CrossRef]
80. Le, O.N.; Rodier, F.; Fontaine, F.; Coppe, J.P.; Campisi, J.; DeGregori, J.; Laverdiere, C.; Kokta, V.; Haddad, E.; Beausejour, C.M. Ionizing radiation-induced long-term expression of senescence markers in mice is independent of p53 and immune status. *Aging Cell* **2010**, *9*, 398–409. [CrossRef]
81. Ungvari, Z.; Tarantini, S.; Hertelendy, P.; Valcarcel-Ares, M.N.; Fulop, G.A.; Logan, S.; Kiss, T.; Farkas, E.; Csiszar, A.; Yabluchanskiy, A. Cerebromicrovascular dysfunction predicts cognitive decline and gait abnormalities in a mouse model of whole brain irradiation-induced accelerated brain senescence. *Geroscience* **2017**, *39*, 33–42. [CrossRef]
82. Hernandez-Segura, A.; de Jong, T.V.; Melov, S.; Guryev, V.; Campisi, J.; Demaria, M. Unmasking Transcriptional Heterogeneity in Senescent Cells. *Curr. Biol.* **2017**, *27*, 2652–2660.e2654. [CrossRef] [PubMed]
83. Coppe, J.P.; Rodier, F.; Patil, C.K.; Freund, A.; Desprez, P.Y.; Campisi, J. Tumor suppressor and aging biomarker p16(INK4a) induces cellular senescence without the associated inflammatory secretory phenotype. *J. Biol. Chem.* **2011**, *286*, 36396–36403. [CrossRef] [PubMed]
84. Hernandez-Segura, A.; Nehme, J.; Demaria, M. Hallmarks of Cellular Senescence. *Trends Cell Biol.* **2018**, *28*, 436–453. [CrossRef] [PubMed]
85. Hayakawa, T.; Iwai, M.; Aoki, S.; Takimoto, K.; Maruyama, M.; Maruyama, W.; Motoyama, N. SIRT1 suppresses the senescence-associated secretory phenotype through epigenetic gene regulation. *PLoS ONE* **2015**, *10*, e0116480. [CrossRef]
86. Palmer, T.D.; Willhoite, A.R.; Gage, F.H. Vascular niche for adult hippocampal neurogenesis. *J. Comp. Neurol.* **2000**, *425*, 479–494. [CrossRef]
87. Remes, T.M.; Suo-Palosaari, M.H.; Koskenkorva, P.K.T.; Sutela, A.K.; Toiviainen-Salo, S.M.; Arikoski, P.M.; Arola, M.O.; Heikkila, V.P.; Kapanen, M.; Lahteenmaki, P.M.; et al. Radiation-induced accelerated aging of the brain vasculature in young adult survivors of childhood brain tumors. *Neurooncol. Pract.* **2020**, *7*, 415–427. [CrossRef]
88. Flanary, B. The role of microglial cellular senescence in the aging and Alzheimer diseased brain. *Rejuvenation Res.* **2005**, *8*, 82–85. [CrossRef]

89. Baxter, P.S.; Hardingham, G.E. Adaptive regulation of the brain's antioxidant defences by neurons and astrocytes. *Free Radic. Biol. Med.* **2016**, *100*, 147–152. [CrossRef]

90. Mata-Garrido, J.; Tapia, O.; Casafont, I.; Berciano, M.T.; Cuadrado, A.; Lafarga, M. Persistent accumulation of unrepaired DNA damage in rat cortical neurons: Nuclear organization and ChIP-seq analysis of damaged DNA. *Acta Neuropathol. Commun.* **2018**, *6*, 68. [CrossRef]

91. De, A.K.; Chipalkatti, S.; Aiyar, A.S. Effects of chronic irradiation on age-related biochemical changes in mice. *Radiat. Res.* **1983**, *95*, 637–645. [CrossRef]

92. Tang, F.R.; Liu, L.; Wang, H.; Ho, K.J.N.; Sethi, G. Spatiotemporal dynamics of gammaH2AX in the mouse brain after acute irradiation at different postnatal days with special reference to the dentate gyrus of the hippocampus. *Aging* **2021**, *13*, 15815–15832. [CrossRef]

93. Fike, J.R.; Rosi, S.; Limoli, C.L. Neural precursor cells and central nervous system radiation sensitivity. *Semin Radiat. Oncol.* **2009**, *19*, 122–132. [CrossRef] [PubMed]

94. Pazzaglia, S.; Briganti, G.; Mancuso, M.; Saran, A. Neurocognitive Decline Following Radiotherapy: Mechanisms and Therapeutic Implications. *Cancers* **2020**, *12*. [CrossRef] [PubMed]

95. Cheng, Z.; Li, Y.Q.; Wong, C.S. Effects of Aging on Hippocampal Neurogenesis after Irradiation. *Int. J. Radiat. Oncol. Biol. Phys.* **2016**, *94*, 1181–1189. [CrossRef]

96. Goldstein, M.; Kastan, M.B. The DNA damage response: Implications for tumor responses to radiation and chemotherapy. *Annu. Rev. Med.* **2015**, *66*, 129–143. [CrossRef] [PubMed]

97. Pollard, J.M.; Gatti, R.A. Clinical radiation sensitivity with DNA repair disorders: An overview. *Int. J. Radiat. Oncol. Biol. Phys.* **2009**, *74*, 1323–1331. [CrossRef] [PubMed]

98. Nahas, S.A.; Gatti, R.A. DNA double strand break repair defects, primary immunodeficiency disorders, and 'radiosensitivity'. *Curr. Opin. Allergy Clin. Immunol.* **2009**, *9*, 510–516. [CrossRef]

99. Andreassen, C.N.; Schack, L.M.; Laursen, L.V.; Alsner, J. Radiogenomics—Current status, challenges and future directions. *Cancer Lett.* **2016**, *382*, 127–136. [CrossRef] [PubMed]

100. Foray, N.; Bourguignon, M.; Hamada, N. Individual response to ionizing radiation. *Mutat. Res. Rev. Mutat. Res.* **2016**, *770*, 369–386. [CrossRef]

101. Borgmann, K.; Kocher, S.; Kriegs, M.; Mansour, W.Y.; Parplys, A.C.; Rieckmann, T.; Rothkamm, K. DNA Repair. *Recent Results Cancer Res.* **2016**, *198*, 1–24. [CrossRef] [PubMed]

102. Wu, Q.; Allouch, A.; Martins, I.; Brenner, C.; Modjtahedi, N.; Deutsch, E.; Perfettini, J.L. Modulating Both Tumor Cell Death and Innate Immunity Is Essential for Improving Radiation Therapy Effectiveness. *Front. Immunol.* **2017**, *8*, 613. [CrossRef]

103. Livingston, K.; Schlaak, R.A.; Puckett, L.L.; Bergom, C. The Role of Mitochondrial Dysfunction in Radiation-Induced Heart Disease: From Bench to Bedside. *Front. Cardiovasc. Med.* **2020**, *7*, 20. [CrossRef]

104. Cabuy, E.; Newton, C.; Joksic, G.; Woodbine, L.; Koller, B.; Jeggo, P.A.; Slijepcevic, P. Accelerated telomere shortening and telomere abnormalities in radiosensitive cell lines. *Radiat. Res.* **2005**, *164*, 53–62. [CrossRef] [PubMed]

105. Castella, M.; Puerto, S.; Creus, A.; Marcos, R.; Surralles, J. Telomere length modulates human radiation sensitivity in vitro. *Toxicol. Lett.* **2007**, *172*, 29–36. [CrossRef]

106. Slijepcevic, P. Is there a link between telomere maintenance and radiosensitivity? *Radiat. Res.* **2004**, *161*, 82–86. [CrossRef]

107. Valko, M.; Leibfritz, D.; Moncol, J.; Cronin, M.T.; Mazur, M.; Telser, J. Free radicals and antioxidants in normal physiological functions and human disease. *Int. J. Biochem. Cell Biol.* **2007**, *39*, 44–84. [CrossRef]

108. Schaue, D.; Kachikwu, E.L.; McBride, W.H. Cytokines in radiobiological responses: A review. *Radiat. Res.* **2012**, *178*, 505–523. [CrossRef]

109. Choi, K.M.; Kang, C.M.; Cho, E.S.; Kang, S.M.; Lee, S.B.; Um, H.D. Ionizing radiation-induced micronucleus formation is mediated by reactive oxygen species that are produced in a manner dependent on mitochondria, Nox1, and JNK. *Oncol. Rep.* **2007**, *17*, 1183–1188. [CrossRef]

110. Kang, M.A.; So, E.Y.; Simons, A.L.; Spitz, D.R.; Ouchi, T. DNA damage induces reactive oxygen species generation through the H2AX-Nox1/Rac1 pathway. *Cell Death Dis.* **2012**, *3*, e249. [CrossRef] [PubMed]

111. Tateishi, Y.; Sasabe, E.; Ueta, E.; Yamamoto, T. Ionizing irradiation induces apoptotic damage of salivary gland acinar cells via NADPH oxidase 1-dependent superoxide generation. *Biochem. Biophys. Res. Commun.* **2008**, *366*, 301–307. [CrossRef]

112. Todd, D.G.; Mikkelsen, R.B. Ionizing radiation induces a transient increase in cytosolic free [Ca^{2+}] in human epithelial tumor cells. *Cancer Res.* **1994**, *54*, 5224–5230.

113. Ohshima, Y.; Tsukimoto, M.; Takenouchi, T.; Harada, H.; Suzuki, A.; Sato, M.; Kitani, H.; Kojima, S. gamma-Irradiation induces P2X(7) receptor-dependent ATP release from B16 melanoma cells. *Biochim. Biophys. Acta* **2010**, *1800*, 40–46. [CrossRef] [PubMed]

114. Tang, D.; Kang, R.; Zeh, H.J., III; Lotze, M.T. High-mobility group box 1, oxidative stress, and disease. *Antioxid. Redox Signal.* **2011**, *14*, 1315–1335. [CrossRef]

115. Veskoukis, A.S.; Tsatsakis, A.M.; Kouretas, D. Dietary oxidative stress and antioxidant defense with an emphasis on plant extract administration. *Cell Stress Chaperones* **2012**, *17*, 11–21. [CrossRef]

116. Harman, D. Aging: A theory based on free radical and radiation chemistry. *J. Gerontol.* **1956**, *11*, 298–300. [CrossRef]

117. Andersen, H.R.; Nielsen, J.B.; Nielsen, F.; Grandjean, P. Antioxidative enzyme activities in human erythrocytes. *Clin. Chem.* **1997**, *43*, 562–568. [CrossRef]

118. Inal, M.E.; Kanbak, G.; Sunal, E. Antioxidant enzyme activities and malondialdehyde levels related to aging. *Clin. Chim. Acta* **2001**, *305*, 75–80. [CrossRef]
119. Harman, D. The biologic clock: The mitochondria? *J. Am. Geriatr. Soc.* **1972**, *20*, 145–147. [CrossRef] [PubMed]
120. Floyd, R.A. Antioxidants, oxidative stress, and degenerative neurological disorders. *Proc. Soc. Exp. Biol. Med.* **1999**, *222*, 236–245. [CrossRef]
121. Riley, P.A. Free radicals in biology: Oxidative stress and the effects of ionizing radiation. *Int. J. Radiat. Biol.* **1994**, *65*, 27–33. [CrossRef] [PubMed]
122. Kasapovic, J.; Stojiljkovic, V.; Gavrilovic, L.; Popovic, N.; Milicevic, Z. Antioxidant protection against curative and palliative doses of ionizing irradiation in human blood decreases with aging. *Sci. World J.* **2012**, *2012*, 982594. [CrossRef]
123. Mattson, M.P.; Gleichmann, M.; Cheng, A. Mitochondria in neuroplasticity and neurological disorders. *Neuron* **2008**, *60*, 748–766. [CrossRef] [PubMed]
124. Green, D.R.; Galluzzi, L.; Kroemer, G. Mitochondria and the autophagy-inflammation-cell death axis in organismal aging. *Science* **2011**, *333*, 1109–1112. [CrossRef]
125. Lin, D.T.; Wu, J.; Holstein, D.; Upadhyay, G.; Rourk, W.; Muller, E.; Lechleiter, J.D. Ca^{2+} signaling, mitochondria and sensitivity to oxidative stress in aging astrocytes. *Neurobiol. Aging* **2007**, *28*, 99–111. [CrossRef]
126. Ghosh, D.; LeVault, K.R.; Barnett, A.J.; Brewer, G.J. A reversible early oxidized redox state that precedes macromolecular ROS damage in aging nontransgenic and 3xTg-AD mouse neurons. *J. Neurosci.* **2012**, *32*, 5821–5832. [CrossRef]
127. Harman, D. The Free Radical Theory of Aging: Effect of Age on Serum Copper Levels. *J. Gerontol.* **1965**, *20*, 151–153. [CrossRef]
128. Hekimi, S.; Lapointe, J.; Wen, Y. Taking a 'good' look at free radicals in the aging process. *Trends Cell Biol.* **2011**, *21*, 569–576. [CrossRef]
129. Kroemer, G.; Galluzzi, L.; Brenner, C. Mitochondrial membrane permeabilization in cell death. *Physiol. Rev.* **2007**, *87*, 99–163. [CrossRef] [PubMed]
130. Raffaello, A.; Rizzuto, R. Mitochondrial longevity pathways. *Biochim. Biophys. Acta* **2011**, *1813*, 260–268. [CrossRef]
131. Santos, R.X.; Correia, S.C.; Zhu, X.; Smith, M.A.; Moreira, P.I.; Castellani, R.J.; Nunomura, A.; Perry, G. Mitochondrial DNA oxidative damage and repair in aging and Alzheimer's disease. *Antioxid. Redox Signal.* **2013**, *18*, 2444–2457. [CrossRef]
132. Stahon, K.E.; Bastian, C.; Griffith, S.; Kidd, G.J.; Brunet, S.; Baltan, S. Age-Related Changes in Axonal and Mitochondrial Ultrastructure and Function in White Matter. *J. Neurosci.* **2016**, *36*, 9990–10001. [CrossRef] [PubMed]
133. Lores-Arnaiz, S.; Lombardi, P.; Karadayian, A.G.; Orgambide, F.; Cicerchia, D.; Bustamante, J. Brain cortex mitochondrial bioenergetics in synaptosomes and non-synaptic mitochondria during aging. *Neurochem. Res.* **2016**, *41*, 353–363. [CrossRef]
134. Nolfi-Donegan, D.; Braganza, A.; Shiva, S. Mitochondrial electron transport chain: Oxidative phosphorylation, oxidant production, and methods of measurement. *Redox Biol.* **2020**, *37*, 101674. [CrossRef] [PubMed]
135. Park, C.B.; Larsson, N.G. Mitochondrial DNA mutations in disease and aging. *J. Cell Biol.* **2011**, *193*, 809–818. [CrossRef] [PubMed]
136. Linnane, A.W.; Marzuki, S.; Ozawa, T.; Tanaka, M. Mitochondrial DNA mutations as an important contributor to ageing and degenerative diseases. *Lancet* **1989**, *1*, 642–645. [CrossRef]
137. Fattal, O.; Budur, K.; Vaughan, A.J.; Franco, K. Review of the literature on major mental disorders in adult patients with mitochondrial diseases. *Psychosomatics* **2006**, *47*, 1–7. [CrossRef] [PubMed]
138. Yoshida, T.; Goto, S.; Kawakatsu, M.; Urata, Y.; Li, T.S. Mitochondrial dysfunction, a probable cause of persistent oxidative stress after exposure to ionizing radiation. *Free Radic. Res.* **2012**, *46*, 147–153. [CrossRef]
139. Limoli, C.L.; Giedzinski, E.; Morgan, W.F.; Swarts, S.G.; Jones, G.D.; Hyun, W. Persistent oxidative stress in chromosomally unstable cells. *Cancer Res.* **2003**, *63*, 3107–3111.
140. Robbins, M.E.; Zhao, W. Chronic oxidative stress and radiation-induced late normal tissue injury: A review. *Int. J. Radiat. Biol.* **2004**, *80*, 251–259. [CrossRef]
141. Wallace, S.S. Enzymatic processing of radiation-induced free radical damage in DNA. *Radiat. Res.* **1998**, *150*, S60–S79. [CrossRef]
142. Chicco, A.J.; Sparagna, G.C. Role of cardiolipin alterations in mitochondrial dysfunction and disease. *Am. J. Physiol. Cell Physiol.* **2007**, *292*, C33–C44. [CrossRef]
143. Dekker, C.J.; Geurts van Kessel, W.S.; Klomp, J.P.; Pieters, J.; De Kruijff, B. Synthesis and polymorphic phase behaviour of polyunsaturated phosphatidylcholines and phosphatidylethanolamines. *Chem. Phys. Lipids* **1983**, *33*, 93–106. [CrossRef]
144. Palm, W.; de Lange, T. How shelterin protects mammalian telomeres. *Annu. Rev. Genet.* **2008**, *42*, 301–334. [CrossRef] [PubMed]
145. Fumagalli, M.; Rossiello, F.; Clerici, M.; Barozzi, S.; Cittaro, D.; Kaplunov, J.M.; Bucci, G.; Dobreva, M.; Matti, V.; Beausejour, C.M.; et al. Telomeric DNA damage is irreparable and causes persistent DNA-damage-response activation. *Nat. Cell Biol.* **2012**, *14*, 355–365. [CrossRef]
146. Hewitt, G.; Jurk, D.; Marques, F.D.; Correia-Melo, C.; Hardy, T.; Gackowska, A.; Anderson, R.; Taschuk, M.; Mann, J.; Passos, J.F. Telomeres are favoured targets of a persistent DNA damage response in ageing and stress-induced senescence. *Nat. Commun.* **2012**, *3*, 708. [CrossRef] [PubMed]
147. Martinez, P.; Blasco, M.A. Role of shelterin in cancer and aging. *Aging Cell* **2010**, *9*, 653–666. [CrossRef]
148. Ilyenko, I.; Lyaskivska, O.; Bazyka, D. Analysis of relative telomere length and apoptosis in humans exposed to ionising radiation. *Exp. Oncol.* **2011**, *33*, 235–238. [PubMed]
149. Wolf, S.A.; Boddeke, H.W.; Kettenmann, H. Microglia in Physiology and Disease. *Annu. Rev. Physiol.* **2017**, *79*, 619–643. [CrossRef]

150. Ojo, J.O.; Rezaie, P.; Gabbott, P.L.; Stewart, M.G. Impact of age-related neuroglial cell responses on hippocampal deterioration. *Front. Aging Neurosci.* **2015**, *7*, 57. [CrossRef] [PubMed]
151. Unryn, B.M.; Hao, D.; Gluck, S.; Riabowol, K.T. Acceleration of telomere loss by chemotherapy is greater in older patients with locally advanced head and neck cancer. *Clin. Cancer Res.* **2006**, *12*, 6345–6350. [CrossRef]
152. Zhang, P.; Furukawa, K.; Opresko, P.L.; Xu, X.; Bohr, V.A.; Mattson, M.P. TRF2 dysfunction elicits DNA damage responses associated with senescence in proliferating neural cells and differentiation of neurons. *J. Neurochem.* **2006**, *97*, 567–581. [CrossRef]
153. Crompton, N.E. Telomeres, senescence and cellular radiation response. *Cell. Mol. Life Sci.* **1997**, *53*, 568–575. [CrossRef] [PubMed]
154. Ayouaz, A.; Raynaud, C.; Heride, C.; Revaud, D.; Sabatier, L. Telomeres: Hallmarks of radiosensitivity. *Biochimie* **2008**, *90*, 60–72. [CrossRef] [PubMed]
155. Mikhel'son, V.M.; Gamalei, I.A. [Telomere shortening is the main mechanism of natural and radiation aging]. *Radiat. Biol. Radioecol.* **2010**, *50*, 269–275. [CrossRef]
156. Bucher, N.; Britten, C.D. G2 checkpoint abrogation and checkpoint kinase-1 targeting in the treatment of cancer. *Br. J. Cancer* **2008**, *98*, 523–528. [CrossRef] [PubMed]
157. Mavragani, I.V.; Nikitaki, Z.; Kalospyros, S.A.; Georgakilas, A.G. Ionizing Radiation and Complex DNA Damage: From Prediction to Detection Challenges and Biological Significance. *Cancers* **2019**, *11*. [CrossRef] [PubMed]
158. Li, M.; You, L.; Xue, J.; Lu, Y. Ionizing Radiation-Induced Cellular Senescence in Normal, Non-transformed Cells and the Involved DNA Damage Response: A Mini Review. *Front. Pharmacol.* **2018**, *9*, 522. [CrossRef]
159. Robello, E.; Bonetto, J.G.; Puntarulo, S. Cellular Oxidative/Antioxidant Balance in gamma-Irradiated Brain: An Update. *Mini-Rev. Med. Chem.* **2016**, *16*, 937–946. [CrossRef]
160. Pateras, I.S.; Havaki, S.; Nikitopoulou, X.; Vougas, K.; Townsend, P.A.; Panayiotidis, M.I.; Georgakilas, A.G.; Gorgoulis, V.G. The DNA damage response and immune signaling alliance: Is it good or bad? Nature decides when and where. *Pharmacol. Ther.* **2015**, *154*, 36–56. [CrossRef]
161. Di Leonardo, A.; Linke, S.P.; Clarkin, K.; Wahl, G.M. DNA damage triggers a prolonged p53-dependent G1 arrest and long-term induction of Cip1 in normal human fibroblasts. *Genes Dev.* **1994**, *8*, 2540–2551. [CrossRef]
162. Campisi, J.; d'Adda di Fagagna, F. Cellular senescence: When bad things happen to good cells. *Nat. Rev. Mol. Cell Biol.* **2007**, *8*, 729–740. [CrossRef]
163. Su, Y.; Ming, G.L.; Song, H. DNA damage and repair regulate neuronal gene expression. *Cell Res.* **2015**, *25*, 993–994. [CrossRef]
164. Gudkov, A.V.; Komarova, E.A. The role of p53 in determining sensitivity to radiotherapy. *Nat. Rev. Cancer* **2003**, *3*, 117–129. [CrossRef] [PubMed]
165. Ciccia, A.; Elledge, S.J. The DNA damage response: Making it safe to play with knives. *Mol. Cell* **2010**, *40*, 179–204. [CrossRef]
166. Ribezzo, F.; Shiloh, Y.; Schumacher, B. Systemic DNA damage responses in aging and diseases. *Semin. Cancer Biol.* **2016**, *37-38*, 26–35. [CrossRef] [PubMed]
167. Rodier, F.; Coppe, J.P.; Patil, C.K.; Hoeijmakers, W.A.; Munoz, D.P.; Raza, S.R.; Freund, A.; Campeau, E.; Davalos, A.R.; Campisi, J. Persistent DNA damage signalling triggers senescence-associated inflammatory cytokine secretion. *Nat. Cell Biol.* **2009**, *11*, 973–979. [CrossRef]
168. Mikhed, Y.; Daiber, A.; Steven, S. Mitochondrial Oxidative Stress, Mitochondrial DNA Damage and Their Role in Age-Related Vascular Dysfunction. *Int. J. Mol. Sci.* **2015**, *16*, 15918–15953. [CrossRef] [PubMed]
169. Shokolenko, I.; Venediktova, N.; Bochkareva, A.; Wilson, G.L.; Alexeyev, M.F. Oxidative stress induces degradation of mitochondrial DNA. *Nucleic Acids Res.* **2009**, *37*, 2539–2548. [CrossRef]
170. Coppe, J.P.; Patil, C.K.; Rodier, F.; Sun, Y.; Munoz, D.P.; Goldstein, J.; Nelson, P.S.; Desprez, P.Y.; Campisi, J. Senescence-associated secretory phenotypes reveal cell-nonautonomous functions of oncogenic RAS and the p53 tumor suppressor. *PLoS Biol.* **2008**, *6*, 2853–2868. [CrossRef]
171. Reinhardt, H.C.; Schumacher, B. The p53 network: Cellular and systemic DNA damage responses in aging and cancer. *Trends Genet.* **2012**, *28*, 128–136. [CrossRef]
172. Ou, H.L.; Schumacher, B. DNA damage responses and p53 in the aging process. *Blood* **2018**, *131*, 488–495. [CrossRef]
173. White, R.R.; Vijg, J. Do DNA Double-Strand Breaks Drive Aging? *Mol. Cell* **2016**, *63*, 729–738. [CrossRef]
174. Shtutman, M.; Chang, B.D.; Schools, G.P.; Broude, E.V. Cellular Model of p21-Induced Senescence. *Methods Mol. Biol.* **2017**, *1534*, 31–39. [CrossRef]
175. Zhang, H. Molecular signaling and genetic pathways of senescence: Its role in tumorigenesis and aging. *J. Cell Physiol.* **2007**, *210*, 567–574. [CrossRef] [PubMed]
176. Mirzayans, R.; Andrais, B.; Hansen, G.; Murray, D. Role of p16(INK4A) in Replicative Senescence and DNA Damage-Induced Premature Senescence in p53-Deficient Human Cells. *Biochem. Res. Int.* **2012**, *2012*, 951574. [CrossRef] [PubMed]
177. Mantovani, A.; Allavena, P.; Sica, A.; Balkwill, F. Cancer-related inflammation. *Nature* **2008**, *454*, 436–444. [CrossRef]
178. Qian, S.; Golubnitschaja, O.; Zhan, X. Chronic inflammation: Key player and biomarker-set to predict and prevent cancer development and progression based on individualized patient profiles. *EPMA J.* **2019**, *10*, 365–381. [CrossRef]
179. Greene-Schloesser, D.; Robbins, M.E. Radiation-induced cognitive impairment—From bench to bedside. *Neuro. Oncol.* **2012**, *14*, iv37–iv44. [CrossRef]
180. Malaquin, N.; Carrier-Leclerc, A.; Dessureault, M.; Rodier, F. DDR-mediated crosstalk between DNA-damaged cells and their microenvironment. *Front. Genet.* **2015**, *6*, 94. [CrossRef] [PubMed]

181. Rodier, F.; Campisi, J. Four faces of cellular senescence. *J. Cell Biol.* **2011**, *192*, 547–556. [CrossRef] [PubMed]
182. Hubackova, S.; Krejcikova, K.; Bartek, J.; Hodny, Z. IL1- and TGFbeta-Nox4 signaling, oxidative stress and DNA damage response are shared features of replicative, oncogene-induced, and drug-induced paracrine 'bystander senescence'. *Aging* **2012**, *4*, 932–951. [CrossRef] [PubMed]
183. Tominaga, K. The emerging role of senescent cells in tissue homeostasis and pathophysiology. *Pathobiol. Aging Age Relat. Dis.* **2015**, *5*, 27743. [CrossRef] [PubMed]
184. Chien, Y.; Scuoppo, C.; Wang, X.; Fang, X.; Balgley, B.; Bolden, J.E.; Premsrirut, P.; Luo, W.; Chicas, A.; Lee, C.S.; et al. Control of the senescence-associated secretory phenotype by NF-kappaB promotes senescence and enhances chemosensitivity. *Genes Dev.* **2011**, *25*, 2125–2136. [CrossRef]
185. Salminen, A.; Kauppinen, A.; Kaarniranta, K. Emerging role of NF-kappaB signaling in the induction of senescence-associated secretory phenotype (SASP). *Cell. Signal.* **2012**, *24*, 835–845. [CrossRef]
186. Acosta, J.C.; Banito, A.; Wuestefeld, T.; Georgilis, A.; Janich, P.; Morton, J.P.; Athineos, D.; Kang, T.W.; Lasitschka, F.; Andrulis, M.; et al. A complex secretory program orchestrated by the inflammasome controls paracrine senescence. *Nat. Cell Biol.* **2013**, *15*, 978–990. [CrossRef]
187. Qiao, Y.; Wang, P.; Qi, J.; Zhang, L.; Gao, C. TLR-induced NF-kappaB activation regulates NLRP3 expression in murine macrophages. *FEBS Lett.* **2012**, *586*, 1022–1026. [CrossRef]
188. Orjalo, A.V.; Bhaumik, D.; Gengler, B.K.; Scott, G.K.; Campisi, J. Cell surface-bound IL-1alpha is an upstream regulator of the senescence-associated IL-6/IL-8 cytokine network. *Proc. Natl. Acad. Sci. USA* **2009**, *106*, 17031–17036. [CrossRef]
189. Meeren, A.V.; Bertho, J.M.; Vandamme, M.; Gaugler, M.H. Ionizing radiation enhances IL-6 and IL-8 production by human endothelial cells. *Mediat. Inflamm.* **1997**, *6*, 185–193. [CrossRef]
190. Hayashi, T.; Morishita, Y.; Kubo, Y.; Kusunoki, Y.; Hayashi, I.; Kasagi, F.; Hakoda, M.; Kyoizumi, S.; Nakachi, K. Long-term effects of radiation dose on inflammatory markers in atomic bomb survivors. *Am. J. Med.* **2005**, *118*, 83–86. [CrossRef]
191. Heissig, B.; Rafii, S.; Akiyama, H.; Ohki, Y.; Sato, Y.; Rafael, T.; Zhu, Z.; Hicklin, D.J.; Okumura, K.; Ogawa, H.; et al. Low-dose irradiation promotes tissue revascularization through VEGF release from mast cells and MMP-9-mediated progenitor cell mobilization. *J. Exp. Med.* **2005**, *202*, 739–750. [CrossRef] [PubMed]
192. Panes, J.; Granger, D.N. Neutrophils generate oxygen free radicals in rat mesenteric microcirculation after abdominal irradiation. *Gastroenterology* **1996**, *111*, 981–989. [CrossRef]
193. Okun, E.; Griffioen, K.J.; Mattson, M.P. Toll-like receptor signaling in neural plasticity and disease. *Trends Neurosci.* **2011**, *34*, 269–281. [CrossRef]
194. Kumar, V. Toll-like receptors in the pathogenesis of neuroinflammation. *J. Neuroimmunol.* **2019**, *332*, 16–30. [CrossRef] [PubMed]
195. Salminen, A.; Kaarniranta, K.; Kauppinen, A. Inflammaging: Disturbed interplay between autophagy and inflammasomes. *Aging* **2012**, *4*, 166–175. [CrossRef]
196. Chaurasia, M.; Bhatt, A.N.; Das, A.; Dwarakanath, B.S.; Sharma, K. Radiation-induced autophagy: Mechanisms and consequences. *Free Radic. Res.* **2016**, *50*, 273–290. [CrossRef] [PubMed]
197. Gorbunov, N.V.; Kiang, J.G. Up-regulation of autophagy in small intestine Paneth cells in response to total-body gamma-irradiation. *J. Pathol.* **2009**, *219*, 242–252. [CrossRef]
198. Kiang, J.G.; Smith, J.T.; Agravante, N.G. Geldanamycin analog 17-DMAG inhibits iNOS and caspases in gamma-irradiated human T cells. *Radiat. Res.* **2009**, *172*, 321–330. [CrossRef]
199. Buytaert, E.; Dewaele, M.; Agostinis, P. Molecular effectors of multiple cell death pathways initiated by photodynamic therapy. *Biochim. Biophys. Acta* **2007**, *1776*, 86–107. [CrossRef]
200. Fulda, S.; Gorman, A.M.; Hori, O.; Samali, A. Cellular stress responses: Cell survival and cell death. *Int. J. Cell Biol.* **2010**, *2010*, 214074. [CrossRef]
201. Kim, I.; Lemasters, J.J. Mitophagy selectively degrades individual damaged mitochondria after photoirradiation. *Antioxid. Redox Signal.* **2011**, *14*, 1919–1928. [CrossRef]
202. Patel, N.H.; Sohal, S.S.; Manjili, M.H.; Harrell, J.C.; Gewirtz, D.A. The Roles of Autophagy and Senescence in the Tumor Cell Response to Radiation. *Radiat. Res.* **2020**, *194*, 103–115. [CrossRef]
203. Leidal, A.M.; Levine, B.; Debnath, J. Autophagy and the cell biology of age-related disease. *Nat. Cell Biol.* **2018**, *20*, 1338–1348. [CrossRef] [PubMed]
204. Wong, S.Q.; Kumar, A.V.; Mills, J.; Lapierre, L.R. Autophagy in aging and longevity. *Hum. Genet.* **2020**, *139*, 277–290. [CrossRef] [PubMed]
205. Menzies, F.M.; Fleming, A.; Rubinsztein, D.C. Compromised autophagy and neurodegenerative diseases. *Nat. Rev. Neurosci.* **2015**, *16*, 345–357. [CrossRef]
206. Niccoli, T.; Partridge, L. Ageing as a risk factor for disease. *Curr. Biol.* **2012**, *22*, R741–R752. [CrossRef]
207. Menzies, F.M.; Fleming, A.; Caricasole, A.; Bento, C.F.; Andrews, S.P.; Ashkenazi, A.; Fullgrabe, J.; Jackson, A.; Jimenez Sanchez, M.; Karabiyik, C.; et al. Autophagy and Neurodegeneration: Pathogenic Mechanisms and Therapeutic Opportunities. *Neuron* **2017**, *93*, 1015–1034. [CrossRef]
208. Plaza-Zabala, A.; Sierra-Torre, V.; Sierra, A. Autophagy and Microglia: Novel Partners in Neurodegeneration and Aging. *Int. J. Mol. Sci.* **2017**, *18*. [CrossRef] [PubMed]

209. Robbins, P.D.; Jurk, D.; Khosla, S.; Kirkland, J.L.; LeBrasseur, N.K.; Miller, J.D.; Passos, J.F.; Pignolo, R.J.; Tchkonia, T.; Niedernhofer, L.J. Senolytic Drugs: Reducing Senescent Cell Viability to Extend Health Span. *Annu. Rev. Pharmacol. Toxicol.* **2021**, *61*, 779–803. [CrossRef]
210. Qing, Y.; Li, H.; Zhao, Y.; Hu, P.; Wang, X.; Yu, X.; Zhu, M.; Wang, H.; Wang, Z.; Guo, Q.; et al. One-Two Punch Therapy for the Treatment of T-Cell Malignancies Involving p53-Dependent Cellular Senescence. *Oxid. Med. Cell Longev.* **2021**, *2021*, 5529518. [CrossRef]
211. Lagoumtzi, S.M.; Chondrogianni, N. Senolytics and senomorphics: Natural and synthetic therapeutics in the treatment of aging and chronic diseases. *Free Radic. Biol. Med.* **2021**, *171*, 169–190. [CrossRef] [PubMed]

Review

Control of Neuroinflammation through Radiation-Induced Microglial Changes

Alexandra Boyd [1], Sarah Byrne [1], Ryan J. Middleton [1], Richard B. Banati [1,2] and Guo-Jun Liu [1,2,*]

1   Australian Nuclear Science and Technology Organisation, Sydney, NSW 2234, Australia; boyda@ansto.gov.au (A.B.); sarahbyrne916@gmail.com (S.B.); rym@ansto.gov.au (R.J.M.); rib@ansto.gov.au (R.B.B.)
2   Discipline of Medical Imaging & Radiation Sciences, Faculty of Medicine and Health, Brain and Mind Centre, University of Sydney, Sydney, NSW 2050, Australia
*   Correspondence: gdl@ansto.gov.au

Abstract: Microglia, the innate immune cells of the central nervous system, play a pivotal role in the modulation of neuroinflammation. Neuroinflammation has been implicated in many diseases of the CNS, including Alzheimer's disease and Parkinson's disease. It is well documented that microglial activation, initiated by a variety of stressors, can trigger a potentially destructive neuroinflammatory response via the release of pro-inflammatory molecules, and reactive oxygen and nitrogen species. However, the potential anti-inflammatory and neuroprotective effects that microglia are also thought to exhibit have been under-investigated. The application of ionising radiation at different doses and dose schedules may reveal novel methods for the control of microglial response to stressors, potentially highlighting avenues for treatment of neuroinflammation associated CNS disorders, such as Alzheimer's disease and Parkinson's disease. There remains a need to characterise the response of microglia to radiation, particularly low dose ionising radiation.

Keywords: neuroinflammation; microglia; TSPO; mitochondria; cytokines; antioxidants

Citation: Boyd, A.; Byrne, S.; Middleton, R.J.; Banati, R.B.; Liu, G.-J. Control of Neuroinflammation through Radiation-Induced Microglial Changes. *Cells* **2021**, *10*, 2381. https://doi.org/10.3390/cells10092381

Academic Editor: Fengru Tang

Received: 28 June 2021
Accepted: 2 September 2021
Published: 10 September 2021

Publisher's Note: MDPI stays neutral with regard to jurisdictional claims in published maps and institutional affiliations.

1. Introduction

Ionising radiation (IR) as a diagnostic tool—such as X-ray, or positron emission tomography (PET)—and therapeutic technique has been widely used for decades in the pursuit of better health outcomes for patients [1]. Typically, these methods use lower doses of ionising radiation, and are prescribed when the potential benefits to receiving the procedure outweigh the risks associated with IR [2]. However, fundamental to this practice is the acceptance of the linear-no-threshold (LNT) model; the understanding that ionising radiation initiates detrimental effects to human health in a manner proportional to dosage [2,3]. This model, developed in the 1950s, arose from the extrapolation of the linear dose-response trend at higher doses and applying it to lower doses, where negative effects have been presumed, but not observed [3,4]. The LNT model has been employed by regulatory bodies and accepted by both scientific and medical communities in the absence of an alternate comprehensively proven model [5].

The acceptance of the LNT model may limit the potential of IR as a therapeutic tool except in instances where the benefits heavily outweigh the perceived risk, for example, radiotherapy. Although high doses of ionising radiation (HDIR) have been shown to have adverse health effects, such as carcinogenesis [6,7], the presumption that low dose ionising radiation (LDIR) would also have negative effects simply to a lesser degree is unfounded [8,9]. In fact, hormetic effects have been demonstrated in numerous aspects of human health; for example, sunlight is essential in vitamin D synthesis. However, high doses or prolonged exposures can result in sunburns and the development of skin cancers [10]. Additionally, data from both large-scale nuclear accidents [11,12], and the atomic bombings of Japan [13], do not fully support the LNT model, and there is a growing body of literature suggesting that exposure to LDIR may enhance putative neuroprotective adaptive cellular pathways, such as increased antioxidant levels and reduced reactive oxygen species, which may reduce inflammation within the CNS [14,15]. The increasing availability and utility of IR, and the growing body of evidence suggesting the invalidity of the LNT model, necessitates the reinvestigation of the current conceptions around the safety and dose limitations of IR.

Microglia, the resident immune cells of the central nervous system (CNS), respond to external stressors such as pathogenic invasion and injury by inducing inflammation [16–18]. During microglial activation, microglia are polarised from the M2 anti-inflammatory state to the M1 pro-inflammatory state [16,19,20]. Alterations in the functional states of microglia in response to stressors are characterised by morphological changes and functional plasticity [19,20]. The immune response that ensues is characterised by increased levels of pro-inflammatory cytokines and reactive oxygen species (ROS) which promote the degradation of damaged tissues and pathogenic invaders [21,22]. The M1 functional state is associated with pro-inflammatory cytokines such as IL-1β, IL-6, and TNF-α, whereas the M2 functional state is associated with anti-inflammatory cytokines such as IL-4 and IL-10 [23]. However, it has been shown that following microglial activation, both pro- and anti-inflammatory genes are upregulated [18].

Notably, the expression of translocator protein (TSPO) is upregulated within the mitochondria of activated microglia, and hence is often used as a biomarker of neuroinflammation [24]. This inflammatory effect though to be beneficial to the body, as it is a protective mechanism against disease. However, neuroinflammation has also been implicated in many CNS diseases, such as Alzheimer's disease [25–28], depression [29], and Parkinson's disease [30,31], indicating inappropriate chronic microglial activation. As microglia appear to play a critical role in the onset and maintenance of neuroinflammation, the physiology behind microglial activation and immune modulation pathways are of interest as potential therapeutic targets [32]. This review will examine the contrasting characteristics of activated microglia when exposed to differing degrees of stressors, with a focus on ionising radiation, to highlight the remaining uncertainties regarding microglial activation.

We acknowledge there is contention around both the term "neuroinflammation" and the diseases it applies to [33–35]. A large portion of the scientific community has embraced the term, applying it to any condition where microglial and astrocytic activation can be observed. This has led to the understanding that diseases such as Alzheimer's disease, Parkinson's disease, and depression are "neuroinflammatory" diseases [29,35,36]. However, gene expression data indicate that these diseases are distinct from other known inflammatory diseases [33–35]. Hence it best to restrictively use the term "neuroinflammation" as a shorthand to describe the presence of microglia whose morphology or RNA or protein expression profile is different from that ordinarily observed in health brain tissue. Since under the term "neuroinflammation" microglial state changes (or "microglial activation") can be the consequence of a wide range of local or systemic immune system responses, "neuroinflammation" should not be used as predictors of specific physiological outcomes [33–35]. However, as the term neuroinflammation continues to be used by many authors, this review, too, will refer all instances of microglial activation as to neuroinflammation in its broader meaning but specify the context within which the term needs to be interpreted.

2. Functional States of Microglia Altered by Stressors

Microglial cells play an important role in inflammation, brain development, and the regulation of neuronal networks [37,38]. Historically referred to as the endogenous macrophages of the CNS, this description is not completely comprehensive as it reflects only one specific functionality of the cell and suggests that the mechanism of action for microglia and macrophages are inherently the same. However, the initiation and maintenance of an immune response is a major aspect of the function of microglial cells.

It is believed that microglia arise from primitive macrophages (myeloid progenitor cells) in the embryonic yolk sac of mammals, before infiltrating the brain where they differentiate and reside for life [39,40]. In the adult CNS, the microglial population does not arise from further myeloid progenitor cells, instead the resident microglia self-renew as needed and can rapidly proliferate in response to neural insults [41]. When a threat is detected, such as a pathogen or radiation injury, microglia undergo morphological transformations as they become "activated". Traditionally, microglia were characterised as either active (M1) or resting (M2); however, it is now understood that a spectrum of microglial functional states exist [42,43]. Generally, active microglia adopt an amoeboid, less ramified morphology, allowing them to become more mobile and phagocytotic, whereas resting microglia have a smaller cell body and are highly ramified, allowing them to survey the microenvironment [43–45]. Morphological changes such as cell area, perimeter and ramification length are still frequently utilised in research as an indicator of the degree of neuroinflammation [46,47].

A third morphology is gaining further scientific attention. Dystrophic microglia tend to be small and de-ramified, with beaded or discontinuous processes [48,49]. The cause of dystrophic microglia remains unclear; however, it has been hypothesised that they are linked to ageing [16,48]. A recent study by Shahidehour et al. (2021) found that hypertrophic (activated) microglial numbers, and not dystrophic microglia numbers, were associated with ageing in the CA1 region of the hippocampus [49]. They found no difference in the percentage of hypertrophic microglia between neurodegenerative pathologies (AD, Lewy body dementia and limbic-predominant age-related TDP-43 encephalopathy) and age matched controls; however, in neurodegenerative pathology, 45% of microglia were dystrophic, compared to 9% in the control [49]. Ethanol exposure has also been shown to both reduce overall microglia numbers, whilst increasing the number dystrophic microglia in the hippocampus [50]. Interestingly, the researchers observed that the microglia following ethanol exposure appeared "activated but not to an M1-like, amoeboid state" highlighting the diversity of microglial functional states and how the M1/M2 classification system may need revision [50].

In addition to morphological changes during microglial activation, there are numerous alterations to bio-cellular pathways which promote an inflammatory immune response. These pathways have primarily been established by exposing microglial cells to lipopolysaccharide (LPS), an endotoxin derived from *Escherichia coli*. LPS, and other stressors, which act on toll-like receptor 4 (TLR4) of microglial cells, initiating an inflammatory response [51,52]. This triggers microglial cells to become phagocytotic; engulfing and degrading foreign materials [53]. It has recently been shown that inhibition of TLR4 in an Alzheimer's cell model promotes an M2 phenotype and improves neurological function [54]. The triggering receptor expressed on myeloid cells 2 (TREM2) has also been implicated in the modulation of microglial phagocytosis, with some studies suggesting that TREM2 function, and hence phagocytotic function, may be impaired in instances of neurodegenerative diseases, particularly Alzheimer's disease [55–57].

TLR4 signalling will lead to the downstream phosphorylation of Nuclear Factor Kappa B (NF-κB) inhibitory protein, promoting the expression of pro-inflammatory genes and therefore the expression of pro-inflammatory proteins [58–62]. Among the more commonly known ones are cytokines IL-1β [63], IL-6 [64], and TFNα [65] which play a role in triggering cell cycle arrest and apoptosis [66], the initiation of neurotoxicity [67], and inflammatory signalling [62,68]. There is also emerging evidence that some pro-inflammatory cytokines may enhance the dopaminergic differentiation of neural stem cells, i.e., they may possess neurogenic properties [69]. The transcription of enzymatic genes, such as *iNOS*, and apoptotic genes, such as *Fas-ligand*, results in the translation of proteins which play an active role in inflammation and cellular death [70–72]. In mice models of neurodegeneration, it has been observed that the downregulation of homeostatic microglial genes correlates with neuronal loss, whereas the upregulation of disease associated microglial genes did not [73]. The one exception to this was the *APOE* gene, which directly correlated with neuronal loss [73]. In fact, *APOE4* is a known genetic risk factor for Alzheimer's disease [74]. The study also found that early AD pathology in the human brain was only associated with a loss of homeostatic genes, but not the gain of any disease related genes. This suggests interspecies variability in microglial gene expression during pathological states, meaning the results of murine studies may not translate into the clinic [73].

Increased reactive oxygen and nitrogen species (RONS) concentration are also a hallmark of neuroinflammation [75]. In microglia, the majority of RONS are reactive oxygen species generated via NADPH oxidase; however, they can also originate from other intra- and extracellular sources [76,77]. Interestingly, some studies show that LPS- and a-Synuclein-induced neuroinflammatory responses are attenuated in NOX2 (an isoform of NADPH oxidase) knockout mice, indicating that NOX2 may play a role in microglial activation [78]. NADPH oxidase has also been implicated in cognitive dysfunction in experimental autoimmune encephalomyelitis, being demonstrated to prevent long term potentiation [79]. The iNOS enzyme in microglia and macrophages is responsible for the production of nitric oxide (NO), a precursor to reactive nitrogen species (RNS) which, in conjunction with reactive oxygen species, result in oxidative damage to lipids and proteins [80]. NO can also prevent cell division, often leading to cell death, by inhibiting an enzyme required for DNA synthesis, or by directly causing double stranded DNA breaks [81–84].

On the other hand, microglia display an array of neuroprotective effects. Immune surveillance is constantly occurring under homeostatic conditions, during which microglial processes and filopodia randomly survey the external environment searching for cues which may trigger an immune response [85,86]. Microglia also have a role in synaptic pruning and promoting neurogenesis [87–91]. Synaptic pruning allows for the removal of weaker synaptic connections, which promotes the development of stronger pathways, allowing for clear and direct signal transduction [92]. The importance of microglia in synaptic pruning is demonstrated through the knockout of Cx3cr1 receptors, a critical receptor for microglial migration. Cx3cr1 knockout in mice results in immature brain circuitry, which possess the electrophysiological hallmarks of undeveloped synaptic function [89]. A transcriptomic analysis also demonstrated that the microglial phagocytosis of human apoptotic cells initiated the expression of neurogenic-related genes, strongly suggesting that microglia modulate the process [93]. Recently, one study found that the supernatant of M2 microglial cells (containing molecule 15-deoxy-Δ12,14-prostaglandin J2) promoted neurogenesis and oligodendrogenesis [94], whereas another demonstrated the direct contact between a microglial cell and neuronal dendrites promoted synaptic formation [95]. It has also been shown that microglia are necessary to learning-dependent synaptic formation, and this plasticity is regulated via brain derived neurotrophic factor [96]. It is therefore clear that microglial have a role in neurogenesis.

3. Impact of Ionising Radiation on Healthy Brains by Altering Microglial Function States

Phenotypic changes to microglia can be achieved by the application of ionising radiation, the effects of which are widely believed to be dependent on the dosage and dose schedule. Knowledge of the effects of ionising radiation, primarily established through large scale nuclear events, indicate that HDIR is detrimental to human health [97,98]. The linear no threshold model arose from extrapolating this knowledge and applying to it LDIR [99]. The LNT model is still widely used by many radiation protection organisations today; however, there is an increasing need to re-evaluate this on the basis of recent evidence highlighting the potentially positive effects of LDIR. Currently, there is no consensus on what dose constitutes LDIR or (HDIR). Often, a low dose is defined as below 100 mSv; however, this does not take into account dose rate, cumulative dose or potential interspecies variability. As such, a variety of doses and their effects have been included.

Aside from the widely accepted risk of carcinogenesis, HDIR has detrimental effects on the human CNS, having been linked to the onset of cognitive dysfunction [100–103], deficits of spatial-temporal learning, and reduced memory (2–54 Gy) [104–108]. Ionising radiation has also been shown to cause demyelination (see review [109]), and to disrupt neurogenesis (see reviews [110,111]). Additionally, a reduction in functional connectivity in the anterior cingulate cortex and right insular region has been observed following radiation therapy in nasopharyngeal carcinoma patients, where 68–70 Gy was administered over 30–33 fractions [112]. These deficits tend to peak around 4 months post radiation treatment [113], and can be irreversible. However, it is important to note that most human epidemiological studies arise from opportunity and, therefore, in most human studies the disease which is being treated may be a confounding factor. Similarly, a lot of long-term effects are unknown in these cases due to the typically shorter life span of the patients.

Cell and animal research continue to help bridge the gap in areas which human studies cannot explore. Various studies have shown a range of microglial responses to high dose ionising radiation; notably that irradiation using a high dosage will elicit a neuroinflammatory response [18,114–118]. A dose of 0.5 Gy has been shown to increase the number of microglia in the hippocampus, compared to control and low dose ionising radiation (0.063 Gy); however, the microglia were less ramified than before [118]. A similar increase in microglial density has been found in the cerebellum following a 6 Gy dose [119]. Osman et al. recently observed that a dose of 8 Gy on the juvenile murine brain induced transient microglial activation, as demonstrated through changes in microglial morphology and density [18]. Microglial activation was also associated with a transient increase in apoptotic cell levels, as well as a simultaneous increase in both pro- and anti-inflammatory genes. Notably, the effects of the ionising radiation tended to peak around 6 hrs, after which they began to decline [18].

Often linked with an increased microglial density is an increase in reactive oxygen and nitrogen species production, which leads to protein oxidation and lipid peroxidation [117,120–124]. Oxidative stress results in an increase in DNA damage such as double stranded breaks and concomitant decrease in DNA repair proteins [125] and diminished antioxidant enzyme activity [126]. However, one study contrasts this, showing that a dose of 200 mGy may reduce lipid peroxidation within the brain, with associated increases in catalase and antioxidant concentrations [127]. Pro-inflammatory cytokines such as IL-1β, TNF-α, and IL-6 have all been shown to be increased after irradiation and play a role in the inhibition of neurogenesis and further promotion of inflammation (refer to Figure 1) [106,117,119,128–131]. Other observed reactions that are of interest include changes in mitochondrial membrane potential and permeability [132], increased blood–brain barrier permeability [119,133], and the induction of pyroptosis [134,135]. Crucially, it has been shown that the transcriptome of microglia that have been exposed to high dose radiation is significantly similar to the M1 classical activation phenotype [136], indicating that HDIR activates microglia. Further evidencing the role of HDIR activated microglial in neuroinflammation are recent studies which demonstrate that acute pharmacological microglial depletion following radiation exposure alter the neuroinflammatory response and can prevent cognitive deficits [102,137].

It has previously been shown that ionising radiation alters the brain architecture in mice. In the hippocampus, ionising radiation has resulted in reductions in dendritic complexity and number, and has altered the concentration of synaptic proteins [138]. Recently, microglia have been associated with radiation-induced synaptic loss. Male mice exposed to 10 Gy IRexhibited a significant decrease in immature spinal density; however, the knockout of complement receptor 3 (CR3) was neuroprotective and prevented this loss [139]. Interestingly, female mice (knockout and wildtype) did not experience radiation induced decreases in spinal density, and also displayed a significantly higher number of intersections basally when compared to male mice, suggesting a sex-dependent effect [139]. This study also observed changes in microglial activation markers following the IR; however, not morphological changes, suggesting that morphological changes may not be a reliable indicator of microglial activation [139]. HDIR may also induce neuronal apoptosis by causing cell cycle arrest at the G2 and M checkpoints [117]. This study also indicated that the effects of HDIR may be delayed, with significant apoptosis and chemokine mRNA expression occurring 1-week post exposure [117].

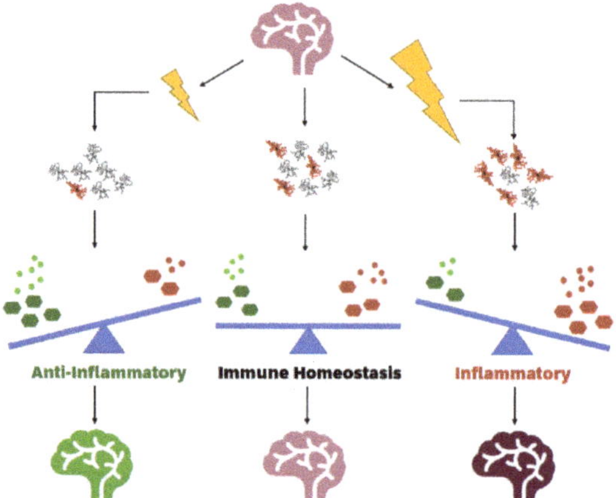

Figure 1. Ionising radiation modulates inflammatory response in a healthy brain by altering microglial functional states. Low dose ionising radiation (**left**) may reduce the number of activated microglia, increasing antioxidants and anti-inflammatory cytokines and thus having a neuroprotective effect when compared to a control brain (**middle**). High dose ionising radiation (**right**) increases the number of activated microglia, which increases oxidants and pro-inflammatory cytokines, creating a neuroinflammatory state.

Additionally, rodent models have allowed the investigation of the CNS effects of prenatal radiation exposure. The paucity of human studies have suggested there is potentially an increased risk of cognitive and health effects on a foetus; however, the Centers for Disease Control and Prevention (CDC, USA) has acknowledged that an acute radiation dose of <100 mGy has no observable non-cancer effects [140–143]. The majority of human studies which involve HDIR come from A-bomb survivors, and these indicate the potential for prenatal IR to cause mental retardation and microcephaly [144]. Animal studies have highlighted dose and pregnancy stage dependent effects of HDIR causing DNA damage, alterations to cell cycle checkpoints and apoptosis in the neocortex [145–147], and have identified behavioural/cognitive changes [145].

The effects of LDIR have been observed to be quite different and conflicting. There is a growing body of evidence which suggests that LDIR may be anti-neuroinflammatory (refer to Figure 1). This hypothesis is termed radiation hormesis. Human cells grown under reduced background radiation manage the stress of acute irradiation at high dose less efficiently than cells cultured under normal background radiation [148], supporting the theory that mild radiation exposure stimulates an adaptive response. Irradiated fruit flies and rodents also have shown enhanced immune systems and extended lifespans compared to non-irradiated controls [149–152]. One murine lifetime study found that an exposure of single 0.063 Gy radiation significantly reduces the risk of the development of many types of tumours, including pheochromocytomas, adenomas, insulinomas, and adenocarcinomas, compared to non-exposed controls [152]. LDIR is also thought to confer protection to cell functioning, molecular structures, synapses and key brain mechanisms such as neurogenesis, as well as inducing reparative functions (see review [153]). These theories have been supported by observed physiological responses LDIR. The suppression of ROS is one such example [154,155]. Studies of occupationally exposed workers found that chronic LDIR (0.1–8.4 mGy per month) was associated with an increased resistance to oxidative stress [156,157]. This effect has also been observed at higher acute doses, with acute 0.2 Gy IR increasing antioxidants in the blood and tissues of rats [127]. Increases in anti-inflammatory cytokines such as IL-4, IL-10, and reductions in inflammatory cytokines such as TFN-α have been observed at 50 mGy [158]; however, this effect has also been observed at higher doses of up to 1 Gy [25,159,160]. Of note, one study found that 100 mGy actually increased inflammatory cytokines [161], and another found that 100 mGy may have a negative effect on cognition, although noted further investigation is required [108]. Notably, 100 mGy appears to disrupt the BBB, which is a key aspect of neuroinflammation [74,162].

A recent study by Ung et al. (2020) specifically examined the effects of a singular radiation event on both murine glial cells and murine behaviour [118]. At the "high" dose of 0.5 Gy, there was an observable decrease in acoustic startle response, exploration, and rearing at 12 months post exposure. At 24 months post exposure, there was a significant increase in the number of microglial cells in the dentate gyrus, and a reduction in both astrocyte number and complex morphology. However, mice exposed to a singular dose of 0.063 Gy radiation had an increased acoustic startle response, increased exploratory behaviour, and increased rearing at 18 months post exposure, and had significant increases in microglial ramification at 24 months (compared to both the control and other irradiated groups of 0.125 and 0.5 Gy) despite no observable increase in microglial density. The astrocyte morphology of this group was also not significantly different to the control group. This study clearly demonstrates the potential for LDIR to be anti-neuroinflammatory in comparison to HDIR [118]. A complementary study by Hladik et al. (2020) observed the effects of a single radiation event on cAMP response element-binding protein (CREB) signalling; a transcription factor involved in memory formation, neuroplasticity and amyloid processing [163]. The CREB pathway, and other associated pathways, were found to be "activated" (as determined by alterations in hippocampal protein levels) by a dose of either 0.063 Gy or 0.125 Gy, whereas these pathways were "deactivated" by a dose of 0.5 Gy. Conditioning, learning, and long-term potentiation were found to be activated by 0.063 Gy or 0.125 Gy and were deactivated at 0.5 Gy. Additionally, a dose of 0.125 Gy was found to deactivate apoptosis. This data further supports the concept that LDIR may allow for neuroprotective cellular adaptive responses in comparison to the detrimental effects of higher doses [163], a premise which is supported by our recent study in press at *Frontiers in Cell and Developmental Biology*, which demonstrated that LDIR (10 mGy) may enhance neuroprotective pathways in the healthy brain [164].

However, radiation hormesis tends to be poorly supported by most human studies. An 11 million person cohort study found that the LDIR dose from a computed tomography (CT) scan (~40 mGy) during childhood and adolescent correlates with an increased incidence of cancers [165]. A recent meta-analysis examining a potential link between low dose ionising radiation exposure in adulthood and cancer similarly found that, after excluding studies with potential biases from the null, there was still a positive risk estimate reported by many studies [166]. Another systematic review found no positive effects of LDIR on neurodevelopment and cognition; however, it indicated that the evidence of adverse effects is "limited to inadequate" [167]. In a population of adults with congenital heart disease, a greater exposure to LDIR from cardiac procedures correlates with an increased cancer incidence [168]; however, the population of "adults without cancer" were significantly younger therefore had less comorbidities than the "adults with cancer population." A review by Lumniczky, Szatmári, and Sáfrány (2017) found that LDIR could result in cognitive defects and other unfavourable outcomes in both human and animal populations, and resulted in the induction of different molecular and cellular mechanisms [169]. The authors called into question the safety of LDIR for diagnostic purposes [169]. Another review by Tang and Loganovsky (2018) concluded that LDIR (<100 mGy) or low dose rate ionising radiation (<6 mSv/hr) may or may not induce cancer, depending on a variety of factors including demographics, lifestyle, and diagnostic accuracy [170]. However, LDIR may increase incidences of vascular diseases, cognitive and mental health disorders, eye diseases, and other pathologies, whilst reducing cancer mortality and mutations, and increasing longevity [170]. Despite more studies beginning to investigate radiosensitivity [171], it remains unclear what effect LDIR truly has on different organs or tissues. One important effect which cannot be ignored is the potential effect of LDIR on cellular senescence. Carbon ion irradiation (1 mGy) can lead to premature senescence in human lung fibroblasts; however, this effect was not observed following 1 mGy gamma irradiation [172].

Given that ionising radiation can induce the polarisation of resting microglia into an activated state, there is the potential to explore different doses of ionising radiation as a tool to trigger the switching between activation states; in particular, from inflammatory to anti-inflammatory state by LDIR. We know that the functional state of microglia is dynamic, and that changing the environment can be a mechanism of manipulating them [173]. It has already been demonstrated that there is potential to convert microglia into an inflammatory, M1 phenotype then alter the environment to trigger a switch back to the M2 phenotype [174]. There is also evidence to suggest that the microglial activation state can be manipulated by repeated challenges with stimuli that influence future microglial behaviour upon subsequent stress [175]. Therefore, future studies should investigate whether challenging microglia with different doses and dose schedules of ionising radiation can enable the control of the microglial functional state, as it is a potential therapeutic tool for neuroinflammation-associated pathologies such as Alzheimer's and Parkinson's disease.

4. Impact of Low Dose Ionising Radiation on Neurodegenerative Diseases

The positive effects of LDIR on models of Alzheimer's disease have been seen in mice, particularly in transgenic AD female mice (refer to Figure 2) [176]. A dose of 100 mGy improved locomotor activity in Alzheimer's-like transgenic (Tg) female mice, improved their grip strength and reduced Aβx-40 levels. The same dose in male Tg mice had few effects but did significantly reduce motor coordination. Interestingly, the higher dose of 500 mGy also improved locomotor activity in Tg females for open maze test and for Tg males in the Y-maze, improved motor coordination in Tg females and reduced Aβx-40 levels. Arguably, the most important takeaway from this study is that, without radiation, Tg female mice have higher Aβ levels than Tg male mice; however, radiation reduced these levels in the female Tg mice, which correlated with decreased microglial activation as determined by CD68 receptor staining [176]. Recently, rat models of AD were also shown to have improved memory performance in response to a higher dose of ionising radiation of 2 Gy/day for 5 days, without increasing neuroinflammation or amyloid load [177]. LDIR has also been shown to promote an M2 morphology in LPS treated mice microglial (BV2) cells [25].

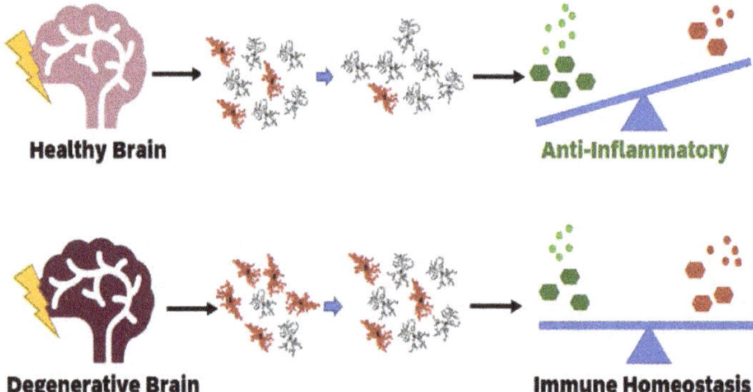

Figure 2. Low dose ionising radiation strengthens immunity in the healthy brain and reduces neurodegenerative disease by changing microglial functional states. A healthy brain exposed to low dose ionising radiation may experience anti-inflammatory effects, whereas in a degenerative brain, such as in Alzheimer's disease, low dose ionising radiation may lessen the severity of neuroinflammation and promote a shift towards a more "normal" brain environment.

Human case studies have used low dose ionising radiation as a treatment for neurodegenerative disorders (refer to Figure 2). The most well-known is a series of articles following a patient with end stage AD. After 2 CT scans (~40 mGy each) to detect anatomical changes, the patient displayed several behavioural changes noticed by her family and carers including signs of old memory return, improved motor function, and short three- to five-word sentence formation [178]. The patient then began to receive a CT scan approximately every 2 weeks. Interestingly, following the fifth CT scan the patient exhibited significant decrease in cognitive and motor abilities, potentially demonstrating the fine balance between both dosage and dose rate on whether radiation has advantageous or deleterious effects. The patient recovered from this setback, and continued to receive frequent, though more spaced out, CT scans [178–180]. Additionally, having observed these positive effects on his wife, the partner of the Alzheimer's patient opted to receive CT scans to treat his Parkinson's disease [179]. He received a CT scan every few months. The patient was able to reduce the dose of his medication, his tremors reduced, he was less constipated and his vision improved [180]. These promising results prompted a small pilot study where CT scans were used on four patients with AD [181]. Minor quantitative changes were observed; however, there were "remarkable improvements" in qualitative measures such as communication and behavioural changes. One of the four patients showed no improvement [181]. Together, this case and pilot study show the positive effects low dose radiation may confer on Alzheimer's disease; however, until we can better control and understand LDIR as a treatment, it is not feasible to use therapeutically.

It is also important to consider the effects of radiation in multiple sclerosis (MS), where the immune system is already under the stress of an autoimmune condition. There is some evidence which may suggest that ionising exposure may represent a greater risk in MS than in healthy individuals. A clinical trial on total lymphoid irradiation (TLI) (19.8 Gy) for the treatment of MS found that irradiation reducing lymphocyte numbers to <900 mm^{-3} slowed the progression of the disease [182]. A follow-up found that any toxicity was "mild and transient" and suggested that the benefits of the treatment outweighed the disadvantages; however, it did indicate that menopause was induced in two patients and a staphylococcal pneumonia infection in another [183]. A further statement from the authors discussed the causes of five deaths among the cohort. All of the deceased were in a high lymphocyte, poor prognosis category, so it is plausible that the deaths may have occurred regardless; however, there is also the possibility that the risk for serious infections is increased after TLI [184]. The patient with the induced staphylococcal pneumonia infection died from aspiration, providing merit to the latter theory [184].

Additionally, a 2013 case report suggested that conventional doses of ionising radiation used to treat meningioma induced the onset of multiple sclerosis in a 43-year-old woman, suggesting a potential connection but not establishing a causal relationship [185]. Other case and cohort studies have indicated a similar trend [186–189]. One cohort study even found that X-ray exposure site was relevant, with chest X-rays, skull X-rays, and brain CT scans all aligning with a higher incidence of MS [190]. Conversely, there are studies that indicate that there is an inverse relationship between IR, specifically ultra-violet B (UVB) exposure, and the development of MS [191,192]. Gamma Knife Radiosurgery (85 Gy) has been used to treat trigeminal neuralgia in MS patients, with 82% of patients reporting a reporting a reduction in pain following one treatment [193]. However, the authors acknowledge this needs to be explored further, with increased radiation toxicity being observed by some studies [194]. One murine study observed that repeated 0.5 Gy doses (to a dose of 10 Gy) of gamma radiation reduced many autoimmune symptoms, including splenomegaly, lymphadenopathy, and proteinuria [195]. It therefore remains unclear the effects of ionising radiation in pre-existing immune conditions.

5. TSPO as a Biomarker for Changes in Microglia

To quantify microglial activation, appropriate biological markers are necessary. These biomarkers can take on a variety of forms, such as cell receptors or cytokines. The most commonly used marker of microglial activation is ionised calcium binding molecule 1 (Iba1), also known as allograft inflammatory factor 1. In the CNS, it is expressed solely by activated microglial cells and is responsible for membrane ruffling and phagocytosis [196]. Iba1 is highly conserved across species and can be easily detected through anti-Iba1 antibodies [197].

Pro-inflammatory and anti-inflammatory cytokines and chemokines can also be examined to elucidate the current activation state of microglia. Increased concentrations of pro-inflammatory molecules such as IL-6, IL-1β, IL-18, TNF-α, IFN-γ, CCL5, or GM-CSF all indicate a greater proportion of activated microglia, and hence likely a greater degree of neuroinflammation [198]. Opposingly, higher concentrations of IL-4, IL-10, TGFβ, or CCL22 all signify an anti-inflammatory environment [198]. However, cytokine and chemokine analyses are often used to complement Iba-1 data, rather than as standalone data, as they are not "neuroinflammation" specific and often require the use of homogenised tissues, leading to spatial information being lost [199]. Therefore, these molecules tend not to be the best biomarker of neuroinflammation.

Mitochondrial translocator protein 18 kDa (TSPO), like Iba-1, has been shown to be upregulated in microglia under conditions of stress and pathology [200–203] and serves as a biomarker of neuroinflammation [204,205], particularly for in vivo molecular imaging such as a PET scan [206]. TSPO has a low basal expression in the central nervous system, predominantly by endothelial cells [207]. Although its exact function remains unclear, TSPO knockout mice have both helped to disprove its role in cholesterol translocation and steroidogenesis [208–211], and highlight a role in microglial activation and mitochondrial function [212]. TSPO is an attractive target for studying the effects of ionising radiation as it has been implicated in ROS production and ROS-mediated oxidative damage [213,214], and the addition of TSPO ligands, such as PK11195 or Midazolam, have been shown to reduce pro-inflammatory gene expression, accompanied by a reduction in activated microglia [215–217]. Manipulating TSPO expression or function will allow the understanding of the link between mitochondrial function and neuroinflammation to strengthen, supporting the future development of therapeutics targeting these elements. The use of a TSPO knockout model to observe the behaviour of TSPO under varying conditions, and the potential impact on the CNS microenvironment, provides a unique tool to characterise microglial response in the presence and absence of TSPO to varying stressors and would be invaluable to future study [209]. Our recent study demonstrates that there are decreases in TSPO and Iba1 mRNA and protein levels in brain, and proinflammatory cytokine IL6 in blood plasma, following 10 mGy IR, and increases in TSPO protein expression at 2 Gy in the brains of healthy mice and in primary cultured microglia [164]. As the levels of neuroinflammation in the healthy brain are minimal, there is little inflammation to be reduced by low dose radiation. However, the clear trend towards downregulation of TSPO and Iba1 expression indicates that LDIR may reduce microglial activation, and hence neuroinflammation. Further investigations should be undertaken in models of neurodegenerative diseases, where elevated levels of neuroinflammation are observed.

One potential downside, elucidated by a recent study, is that neuronal activity increases TSPO levels within the brain, specifically in neurons, meaning it may be an unreliable measure of glial activation [218]. Additionally, the post-mortem brains of late stage AD and Dementia with Lewy Bodies have shown similar TSPO levels to age matched controls, and even showed a reduction in some areas such as the substantia nigra [219]. Here, a reduction in TSPO may not indicate a reduction in neuroinflammation, rather may reflect "dystrophy, senescence and death, or dysfunction of mitochondria in the microglia" [219]. Another recent study concluded that whilst TSPO effectively marks activated microglia, it is not a predictor of neuronal loss, and as such only marks neuroinflammation and not neurodegeneration [220]. Finally, there is evidence to suggest that TSPO ligands may bind to plasma proteins, and therefore are unavailable to bind to TSPO, affecting the accuracy of PET scan results regarding neuroinflammation [221].

6. Conclusions and Future Directions

As the innate immune cells of the central nervous system, microglia facilitate the initiation and maintenance of basic immunity and neuroinflammation. The activation of microglia through stressors such as radiation prompts the transcription of pro-inflammatory genes, leading to the release of molecules such as IL-6 and TFN-α and the adoption of an ameboid, phagocytotic morphology. However, there is a growing body of evidence that LDIR may not act as a stressor, rather that LDIR may confer neuroprotection. Whilst the molecular mechanisms behind this are largely unknown, many cellular and animal studies have found LDIR promotes longevity, neurogenesis, and cognition, while decreasing ROS production. There is paucity of human studies on the effects of LDIR on microglia, and hence neuroinflammation. As such, the current understanding of radiation hormesis is poor. As highlighted throughout the review, the dosage which divides the positive and negative effects of radiation is hard to determine, with many studies providing conflicting results. Furthermore, the duration (transient or indefinite) of any potential positive outcomes is undetermined, and the interactions between different doses and exposure frequencies are under-explored. It may be that case that repeated "low dose" treatments will accrue and transform a positive effect into a detrimental one, as seen in the case study where CT scans were used to treat a patient with AD [178]. Whilst the case studies of a woman with AD and a man with PD show promise in the potential treatment of neuroinflammatory conditions [179], the large cohort study of people who received a CT scan in childhood or adolescence indicated an increased cancer risk [165]. It remains unclear whether the effects of LDIR are beneficial to or adversely impact human health. Further animal and cell work is required to elucidate the mechanisms behind the observed microglial mediated neuroprotection, and research needs to be undertaken regarding dose-rate, age when exposed (adulthood vs adolescence) and whether the impact is not only beneficial in disease states but also to a healthy population.

Author Contributions: Conceptualization, G.-J.L.; writing—original draft preparation, A.B. and S.B.; writing—review and editing, G.-J.L., R.J.M., and R.B.B. All authors have read and agreed to the published version of the manuscript.

Funding: This research received no external funding.

Institutional Review Board Statement: Not applicable.

Informed Consent Statement: Not applicable.

Acknowledgments: Not applicable.

Conflicts of Interest: The authors declare no conflict of interest.

References

1. Furdui, C.M. Ionizing radiation: Mechanisms and therapeutics. *Antioxid. Redox Signal.* **2014**, *21*, 218–220. [CrossRef] [PubMed]
2. Lin, E.C. Radiation Risk From Medical Imaging. *Mayo Clin. Proc.* **2010**, *85*, 1142–1146. [CrossRef] [PubMed]
3. Seong, K.M.; Seo, S.; Lee, D.; Kim, M.; Lee, S.; Parl, S.; Jin, Y.W. Is the Linear No-Threshold Dose-Response Paradigm Still Necessary for the Assessment of Health Effects of Low Dose Radiation? *J. Korean Med. Sci.* **2016**, *31*, S10–S23. [CrossRef] [PubMed]
4. Calabrese, E.J.; O'Connor, M.K. Estimating risk of low radiation doses—A critical review of the BEIR VII report and its use of the linear no-threshold (LNT) hypothesis. *Radiat. Res.* **2014**, *182*, 463–474. [CrossRef] [PubMed]
5. Harbron, R.W. Cancer risks from low dose exposure to ionising radiation—Is the linear no-threshold model still relevant? *Radiography* **2012**, *18*, 28–33. [CrossRef]
6. Nakamura, N. A hypothesis: Radiation carcinogenesis may result from tissue injuries and subsequent recovery processes which can act as tumor promoters and lead to an earlier onset of cancer. *Br. J. Radiol.* **2020**, *93*, 20190843. [CrossRef]
7. Williams, D. Radiation carcinogenesis: Lessons from Chernobyl. *Oncogene* **2008**, *27* (Suppl. 2), S9–18. [CrossRef]
8. Boice, J.D.; Held, K.D.; Shore, R.E. Radiation epidemiology and health effects following low-level radiation exposure. *J. Radiol. Prot.* **2019**, *39*, 14–27. [CrossRef] [PubMed]
9. Tharmalingam, S.; Sreetharan, S.; Brooks, A.L.; Boreham, D.R. Re-evaluation of the linear no-threshold (LNT) model using new paradigms and modern molecular studies. *Chem Biol. Interact.* **2019**, *301*, 54–67. [CrossRef] [PubMed]
10. Vaiserman, A.; Koliada, A.; Zabuga, O.; Socol, Y. Health Impacts of Low-Dose Ionizing Radiation: Current Scientific Debates and Regulatory Issues. *Dose-Reponse* **2018**, *16*, 1–27. [CrossRef] [PubMed]

11. Socol, Y. Reconsidering Health Consequences of the Chernobyl Accident. *Dose-Response* **2015**, *13*. [CrossRef] [PubMed]
12. *UNSCEAR 2020 REPORT: SOURCES, EFFECTS AND RISKS OF IONIZING RADIATION*; United Nations: New York, NY, USA, 2020.
13. Sutou, S. Black rain in Hiroshima: A critique to the Life Span Study of A-bomb survivors, basis of the linear no-threshold model. *Genes Environ.* **2020**, *42*, 1–11. [CrossRef]
14. Calabrese, E.J. Key studies used to support cancer risk assessment questioned. *Environ. Mol. Mutagenesis* **2011**, *52*, 595–606. [CrossRef] [PubMed]
15. Calabrese, E.J. Muller's Nobel Prize Lecture: When ideology prevailed over science. *Toxicol. Sci.* **2012**, *126*, 1–4. [CrossRef]
16. Luo, X.-G.; Chen, S.-D. The changing phenotype of microglia from homeostasis to disease. *Transl. Neurodegener.* **2012**, *1*, 9. [CrossRef] [PubMed]
17. Donat, C.K.; Scott, G.; Gentleman, S.M.; Sastre, M. Microglial Activation in Traumatic Brain Injury. *Front Aging Neurosci.* **2017**, *9*, 208. [CrossRef]
18. Osman, A.M.; Sun, Y.; Burns, T.C.; He, L.; Kee, N.; Oliva-Vilarnau, N.; Alevyzaki, A.; Zhou, K.; Louhivuori, L.; Uhlen, P.; et al. Radiation Triggers a Dynamic Sequence of Transient Microglial Alterations in Juvenile Brain. *Cell Rep* **2020**, *31*, 107699. [CrossRef] [PubMed]
19. Banati, R.B.; Egensperger, R.; Maassen, A.; Hager, G.; Kreutzberg, G.W.; Graeber, M.B. Mitochondria in activated microglia in vitro. *J. Neurocytol.* **2004**, *33*, 535–541. [CrossRef] [PubMed]
20. Moneta, M.E.; Gehrmann, J.; Töpper, R.; Banati, R.B.; Kreutzberg, G.W. Cell adhesion molecule expression in the regenerating rat facial nucleus. *J. Neuroimmunol.* **1993**, *45*, 203–206. [CrossRef]
21. Banati, R.B.; Gehrmann, J.; Schubert, P.; Kreutzberg, G.W. Cytotoxicity of microglia. *Glia* **1993**, *7*, 111–118. [CrossRef] [PubMed]
22. Ye, J.; Jiang, Z.; Chen, X.; Liu, M.; Li, J.; Liu, N. Electron transport chain inhibitors induce microglia activation through enhancing mitochondrial reactive oxygen species production. *Exp. Cell Res.* **2016**, *340*, 315–326. [CrossRef] [PubMed]
23. Martinez, F.O.; Gordon, S. The M1 and M2 paradigm of macrophage activation: Time for reassessment. *F1000Prime Rep.* **2014**, *6*, 13. [CrossRef] [PubMed]
24. Lee, Y.; Park, Y.; Nam, H.; Lee, J.-W.; Yu, S.-W. Translocator protein (TSPO): The new story of the old protein in neuroinflammation. *Biochem. Mol. Biol. Rep.* **2020**, *53*, 20–27. [CrossRef]
25. Kim, S.; Chung, H.; Mai, H.N.; Nam, Y.; Shin, S.J.; Park, Y.H.; Chung, M.J.; Lee, J.K.; Rhee, H.Y.; Jahng, G.-H.; et al. Low-Dose Ionizing Radiation Modulates Microglia Phenotypes in the Models of Alzheimer's Disease. *Int. J. Mol. Sci.* **2020**, *21*, 4532. [CrossRef] [PubMed]
26. Tournier, B.; Tsartsalis, S.; Ceyzériat, K.; Garibotto, V.; Millet, P. In Vivo TSPO Signal and Neuroinflammation in Alzheimer's Disease. *Cells* **2020**, *9*, 1941. [CrossRef] [PubMed]
27. Nogueria, M.L.; Epelbaum, S.; Steyaert, J.M.; Dubois, B.; Schwartz, L. Mechanical tress models of Alzheimer's disease pathology. *Alzheimers Dement.* **2016**, *12*, 324–333. [CrossRef] [PubMed]
28. Jiang, T.; Zhang, Y.D.; Chen, Q.; Gao, Q.; Zhu, X.C.; Zhou, J.S.; Shi, J.Q.; Lu, H.; Tan, L.; Yu, J.T. TREM2 modifies microglial phenotype and provides neuroprotection in P301S tau transgenic mice. *Neuropharmacology* **2016**, *105*, 196–206. [CrossRef] [PubMed]
29. Troubat, R.; Barone, P.; Leman, S.; Desmidt, T.; Cressant, A.; Atanasova, B.; Brizard, B.; El Hage, W.; Surget, A.; Belzung, C.; et al. Neuroinflammation and depression: A review. *Eur. J. Neurosci.* **2020**, *53*, 151–171. [CrossRef] [PubMed]
30. McGeer, P.L.; Itagaki, S.; Boyes, B.E.; McGeer, E.G. Reactive microglia are positive for HLA-DR in the substantia nigra of Parkinson's and Alzheimer's disease brains. *Neurology* **1988**, *38*, 1285–1291. [CrossRef]
31. Terada, T.; Yokokura, M.; Yoshikawa, E.; Futatsubashi, M.; Kono, S.; Konishi, T.; Miyajima, H.; Hashizume, T.; Ouchi, Y. Extrastriatal spreading of microglial activation in Parkinson's disease: A positron emission tomography study. *Ann. Nucl. Med.* **2016**, *30*, 579–587. [CrossRef]
32. Akhmetzyanova, E.; Kletenkov, K.; Mukhamedshina, Y.; Rizvanov, A. Different Approaches to Modulation of Microglia Phenotypes After Spinal Cord Injury. *Front Syst. Neurosci.* **2019**, *13*, 37. [CrossRef]
33. Filiou, M.D.; Arefin, A.S.; Moscato, P.; Graeber, M.B. 'Neuroinflammation' differs categorically from inflammation: Transcriptomes of Alzheimer's disease, Parkinson's disease,schizophrenia and inflammatory diseases compared. *Neurogenetics* **2014**, *15*, 201–212. [CrossRef] [PubMed]
34. Graeber, M.B.; Li, W.; Rodriguez, M.L. Role of microglia in CNS inflammation. *FEBS Lett.* **2011**, *585*, 3798–3805. [CrossRef] [PubMed]
35. Andoh, M.; Koyama, R. Microglia regulate synaptic development and plasticity. *Dev. Neurbiology* **2021**. [CrossRef] [PubMed]
36. Guzman-Martinez, L.; Maccioni, R.B.; Andrade, V.; Navarrete, L.P.; Pastor, M.G.; Ramos-Escobar, N. Neuroinflammation as a Common Feature of Neurodegenerative Disorders. *Front Pharm.* **2019**, *10*, 1008. [CrossRef] [PubMed]
37. Brown, G.C.; Vilalta, A. How Microglia kill neurons. *Brain Reserach* **2015**, *1628*, 288–297. [CrossRef]
38. York, E.M.; Bernier, L.-P.; MacVicar, B.A. Microglial modulation of neuronal activity in the healthy brain. *Dev. Neurobiol.* **2017**, *78*, 593–603. [CrossRef] [PubMed]
39. Ginhoux, F.; Greter, M.; Leboeuf, M.; Nandi, S.; See, P.; Gokhan, S.; Mehler, M.F.; Conway, S.J.; Ng, L.G.; Stanley, E.R.; et al. Fate Mapping Analysis Reveals That Adult Microglia Derive from Primitive Macrophages. *Science* **2010**, *330*, 841–845. [CrossRef] [PubMed]

40. Kierdorf, K.; Erny, D.; Goldmann, T.; Sander, V.; Schulz, C.; Perdiguero, E.G.; Wieghofer, P.; Heinrich, A.; Riemke, P.; Hölscher, C.; et al. Microglia emerge from erythromyeloid precursors via Pu.1- and Irf8-dependent pathways. *Nat. Neurosci.* **2013**, *16*, 273–280. [CrossRef] [PubMed]

41. Huang, Y.; Xu, Z.; Xiong, S.; Sun, F.; Qin, G.; Hu, G.; Wang, J.; Zhao, L.; Liang, Y.-X.; Wu, T.; et al. Repopulated microglia are solely derived from the proliferation of residual microglia after acute depletion. *Nat. Neurosci.* **2018**, *21*, 530–540. [CrossRef] [PubMed]

42. Bachiller, S.; Jiménez-Ferrer, I.; Paulus, A.; Yang, Y.; Swanberg, M.; Deierborg, T.; Boza-Serrano, A. Microglia in Neurological Diseases: A Road Map to Brain-Disease Dependent-Inflammatory Response. *Front. Cell. Neurosci.* **2018**, *12*, 488. [CrossRef] [PubMed]

43. Boche, D.; Perry, V.H.; Nicoll, J.A. Review: Activation patterns of microglia and their identification in the human brain. *Neuropathol. Appl. Neurobiol.* **2013**, *39*, 3–18. [CrossRef] [PubMed]

44. Morrison, H.; Young, K.; Qureshi, M.; Rowe, R.K.; Lifshitz, J. Quantitative microglia analyses reveal diverse morphologic responses in the rat cortex after diffuse brain injury. *Sci. Rep.* **2017**, *7*, 13211. [CrossRef] [PubMed]

45. Fernández-Arjona, M.d.M.; Grondona, J.M.; Granados-Durán, P.; Fernández-Llebrez, P.; Fernández-Llebrez, M.D. Microglia Morphological Categorization in a Rat Model of Neuroinflammation by Hierarchical Cluster and Principal Components Analysis. *Front. Cell. Neurosci.* **2017**, *11*, 1–22. [CrossRef] [PubMed]

46. Adeluyi, A.; Guerin, L.; Fisher, M.L.; Galloway, A.; Cole, R.D.; Chan, S.S.L.; Wyatt, M.D.; Davis, S.W.; Freeman, L.R.; Ortinski, P.I.; et al. Microglia morphololgy and proinflammatory signalling in the nucleus accumbens during nictoine withdrawal. *Sci. Adv.* **2019**, *5*, eaax7031. [CrossRef] [PubMed]

47. Davies, D.S.; Ma, J.; Jegathees, T.; Goldsbury, C. Microglia show altered morphology and reduced arborization in human brain during aging and Alzheimer's disease. *Brain Pathol.* **2017**, *27*, 795–808. [CrossRef] [PubMed]

48. Angelova, D.M.; Brown, D.R. Microglia and the aging brain: Are senescent microglia the key to neurodegeneration? *J. Neuer.* **2019**, *151*, 676–688. [CrossRef] [PubMed]

49. Shahidehpour, R.K.; Higdon, R.E.; Crawford, N.G.; Nelter, J.H.; Ighodaro, E.T.; Patel, E.; Price, D.; Nelson, P.T.; Bachstetter, A.D. Dystrophic microglia are a disease associated microglia morphology in the human brain. *Neurobiol. Aging* **2021**, *22*, 19–27. [CrossRef]

50. Marshall, S.A.; McClain, J.A.; Wooden, J.I.; Nixon, K. Microglia Dystrophy Following Binge-Like Alcohol Exposure in Adolescent and Adult Male Rats. *Front. Neuroanat.* **2020**, *14*, 52. [CrossRef] [PubMed]

51. Shimazu, R.; Akashi, S.; Ogata, H.; Nagai, Y.; Fukudome, K.; Miyake, K.; Masao, K. MD-2, a Molecule that Confers Lipopolysaccharide Responsiveness on Toll-like Receptor 4. *J. Exp. Med.* **1999**, *189*, 1777–1782. [CrossRef]

52. Fernández-Arjona, M.d.M.; Grondona, J.M.; Fernández-Llebrez, P.; López-Ávalos, M.D. Microglial activation by microbial neuraminidase through TLR2 and TLR4 receptors. *J. Neuroinflammation* **2019**, *16*, 254. [CrossRef] [PubMed]

53. Galloway, D.A.; Phillips, A.E.M.; Owen, D.R.J.; Moore, C.S. Phagocytosis in the Brain: Homeostasis and Disease. *Front. Immunol.* **2019**, *10*, 790. [CrossRef] [PubMed]

54. Cui, W.; Sun, C.; Ma, Y.; Wang, S.; Wang, X.; Zang, Y. Inhibition of TLR4 Induces M2 Microglial Polarization and Provides Neuroprotection via the NLRP3 Inflammasome in Alzheimer's Disease. *Front. Neurosci.* **2020**, *14*, 444. [CrossRef] [PubMed]

55. Lucin, K.M.; O'Brien, C.E.; Bieri, G.; Czirr, E.; Mosher, K.I.; Abbey, R.J.; Mastroeni, D.F.; Jospeh, R.; Spencer, B.; Masliah, E.; et al. Microglial Beclin 1 Regulates Retromer Trafficking and Phagocytosis and Is Impaired in Alzheimer's Disease. *Neuron* **2013**, *79*, 873–886. [CrossRef] [PubMed]

56. Krasemann, S.; Madore, C.; Cialic, R.; Baufeld, C.; Calcagno, N.; El Fatimy, R.; Beckers, L.; O'Loughlin, E.; Xu, Y.; Fanek, Z.; et al. The TREM2-APOE Pathway Drives the Transcriptional Phenotype of Dysfunctional Microglia in Neurodegenerative Diseases. *Immunity* **2017**, *47*, 566–581. [CrossRef]

57. Ren, M.; Guo, Y.; Wei, X.; Yan, S.; Qin, Y.; Zhang, X.; Jiang, F.; Lou, H. TREM2 overexpression attenuates neuroinflammation and protects dopaminergic neurons in experimental models of Parkinson's disease. *Exp. Neurol.* **2018**, *302*, 205–213. [CrossRef] [PubMed]

58. Ghosh, S.; Karin, M. Missing Pieces in the NF-κB Puzzle. *Cell* **2002**, *109*, 81–96. [CrossRef]

59. Frakes, A.E.; Ferraiuolo, L.; Haidet-Phillips, A.M.; Schmelzer, L.; Braun, L.; Miranda, C.J.; Ladner, K.J.; Bevan, A.K.; Foust, K.D.; Godbout, J.P.; et al. Microglia induce motor neuron death via the classical NF-κB pathway in amyotrophic lateral sclerosis. *Neuron* **2015**, *81*, 1009–1023. [CrossRef]

60. Cunha, C.; Gomes, C.; Vaz, A.R.; Brites, D. Exploring New Inflammatory Biomarkers and Pathways during LPS-Induced M1 Polarization. *Mediat. Inflamm.* **2016**, *2016*, 6986175. [CrossRef] [PubMed]

61. Brás, J.P.; Bravo, J.; Freitas, J.; Barbosa, M.A.; Santos, S.G.; Summavieele, T.; Almeida, M.I. TNF-alpha-induced microglia activation requires miR-342: Impact on NF-kB signaling and neurotoxicity. *Cell Death Dis.* **2020**, *11*, 415. [CrossRef]

62. Liu, T.; Zhang, L.; Joo, D.; Sun, S.-C. NF-κB signaling in inflammation. *Signal Transduct. Target Ther.* **2017**, *2*, 17023. [CrossRef]

63. Hiscott, J.; Marois, J.; Garoufalis, J.; D'Addario, M.; Roulston, A.; Kwan, I.; Pepin, N.; Lacoste, J.; Nguyen, H.; Bensi, G. Characterization of a functional NF-kappa B site in the human interleukin 1 beta promoter: Evidence for a positive autoregulatory loop. *Mol. Cell. Biol.* **1993**, *13*. [CrossRef]

64. Son, Y.-H.; Jeong, Y.-T.; Lee, K.-A.; Choi, K.-H.; Kim, S.-M.; Rhim, B.-Y.; Kim, K. Roles of MAPK and NF-kB in interleukin-6 induction by lipopolysaccharide in vascular smooth muscle cells. *J. Cardiovasc. Pharmacol.* **2008**, *51*, 71–77. [CrossRef]

65. Baeuerle, P.; Vassalli, P.; Collart, M.A. Regulation of tumor necrosis factor alpha transcription in macrophages: Involvement of four kappa B-like motifs and of constitutive and inducible forms of NF-kappa B. *Mol. Cell. Biol.* **1990**, *10*, 1498–1506.
66. Guadagno, J.; Swan, P.; Cregan, S.P. Microglia-derived IL-1β triggers p53-mediated cell cycle arrest and apoptosis in neural precursor cells. *Cell Death Dis.* **2015**, *6*, e1779. [CrossRef]
67. Ye, L.; Huang, Y.; Zhao, L.; Li, Y.; Sun, L.; Zhou, Y.; Qian, G.; Zheng, J.C. IL-1β and TNF-α induce neurotoxicity through glutamate production: A potential role for neuronal glutaminase. *J. Neurochem.* **2013**, *125*, 897–908. [CrossRef] [PubMed]
68. Reeh, H.; Rudolph, N.; Billing, U.; Christen, H.; Streif, S.; Bullinger, E.; Schliemann-Bullinger, M.; Findeisen, R.; Schaper, F.; Huber, H.J.; et al. Response to IL-6 trans- and IL-6 classic signalling is determined by the ratio of the IL-6 receptor α to gp130 expression: Fusing experimental insights and dynamic modelling. *Cell Commun. Signal.* **2019**, *17*, 46. [CrossRef] [PubMed]
69. Schmidt, S.I.; Bogetofte, H.; Ritter, L.; Agergaard, J.B.; Hammerick, D.; Kabiljagic, A.A.; Wlodarczyk, A.; Lopez, S.G.; Sørensen, M.D.; Jørgensen, M.L.; et al. Microglia-Secreted Factors Enhance Dopaminergic Differentiation of Tissueand iPSC-Derived Human Neural Stem Cells. *Stem Cell Rep.* **2021**, *16*, 281–294. [CrossRef] [PubMed]
70. Ciesielski-Treska, J.; Ulrich, G.; Chasserot-Golaz, S.; Zwiller, J.; Revel, M.-O.; Aunis, D.; Bader, M.-F. Mechanisms Underlying Neuronal Death Induced by Chromogranin A-activated Microglia. *J. Biol. Chem.* **2001**, *276*, 13113–13120. [CrossRef] [PubMed]
71. Taylor, D.L.; Jones, F.; Kubota, E.S.F.C.S.; Pocock, J.M. Stimulation of Microglial Metabotropic Glutamate Receptor mGlu2 Triggers Tumor Necrosis Factor α-Induced Neurotoxicity in Concert with Microglial-Derived Fas Ligand. *J. Neurosci.* **2005**, *25*, 2952–2964. [CrossRef]
72. Förstermann, U.; Sessa, W.C. Nitric oxide synthases: Regulation and function. *Eur. Heart J.* **2012**, *33*, 829–837. [CrossRef]
73. Sobue, A.; Komine, O.; Hara, Y.; Endo, F.; Mizoguchi, H.; Watanabe, S.; Murayama, S.; Saito, T.; Saido, T.C.; Sahara, N.; et al. Microglial gene signature reveals loss of homeostatic microglia associated with neurodegeneration of Alzheimer's disease. *Acta Neuropathol. Commun.* **2021**, *9*, 1. [CrossRef] [PubMed]
74. Emrani, S.; Arain, H.A.; DeMarshall, C.; Nuriel, T. APOE4 is associated with cognitive and pathological heterogeneity in patients with Alzheimer's disease: A systematic review. *Alzheimer's Res. Ther.* **2020**, *12*, 141. [CrossRef]
75. Wa, M.W.; Wang, J.; Zhang, Q.; Wang, R.; Dhandapani, K.M.; Vadlamudi, R.K.; Brann, D.W. NADPH oxidase in brain injury and neurodegenerative disorders. *Mol. Neurodegener.* **2017**, *12*, 7.
76. Qin, L.; Liu, Y.; Wang, T.; Wei, S.-J.; Block, M.L.; Wilson, B.; Liu, B.; Hong, J.-S. NADPH Oxidase Mediates Lipopolysaccharide-induced Neurotoxicity and Proinflammatory Gene Expression in Activated Microglia. *J. Biol. Chem.* **2004**, *279*, 1415–1421. [CrossRef] [PubMed]
77. Cooney, S.J.; Bermudez-Sabogal, S.L.; Brynes, K.R. Cellular and temporal expression of NADPH oxidase (NOX) isotypes after brain injury. *J. Neuroinflammation* **2013**, *10*, 917. [CrossRef] [PubMed]
78. Zhang, W.; Gao, J.-H.; Yan, Z.-F.; Huang, X.-Y.; Guo, P.; Sun, L.; Liu, Z.; Hu, Y.; Yu, S.-Y.; Cao, C.-J.; et al. Minimally toxic dose of lipopolysaccharide and α-Synuclein oligomer elicit synergistic dopaminergic neurodegeneration: Role and mechanism of microglial NOX2 activation. *Mol. Neurobiol.* **2018**, *55*, 619–632. [CrossRef] [PubMed]
79. Di Filippo, M.; de Iure, A.; Giampà, C.; Chiasserini, D.; Tozzi, A.; Orvietani, P.L.; Ghiglieri, V.; Tantucci, M.; Durante, V.; Quiroga-Varela, A.; et al. Persistent activation of microglia and NADPH oxidase drive hippocampal dysfunction in experimental multiple sclerosis. *Sci. Rep.* **2016**, *6*, 20926. [CrossRef] [PubMed]
80. Solleiro-Villavicencio, H.; Rivas-Arancibia, S. Effect of Chronic Oxidative Stress on Neuroinflammatory Response Mediated by CD4+T Cells in Neurodegenerative Diseases. *Front. Cell. Neurosci.* **2018**, *12*, 114. [CrossRef] [PubMed]
81. Baydoun, H.H.; Cherian, A.M.; Green, P.; Lee, R. Inducible nitric oxide synthase mediates DNA double strand breaks in Human T-Cell Leukemia Virus Type 1-induced leukemia/lymphoma. *Retrovirology* **2015**, *12*, 71. [CrossRef]
82. Folkes, L.K.; O'Neill, P. DNA damage induced by nitric oxide during ionizing radiation is enhanced at replication. *Nitric Oxide Chem. Biol.* **2013**, *34*, 47–55. [CrossRef]
83. Clemons, N.J.; McColl, K.E.L.; Fitzgerald, R.C. Nitric oxide and acid induce double-strand DNA breaks in Barrett's esophagus carcinogenesis via distinct mechanisms. *Gastroentrerology* **2007**, *133*, 1198–1209. [CrossRef] [PubMed]
84. Lepoivre, M.; Chenais, B.; Yapo, A.; Lemaire, G.; Thelander, L.; Tenu, J.-P. Alterations of Ribonucleotide Reductase Activity Following Induction of the Nitrite-generating Pathway in Adenocarcinoma Cells. *J. Biol. Chem.* **1990**, *265*, 14143–14149. [CrossRef]
85. Bernier, L.-P.; Bohlen, C.J.; York, E.M.; Choi, H.B.; Kamyabi, A.; Dissing-Olesen, L.; Hefendehl, J.K.; Collins, H.Y.; Stevens, B.; Barres, B.A.; et al. Nanoscale Surveillance of the Brain by Microglia via cAMP-Regulated Filopodia. *Cell Rep.* **2019**, *27*, 2895–2908. [CrossRef] [PubMed]
86. Nimmerjahn, A.; Kirchhoff, F.; Helmchen, F. Resting Microglial Cells Are Highly Dynamic Surveillants of Brain Parenchyma in Vivo. *Science* **2005**, *308*, 1314–1318. [CrossRef] [PubMed]
87. Bialas, A.R.; Stevens, B. TGF-β Signaling Regulates Neuronal C1q Expression and Developmental Synaptic Refinement. *Nat. Neurosci.* **2013**, *16*, 1773–1782. [CrossRef]
88. Chen, Z.; Trapp, B.D. Microglia and neuroprotection. *J. Neurochem.* **2015**, *136*, 10–17. [CrossRef] [PubMed]
89. Paolicelli, R.C.; Bolasco, G.; Pagani, F.; Maggi, L.; Scianni, M.; Panzanelli, P.; Giustetto, M.; Ferreira, T.A.; Guiducci, E.; Dumas, L.; et al. Synaptic Pruning by Microglia is Necessary for Normal Brain Development. *Science* **2011**, *333*, 1456–1458. [CrossRef]
90. Walton, N.M.; Sutter, B.M.; Laywell, E.D.; Levkoff, L.H.; Kearns, S.M.; Marshall, G.P.; Scheffler, B.; Steindler, D.A. Microglia instruct subventricular zone neurogenesis. *Glia* **2006**, *54*, 815–825. [CrossRef] [PubMed]

91. Lenz, K.M.; Nelson, L.H. Microglia and Beyond: Innate Immune Cells As Regulators of Brain Development and Behavioral Function. *Front. Immunol.* **2018**, *9*, 698. [CrossRef]

92. Schafer, D.P.; Stevens, B. Phagocytic glial cells: Sculpting synaptic circuits in the developing nervous system. *Curr. Opin. Neurobiol.* **2013**, *23*, 1034–1040. [CrossRef] [PubMed]

93. Diaz-Aparicio, I.; Paris, I.; Sierra-Torre, V.; Plaza-Zabala, A.; Rodríguez-Iglesias, N.; Márquez-Ropero, M.; Beccari, S.; Huguet, P.; Abiega, O.; Alberdi, E.; et al. Microglia Actively Remodel Adult Hippocampal Neurogenesis through the Phagocytosis Secretome. *J. Neurosci.* **2020**, *40*, 1453–1482. [CrossRef] [PubMed]

94. Yuan, J.; Ge, H.; Liu, W.; Zhu, H.; Chen, Y.; Zhang, X.; Yang, Y.; Yin, Y.; Chen, W.; Wu, W.; et al. M2 microglia promotes neurogenesis and oligodendrogenesis from neural stem/progenitor cells via the PPARγ signaling pathway. *Oncotarget* **2017**, *8*, 19855–19865. [CrossRef] [PubMed]

95. Miyamoto, A.; Wake, H.; Ishikawa, A.W.; Eto, K.; Shibata, K.; Murakoshi, H.; Koizumi, S.; Moorhouse, A.J.; Yoshimura, Y.; Nabekura, J. Microglia contact induces synapse formation in developing somatosensory cortex. *Nat. Commun.* **2016**, *7*, 12540. [CrossRef] [PubMed]

96. Parkhurst, C.N.; Yang, G.; Ninan, I.; Savas, J.N.; Yates, J.R.; Lafaille, J.J.; Hempstead, B.L.; Littman, D.R.; Gan, W.-B. Microglia Promote Learning-Dependent Synapse Formation through Brain-Derived Neurotrophic Factor. *Cell* **2013**, *155*, 1596–1609. [CrossRef] [PubMed]

97. Douple, E.B.; Mabuchi, K.; Cullings, H.M.; Preston, D.L.; Kodama, K.; Shimizu, Y.; Fujiwara, S.; Shore, R.E. Long-term Radiation-Related Health Effects in a Unique Human Population: Lessons Learned from the Atomic Bomb Survivors of Hiroshima and Nagasaki. *Disaster Med. Public Health Prep.* **2011**, *5*, S122–133. [CrossRef] [PubMed]

98. Hasegawa, A.; Tanigawa, K.; Ohtsuru, A.; Yabe, H.; Maeda, M.; Shigemura, J.; Ohira, T.; Tominaga, T.; Akashi, M.; Hirohashi, N.; et al. Health effects of radiation and other health problems in the aftermath of nuclear accidents, with an emphasis on Fukushima. *Lancet* **2015**, *386*, 479–488. [CrossRef]

99. Tran, L.; Seeram, E. Current Perspectives on the Use of the Linear Non-Threshold (LNT) Model. *Int. J. Radiol. Med Imaging* **2017**, *3*, 1–8.

100. Cramer, C.K.; McKee, N.; Case, L.D.; Chan, M.D.; Cummings, T.L.; Lesser, G.J.; Shaw, E.G.; Rapp, S.R. Mild cognitive impairment in long-term brain tumor survivors following brain irradiation. *J. Neuro-Oncol.* **2019**, *141*, 235–244. [CrossRef] [PubMed]

101. Barani, I.J.; Cuttino, L.W.; Benedict, S.H.; Todor, D.; Bump, E.A.; Wu, Y.; Chund, T.D.; Broaddus, W.C.; Lin, P.-S. Neural stem cell-preserving external-beam radiotherapy of central nervous system malignancies. *Int. J. Radiat. Oncol. Biol. Phys.* **2007**, *68*, 978–985. [CrossRef]

102. Acharya, M.M.; Green, K.N.; Allen, B.D.; Najafi, A.R.; Syage, A.; Minasyan, H.; Le, M.T.; Kawashita, T.; Giedzinski, E.; Parihar, V.K.; et al. Elimination of microglia improves cognitive function following cranial irradiation. *Sci. Rep.* **2016**, *6*, 31545. [CrossRef]

103. Farjam, R.; Pramanik, P.; Aryal, M.P.; Srinivasan, A.; Chapman, C.H.; Tsien, C.I.; Lawrence, T.S.; Cao, Y. A radiation-induced hippocampal vascular injury surrogate marker predicts late neurocognitive dysfunction. *Int. J. Radiat. Oncol. Biol. Phys.* **2015**, *93*, 908–915. [CrossRef]

104. Raber, J.; Rola, R.; LeFevour, A.; Morhardt, D.; Curley, J.; Mizumatsu, S.; VandenBerg, S.R.; Fike, J.R. Radiation-induced cognitive impairments are associated with changes in indicators of hippocampal neurogenesis. *Radiat. Res.* **2004**, *162*, 39–47. [CrossRef]

105. Rola, R.; Raber, J.; Rizk, A.; Otsuka, S.; VandenBerg, S.R.; Morhardt, D.R.; Fike, J.R. Radiation-induced impairment of hippocampal neurogenesis is associated with cognitive deficits in young mice. *Exp. Neurol.* **2004**, *188*, 316–330. [CrossRef] [PubMed]

106. Mizumatsu, S.; Monje, M.L.; Morhardt, D.R.; Rola, R.; Palmer, T.D.; Fike, J.R. Extreme Sensitivity of Adult Neurogenesis to Low Doses of X-Irradiation. *Cancer Res.* **2003**, *63*, 4021–4027.

107. Tallet, A.V.; Azria, D.; Barlesi, F.; Spano, J.-P.; Carpentier, A.F.; Goncalves, A.; Metellus, P. Neurocognitive function impairment after whole brain radiotherapy for brain metastases: Actual assessment. *Radiat. Oncol.* **2012**, *7*, 77. [CrossRef] [PubMed]

108. Casciati, A.; Dobos, K.; Antonelli, F.; Benedek, A.; Kempf, S.J.; Belles, M.; Balogh, A.; Tanori, M.; Heredia, L.; Atkinson, M.J.; et al. Age-related effects of X-ray irradiation on mouse hippocampus. *Oncotarget* **2016**, *7*, 28040–28058. [CrossRef] [PubMed]

109. Helson, L. Radiation-induced Demyelination and Remyelination in the Central Nervous System: A Literature Review. *Anticancer Res.* **2018**, *38*, 4999–5002. [CrossRef] [PubMed]

110. Monje, M. Cranial Radiation Therapy and Damage to Hippocampal Neurogenesis. *Dev. Disabil. Res. Rev.* **2008**, *14*, 238–242. [CrossRef]

111. Andres-Mach, M.; Rola, R.; Fike, J.R. Radiation effects on neural precursor cells in the dentate gyrus. *Cell Tissue Res.* **2008**, *331*, 251–262. [CrossRef]

112. Qiu, Y.; Guo, Z.; Han, L.; Yang, Y.; Li, J.; Liu, S.; Lv, X. Network-level dysconnectivity in patients with nasopharyngeal carcinoma (NPC) early post-radiotherapy: Longitudinal resting state fMRI study. *Brain Imaging Behav.* **2018**, *12*, 1279–1289. [CrossRef] [PubMed]

113. Armstrong, C.; Ruffer, J.; Corn, B.; DeVries, K.; Mollman, J. Biphasic patterns of memory deficits following moderate-dose partial-brain irradiation: Neuropsychologic outcome and proposed mechanisms. *J. Clin. Oncol.* **1995**, *13*, 2263–2271. [CrossRef] [PubMed]

114. Rana, P.; Khan, A.R.; Modi, S.; Kumar, B.S.H.; Javed, S.; Tripathi, R.P.; Kushu, S. Altered brain metabolism after whole body irradiation in mice: A preliminary in vivo 1H MRS study. *Int. J. Radiat. Biol.* **2013**, *89*, 212–218. [CrossRef] [PubMed]

115. Jenrow, K.A.; Brown, S.L.; Lapanowski, K.; Naei, H.; Kolozsvary, A.; Kim, J. Selective inhibition of microglia-mediated neuroinflammation mitigates radiation-induced cognitive impairment. *Radiat. Res.* **2013**, *179*, 549–556. [CrossRef]
116. Hua, K.; Schindler, M.K.; McQuail, J.A.; Forbes, M.E.; Riddle, D.R. Regionally distinct responses of microglia and glial progenitor cells to whole brain irradiation in adult and aging rats. *PLoS ONE* **2012**, *7*, e52728. [CrossRef]
117. Chen, H.; Chong, Z.Z.; De Toledo, S.M.; Azzam, E.I.; Elkabes, S.; Souayah, N. Delayed activation of human microglial cells by high dose ionizing radiation. *Brain Res.* **2016**, *1646*, 193–198. [CrossRef]
118. Ung, M.-C.; Garrett, L.; Dalke, C.; Leitner, V.; Dragosa, D.; Hladik, D.; Neff, F.; Wagner, F.; Zitzelsberger, H.; Miller, G.; et al. Dose-dependent long-term effects of a single radiation event on behaviour and glial cells. *Int. J. Radiat. Biol.* **2020**, *97*, 156–169. [CrossRef] [PubMed]
119. Zhou, K.; Boström, M.; Ek, C.J.; Li, T.; Xie, C.; Xu, Y.; Sun, Y.; Blomgren, K.; Zhu, C. Radiation induces progenitor cell death, microglia activation, and blood-brain barrier damage in the juvenile rat cerebellum. *Sci. Rep.* **2017**, *7*, 46181. [CrossRef]
120. Ismail, A.F.M.; El-Sonbaty, S.M. Fermentation enhances Ginkgo biloba protective role on gamma-irradiation induced neuroinflammatory gene expression and stress hormones in rat brain. *J. Photochem. Photobiol. B: Biol.* **2016**, *158*, 154–163. [CrossRef]
121. Limoli, C.L.; Giedzinski, E.; Rola, R.; Otsuka, S.; Palmer, T.D.; Fike, J.R. Radiation response of neural precursor cells: Linking cellular sensitivity to cell cycle checkpoints, apoptosis and oxidative stress. *Radiat. Res.* **2004**, *161*, 17–27. [CrossRef]
122. Kam, W.W.-Y.; Banati, R.B. Effects of ionizing radiation on mitochondria. *Free Radic. Biol. Med.* **2013**, *65*, 607–619. [CrossRef] [PubMed]
123. Huang, T.-T.; Leu, D.; Zou, Y. Oxidative stress and redox regulation on hippocampal-dependent cognitive functions. *Arch. Biochem. Biophys.* **2015**, *576*, 2–7. [CrossRef] [PubMed]
124. Sridharan, D.M.; Asaithamby, A.; Bailey, S.M.; Costes, S.V.; Doestch, P.W.; Dynan, W.S.; Kronenberg, A.; Rithidech, K.N.; Saha, J.; Snijders, A.M. Understanding cancer development processes after HZE-particle exposure: Roles of ROS, DNA damage repair and inflammation. *Radiat. Res.* **2015**, *183*, 1–26. [CrossRef] [PubMed]
125. Suman, S.; Rodriguez, O.C.; Winters, T.A.; Fornace, A.J.; Albanese, C.; Datta, K. Therapeutic and space radiation exposure of mouse brain causes impaired DNA repair response and premature senescence by chronic oxidant production. *Aging* **2013**, *5*, 607–622. [CrossRef] [PubMed]
126. Fishman, K.; Baure, J.; Zou, Y.; Huang, T.-T.; Andres-Mach, M.; Rola, R.; Suarez, T.; Acharya, M.; Limoli, C.L.; Lamborn, K.R.; et al. Radiation-induced reductions in neurogenesis are ameliorated in mice deficient in CuZnSOD or MnSOD. *Free Radic. Biol. Med.* **2009**, *47*, 1459–1467. [CrossRef]
127. Sharma, S.; Singla, N.; Chadha, V.D.; Dhawan, D.K. A concept of radiation hormesis. Stimulation of antioxidant machinery in rats by low dose ionizing radiation. *Hell. J. Nucl. Med.* **2019**, *22*, 43–48.
128. Monje, M.L.; Toda, H.; Palmer, T.D. Inflammatory blockade restores adult hippocampal neurogenesis. *Science* **2003**, *302*, 1760–1765. [CrossRef] [PubMed]
129. Morganti, J.M.; Jopson, T.D.; Liu, S.; Gupta, N.; Rosi, S. Cranial irradiation alters the brain's microenvironment and permits CCR2+ macrophage infiltration. *PLoS ONE* **2014**, *9*, e93650. [CrossRef]
130. Acharya, M.M.; Patel, N.H.; Craver, B.M.; Tran, K.K.; Giedzainski, E.; Tseng, B.P.; Parihar, V.K.; Limoli, C.L. Consequences of low dose ionizing radiation exposure on the hippocampal microenvironment. *PLoS ONE* **2015**, *10*, e1028316. [CrossRef] [PubMed]
131. Kalm, M.; Fukuda, A.; Fukuda, H.; Ohrfelt, A.; Lannering, B.; Björk-Eriksson, T.; Blennow, K.; Márky, I.; Blomgran, K. Transient inflammation in neurogenic regions after irradiation of the developing brain. *Radiat. Res.* **2009**, *171*, 66–76. [CrossRef]
132. Leach, J.K.; Van Tuyle, G.; Lin, P.-S.; Schmidt-Ullrich, R.; Mikkelsen, R.B. Ionizing radiation-induced, mitochondria-dependent generation of reactive oxygen/nitrogen. *Cancer Res.* **2001**, *61*, 3894–3901. [PubMed]
133. Takács, S.F.; Benedek, A.; Mán, I.; Ozsvári, B.; Puskás, L.G.; Neefs, M.; Benotmane, M.A.; Sáfrány, G.; Lumniczky, K. Analysis of radiation-induced blood-brain barrier damage in mice by in vivo bio-imaging technique. *Cent. Eur. J. Occup. Environ. Med.* **2015**, *21*, 87–95.
134. Liao, H.; Wang, H.; Rong, X.; Li, E.; Xu, R.-H.; Peng, Y. Mesenchymal Stem Cells Attenuate Radiation-Induced Brain Injury by Inhibiting Microglia Pyroptosis. *BioMed Res. Int.* **2017**, *2017*, 1948985. [CrossRef] [PubMed]
135. McKenzie, B.A.; Fernades, J.P.; Doan, M.A.L.; Schmitt, L.M.; Branton, W.G.; Power, C. Activation of the executioner caspases-3 and -7 promotes microglial pyroptosis in models of multiple sclerosis. *J. Neuroinflammation* **2020**, *17*, 253. [CrossRef]
136. Li, M.D.; Burns, T.C.; Kumar, S.; Morgan, A.A.; Sloan, S.A.; Palmer, T.D. Aging-like Changes in the Transcriptome of Irradiated Microglia. *Glia* **2015**, *63*, 754–767. [CrossRef] [PubMed]
137. Krukowski, K.; Feng, X.; Paladini, M.S.; Chou, A.; Sacramento, K.; Grue, K.; Riparip, L.-K.; Jones, T.; Campbell-Beachler, M.; Nelson, G.; et al. Temporary microglia-depletion after cosmic radiation modifies phagocytic activity and prevents cognitive deficits. *Sci. Rep. -Nat.* **2018**, *8*, 7857. [CrossRef]
138. Parihar, V.K.; Limoli, C.L. Cranial irradiation compromises neuronal architecture in the hippocampus. *Proc. Natl. Acad. Sci. USA* **2013**, *110*, 12822–12827. [CrossRef] [PubMed]
139. Hinkle, J.J.; Olschowka, J.A.; Love, T.M.; Williams, J.P.; O'Banion, M.K. Cranial irradiation mediated spine loss is sex-specific and complement receptor-3 dependent in male mice. *Sci. Rep.* **2019**, *9*, 18800. [CrossRef]
140. Centers for Disease Control and Prevention (CDC). *Radiation and Pregnancy: Infomation for Clinicians*; CDC: Altanta, GA, USA, 2019.

141. Preston, D.L.; Cullings, H.; Suyama, A.; Funamoto, S.; Nishi, N.; Soda, M.; Mabuchi, K.; Kodama, K.; Kasagi, F.; Shore, R.E. Solid Cancer Incidence in Atomic Bomb Survivors Exposed In Utero or as Young Children. *J. Natl. Cancer Inst.* **2008**, *100*, 428–436. [CrossRef] [PubMed]
142. De Santis, M.; Cesari, E.; Nobili, E.; Straface, G.; Cavaliere, A.F.; Caruso, A. Radiation effects on development. *Birth Defects Res. Part C* **2007**, *81*, 177–182. [CrossRef]
143. Guilbaud, L.; Beghin, D.; Dhombres, F.; Blondiaux, E.; Friszer, S.; Le Pointe, H.D.; Éléfant, E.; Jouannic, J.-M. Pregnancy outcome after first trimester exposure to ionizing radiations. *Eur. J. Obstet. Gynecol. Reprod. Biol.* **2019**, *232*, 18–21. [CrossRef]
144. Schull, W.J.; Otake, M. Cognitive Function and Prenatal Exposure to Ionising Radiation. *Teratology* **1999**, *59*, 222–226. [CrossRef]
145. Verreet, T.; Rangaraja, J.R.; Quintens, R.; Verslegers, M.; Lo, A.C.; Govaerts, K.; Neefs, M.; Leysen, L.; Baatout, S.; Maes, F.; et al. Persistent Impact of In utero Irradiation on Mouse Brain Structure and Function Characterized by MR Imaging and Behavioral Analysis. *Front. Behav. Neurosci.* **2016**, *10*, 83. [CrossRef] [PubMed]
146. Barazzuol, L.; Rickett, N.; Ju, L.; Jeggo, P.A. Low levels of endogenous or X-ray-induced DNA double-strand breaks activate apoptosis in adult neural stem cells. *J. Cell Sci.* **2015**, *128*, 3597–3606. [PubMed]
147. Roque, T.; Haton, C.; Etienne, O.; Chicheportiche, A.; Rousseau, L.; Martin, L.; Mouthon, M.A.; Boussin, F.D. Lack of a p21waf1/cip-Dependent G1/S Checkpoint in Neural Stem and Progenitor Cells After DNA Damage In Vivo. *Stem Cells* **2011**, *30*, 537–547. [CrossRef] [PubMed]
148. Carbone, M.C.; Pinto, M.; Antonelli, F.; Amicarelli, F.; Balata, M.; Belli, M.; Conti Devirgiliis, L.; Ioannucci, L.; Nisi, S.; Sapora, O.; et al. The Cosmic Silence experiment: On the putative adaptive role of environmental ionizing radiation. *Radiat. Environ. Biophys.* **2009**, *48*, 189–196. [CrossRef] [PubMed]
149. Seong, K.M.; Kim, C.S.; Lee, B.-S.; Nam, S.Y.; Kang, K.H.; Kim, J.-Y.; Park, J.-J.; Min, K.-J.; Jin, Y.-W. Low-dose radiation induces Drosophila innate immunity through Toll pathway activation. *J. Radiat. Res.* **2012**, *53*, 242–249. [CrossRef] [PubMed]
150. Seong, K.M.; Kim, C.S.; Seo, S.-W.; Jeon, H.Y.; Lee, B.-S.; Nam, S.Y.; Yang, K.H.; Kim, J.-Y.; Kim, C.-S.; Min, K.-J.; et al. Genome-wide analysis of low-dose irradiated male Drosophila melanogaster with extended longevity. *Biogerontology* **2011**, *12*, 93–107. [CrossRef]
151. Calabrese, E.J.; Iavicoli, I.; Calabrese, V. Hormesis: Why it is important to biogerontologists. *Biogerontology* **2012**, *13*, 215–235. [CrossRef]
152. Dalke, C.; Neff, F.; Bains, S.K.; Bright, S.; Lord, D.; Reitmeir, P.; Rößler, U.; Samaga, D.; Unger, K.; Braselmann, H.; et al. Lifetime study in mice after acute low-dose ionizing radiation: A multifactorial study with special focus on cataract risk. *Radiat. Environ. Biophys.* **2018**, *57*, 99–113. [CrossRef]
153. Gori, T.; Münzel, T. Biological effects of low-dose radiation: Of harm and hormesis. *Eur. Heart J.* **2012**, *33*, 292–295. [CrossRef] [PubMed]
154. Chien, L.; Chen, W.-K.e.; Liu, S.-T.; Chang, C.-R.; Kao, M.-C.; Chen, K.-W.; Chiu, S.-C.; Hsu, M.-L.; Hsiang, I.-C.; Chen, Y.-J.; et al. Low-dose ionizing radiation induces mitochondrial fusion and increases expression of mitochondrial complexes I and III in hippocampal neurons. *Oncotarget* **2015**, *6*, 30628–30629. [CrossRef]
155. Baulch, J.E.; Craver, B.M.; Tran, K.K.; Yu, L.; Chmielewski, N.; Allen, B.D.; Limoli, C.L. Persistent oxidative stress in human neural stem cells exposed to low fluences of charged particles. *Redox. Biol.* **2015**, *5*, 24–32. [CrossRef] [PubMed]
156. Russo, G.L.; Tedesco, I.; Russo, M.; Cioppa, A.; Anreassi, M.G.; Picano, E. Cellular adaptive response to chronic radiation exposure in interventional cardiologists. *Eur. Heart J.* **2012**, *33*, 408–414. [CrossRef] [PubMed]
157. Eken, A.; Aydin, A.; Erdem, O.; Akay, C.; Sayal, A.; Somuncu, I. Induced antioxidant activity in hospital staff occupationally exposed to ionizing radiation. *Int. J. Radiat. Biol.* **2012**, *88*, 648–653. [CrossRef] [PubMed]
158. Cho, S.-J.; Kang, H.; Hong, E.-H.; Kim, J.Y.; Nam, S.Y. Transcriptome analysis of low-dose ionizing radiation-impacted genes in CD4+ T-cells undergoing activation and regulation of their expression of select cytokines. *J. Immunotoxicol.* **2018**, *15*, 137–146. [CrossRef] [PubMed]
159. Tsukimoto, M.; Nakatsukasa, H.; Sugawara, K.; Yamashita, K.; Kojima, S. Repeated 0.5-Gy gamma irradiation attenuates experimental autoimmune encephalomyelitis with up-regulation of regulatory T cells and suppression of IL17 production. *Radiat. Res.* **2008**, *170*, 429–436. [CrossRef]
160. Shimura, N.; Kojima, S. Effects of low-dose-gamma rays on the immune system of different animal models of disease. *Dose Response* **2014**, *12*, 429–465. [CrossRef]
161. Sekihara, K.; Saitoh, K.; Yang, H.; Kawashima, H.; Kazuno, S.; Kikkawa, M.; Arai, H.; Miida, T.; Hayashi, N.; Sasai, K.; et al. Low-dose ionizing radiation exposure represses the cell cycle and protein synthesis pathways in in vitro human primary keratinocytes and U937 cell lines. *PLoS ONE* **2018**, *13*, e0199117.
162. Sandor, N.; Walter, F.R.; Bocsik, A.; Santha, P.; Schilling-Toth, B.; Lener, V.; Varga, Z.; Kahan, Z.; Deli, M.A.; Sáfrány, G.; et al. Low dose cranial irradiation-induced cerebrovascular damage is reversible in mice. *PLoS ONE* **2014**, *9*, e112397. [CrossRef] [PubMed]
163. Hladik, D.; Dalke, C.; von Toerne, C.; Hauck, S.M.; Azimzadeh, O.; Philipp, J.; Ung, M.-C.; Schlattl, H.; Rößler, U.; Graw, J.; et al. CREB Signaling Mediates Dose-Dependent Radiation Response in the Murine Hippocampus Two Years after Total Body Exposure. *J. Proteome Res.* **2020**, *19*, 337–345. [CrossRef] [PubMed]
164. Betlazar, C.; Middleton, R.J.; Howell, N.R.; Storer, B.; Davis, E.; Davies, J.; Banati, R.B.; Liu, G.-J. Mitochondrial translocator protein (TSPO) expression in the brain after whole body gamma irradiation. *Front. Cell Dev. Biol.* **2021**, in press.

165. Matthews, J.D.; Forsythe, A.V.; Brady, Z.; Bulter, M.W.; Goegen, S.K.; Byrnes, G.B.; Giles, G.G.; Wallace, A.B.; Anderson, P.R.; Guiver, T.A.; et al. Cancer risk in 680 000 people exposed to computed tomography scans in childhood or adolescence: Data linkage study of 11 million Australians. *Br. Med. J.* **2013**, *346*, f2360. [CrossRef] [PubMed]

166. Hauptmann, M.; Daniels, R.D.; Cardis, E.; Cullings, H.M.; Kendall, G.; Laurier, D.; Linet, M.S.; Little, M.P.; Lubin, J.H.; Preston, D.L.; et al. Epidemiological Studies of Low-Dose Ionizing Radiation and Cancer: Summary Bias Assessment and Meta-Analysis. *J. Natl. Cancer Inst. Monogr.* **2020**, *2020*, 188–200. [CrossRef] [PubMed]

167. Pasqual, E.; de Basea, M.B.; López-Vicente, M.; Thierry-Chef, I.; Cardis, E. Neurodevelopmental effects of low dose ionizing radiation exposure: A systematic review of the epidemiological evidence. *Environ. Int.* **2020**, *136*, 105371. [CrossRef] [PubMed]

168. Cohen, S.; Liu, A.; Gurvitz, M.; Guo, L.; Therrien, J.; Laprise, C.; Kaufman, J.S.; Abrahamowicz, M.; Marelli, A.J. Exposure to Low-Dose Ionizing Radiation From Cardiac Procedures and Malignancy Risk in Adults With Congenital Heart Disease. *Circulation* **2018**, *137*, 1334–1345. [CrossRef]

169. Lumniczky, K.; Szatmari, T.; Safrany, G. Ionizing Radiation-Induced Immune and Inflammatory Reactions in the Brain. *Front Immunol* **2017**, *8*, 517. [CrossRef]

170. Tang, F.R.; Loganovsky, K. Low dose or low dose rate ionizing radiation-induced health effect in the human. *J. Environ. Radioact.* **2018**, *192*, 32–47. [CrossRef]

171. Safrany, G.; Lumniczky, K.; Manti, L. New Discoveries in Radiation Science. *Cancers* **2021**, *13*, 1034. [CrossRef] [PubMed]

172. Okada, M.; Okabe, A.; Uchihori, Y.; Kitamura, H.; Sekine, E.; Ebisawa, S.; Suzuki, M.; Okayasu, R. Single extreme low dose/low dose rate irradiation causes alteration in lifespan and genome instability in primary human cells. *Br. J. Cancer* **2007**, *96*, 1707–1710. [CrossRef]

173. Davis, M.J.; Tsang, T.M.; Qiu, Y.; Dayrit, J.K.; Freji, J.B.; Huffnagle, G.B.; Olszewski, M.A. Macrophage M1/M2 polarization dynamically adapts to changes in cytokine microenvironments in Cryptococcus neoformans infection. *MBio* **2013**, *4*, e00264-13. [CrossRef]

174. Italiani, P.; Mazza, E.M.C.; Lucchesi, D.; Cifola, I.; Gemelli, C.; Grande, A.; Battaglia, C.; Bicciato, S.; Boraschi, D. Transcriptomic Profiling of the Development of the Inflammatory Response in Human Monocytes In Vitro. *PLoS ONE* **2014**, *9*, e87680. [CrossRef] [PubMed]

175. Orihuela, R.; McPherson, C.A.; Harry, G.J. Microglial M1/M2 polarization and metabolic states. *Br. J. Pharmacol.* **2016**, *173*, 649–665. [CrossRef] [PubMed]

176. Liu, B.; Hinshaw, R.G.; Le, K.X.; Park, M.-A.; Wang, S.; Belanger, A.P.; Dubey, S.; Frost, J.L.; Shi, Q.; Holton, P.; et al. Space-like 56Fe irradiation manifests mild, early sex-specific behavioral and neuropathological changes in wildtype and Alzheimer's-like transgenic mice. *Nat. Sci. Rep.* **2019**, *9*, 12118. [CrossRef]

177. Ceyzériat, K.; Zilli, T.; Fall, A.B.; Millet, P.; Koutsouvelis, N.; Dipasquale, G.; Frisoni, G.B.; Tournier, B.B.; Garibotto, V. Treatment by low-dose brain radiation therapy improves memory performances without changes of the amyloid load in the TgF344-AD rat model. *Neurobiol. Aging* **2021**, *103*, 117–127. [CrossRef] [PubMed]

178. Cuttler, J.M.; Moore, E.R.; Hosfeld, V.D.; Nadolski, D.L. Treatment of Alzheimer Disease With CT Scans: A Case Report. *Dose-Response* **2016**, *14*, 1559325816640073. [CrossRef] [PubMed]

179. Cuttler, J.M.; Moore, E.R.; Hosfeld, V.D.; Nadolski, D.L. Update on a Patient With Alzheimer Disease Treated With CT Scans. *Dose-Response* **2017**, *15*, 1559325817693167. [CrossRef]

180. Cuttler, J.M.; Moore, E.R.; Hosfeld, V.D.; Nadolski, D.L. Second Update on a Patient With Alzheimer Disease Treated by CT Scans. *Dose-Response* **2018**, *16*, 1559325818756461. [CrossRef]

181. Cuttler, J.M.; Abdellah, E.; Goldberg, Y.; Al-Shamaa, S.; Symons, S.P.; Black, S.E.; Freedman, M. Low Doses of Ionizing Radiation as a Treatment for Alzheimer's Disease: A Pilot Study. *J. Alzheimers Dis.* **2021**, *80*, 1119–1128. [CrossRef] [PubMed]

182. Cook, S.D.; Devereux, C.; Troiano, R.; Zito, G.; Hafstein, M.; Lavenhar, M.; Hernandez, E.; Dowling, P.C. Total lymphoid irradiation in multiple sclerosis: Blood lymphocytes and clinical course. *Ann. Neurol.* **1987**, *22*, 634–638. [CrossRef]

183. Devereux, C.K.; Vidaver, R.; Hafstein, M.P.; Zito, G.; Troiano, R.; Dowling, P.C.; Cook, S.D. Total lymphoid irradiation for multiple sclerosis. *Int. J. Radiat. Oncol. Biol. Phys.* **1988**, *14*, 197–203. [CrossRef]

184. Cook, S.D.; Troiano, R.; Zito, G.; Rohowsky-Kochan, C.; Sheffit, A.; Dowling, P.C.; Devereux, C.K. Deaths after Total Lymphoid Irradiation for Multiple Sclerosis. *Lancet* **1989**, *334*, 277–278. [CrossRef]

185. Shaygannejad, V.; Zare, M.; Maghzi, H.; Emami, P. Brain radiation and possible presentation of multiple sclerosis. *J. Res. Med Sci.* **2013**, *18*, S93–S95. [PubMed]

186. Murphy, C.B.; Hashimoto, S.A.; Graeb, D.; Thiessen, B.A. Clinical Exacerbation of Multiple Sclerosis Following Radiotherapy. *Achives Neurol.* **2003**, *60*, 273–275. [CrossRef]

187. McMeekin, R.R.; Hardman, J.M.; Kempe, L.G. Multiple Sclerosis After X-radiation. *Achives Otolaryngol. -Head Neck Surg.* **1969**, *90*, 617–621. [CrossRef] [PubMed]

188. Axelson, O.; Landtblom, A.-M.; Flodin, U. Multiple sclerosis and ionizing radiation. *Neuroepidemiology* **2001**, *20*, 175–178. [CrossRef] [PubMed]

189. Miller, R.C.; Lachance, D.H.; Lucchinetti, C.F.; Keegan, B.M.; Gavrilova, R.H.; Brown, P.D.; Weinshenker, B.G.; Rodriguez, M. Multiple sclerosis, brain radiotherapy, and risk of neurotoxicity: The Mayo Clinic experience. *Int. J. Radiat. Oncol. -Biol. -Phsyics* **2006**, *66*, 1178–1186. [CrossRef] [PubMed]

190. Motamed, M.R.; Fereshtehnejad, S.-M.; Abbasi, M.; Sanei, M.; Abbaslou, M.; Meysami, S. X-ray radiation and the risk of multiple sclerosis: Do the site and dose of exposure matter? *Med. J. Islamic Repub. Iran* **2014**, *28*, 145.
191. DeLuca, H.F.; Plum, L. UVB radiation, vitamin D and multiple sclerosis. *Photochem. Photobiol. Sci.* **2017**, *16*, 411–415. [CrossRef]
192. Irving, A.A.; Marling, S.J.; Seeman, J.; Plum, L.A.; DeLuca, H.F. UV light suppression of EAE (a mouse model of multiple sclerosis) is independent of vitamin D and its receptor. *Proc. Natl. Acad. Sci. USA* **2019**, *116*, 22552–22555. [CrossRef] [PubMed]
193. Helis, C.A.; McTyre, E.; Munley, M.T.; Bourland, J.D.; Lucas, J.T.; Cramer, C.K.; Tatter, S.B.; Laxton, A.W.; Chan, M.D. Gamma Knife Radiosurgery for Multiple Sclerosis-Associated Trigeminal Neuralgia. *Neurosurgery* **2019**, *85*, E933–939. [CrossRef] [PubMed]
194. Kemp, S.; Allan, R.S.; Patanjali, N.; Barnett, M.H.; Jonker, B.P. Neurological deficit following stereotactic radiosurgery for trigeminal neuralgia. *J. Clin. Neurosci.* **2016**, *34*, 229–231. [CrossRef] [PubMed]
195. Tanaka, T.; Tago, F.; Fang, S.-P.; Shimura, N.; Kojima, S. Repeated 0.5-Gy γ-ray irradiation attenuates autoimmune manifestations in MRL-lpr/lpr mice. *Int. J. Radiat. Biol.* **2005**, *81*, 731–740. [CrossRef]
196. Ohsawa, K.; Imai, Y.; Sasaki, Y.; Kohsaka, S. Microglia/macrophage-specific protein Iba1 binds to fimbrin and enhances its actin-bundling activity. *J. Neurochem.* **2004**, *88*, 844–856. [CrossRef] [PubMed]
197. Korzhevskii, D.E.; Kirik, O.V. Brain Microglia and Microglial Markers. *Neurosci. Behav. Physiol.* **2015**, *46*, 284–290. [CrossRef]
198. Jurga, A.M.; Paleczna, M.; Kuter, K.Z. Overview of General and Discriminating Markers of Differential Microglia Phenotypes. *Front. Cell. Neurosci.* **2020**, *14*, 198. [CrossRef] [PubMed]
199. Woodcock, T.; Morganti-Kossmann, M.C. The role of markers of inflammation in traumatic brain injury. *Front. Neurol.* **2013**, *4*, 18. [CrossRef]
200. Wang, M.; Wang, X.; Zhao, L.; Ma, W.; Rodriguez, I.R.; Fariss, R.N.; Wong, W.T. Macroglia-microglia interactions via TSPO signaling regulates microglial activation in the mouse retina. *J. Neurosci.* **2014**, *34*, 3793–3806. [CrossRef] [PubMed]
201. Banati, R.B.; Goerres, G.W.; Myers, R.; Gunn, R.N.; Turkheimer, F.E.; Kreutzberg, G.W.; Brooks, D.J.; Jones, T.; Duncan, J.S. [11C](R)-PK11195 positron emission tomography imaging of activated microglia in vivo in Rasmussen's encephalitis. *Neurology* **1999**, *53*, 2199–2203. [CrossRef] [PubMed]
202. Pappata, S.; Levasseur, M.; Gunn, R.N.; Myers, R.; Crouzel, C.; Syrota, A.; Jones, T.; Kreatuzberg, G.T.; Banati, R.B. Thalamic microglial activation in ischemic stroke detected in vivo by PET and [(11)C]PK11195. *Neurology* **2000**, *55*, 1052–1054. [CrossRef]
203. Cagnin, A.; Kassiou, M.; Meikle, S.R.; Banati, R.B. In vivo evidence for microglial activation in neurodegenerative dementia. *Acta Neurol. Scand.* **2006**, *114*, 107–114. [CrossRef] [PubMed]
204. Cagnin, A.; Gerhard, A.; Banati, R.B. In vivo imaging of neuroinflammation. *Eur. Neuropsychopharmacol.* **2002**, *12*, 581–586. [CrossRef]
205. Banati, R.B. Visualising microglial activation in vivo. *Glia* **2002**, *40*, 206–217. [CrossRef] [PubMed]
206. Pannell, M.; Economopoulos, V.; Wilson, T.C.; Kersemans, V.; Isenegger, P.G.; Larkin, J.R.; Smart, S.; Gilchrist, S.; Gouverneur, V.; Sibson, N.R. Imaging of translocator protein upregulation is selective for pro-inflammatory polarized astrocytes and microglia. *Glia* **2019**, *68*, 280–297. [CrossRef] [PubMed]
207. Betlazar, C.; Harrison-Brown, M.; Middleton, R.J.; Banati, R.; Liu, G.-J. Cellular Sources and Regional Variations in the Expression of the Neuroinflammatory Marker Translocator Protein (TSPO) in the Normal Brain. *Int. J. Mol. Sci.* **2018**, *19*, 2707. [CrossRef] [PubMed]
208. Tu, L.N.; Morohaku, K.; Manna, P.R.; Pelton, S.H.; Butler, W.R.; Stocco, D.M.; Selvaraj, V. Peripheral Benzodiazepine Receptor/Translocator Protein Global Knock-out Mice Are Viable with No Effects on Steroid Hormone Biosynthesis. *J. Biol. Chem.* **2014**, *289*, 27444–27454. [CrossRef]
209. Banati, R.B.; Middleton, R.J.; Chan, R.; Hatty, C.R.; Kam, W.W.-Y.; Quin, C.; Graeber, M.B.; Parmar, A.; Zahra, D.; Callaghan, P.; et al. Positron emission tomography and functional characterization of a complete PBR/TSPO knockout. *Nat. Commun.* **2014**, *5*, 5452. [CrossRef]
210. Liu, G.J.; Middleton, R.J.; Hatty, C.R.; Kam, W.W.; Chan, R.; Pham, T.; Harrison-Brown, M.; Dodson, E.; Veale, K.; Banati, R.B. The 18 kDa translocator protein, microglia and neuroinflammation. *Brain Pathol.* **2014**, *24*, 631–653. [CrossRef]
211. Betlazar, C.; Middleton, R.J.; Banati, R.B.; Liu, G.J. The impact of high and low dose ionising radiation on the central nervous system. *Redox. Biol.* **2016**, *9*, 144–156. [CrossRef]
212. Yao, R.; Pan, R.; Shang, C.; Li, X.; Cheng, J.; Xu, J.; Li, Y. Translocator Protein 18 kDa (TSPO) Deficiency Inhibits Microglial Activation and Impairs Mitochondrial Function. *Front. Pharmacol.* **2020**, *11*, 986. [CrossRef]
213. Meng, Y.; Tian, M.; Yin, S.; Lai, S.; Zhou, Y.; Chen, J.; He, M.; Liao, Z. Downregulation of TSPO expression inhibits oxidative stress and maintains mitochondrial homeostasis in cardiomyocytes subjected to anoxia/reoxygenation injury. *Biomed. Pharmacother.* **2020**, *121*, 109588. [CrossRef] [PubMed]
214. Guilarte, T.R.; Loth, M.K.; Guariglia, S.R. TSPO Finds NOX2 in Microglia for Redox Homeostasis. *Trends Pharmacol. Sci.* **2016**, *37*, 334–343. [CrossRef] [PubMed]
215. Scholz, R.; Caramoy, A.; Bhuckory, M.B.; Rashid, K.; Chen, M.; Xu, H.; Grimm, C.; Langmann, T. Targeting translocator protein (18 kda) (TSPO) dampens pro-inflammatory microglia reactivity in the retina and protects from degeneration. *J. Neuroinflammation* **2015**, *12*, 201. [CrossRef] [PubMed]
216. Monga, S.; Nagler, R.; Amara, R.; Weizman, A.; Gavish, M. Inhibitory Effects of the Two Novel TSPO Ligands 2-Cl-MGV-1 and MGV-1 on LPS-induced Microglial Activation. *Cells* **2019**, *8*, 486. [CrossRef] [PubMed]

217. Feng, H.; Liu, Y.; Zhang, R.; Liang, Y.; Lan, N.; Ma, B. TSPO Ligands PK11195 and Midazolam Reduce NLRP3 Inflammasome Activation and Proinflammatory Cytokine Release in BV-2 Cells. *Front. Cell. Neurosci.* **2020**, *14*, 544431. [CrossRef] [PubMed]
218. Notter, T.; Schalbetter, S.M.; Clifton, N.E.; Mattei, D.; Richetto, J.; Thomas, K.; Meyer, U.; Hall, J. Neuronal activity increases translocator protein (TSPO) levels. *Mol. Psychiatry* **2020**. [CrossRef] [PubMed]
219. Xu, J.; Sun, J.; Perrin, R.J.; Mach, R.H.; Bales, K.R.; Morris, J.C.; Benzinger, T.L.S.; Holtzman, D.M. Translocator protein in late stage Alzheimer's disease and Dementia with Lewy bodies brains. *Ann. Clin. Transl. Neurol.* **2019**, *6*, 1423–1434. [CrossRef] [PubMed]
220. Beckers, L.; Ory, D.; Geric, I.; Declercq, L.; Koole, M.; Kassiou, M.; Bormans, G.; Baes, M. Increased Expression of Translocator Protein (TSPO) Marks Pro-inflammatory Microglia but Does Not Predict Neurodegeneration. *Mol. Imaging Biol.* **2018**, *20*, 94–102. [CrossRef] [PubMed]
221. Nettis, M.A.; Veronese, M.; Nikkheslat, N.; Mariani, N.; Lombardo, G.; Sforzini, L.; Enache, D.; Harrison, N.A.; Turkheimer, F.E.; Modelli, V.; et al. PET imaging shows no changes in TSPO brain density after IFN-α immune challenge in healthy human volunteers. *Transl. Psychiatry* **2020**, *10*, 89. [CrossRef]

Article

The Central Domain of MCPH1 Controls Development of the Cerebral Cortex and Gonads in Mice

Yaru Wang [1], Wen Zong [1], Wenli Sun [1], Chengyan Chen [1], Zhao-Qi Wang [2,3,*] and Tangliang Li [1,4,*]

1   State Key Laboratory of Microbial Technology, Shandong University, Qingdao 250100, China
2   Leibniz Institute on Aging—Fritz Lipmann Institute (FLI), 07745 Jena, Germany
3   Faculty of Biological Sciences, Friedrich-Schiller University of Jena, 07743 Jena, Germany
4   Department of Pathology and Pathophysiology, School of Basic Medical Sciences, Hangzhou Normal University, Hangzhou 311121, China
*   Correspondence: zhao-qi.wang@leibniz-fli.de (Z.-Q.W.); li.tangliang@sdu.edu.cn (T.L.); Tel.: +49-3641-656415 (Z.-Q.W.); +86-532-5863-2368 (T.L.); Fax: +49-3641-656413 (Z.-Q.W.)

Abstract: *MCPH1* is the first gene identified to be responsible for the human autosomal recessive disorder primary microcephaly (MCPH). Mutations in the N-terminal and central domains of MCPH1 are strongly associated with microcephaly in human patients. A recent study showed that the central domain of MCPH1, which is mainly encoded by exon 8, interacts with E3 ligase βTrCP2 and regulates the G2/M transition of the cell cycle. In order to investigate the biological functions of MCPH1's central domain, we constructed a mouse model that lacked the central domain of MCPH1 by deleting its exon 8 (designated as *Mcph1-Δe8*). *Mcph1-Δe8* mice exhibited a reduced brain size and thinner cortex, likely caused by a compromised self-renewal capacity and premature differentiation of *Mcph1-Δe8* neuroprogenitors during corticogenesis. Furthermore, *Mcph1-Δe8* mice were sterile because of a loss of germ cells in the testis and ovary. The embryonic fibroblasts of *Mcph1-Δe8* mice exhibited premature chromosome condensation (PCC). All of these findings indicate that *Mcph1-Δe8* mice are reminiscent of MCPH1 complete knockout mice and *Mcph1-ΔBR1* mice. Our study demonstrates that the central domain of MCPH1 represses microcephaly, and is essential for gonad development in mammals.

Keywords: microcephaly; MCPH1; central domain; brain development; gonad development

Citation: Wang, Y.; Zong, W.; Sun, W.; Chen, C.; Wang, Z.-Q.; Li, T. The Central Domain of MCPH1 Controls Development of the Cerebral Cortex and Gonads in Mice. *Cells* **2022**, *11*, 2715. https://doi.org/10.3390/cells11172715

Academic Editor: Markus Fendt

Received: 20 July 2022
Accepted: 26 August 2022
Published: 31 August 2022

Publisher's Note: MDPI stays neutral with regard to jurisdictional claims in published maps and institutional affiliations.

1. Introduction

Autosomal recessive primary microcephaly (MCPH) is a rare heterogenous neurodevelopmental disorder that is characterized by a pronounced reduction in the brain size. It is caused by defects in neuroprogenitors during neurodevelopment, and has an incidence of 1:30,000 to 1:250,000 in live births in the general population [1–3]. MCPH patients exhibit a small brain with simplified gyri, and by marked reduction in the cerebral cortex, although the brain architecture is grossly normal [4]. Currently, mutations in 29 genes have been identified as causes of MCPH [5–7]. *MCPH1* (*or BRIT1*) was the first gene reported to be causative for primary microcephaly type 1 (MCPH1, OMIM251200) [1]. In addition to brain developmental abnormalities, MCPH mutations cause premature chromosome condensation (PCC) syndrome (OMIM 606858) [8,9]. MCPH patient cells show defective chromosome condensation, with a high proportion of prophase-like cells (PLCs) in late G2-phase with delayed de-condensation post-mitosis [8,10].

The *MCPH1* gene encodes a multifunctional protein that plays an important role in chromosome condensation, DNA damage response (DDR), cell cycle progression, chromatin remodeling and tumorigenesis [11,12]. These functions enable MCPH1 to play an important role in brain development, gonad formation and tumorigenesis [7,13,14]. MCPH1 contains three functional domains: the N-terminal BRCT domain (BRCT1) is necessary for prevention of PCC and contributes to the centrosome localization of MCPH1 [15,16]; the two C-terminal BRCT domains (BRCT2 and BRCT3) bind to phosphorylated proteins in DDR, and are crucial for its localization to DNA damage sites by interacting with γH2AX after ionizing radiation [17,18]. The central domain of MCPH1 (amino acids 381–435) is primarily responsible for the condensin II–MCPH1 interaction [19].

Currently, numerous loss-of-function mutations and variants of *MCPH1* have been associated with MCPH [12,20]. Most missense mutations of *MCPH1* are located in exons 2 and 3 that encode the N-terminal BRCT domain. In accordance to human MCPH1, complete knockout of *Mcph1* (*Mcph1*-del) by deletion of exons 4 and 5 in mice results in primary microcephaly, recapitulating human MCPH1 [21]. Mechanistically, MCPH1 loss reduces Chk1 in the centrosome, leading to aberrant Cdk1 activation and premature mitotic entry of neuroprogenitors [21]. Via this mechanism, MCPH1 regulates the division mode of neuroprogenitors. Of note, mice expressing MCPH1 that is missing the first BRCT domain at the N terminus (*Mcph1*-ΔBRCT1, or *Mcph1*-ΔBR1) reproduce the primary microcephaly phenotype as seen in MCPH1 knockout mice and MCPH patients [13].

Recently, novel MCPH1 mutations have been found to be linked to disruptions in the MCPH1 middle domain. For example, two heterozygous missense mutations (C.982G > A and C.1273T > A) in exon 8 of MCPH1 were found in microcephaly individuals of a Saudi family [22]. A recent finding reports a novel frameshift deletion mutation c.1254delT in exon 8 of the *MCPH1* gene which disrupts the conserved central domain of MCPH1 [23]. These findings suggest an important function of the central domain of MCPH1 in brain development. Our recent molecular biological study conducted in human 293T and Hela cells showed that MCPH1, via its central domain, modulated the dimerization of βTrCP2 to regulate Cdc25A turnover [24,25]. Thus, MCPH1 controls the G2/M transition and determines the mitotic fate of neuroprogenitors [24]. In the current study, we aimed to investigate the physiological function of the central domain of MCPH1, and generated a knockout mouse model lacking exon 8 of *Mcph1* (designated as *Mcph1*-Δe8), which expresses a truncated MCPH1 without its central domain. Our analysis showed that the central domain of MCPH1 is essential for corticogenesis and gonad development in mice.

2. Materials and Methods

2.1. Mice and Mating Scheme

MCPH1's central domain conditional knockout mouse (*Mcph1*$^{\text{flox-e8/flox-e8}}$) was produced by a commercial service provided by Cyagen Biosciences Inc., Suzhou, China. The targeting vector was designed as follows: homology arms (a 2.9-kb 5′ arm and 2.6-kb 3′ arm) were generated using BAC clones from the C57BL/6J RPCI-23 BAC library; the 5′ LoxP site was inserted into intron 7 of *MCPH1* together with a self-excision Neo cassette flanked by Rox sites that can be cleaved by the testes-specific Dre recombinase (SDA–Neo–SDA–LoxP cassette) (Figure 1A); the second LoxP site (3′ LoxP) was inserted into intron 8. The negative selection marker diphtheria toxin A (DTA) cassette was placed upstream of the 5′ homology arm. The linearized targeting vector was introduced into Cyagen's proprietary TurboKnockout ES cells (on a C57BL/6N background) by electroporation. Homologous recombinant clones were isolated using positive (neomycin) and negative (DTA) selection. There were 188 neo-resistant ES clones that were analyzed by Southern blotting in order to identify the gene targeting events. Three probes for Southern blotting were designed to screen the correctly targeted ES clones. The probe 1 (P1) in intron 7 within the 5′ homology arm detected a fragment of 8.44 kb in wild-type (WT) allele and 3.39 kb in targeted (Tg) allele after KpnI digestion (Figure 1A). The probe 2 (P2) was located in intron 8, which detected a fragment of 13.91 kb in WT allele and 10.76 kb in Tg allele after AvrII digestion (Figure 1A). The probe Neo detected a fragment of 9.06 kb after genomic DNA digestion with KpnI, which was used to rule out any random insertion of the targeting vector into other parts of the genome (Supplementary Figure S1A). In total, six ES clones were confirmed to have correct gene targeting events. The targeted ES cell clone 1B1 was introduced into blastocysts by microinjection and then surgically transferred into pseudo-pregnant (surrogate) mothers in order to generate chimeric offspring, which were further crossed with C57BL/6N females to generate heterozygous floxed mice (*Mcph1*^flox-e8/+^). The *Mcph1*^flox-e8/flox-e8^ mouse was bred with the EIIa-Cre transgenic mouse to remove the exon 8 and to generate a conventional central domain knockout mouse for MCPH1 (*Mcph1*-Δe8) (see the Results section; Figure 1 and Figure S1).

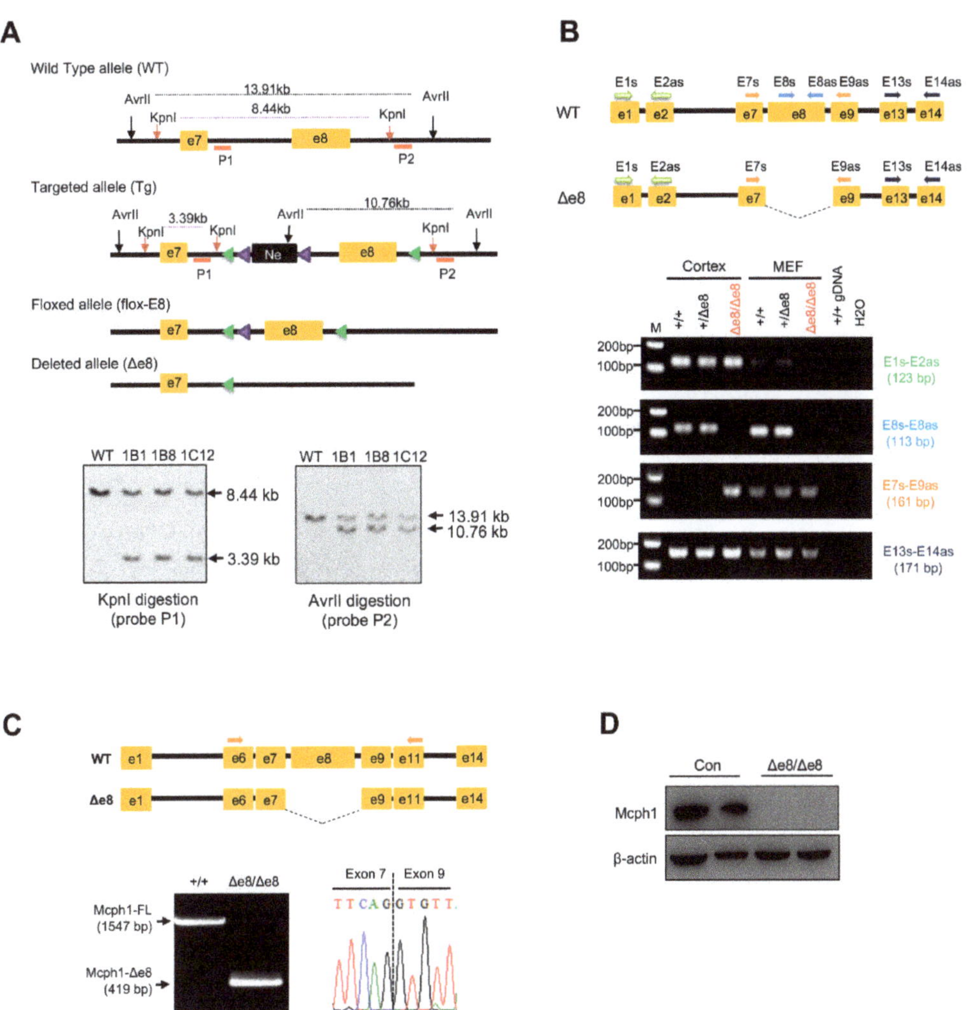

Figure 1. Generation of the MCPH1 central domain knockout mouse (*Mcph1*-Δe8). (**A**) Gene targeting strategy to generate MCPH1 central domain knockout mice. Exon 8 of *Mcph1* that encodes the MCPH1 central domain was floxed by two LoxP sites. The expected alleles before and after gene targeting are shown. The expected sizes of wild-type allele (WT), targeted allele (Tg), floxed allele (Flox) and exon 8 deleted allele, after respective enzyme digestion in Southern blotting, are shown. The locations of the Southern blotting probes (P1 for the 5′ homology arm; P2 for the 3′ homology arm) are marked under their respective alleles. Lower panel: Southern blotting to identify the correct homologous recombination events in ES cell clones after digestion with the indicated restriction enzymes and hybridization with the indicated probes. (**B**) RT-PCR analysis to validate exon 8 deletion in the *Mcph1*

transcript of the *Mcph1-Δe8* cortex and MEFs. Top panel: schematic presentation of localization of PCR primers in the WT and *Mcph1-Δe8* (Δe8) transcripts. Lower panels: RT-PCR results by the indicated primers of the cortex and MEF samples from WT (+/+), heterozygote (+/Δe8) and homozygote (Δe8/Δe8) mice. (**C**) cDNA sequencing to validate exon 8 deletion in *Mcph1-Δe8* mice. PCRs using primers located on *Mcph1* exon 6 and exon 11 produced a 1547-base-pair fragment in WT cDNA and a 419-base-pair fragment in Δe8 cDNA. Right lower panel shows a loss of exon 8 and the only junction between exon 7 and exon 9 of *Mcph1-Δe8* transcripts. (**D**) Western blot analysis of the expression of MCPH1 protein in neurospheres derived from control (Con, wild-type) and *Mcph1-Δe8* homozygous mice (Δe8/Δe8) using an anti-MCPH1 antibody that recognizes residues in the central domain of MCPH1. β-actin was used as a loading control.

The *MCPH1* alleles were genotyped using the following primers: F1: GGTCTGAG-TAATGACCACAGGTTC; R1: GTGGGTAAACACAACTCATCCTTC; R2: TGGGTACTC-CTGCTAGCCTC (wild-type allele, 193 bps; flox allele, 327 bps; and exon 8 deletion allele, 543 bps). EIIa-Cre transgene was genotyped by the primers Cre1, 5′-CATATTGGCAGAACG AAAACGC-3′; and Cre-2, 5′-CCTGTTTCACTATCCAGGTTACGG-3′; the amplicon size was 413 bps.

Mouse strains, including *Mcph1*$^{flox-e8/flox-e8}$, *Mcph1-Δe8* and EIIa-Cre mice, were maintained under specific pathogen-free conditions at the animal facility of Shandong University, Qingdao, P. R. China. Animal care and experiments were performed in accordance with the ethics committee's guidelines (License number: SYDWLL-2022-064).

2.2. Cell Culture

Mouse embryonic fibroblasts (MEFs) were isolated from E13.5 embryos following previously published protocols [26]. Neuroprogenitors were isolated from E14.5 embryonic cortex tissues and maintained in the following neural stem cell medium: DMEM/F12 (Gibco) supplemented with B-27 (Gibco), penicillin/streptomycin (Gibco), 10 ng/mL EGF (Peprotech) and 20 ng/mL bFGF (Peprotech) [27]. All of the cell cultures were maintained at 37 °C in a CO_2 incubator.

2.3. RNA Extraction and RT-PCR Analysis

Total RNAs were extracted from cells and mouse tissues (cerebral cortex) using an RNA extraction kit (Accurate Biotechnology, Changsha, China). cDNA was synthesized using PrimeScript II 1st Strand cDNA Synthesis Kit (Takara, Kusatsu, Japan). RT-PCRs for different fragments of *Mcph1* were conducted using PrimerSTAR® HS premix (Takara, Kusatsu, Japan). The primers used to characterize different fragments of *Mcph1* are listed in Supplementary Table S1. PCR products were isolated with the Gel Extraction Kit (Sangon Biotech, Shanghai, China) and sequenced through services provided by Tsingke.

2.4. Histological Analysis

For paraffin sections, mouse brains were fixed with 4% paraformaldehyde at 4 °C for 24–48 h. Testicular and ovarian tissues were fixed with Bouin's solution for 24 h. Tissue samples were further processed and embedded in paraffin. Sections of 4 μm were used in this study. The sections were stained with hematoxylin and eosin solutions. The histological images were scanned and processed with an Olympus VS200 microscope.

2.5. Immunofluorescence (IF) Staining on Cells and Brain Sections

For IF analysis of cells, primary MEF cells were fixed with 4% PFA for 15 min at room temperature. Then, MEFs were stained with pS10-histone 3 antibody (1:400, 9706S, Cell Signaling, Danvers, MA, USA) in order to investigate the mitotic cells. At least 250 prophase cells from each cell line were counted manually.

For IF in brain tissues, embryonic brains at indicated developmental stages were fixed with 4% PFA at 4 °C for 24–48 h, then transferred to 30% sucrose. Brains were then embedded with OCT (Thermo Fisher, Waltham, MA, USA) and stored at −80 °C. A 10-micrometer section was used for cryosection. For immunostaining, a previously published protocol was followed [27]. The following primary antibodies were used: rabbit anti-Sox2 (1:400, Ab97959, Abcam, Cambridge, UK), rabbit anti-Tbr2 (1:400, Ab183991, Abcam, Cambridge, UK), rabbit anti-Ki67 (1:400, 9129S, Cell Signaling), rabbit anti-cleaved caspase-3 (Asp175) (1:400, 9579S, Cell Signaling, Danvers, MA, USA), mouse anti-pS10-H3 (1:400, 9706S, Cell Signaling, Danvers, MA, USA) and rat anti-pS28-H3 (1:400, Ab10543, Abcam, Cambridge, UK). The following secondary antibodies were used: Goat Anti-Mouse IgG Alexa Fluor® 594 (1:1000, Ab150113, Abcam, Cambridge, UK), Goat Anti-Rabbit IgG Alexa Fluor® 594 (1:1000, Ab150080, Abcam, Cambridge, UK), and Goat Anti-Rat IgG Alexa Fluor®594 (1:1000, Ab150160, Abcam, Cambridge, UK). The antibodies used for IF analysis are summarized in supplementary files (Supplementary Tables S2 and S3).

All of the IF images from cells and tissues were captured with OLYMPUS cellSens software (Standard Version, Shinjuku City, Tokyo, Japan) on an Olympus microscope BX53 installed with a DP80 camera. The images were further analyzed and processed with OLYMPUS OlyVIA software (Version 3.2, Shinjuku City, Tokyo, Japan).

2.6. Western Blotting

In order to detect MCPH1 expression, cell or tissue samples were lysed with RIPA buffer supplements with protease/phosphatase inhibitors (APExBIO). An amount of 40–60 ug total protein was separated with SDS-PAGE gels (10%), and further transferred onto PVDF membranes. The primary antibodies used for this study were the following: rabbit anti-MCPH1 (1:1000, Cell Signaling, Danvers, MA, USA); and mouse anti-β-actin (1:10,000, Sigma-Aldrich, St. Louis, MO, USA). For the DNA damage assay, MEFs were treated with hydroxyurea (HU, 2 mM) for 3 h, and then recovered for 0, 3 and 6 h. Protein samples were harvested at different time-points. In order to investigate the DNA repair dynamics, the following antibodies were used: mouse anti-phospho histone H2A.X (Ser139) (1:2000, 05-636, EMD Millipore, Burlington, MA, USA), rabbit anti-phospho Chk1 (Ser317) (1:1000,12302S, Cell Signaling, Danvers, MA, USA) and mouse anti-β-actin (1:10,000, A5441, Sigma-Aldrich, St. Louis, MO, USA). The secondary antibodies used were: HRP-conjugated goat anti-rabbit IgG and HRP-conjugated goat anti-mouse IgG (1:2000; Proteintech, Tokyo, Japan). The blotting result was developed using BeyoECL Plus substrates (Beyotime, P0018S, Shanghai, China) and quantified with Evolution-Capt Solo S 17.00 software. The antibodies used for WB are summarized in supplementary files (Supplementary Tables S2 and S3).

2.7. Statistical Analysis

For the quantitative analysis conducted in this study, at least 3 biological samples from each genotype were used. Unpaired Student's *t*-test was employed. The statistical analysis in this study was performed with GraphPad Prism (Ver 6.00, GraphPad Software, San Diego, CA, USA) and graphed with the format of mean $\pm$ SD. Statistical significance between genotypes was used as follows: n.s, $p > 0.05$; *, $p < 0.05$; **, $p < 0.01$; ***, $p < 0.001$.

3. Results

3.1. Generation of Mcph1-Δe8 Mice

The central domain of mouse MCPH1 is mainly encoded by exon 8 of the *Mcph1* gene. Thus, we generated an Mcph1 exon 8 conditional knockout mouse by introducing LoxP sites that flanked exon 8 (Figure 1A, please see the Materials and Methods section for details). In order to produce conventional knockout mice with MCPH1's central domain deletion, $Mcph1^{flox-e8/+}$ mice were further crossed with EIIa-Cre mice to delete exon 8 of *Mcph1* in the germline. Heterozygous $Mcph1^{\Delta e8/+}$ mice were further intercrossed to generate $Mcph1^{\Delta e8/\Delta e8}$ mice (designated as *Mcph1-Δe8*) (Supplementary Figure S1B). In order to confirm the successful generation of the *Mcph1-Δe8* mouse, we performed PCR genotyping and found that the Δe8 mutant allele (Δe8) produced an expected 543-base-pair PCR product (Supplementary Figure S1C). RT-PCR experiments revealed the deletion of exon 8 in *MCPH1* transcripts in *Mcph1-Δe8* mouse embryos (Figure 1B, primer set E8-E8as), which was further confirmed by sequencing (Figure 1C). Of note, the mouse *Mcph1* exon 8 is 1128 bps, deletion of which resulted in in-frame deletion of the central domain in MCPH1. Western blotting using an MCPH1 antibody (D38G5, #4120, Cell Signaling Technology) that recognizes the MCPH1's central domain confirmed that *Mcph1-Δe8* indeed was missing this middle domain (Figure 1D). All of these data indicate that the *Mcph1-Δe8* mouse was successfully generated.

3.2. Mcph1-Δe8 Mice Develop Microcephaly

Intercross of heterozygous *Mcph1-Δe8* mice (*Mcph1*$^{\Delta e8/+}$) generated homozygous *Mcph1-Δe8* mice at birth with normal Mendelian ratios (data not shown). Interestingly, we noticed that *Mcph1-Δe8* mice had a smaller brain as well as a reduction in the brain weight as compared to control littermates at postnatal day 0 (P0) (Figure 2A,B), which is similar to *MCPH1* complete knockout mice [21] and *Mcph1-ΔBR1* mice [13]. In addition to the cortex, the middle brain of *Mcph1-Δe8* mice at P0 seemed smaller than that of control mice (Figure 2A). Histological analysis of coronal sections of P0 brains revealed a thinner cerebral cortex in *Mcph1-Δe8* as compared to control mice (Figure 2C,D), indicating that *Mcph1-Δe8* mice are microcephalic.

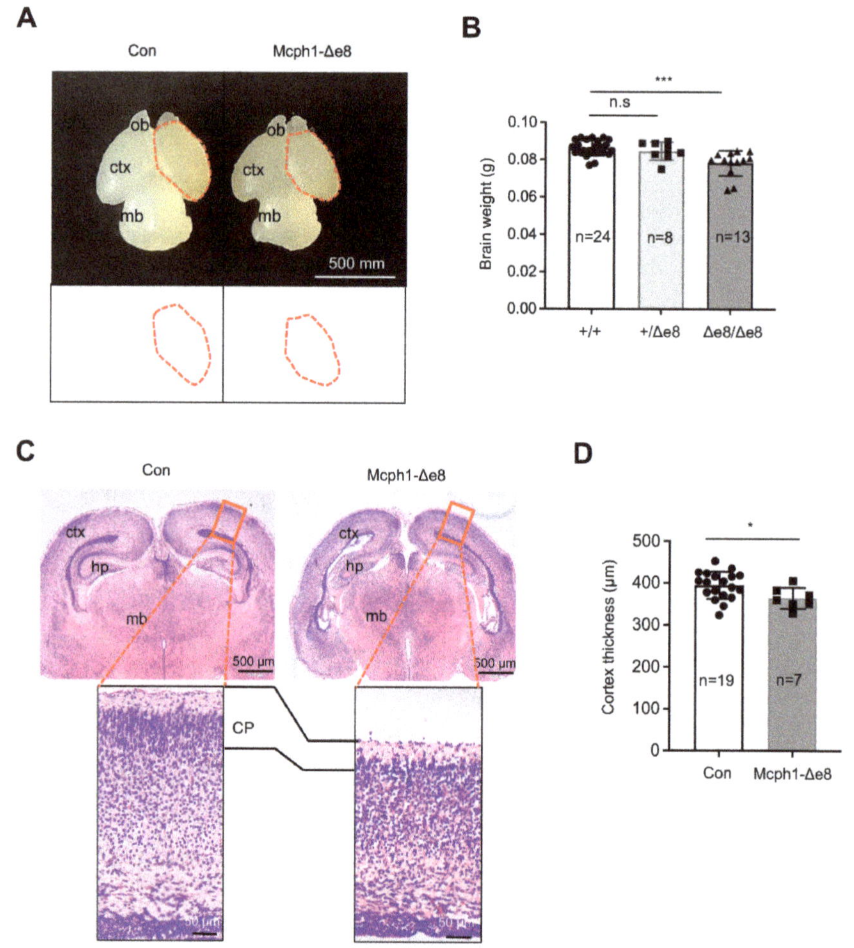

Figure 2. *Mcph1-Δe8* mice develop primary microcephaly. (**A**) Macroscopic dorsal view of the mouse

brain of wild-type control (Con) and homozygous *Mcph1-Δe8* (Δe8/Δe8) mice at birth (P0). Schematic views of the cortical areas from control and *Mcph1-Δe8* brains are shown in the lower panel. ob: olfactory bulb; ctx: cerebral cortex; mb: midbrain. (**B**) Quantification of the brain weights of the indicated genotypes of mice at P0. (**C**) H & E staining of coronal sections of control and *Mcph1-Δe8* brains at P0. Overview of coronal brain sections and enlarged view of the forebrain cortex from the rectangular areas of the upper panel are shown. ctx: cerebral cortex; hp: hippocampus; mb: midbrain. (**D**) Quantification of the thickness of the cortex of control and homozygous Δe8/Δe8 mice. The cortex thickness of control: 394.22 μm and *Mcph1-Δe8*: 363.75 μm. *n*: the number of mice analyzed. Unpaired Student's *t*-test was used for statistical analysis. *, $p < 0.05$; ***, $p < 0.001$; n.s, not significant.

3.3. Mcph1-Δe8 Neuroprogenitors Have Self-Renewal Defect

In order to study the reasons behind the primary microcephaly phenotype of *Mcph1-Δe8* mice, we analyzed the embryonic cortex at the middle stage of neurogenesis, namely E15.5. By immunostaining with antibodies against Sox2, a marker for radial glial cells (RGCs) in the ventricular zone (VZ), and Tbr2, a marker for intermediate progenitors (IPs) in the subventricular zone (SVZ), we found less Sox2-positive (Sox2+) cells in the *Mcph1-Δe8* cortex compared to controls (Figure 3A,B); however, we found a normal number of Tbr2-positive (Tbr2+) cells in the SVZ (Figure 3A,B), indicating that RGCs, but not intermediate progenitors, are affected by the mutation.

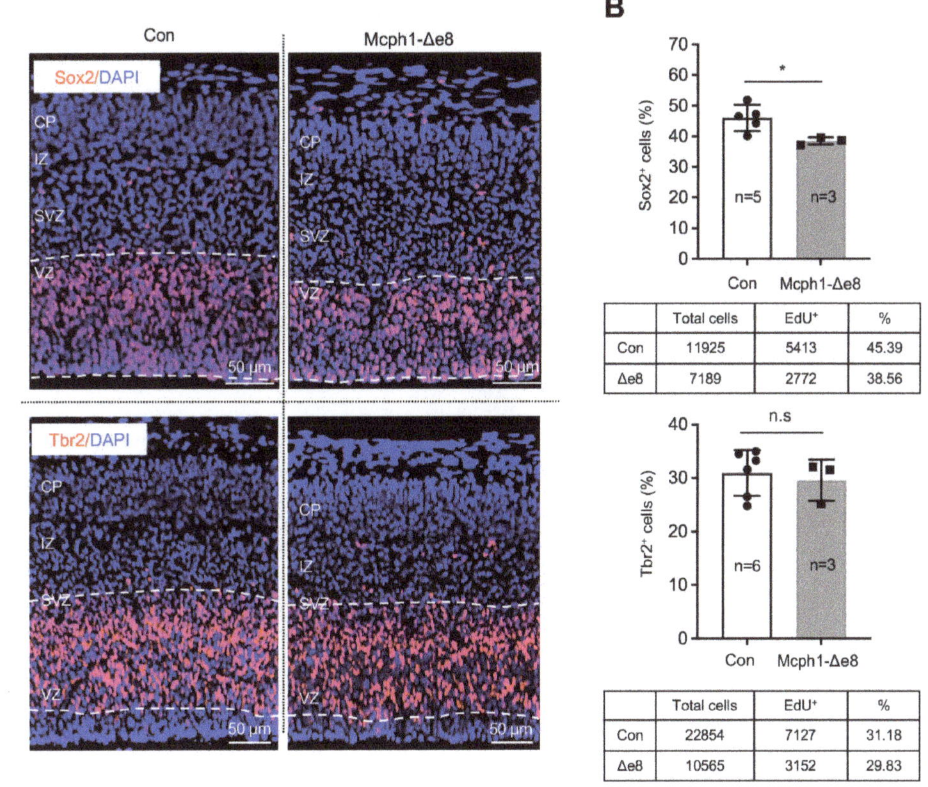

	Total cells	EdU+	%
Con	11925	5413	45.39
Δe8	7189	2772	38.56

	Total cells	EdU+	%
Con	22854	7127	31.18
Δe8	10565	3152	29.83

Figure 3. Reduction of neuroprogenitors in *Mcph1-Δe8* embryonic cortex. (**A**) Immunofluorescence

staining of sagittal sections of the E15.5 brain with Sox2 and Tbr2 antibodies. (**B**) Quantification of Sox2$^+$ and Tbr2$^+$ cells. N: the number of mice analyzed. The total numbers of cells of control (Con) and mutant (*Mcph1-Δe8*) scored are summarized under the respective graphs. Unpaired Student's *t*-test was used for statistical analysis. *, $p < 0.05$; n.s, not significant.

In order to test whether the fewer Sox2$^+$ RGCs observed were caused by an increase in apoptosis, we performed immunofluorescence staining for cleaved-Caspase3 (a classical assay to detect apoptotic cells) and found there was no obvious difference in the apoptotic index between the control's and *Mcph1-Δe8* genotype's cerebral cortex tissues (Supplementary Figure S2A,B). We next analyzed the proliferation of the neuroprogenitors. In vivo pulse labeling (1 h) by EdU incorporation into E15.5 embryos revealed that the ratio of EdU$^+$ cells in the VZ and SVZ of the *Mcph1-Δe8* cortex was significantly reduced as compared to control littermates (Figure 4A,B). We next examined whether the low number of EdU-positive cells affects cell division by measuring the number of mitotic cells after immunostaining against mitosis maker pS28-H3 (Figure 4A). Indeed, the number of pS28-H3-positive cells was much less in the *Mcph1-Δe8* cortex compared to control littermates (Figure 4B). Thus, the middle domain of MCPH1 is required for the self-renewal and maintenance of neuroprogenitors [22–24], but dispensable for apoptosis.

Human cells with siRNA knockdown of MCPH1, or MCPH patient lymphoblast cells, exhibit a defective DDR that involves the ATR-Chk1-mediated cell cycle checkpoint [28,29]. We next investigated if *Mcph1-Δe8* primary MEF cells had altered Atr-Chk1 signaling using Western blotting. Hydroxyurea (HU) inhibits DNA synthesis, and has been widely used to investigate Atr-Chk1 -mediated DDR [26]. To this end, we treated the control and *Mcph1-Δe8* MEFs with or without HU, and analyzed the expressions of p-Chk1 and γH2AX at different time-points. p-Chk1 levels indicate an activation of Atr signaling, while γH2AX levels mark DNA damage accumulation in the cells [26]. We found that *Mcph1-Δe8* MEF cells had higher basal as well as HU-induced levels of p-Chk1 and γH2AX compared to those of controls (Figure 5C,D), indicating that these mutant cells have malfunctional Atr signaling.

3.4. Defects in Gonad Development in Mcph1-Δe8 Mice

We crossed *Mcph1-Δe8* mice, both males and females, with wild-type mice for 4 months, but all mutant mice failed to produce any offspring (data not shown), indicating that *Mcph1-Δe8* mice were infertile, similarly to the sterility of *Mcph1-del* and *Mcph1-ΔBR1* mice [13,21]. Macroscopically, the *Mcph1-Δe8* testis was significantly smaller than that of the control (Figure 6A). The ratio of testis weight (TW) to body weight (BW) was also significantly smaller in *Mcph1-Δe8* mice (Figure 6A). Histological analysis after H&E staining showed that *Mcph1-Δe8* testes were devoid of round spermatids and elongated spermatids, but still possessed some pachytene spermatocytes within the seminiferous tubules (Figure 6B). In addition, we performed H&E staining of the epididymis and did not detect any spermatozoa-filled tubules in the *Mcph1-Δe8* epididymis (Figure 6B). The diameters of *Mcph1-Δe8* seminiferous tubules in testes were smaller (Figure 6C). A significantly higher frequency of vacuolized seminiferous tubules was found in *Mcph1-Δe8* testes (Figure 6D). These data suggest that spermatogenesis in *Mcph1-Δe8* mice was arrested at an early stage of development. In female *Mcph1-Δe8* mice, the size of the ovary was smaller and the uterine wall was thinner compared to controls (Figure 6E). H&E staining revealed that ovarian follicles were invisible in *Mcph1-Δe8* female mice (Figure 6F). We conclude that the central domain of MCPH1 is essential for gonad development.

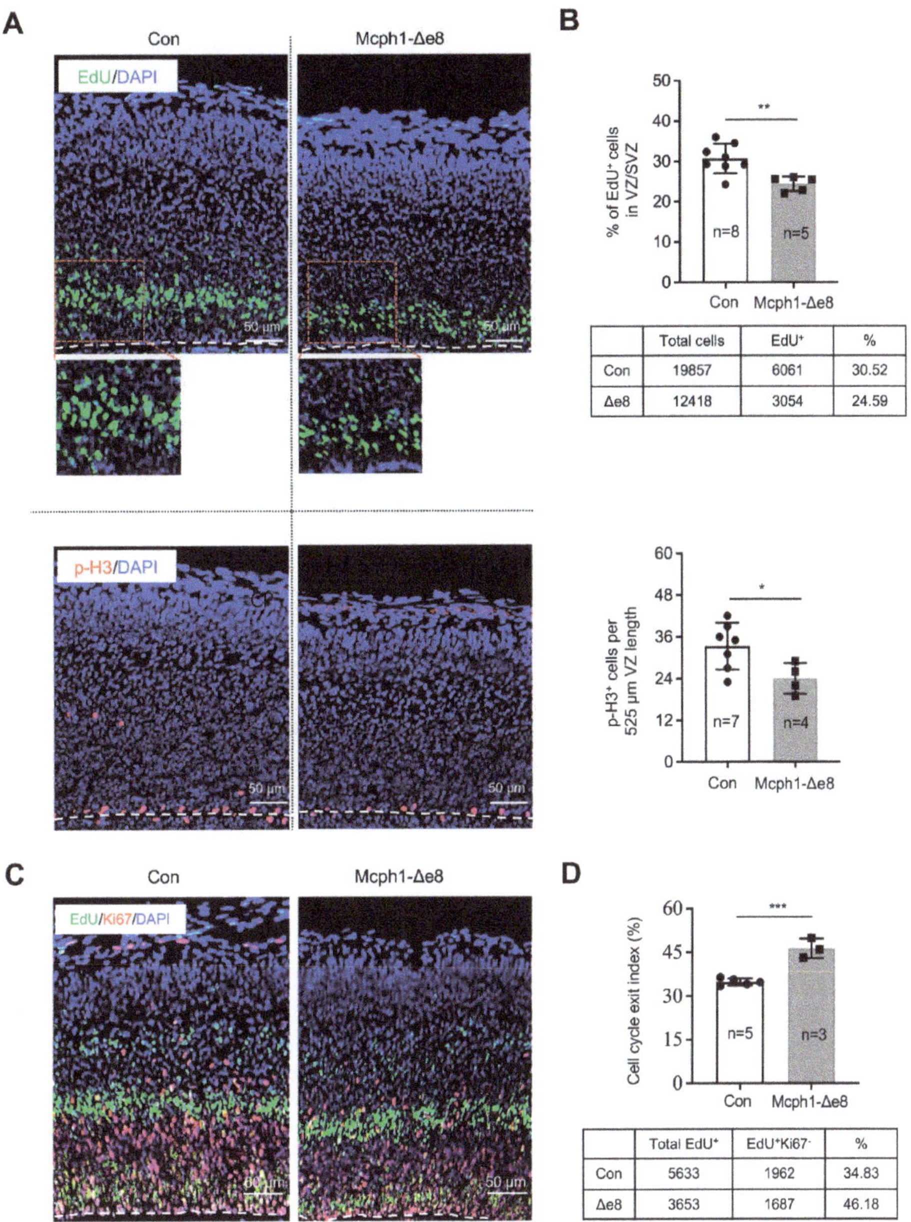

	Total cells	EdU+	%
Con	19857	6061	30.52
Δe8	12418	3054	24.59

	Total EdU+	EdU+Ki67-	%
Con	5633	1962	34.83
Δe8	3653	1687	46.18

Figure 4. Proliferation defects of neuroprogenitors in the *Mcph1*-Δe8 embryonic cortex. (**A**) Immunostaining of the E15.5 embryonic cortex after EdU pulse labeling for 1 hr using antibodies against EdU (green) and pS28-H3 (a mitotic marker, red). The nucleus is counterstained by DAPI (blue). (**B**) Quantification of the percentages of EdU$^+$ and pS28-H3$^+$ cells among the total DAPI$^+$ cells. (**C**) Double staining of the E15.5

brain cortex using antibodies against EdU (green) and Ki67 (red). EdU was injected on E14.5 of embryonic development and embryos were harvested at 24 hrs later (E15.5). (**D**) Cell cycle exit index was calculated by the ratio of EdU$^+$Ki67$^-$ vs. total EdU$^+$ cells. The numbers of control (Con) and mutant (*Mcph1-Δe8*) cells scored are indicated in the tables below their respective graphs. *n*: the number of mice analyzed. Unpaired Student's *t*-test was used for statistical analysis. *, $p < 0.05$; **, $p < 0.01$; ***, $p < 0.001$.

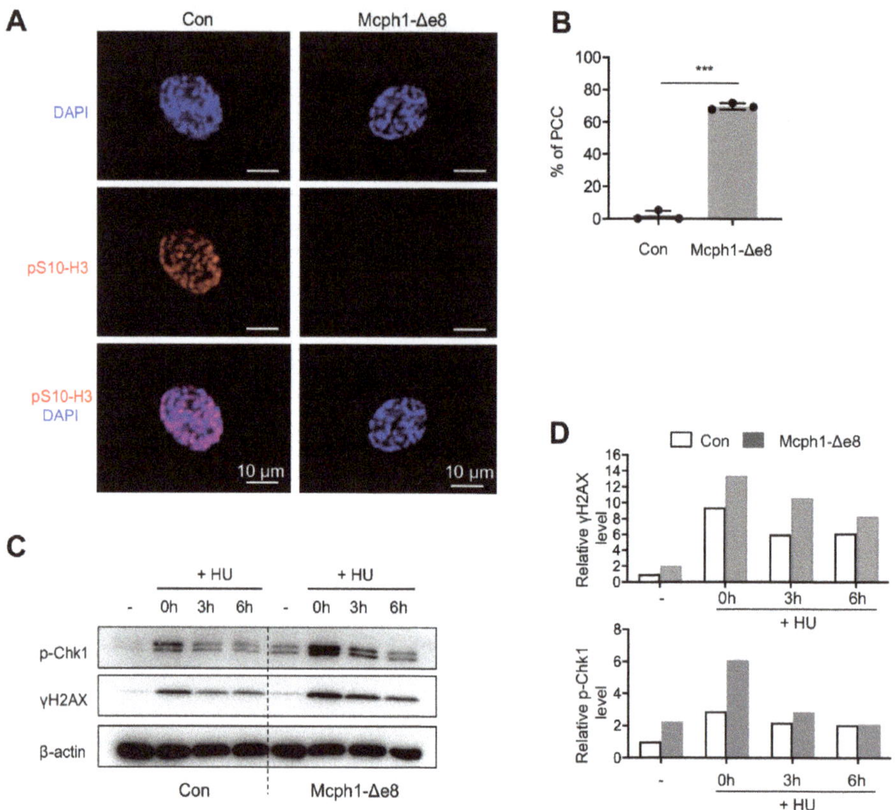

Figure 5. *Mcph1-Δe8* cells exhibit PCC and defective DDR. (**A**) Representative images of PCC in *Mcph1-Δe8* MEFs. Primary MEF cells were stained with a pS10-H3 antibody (red) and counterstained with DAPI (blue). (**B**) Quantification of the percentages of PCC cells (prophase cells lacking the pS10-H3 signal) in control and *Mcph1-Δe8* primary MEF cells. More than 250 prophase cells were scored in each group of the indicated genotype. (**C**) Western blot analysis of p-Chk1 and γH2AX in control and *Mcph1-Δe8* MEFs with or without HU treatment. β-actin was used as a loading control. (**D**) The quantification of the indicated protein intensities from panel (**C**). Unpaired Student's *t*-test was used for statistical analysis. ***, $p < 0.001$.

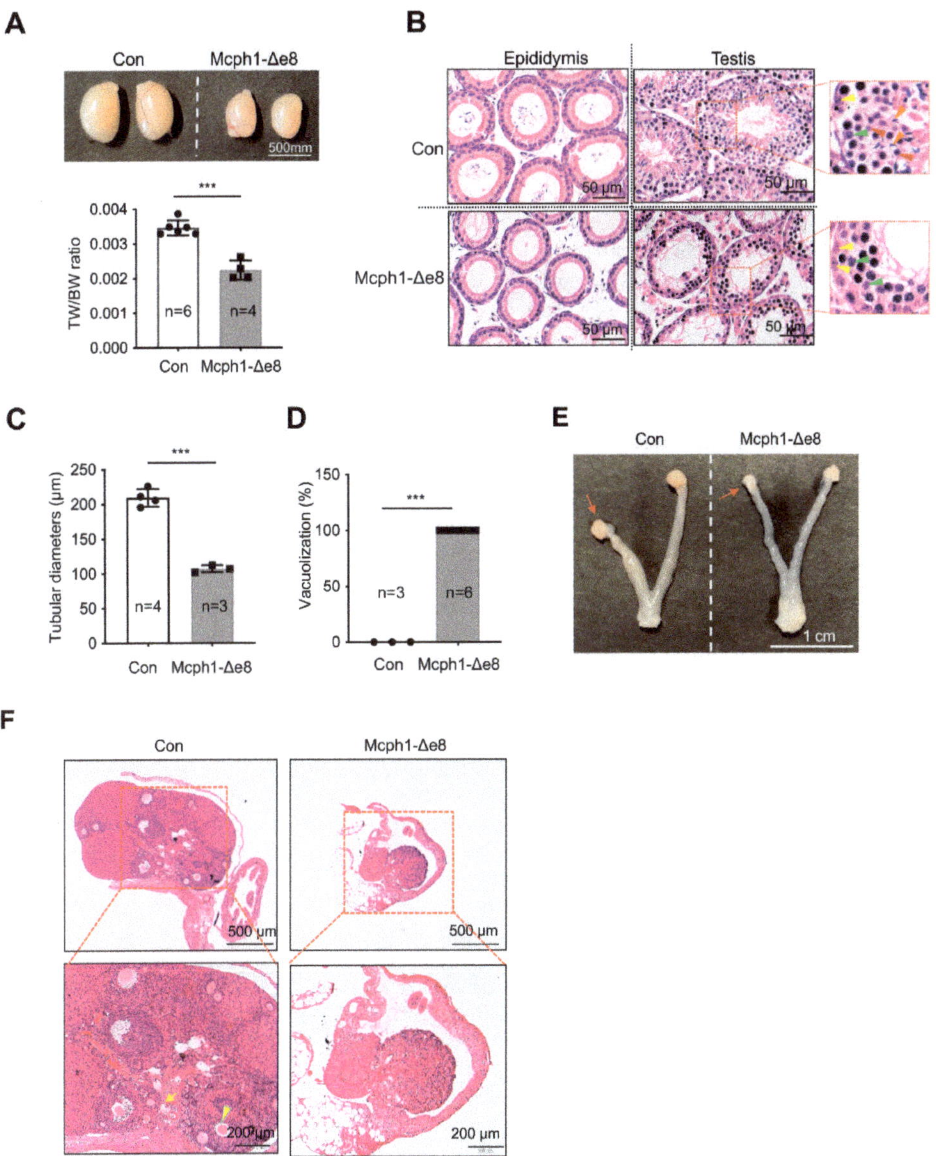

Figure 6. Defective gonad development in *Mcph1-*Δe8 mice. (**A**) Macroscopic view of testes from a control and a *Mcph1-*Δe8 mouse (6 weeks of age). The ratios of the testis weight (TW) to body weight (BW) from 6-week-old control and *Mcph1-*Δe8 mice are shown in the lower panel. (**B**) H&E staining of testis sections from 6-week-old control and *Mcph1-*Δe8 male mice. The epididymis (left panel) and seminiferous tubules (right panel) are shown. Yellow, green, red and blue arrowheads mark spermatogonia, pachytene spermatocytes, round spermatids and elongated spermatids, respectively.

(**C**) Quantification of the diameters of seminiferous tubules of *Mcph1-Δe8* male mice at P100. (**D**) Quantification of tubules lacking spermatocytes (vacuolization) as a percentage of the total tubules of 6-week-old *Mcph1-Δe8* male mice. (**E**) Representative image of ovaries from a 6-week-old control and a *Mcph1-Δe8* mouse. Arrows mark ovaries. (**F**) H&E staining of ovary sections of 6-week-old control and *Mcph1-Δe8* mice. The yellow arrow and arrowhead point to a primary and secondary follicle, respectively. *n*: the number of mice analyzed. Unpaired Student's *t*-test was used for statistical analysis. ***, $p < 0.001$.

4. Discussion

In humans, MCPH1 mutations cause primary microcephaly. Genetic studies identify that many mutations and genetic variants of MCPH1 are prominently located in the N-terminal and central domains [7,20,22,23,30]. We previously generated MCPH1 conventional knockout mice (*Mcph1*-del) [21], MCPH1 N-terminal BRCT domain knockout mice (*Mcph1*-ΔBR1) [13], and MCPH1 fetal brain specific knockout mice [21,31], and revealed that the whole MCPH1 protein and the first BRCT domain at the MCPH1 N′ terminus are essential for neuroprogenitor self-renewal and differentiation during corticogenesis. The first BRCT domain of MCPH1 is required for its centromere localization, which has an impact on choices of symmetric and asymmetric division of neuroprogenitors. Thus, the altered cell division mode in neuroprogenitors from *Mcph1*-del and *Mcph1*-ΔBR1 embryonic brains is responsible for the microcephaly phenotype in these models [21,24].

Recently, we found that MCPH1, via its central domain, directly interacts with βTrCP2 to promote its activity to degrade Cdc25A during the G2/M transition in cell cycle control. Ectopic expression of βTrCP2 or the Cdc25A knockdown remedied the premature differentiation of MCPH1-deficient neuroprogenitors [24]. Therefore, a stabilization role of the MCPH1 central domain on βTrCP2 is presumably important for brain development. We showed here, using a mouse model in which the central domain of MCPH1 is deleted (by deleting exon 8), that the central domain of MCPH1 is required for preventing microcephaly, and maintains a proper balance of the self-renewal and differentiation capacity of neuroprogenitors during brain development. *Mcph1-Δe8* mice show a reduction in cortical thickness that is likely due to impaired proliferation and premature differentiation of neuroprogenitors during the early stages of embryonic neurogenesis, ultimately leading to a reduction in the neuroprogenitor pool and a small brain size.

In addition to brain developmental defects, *Mcph1-Δe8* mice show severe gonad atrophy in both males and females, similarly to *MCPH1* complete knockout and *Mcph1*-ΔBR1 mice [13,14,21]; this indicates that the central domain of MCPH1 is essential for gonadal development. It might be surprising that all of these *Mcph1* mutant mice are always associated with defects in gonad development and infertility. It strongly suggests that at least in mice, MCPH1 is critical for the normal development of the reproductive system [13,14,21]. Currently, however, there is no case reports of MCPH patients with testicular or ovarian atrophy. How MCPH1 regulates spermatogenesis or ovary formation is currently unclear. One possible explanation is that defective DNA damage repair (as judged by a high level of p-Chk1 and γH2AX shown in *Mcph1-Δe8* MEFs) may contribute to infertility phenotypes in *Mcph1-Δe8* testes and ovaries. It was found that the recruitment of DNA repair proteins BRCA2-RAD51 was impaired in MCPH1 mutant mice, leading to meiosis arrest and apoptosis of spermatocytes, as well as a completely loss of pachytene and post-meiosis spermatocytes [14].

The central domain of MCPH1 has been shown to interact with condensin II, which is believed to be responsible for chromosome condensation [16,19]. Similarly to *Mcph1*-deficient [14,21] and *Mcph1*-ΔBR1 cells [13], *Mcph1-Δe8* MEF cells exhibit a high incidence of PCC, indicating that the central domain is involved in the regulation of pre-mitotic chromosomal states. Of note, previous studies found that *Mcph1*-ΔBR1 MEFs have around 40% of PCC, while complete knockout (*Mcph1*-del) MEFs have around 30% PCC [13,21]. The difference in PCC index in each mouse line could be caused by an MEF isolation process, cell culture conditions, or molecular markers used to define metaphase cells.

In summary, we demonstrated that analogously to *Mcph1*-complete knockout [14,21] and *Mcph1*-ΔBR1 [13] mice, the deletion of the central domain of MCPH1 in mice recapitulates phenotypes of MCPH patients, i.e., primary microcephaly and the PCC. The three MCPH1 mutant mouse lines generated in our lab [13,21], together with three additional MCPH1 mutants produced by other research groups [14,32,33], allowed us to conclude that the decisive function of MCPH1 in brain development and fertility is located in the N-terminal and the central domains of MCPH1. Since MCPH1 has no enzymatic activity, it is plausible that MCPH1 participates in various cellular activities through interactions with different partners. For example, MCPH1 interacts with βTrCP2 through its central domain and participates in the regulation of the G2/M cell cycle checkpoint [24]. Therefore, finding proteins that interact with specific domains of MCPH1 could provide further hints to decipher the biological functions of MCPH1 in all of these physiological and developmental processes.

Supplementary Materials: The following supporting information can be downloaded at: https://www.mdpi.com/article/10.3390/cells11172715/s1, Figure S1: Characterization of Mcph1-Δe8 mice; Figure S2: Cell death in Mcph1-Δe8 embryonic cortex; Table S1: List of primers; Table S2: List of primary antibodies used in this study; Table S3: List of secondary antibodies used in this study.

Author Contributions: Y.W., W.Z., W.S. and C.C., methodology and experiments, data acquisition and validation; Y.W., W.S., W.Z., Z.-Q.W. and T.L., data analysis; Y.W., W.Z., Z.-Q.W. and T.L., writing manuscript; Z.-Q.W. and T.L., conceptualization, supervision and administration of the project; Z.-Q.W. and T.L., funding acquisition. All authors have read and agreed to the published version of the manuscript.

Funding: This research was supported by the DFG grant to Z.-Q.W. (WA2627/2-2). T.L is supported by Grant Nos. KF2020005 and KF2021008 from NHC Key Laboratory of Birth Defect for Research and Prevention (Hunan Provincial Maternal and Child Health Care Hospital, Changsha, China), Grant No. 31770871 from the National Natural Science Foundation of China, and Qilu Youth Scholar Startup Funding of Shandong University.

Institutional Review Board Statement: The animal study protocol was approved by Ethics Committee of Shandong University (License number: SYDWLL-2022-064).

Informed Consent Statement: Not applicable.

Data Availability Statement: All datasets presented can be requested from the corresponding authors.

Acknowledgments: We are grateful to the members of Wang and Li laboratories for their helpful discussions and critical comments.

Conflicts of Interest: The authors declare no competing interest.

References

1. Jackson, A.P.; Eastwood, H.; Bell, S.M.; Adu, J.; Toomes, C.; Carr, I.M.; Roberts, E.; Hampshire, D.J.; Crow, Y.J.; Mighell, A.J.; et al. Identification of microcephalin, a protein implicated in determining the size of the human brain. *Am. J. Hum. Genet.* **2002**, *71*, 136–142. [CrossRef] [PubMed]
2. Kumar, A.; Markandaya, M.; Girimaji, S.C. Primary microcephaly: Microcephalin and ASPM determine the size of the human brain. *J. Biosci.* **2002**, *27*, 629–632. [CrossRef] [PubMed]
3. Woods, C.G.; Bond, J.; Enard, W. Autosomal recessive primary microcephaly (MCPH): A review of clinical, molecular, and evolutionary findings. *Am. J. Hum. Genet.* **2005**, *76*, 717–728. [CrossRef]
4. Desir, J.; Cassart, M.; David, P.; Van Bogaert, P.; Abramowicz, M. Primary microcephaly with ASPM mutation shows simplified cortical gyration with antero-posterior gradient pre- and post-natally. *Am. J. Med. Genet. Part A* **2008**, *146*, 1439–1443. [CrossRef]
5. Siskos, N.; Stylianopoulou, E.; Skavdis, G.; Grigoriou, M.E. Molecular Genetics of Microcephaly Primary Hereditary: An Overview. *Brain Sci.* **2021**, *11*, 581. [CrossRef] [PubMed]
6. Jean, F.; Stuart, A.; Tarailo-Graovac, M. Dissecting the Genetic and Etiological Causes of Primary Microcephaly. *Front. Neurol.* **2020**, *11*, 570830. [CrossRef]
7. Kristofova, M.; Ori, A.; Wang, Z.Q. Multifaceted Microcephaly-Related Gene MCPH1. *Cells* **2022**, *11*, 275. [CrossRef]
8. Trimborn, M.; Bell, S.M.; Felix, C.; Rashid, Y.; Jafri, H.; Griffiths, P.D.; Neumann, L.M.; Krebs, A.; Reis, A.; Sperling, K.; et al. Mutations in microcephalin cause aberrant regulation of chromosome condensation. *Am. J. Hum. Genet.* **2004**, *75*, 261–266. [CrossRef]
9. Pfau, R.B.; Thrush, D.L.; Hamelberg, E.; Bartholomew, D.; Botes, S.; Pastore, M.; Tan, C.; del Gaudio, D.; Gastier-Foster, J.M.; Astbury, C. MCPH1 deletion in a newborn with severe microcephaly and premature chromosome condensation. *Eur. J. Med. Genet.* **2013**, *56*, 609–613. [CrossRef]

10. Trimborn, M.; Schindler, D.; Neitzel, H.; Hirano, T. Misregulated chromosome condensation in MCPH1 primary microcephaly is mediated by condensin II. *Cell Cycle* **2006**, *5*, 322–326.

11. Barbelanne, M.; Tsang, W.Y. Molecular and cellular basis of autosomal recessive primary microcephaly. *BioMed Res. Int.* **2014**, *2014*, 547986. [CrossRef] [PubMed]

12. Liu, X.; Zhou, Z.W.; Wang, Z.Q. The DNA damage response molecule MCPH1 in brain development and beyond. *Acta Biochim. Biophys. Sin.* **2016**, *48*, 678–685. [CrossRef] [PubMed]

13. Liu, X.; Schneble-Lohnert, N.; Kristofova, M.; Qing, X.; Labisch, J.; Hofmann, S.; Ehrenberg, S.; Sannai, M.; Jorss, T.; Ori, A.; et al. The N-terminal BRCT domain determines MCPH1 function in brain development and fertility. *Cell Death Dis.* **2021**, *12*, 143. [CrossRef] [PubMed]

14. Liang, Y.; Gao, H.; Lin, S.Y.; Peng, G.; Huang, X.; Zhang, P.; Goss, J.A.; Brunicardi, F.C.; Multani, A.S.; Chang, S.; et al. BRIT1/MCPH1 is essential for mitotic and meiotic recombination DNA repair and maintaining genomic stability in mice. *PLoS Genet.* **2010**, *6*, e1000826. [CrossRef]

15. Jeffers, L.J.; Coull, B.J.; Stack, S.J.; Morrison, C.G. Distinct BRCT domains in Mcph1/Brit1 mediate ionizing radiation-induced focus formation and centrosomal localization. *Oncogene* **2008**, *27*, 139–144. [CrossRef]

16. Wood, J.L.; Liang, Y.; Li, K.; Chen, J. Microcephalin/MCPH1 associates with the Condensin II complex to function in homologous recombination repair. *J. Biol. Chem.* **2008**, *283*, 29586–29592. [CrossRef]

17. Manke, I.A.; Lowery, D.M.; Nguyen, A.; Yaffe, M.B. BRCT repeats as phosphopeptide-binding modules involved in protein targeting. *Science* **2003**, *302*, 636–639. [CrossRef]

18. Wood, J.L.; Singh, N.; Mer, G.; Chen, J. MCPH1 functions in an H2AX-dependent but MDC1-independent pathway in response to DNA damage. *J. Biol. Chem.* **2007**, *282*, 35416–35423. [CrossRef]

19. Yamashita, D.; Shintomi, K.; Ono, T.; Gavvovidis, I.; Schindler, D.; Neitzel, H.; Trimborn, M.; Hirano, T. MCPH1 regulates chromosome condensation and shaping as a composite modulator of condensin II. *J. Cell Biol.* **2011**, *194*, 841–854. [CrossRef]

20. Pulvers, J.N.; Journiac, N.; Arai, Y.; Nardelli, J. MCPH1: A window into brain development and evolution. *Front. Cell. Neurosci.* **2015**, *9*, 92. [CrossRef]

21. Gruber, R.; Zhou, Z.; Sukchev, M.; Joerss, T.; Frappart, P.O.; Wang, Z.Q. MCPH1 regulates the neuroprogenitor division mode by coupling the centrosomal cycle with mitotic entry through the Chk1-Cdc25 pathway. *Nat. Cell Biol.* **2011**, *13*, 1325–1334. [CrossRef] [PubMed]

22. Naseer, M.I.; Rasool, M.; Abdulkareem, A.A.; Bassiouni, R.I.; Algahtani, H.; Chaudhary, A.G.; Al-Qahtani, M.H. Novel compound heterozygous mutations in MCPH1 gene causes primary microcephaly in Saudi family. *Neurosciences* **2018**, *23*, 347–350. [CrossRef]

23. Khan, N.M.; Masoud, M.S.; Baig, S.M.; Qasim, M.; Chang, J. Identification of Pathogenic Mutations in Primary Microcephaly- (MCPH-) Related Three Genes CENPJ, CASK, and MCPH1 in Consanguineous Pakistani Families. *BioMed Res. Int.* **2022**, *2022*, 3769948. [CrossRef]

24. Liu, X.; Zong, W.; Li, T.; Wang, Y.; Xu, X.; Zhou, Z.W.; Wang, Z.Q. The E3 ubiquitin ligase APC/C(C)(dh1) degrades MCPH1 after MCPH1-βTrCP2-Cdc25A-mediated mitotic entry to ensure neurogenesis. *EMBO J.* **2017**, *36*, 3666–3681. [CrossRef] [PubMed]

25. Busino, L.; Donzelli, M.; Chiesa, M.; Guardavaccaro, D.; Ganoth, D.; Dorrello, N.V.; Hershko, A.; Pagano, M.; Draetta, G.F. Degradation of Cdc25A by beta-TrCP during S phase and in response to DNA damage. *Nature* **2003**, *426*, 87–91. [CrossRef]

26. Li, T.; Wang, Z.Q. Point mutation at the Nbs1 Threonine 278 site does not affect mouse development, but compromises the Chk2 and Smc1 phosphorylation after DNA damage. *Mech. Ageing Dev.* **2011**, *132*, 382–388. [CrossRef]

27. Guerra, G.M.; May, D.; Kroll, T.; Koch, P.; Groth, M.; Wang, Z.Q.; Li, T.L.; Grigaravicius, P. Cell Type-Specific Role of RNA Nuclease SMG6 in Neurogenesis. *Cells* **2021**, *10*, 3365. [CrossRef]

28. Alderton, G.K.; Galbiati, L.; Griffith, E.; Surinya, K.H.; Neitzel, H.; Jackson, A.P.; Jeggo, P.A.; O'Driscoll, M. Regulation of mitotic entry by microcephalin and its overlap with ATR signalling. *Nat. Cell Biol.* **2006**, *8*, 725–733. [CrossRef]

29. Rai, R.; Dai, H.; Multani, A.S.; Li, K.; Chin, K.; Gray, J.; Lahad, J.P.; Liang, J.; Mills, G.B.; Meric-Bernstam, F.; et al. BRIT1 regulates early DNA damage response, chromosomal integrity, and cancer. *Cancer Cell* **2006**, *10*, 145–157. [CrossRef]

30. Ghafouri-Fard, S.; Fardaei, M.; Gholami, M.; Miryounesi, M. A case report: Autosomal recessive microcephaly caused by a novel mutation in MCPH1 gene. *Gene* **2015**, *571*, 149–150. [CrossRef]

31. Journiac, N.; Gilabert-Juan, J.; Cipriani, S.; Benit, P.; Liu, X.; Jacquier, S.; Faivre, V.; Delahaye-Duriez, A.; Csaba, Z.; Hourcade, T.; et al. Cell Metabolic Alterations due to Mcph1 Mutation in Microcephaly. *Cell Rep.* **2020**, *31*, 107506. [CrossRef] [PubMed]

32. Trimborn, M.; Ghani, M.; Walther, D.J.; Dopatka, M.; Dutrannoy, V.; Busche, A.; Meyer, F.; Nowak, S.; Nowak, J.; Zabel, C.; et al. Establishment of a mouse model with misregulated chromosome condensation due to defective Mcph1 function. *PLoS ONE* **2010**, *5*, e9242. [CrossRef] [PubMed]

33. Chen, J.; Ingham, N.; Clare, S.; Raisen, C.; Vancollie, V.E.; Ismail, O.; McIntyre, R.E.; Tsang, S.H.; Mahajan, V.B.; Dougan, G.; et al. Mcph1-deficient mice reveal a role for MCPH1 in otitis media. *PLoS ONE* **2013**, *8*, e58156. [CrossRef]

Article

Enhanced Cognition and Neurogenesis in miR-146b Deficient Mice

Keerthana Chithanathan [1], Kelli Somelar [2], Monika Jürgenson [2], Tamara Žarkovskaja [2], Kapilraj Periyasamy [3], Ling Yan [1], Nathaniel Magilnick [4], Mark P. Boldin [4], Ana Rebane [3], Li Tian [1] and Alexander Zharkovsky [2,*]

1 Department of Physiology, Institute of Biomedicine and Translational Medicine, University of Tartu, Ravila 19, 50411 Tartu, Estonia; keerthana.chithanathan@ut.ee (K.C.); ling.yan@ut.ee (L.Y.); li.tian@ut.ee (L.T.)
2 Department of Pharmacology, Institute of Biomedicine and Translational Medicine, University of Tartu, Ravila 19, 50411 Tartu, Estonia; kelli.somelar@ut.ee (K.S.); monika.jurgenson@ut.ee (M.J.); tamara.zarkovskaja@ut.ee (T.Ž.)
3 Department of Biomedicine, Institute of Biomedicine and Translational Medicine, University of Tartu, Ravila 19, 50411 Tartu, Estonia; kapilraj.periyasamy@ut.ee (K.P.); ana.rebane@ut.ee (A.R.)
4 Department of Molecular and Cellular Biology, Beckman Research Institute of City of Hope National Medical Center, Duarte, CA 91010, USA; n.magilnick@gmail.com (N.M.); mboldin@coh.org (M.P.B.)
* Correspondence: aleksandr.zarkovski@ut.ee

Abstract: The miR-146 family consists of two microRNAs (miRNAs), miR-146a and miR-146b, which are both known to suppress a variety of immune responses. Here in this study, we show that miR-146b is abundantly expressed in neuronal cells, while miR-146a is mainly expressed in microglia and astroglia of adult mice. Accordingly, miR-146b deficient (*Mir146b-/-*) mice exhibited anxiety-like behaviors and enhanced cognition. Characterization of cellular composition of *Mir146b-/-* mice using flow cytometry revealed an increased number of neurons and a decreased abundancy of astroglia in the hippocampus and frontal cortex, whereas microglia abundancy remained unchanged. Immunohistochemistry showed a higher density of neurons in the frontal cortex of *Mir146b-/-* mice, enhanced hippocampal neurogenesis as evidenced by an increased proliferation, and survival of newly generated cells with enhanced maturation into neuronal phenotype. No microglial activation or signs of neuroinflammation were observed in *Mir146b-/-* mice. Further analysis demonstrated that miR-146b deficiency is associated with elevated expression of glial cell line-derived neurotrophic factor (*Gdnf*) mRNA in the hippocampus, which might be at least in part responsible for the observed neuronal expansion and the behavioral phenotype. This hypothesis is partially supported by the positive correlation between performance of mice in the object recognition test and *Gdnf* mRNA expression in *Mir146b-/-* mice. Together, these results show the distinct function of miR-146b in controlling behaviors and provide new insights in understanding cell-specific function of miR-146b in the neuronal and astroglial organization of the mouse brain.

Keywords: miR-146b; cognition; anxiety; astrocytes; microglia; neurogenesis; neuronal development; *Gdnf*

Citation: Chithanathan, K.; Somelar, K.; Jürgenson, M.; Žarkovskaja, T.; Periyasamy, K.; Yan, L.; Magilnick, N.; Boldin, M.P.; Rebane, A.; Tian, L.; et al. Enhanced Cognition and Neurogenesis in miR-146b Deficient Mice. *Cells* **2022**, *11*, 2002. https://doi.org/10.3390/cells11132002

Academic Editor: FengRu Tang

Received: 1 June 2022
Accepted: 20 June 2022
Published: 22 June 2022

Publisher's Note: MDPI stays neutral with regard to jurisdictional claims in published maps and institutional affiliations.

1. Introduction

MicroRNAs (miRNAs) are a family of small, endogenous, noncoding RNAs which are approximately 22 nucleotides long and act at post-transcriptional level to regulate gene expression via binding to partially complementary mRNAs [1–3]. The brain expresses more distinct and a relatively high number of miRNAs than any other tissue in vertebrates [4]. Early microarray profiling analyses have shown that miRNAs are differentially expressed in brain regions [5] and in a cell-specific manner [6]. As miRNAs can influence expression of many target mRNAs, they are capable of modulating various physiological processes in the brain, such as neural differentiation [7] and synaptic plasticity [8], and are thereby proposed to regulate to regulate complex animal behaviors including cognition [9] and anxiety [10]. In addition, miRNAs are involved in the several brain pathologies including neurodegenerative and psychiatric disorders [11–13].

The miR-146b belongs to the miR-146 family consisting of two miRNAs, miR-146a and miR-146b (miR-146a/b), which are encoded by two distinct genes located on different chromosomes and regulated by different pathways. Accordingly, the expression of miR-146a is upregulated through the nuclear factor kappa B (NF-κB) pathway [14], while miR-146b expression has been shown to be activated by signal transducer and activator of transcription (STAT) proteins [15,16]. As miR-146a/b differ only in two nucleotides, they probably target the same or a very similar set of genes [17,18]. The well-known target genes for miR-146a/b are TNF receptor-associated factor 6 (TRAF6) and interleukin-1 receptor-associated kinase 1 (IRAK1), both of which are from the NF-κB pathway and have been characterized already by an initial study on miR-146a/b [19]. Accordingly, there is ample evidence concerning the participation of the miR146 family in negative regulation of immune responses [14–18,20].

In the central nervous system, miR-146a has been reported to have immune suppressive function in microglia [21,22]. In addition, miR-146a has been related to autism, neural lineage determination, neurite outgrowth [23], differentiation of neural stem cells and hippocampal-dependent memory impairments [24]. The functions of miR-146b in the nervous system have been characterized by some existing studies. For example, one study showed that overexpression of miR-146b inhibited inflammatory responses via suppression of the activation of the NF-κB signaling in the brain in the rat encephalopathy models [25]. Another study in rats has shown that miR-146b overexpression with lentivirus vector could inhibit the proliferation of primary hippocampal neural stem cells [26]. A recent study has identified miR-146b as a candidate modulator of microglial activation [27]. In addition, miR-146b has also been reported to be a prognostic biomarker of gliomas, as it inhibits glioma cell proliferation and migration and induces apoptosis [28–31].

In this study, we used miR-146b deficient and wild type mice to explore the role of miR-146b in brain functions. We detected high expression of miR-146b in neuronal cells of the brain and observed that loss of miR-146b facilitates learning abilities and induces anxiety-like behaviors. Next, we found that lack of miR-146b affects cellular composition of the brain, reflected by the higher number of neurons, reduced number of astrocytes and increased hippocampal neurogenesis. In addition, we detected that miR-146b target GDNF mRNA is upregulated in *Mir146b-/-* mice in the hippocampus. No microglial activation or signs of neuroinflammation were observed in *Mir146b-/-* mice. Together, these results highlight the function of miR-146b in controlling cellular organization and behaviors in mice.

2. Materials and Methods

2.1. Animals and Experimental Design

Mir146b-/- mouse line on C57BL/6J background was generated by deletion of miR-146b encoding gene from mouse chromosome 19 as previously described in [17]. *Mir146b-/-* and corresponding *WT* mice used for this study were obtained by crossing *Mir146b+/-* heterozygous mice maintained and bred in the animal facility at the Laboratory Animal Centre at the Institute of Biomedicine and Translational Medicine, University of Tartu, according to the Institute's regulations. The generated animals were genotyped using the primer sequence:

146b locus 5′ forward primer- 5′ CTCACACTCTTGTTCTTACCCAGTTCTT 3′;
146b locus 3′ reverse primer- 5′ CAAACAAACAAACAAAAGGTTCAGCTAAG 3′;
146b locus internal reverse primer-5′ACACACAGGGCATATGAGATCAGTTGGTT 3′ and same generation littermates were used in experiments. Two–three months old male mice were used for all experiments, which were undertaken in agreement with the guidelines established in the principles of laboratory animal care (Directive 2010/63/EU). All the mice were group-housed with a 12-h light/dark cycle with food and water available ad libitum. The Animal Experimentation Committee at the Estonian Ministry of Agriculture

(no. 183, 2021) approved the experimental protocol. The sequence of the experiments performed is summarized in Table 1. The age of animals at the start of the experiments was approximately 2 months old. The age of animals at the time of sacrifice was approximately 3 months old.

Table 1. Sequence of the experiments.

Time Point	Task Assigned
Cohort 1 (*WT* and *Mir146b-/-*)	
Day 0	Open field test
Day 4	Elevated plus maze
Day 8–10	Novel object recognition test
Day 13–14	Social dominance test
Day 17	Tail suspension test
Day 22	Sacrifice the animals and collect the tissues for flow cytometry, qPCR and immunohistochemistry
Cohort 2 (*WT* and *Mir146b-/-*)	
Day 0–7	Contextual fear conditioning
Day 10	BrdU injections (300 mg/kg)
Day 31	Sacrifice animals for immunohistochemistry and qPCR

2.2. Novel Object Recognition Test (NORT)

NORT was performed as described by [32] in open chamber 50 cm × 50 cm × 50 cm (L × W × H), made up of brown wood. The objects were opaque glass cups of similar textures and colors but different sizes and shapes, and were heavy enough to prevent the mice from moving them. The experiment consisted of three phases: habituation, training and retention. During habituation phase, the animals were allowed to explore the empty arena without presence of any object for 5 min. Twenty-four hours later, in training phase, two identical objects were placed on a diagonal (both 10 cm from the corner) and each mouse was allowed to explore in the field for 5 min. The amount of time each mouse spent exploring both objects was recorded. Either 2 or 24 h later (retention phase), the mice explored the arena with presence of one familiar object and one novel object to measure their short-term recognition memory (STM) and long-term recognition memory (LTM), respectively. A preference ratio for each mouse was expressed as percentage of time spent exploring the new object (Tnew ×100)/(Tf + Tnew), where Tf and Tnew are the times spent exploring the familiar object and the novel object, respectively. The time spent exploring each object was scored by an observer "blind" to genotypes and in between trials. All of the objects were cleaned with 5% ethanol solution after each trial. Exploration was defined as sniffing or touching the object with the nose or forepaws.

2.3. Contextual Fear Conditioning (CFC) and Tone Fear Recall

The procedure was adapted from [24,33]. The setup was an experimental chamber 22 cm × 22 cm × 35 cm (L × W × H) located inside a larger noise-attenuating box, and a built-in ventilation fan provided a background noise. The floor of the box was made of stainless-steel rods designed for mice and connected to a scrambled shock generator (TSE Systems) containing a speaker for audible tone. The CFC experiment took place on seven consecutive days. On day 0, the mice were allowed to freely explore the conditioning chamber for 3 min and baseline freezing was measured. Immediately after that, conditioned stimuli tone (75 dB, 2 kHz, 30 s) paired with unconditioned stimuli foot shock (1 s, 0.50 mA, constant electric current) was automatically delivered with 1 min intervals for three times through a grid floor. After completing the conditioning session, the mice were returned to their home cage. On day 1, contextual fear retention was assessed at 24 h after the conditioning session by placing animals into conditioned context for 3 min in the absence of tone and foot shock, during which the duration of freezing time (absence of any movement other than that due to respiration) was measured. The extinction of contextual fear memory was measured from day 2 to day 6. Each day, animals were placed in the same context for 3 min and the freezing time was recorded. On day 7, tone fear recall was assessed by placing the mice in the novel context for 3 min and baseline freezing was measured. Immediately after that, tone (75 dB, 2 kHz, 30 s) was presented and freezing time was measured within the next three minutes.

2.4. Open Field Test (OFT)

OFT was used to assess anxiety and locomotion was performed according to [34]. Mice were placed in an experimental room for about 1 h before starting the experiment for their habituation. Each mouse was placed in the center of OFT chamber (45 cm × 45 cm × 45 cm) and allowed to explore freely for 30 min. The light luminosity was set to 500 lux throughout the box. During this time, the mice were monitored and data were collected and recorded by an analytical system (TSE Systems, Chesterfield, VA, USA). Anxiety was quantified by measuring the time spent by mice in central sector of the open field, while locomotor activity was measured by estimating total distance travelled.

2.5. Elevated Plus Maze (EPM)

The EPM test measuring anxiety was performed according to [32]; it was carried out in a plus-maze with setup consisting of a central zone (5 cm × 5 cm), two open arms (45 cm × 10 cm) and two arms closed by walls (45 cm × 10 cm and 15 cm in height). The maze is elevated 60 cm above the floor level. The animals were placed on the central zone of the EPM and allowed to explore plus-maze freely for 5 min with light luminosity set to 40 lux. A live video-tracking system (Noldus, with EthoVision XT version 8 software, Wageningen, the Netherlands) was used for automated animal tracking and data collection to measure the total number of entries, number of entries onto the open arms and time spent on the open arms. An observer "blind" to genotypes also scored the behavior. The level of anxiety was calculated as percentage of entries onto open arms and percentage of time spent on the open arms.

2.6. Tail Suspension Test (TST)

TST was performed as previously described by [35]. This test was used to measure learned hopelessness in mice, where the animals are placed to an inescapable stressful situation by hanging them in separate sections of the test apparatus on a wooden bar by the tip of their tail using an adhesive tape. During the 6 min test period, the behavior was recorded with a camera and the duration of immobility during the testing period was measured. Immobility was defined as a complete lack of movement other than respiration. However, small movements of forefeet and swinging caused by earlier movements were also scored as immobility. An observer "blind" to genotypes scored behavior.

2.7. Social Dominance Test (SDT)

SDT was adapted from [36]. This test was performed by placing the mice of different genotypes simultaneously into the opposite ends of a transparent plastic tube (30 cm long, 4.0 cm inner diameter). When the animals interacted in the tube, the more dominant animal forced its opponent out of the tube. The animal with four paws out of the tube was declared as loser, while the animal remaining inside the tube was considered the winner. Each match was set within 2 min. Matches lasting > 2 min were scored as "even". Animal pairs were decided according to their matched body weight and each animal was encountered with the opponent for three rounds. Numbers of losses, wins and evens were counted and an average of three rounds was taken as percentage of win to assess the social dominance.

2.8. Flow Cytometry

Mice were euthanized with CO2. Dissected brain tissues (hippocampus and frontal cortex) were mechanically dissociated through 70 µm cell strainers (352350, BD Bioscience, San Jose, CA, USA) in ice-cold flow buffer (phosphate buffered saline (PBS) with 1% fetal calf serum). Isolated cells were then blocked with 10% rat serum in ice-cold flow buffer for 1 h at 4 °C. The cells were stained with 0.5 µL of the following antibodies: anti-mouse CD11b-BV421 (101251, Biolegend, San Diego, CA, USA), CD45-Brilliant Violet 650 (103151, Biolegend, San Diego, CA, USA), MHCII-Brilliant Violet 711 (cat no. 107643, Biolegend, San Diego, CA, USA), GLAST-APC (130-123-555, Miltenyi, Bergisch Gladbach, Germany) and O4-PE (130-117-357, Miltenyi, Bergisch Gladbach, Germany) with the corresponding isotype control antibodies rat IgG2b-BV421 (400639), rat IgG2b-BV650 (400651), rat IgG2b-BV711 (400653), mouse IgG2a-APC (400219) and mouse IgM-PE (401611) (all from Biolegend, San Diego, CA, USA) in flow buffer for 1 h. After staining, cells were fixed by 4% paraformaldehyde (PFA), permeabilized with PBS containing 0.05% TritonX-100 at 4 °C for 30 min and incubated with an anti-VGLUT2 mAb-Alexa488 (MAB5504A4 Millipore, Burlington, MA, USA) or isotype control mAb-Alexa488 (400132 Biolegend) for 1 h at 4 °C. The cells were washed with PBS, resuspended and acquired with Fortessa flow cytometer (BD Bioscience, San Jose, CA, USA). In total, 100,000 events were recorded in all the samples. Data were analyzed by Kaluza v2.1 software (Beckman Coulter, Indianapolis, IN, USA). GLAST was used to measure number of astrocytes among total brain cells, CD11b was used to detect number of microglial cells, O4 was used to measure number of oligodendrocyte precursor cells (OPC), and the negative selection was used to measure the number of neurons among non-astrocytes. Number of VGLUT2+ cells was measured under the neuron gate. For measuring M1 microglial polarization, CD11b and CD45 was used to collect total microglial cells and MHCII was used to measure the percentage of M1 type of microglia.

2.9. Brain Volume Assessment and Immunohistochemistry

After behavioral experiments, animals were deeply anesthetized with chloral hydrate (300 mg/kg, i.p.) and transcardially perfused using 0.9% saline and then with 4% PFA in PBS (pH = 7.4). After fixation of the brain in PFA for 24 h, 40 µm-thick sections were cut on a Leica VT1000S vibro-microtome (Leica Microsystems Pvt Ltd., Wetzlar, Germany) and stored at −20 °C in the cryo-protectant (30% ethylene glycol, 30% glycerol in PBS; pH 7.4).

For measuring the volume of the whole brain and hippocampus, every sixth section was selected and was incubated in a 0.1M TRIS HCl buffer containing 0.025% trypsin and 0.1% CaCl2 for 10 min followed by washing with PBS. The sections were then incubated with Triton X-100 (0.25%) for 1 h and washed with PBS. Hematoxylin solution was first added to the section for about 30 s, followed by incubation with acidic alcohol solution (HCl 1% in ethanol 70%) for 10 s and washing with tap water. Eosin solution was added for 10 s and the sections were placed on the glass with water-based mounting medium (Vector Laboratories, Newark, CA, USA) and cover-slipped. An average of 6–8 sections per animal were analyzed. For the analysis of the volume, sections were scanned using Leica SCN400 scanner (Leica Microsystems Pty Ltd., Wetzlar, Germany). The volumes of the areas of interest were calculated from the surface area, measured by Aperio Imagescope (v12.4.3.5008), and multiplied by the thickness of the sections and distance between sections.

For Ki67, NeuN and Iba1 staining, sections were washed three times in PBS and treated with 2% H2O2 solution for 20 min followed by incubation in 0.01 M citrate buffer (pH 6.0) at 85 °C for 30 min in water bath and then stood for 30 min at room temperature. Sections were then washed two times in PBS and once in PBS containing 0.1% Triton X-100. Blocking was done with solution containing 5% goat serum, 0.5% Tween-20, 0.25% Triton X-100 in 100 mM PBS for 1 h. Ki67 primary antibody (1:200, rabbit monoclonal antibody (SP6), ab16667, Abcam, Cambridge, UK) was added for 24 h, NeuN primary antibody (1:200, rabbit anti-NeuN, D4G40, Cell signaling technology, Danvers, MA, USA) was added for 48 h, and Iba1 primary antibody (1: 700, rabbit anti-Iba1, CAF6806, FUJIFILM Wako Chemicals Europe GmbH, Neuss, Germany) was added for 72 h. All antibodies were added in blocking buffer and incubation was carried out at 4 °C. After being washed three times, the sections were incubated with secondary antibody (1:400 or 1:700, affinity purified goat anti-rabbit biotinylated IgG (H+L), Vector Laboratories) in blocking buffer at room temperature for 1 h. Ki67-, NeuN- and Iba1-positive cells were visualized using peroxydase method (ABC system and diaminobenzidine as chromogen, Vector Laboratories). The sections were dried, cleared with xylol and cover-slipped with mounting medium (Vector Laboratories, Newark, CA, USA).

For quantifying morphological characteristics of microglia (cell size, cell body size, size dendritic processes), the images from sections stained for Iba1-positive cells were analyzed using image analysis software (ImageJ 1.48v, http://imagej.nih.gov/ij (accessed on 1 June 2022), National Institutes of Health, Rokville Pike, Bethesda, USA). To quantify morphological characteristics of Iba-1-positive cells, cell size, cell body size, size of dendritic processes and cell body size to cell ratio, an algorithm described in [37] was used. Briefly, images were converted into 8-bit format, before "adjusted threshold" and "analyze particles" functions were used to apply intensity thresholds and size filter. To measure the total cell size, the threshold was maintained at the level that was automatically provided by the ImageJ program, and size filter of 150 pixels was applied. To measure the total cell body size, the threshold was lowered 40 points and no size filter was applied.

The counts of Iba-1- or NeuN-positive cells were obtained from images according to the algorithm described previously in [38]. Briefly, images were converted to 8-bit, background was subtracted, and then obtained images were thresholded, binarized and counted using "analyze particles" command in ImageJ software.

2.10. Assessment of Neurogenesis in the Adult Mouse Dentate Gyrus

Cell proliferation in the dentate gyrus was assessed using immunohistochemical detection of Ki67 as endogenous marker of proliferating cells as described above.

Cell survival assessment was performed according to [39]. Briefly, mice received three BrdU injections (100 mg/kg, i.p, Sigma Aldrich, Berlington, MA, USA) in a total dosage of 300 mg/kg separated by the intervals of 2 h. After three weeks of injection, the animals were sacrificed and the brains were sectioned and stored in the cryo-protectant.

For BrdU immunohistochemistry, sections were incubated in 0.3% H2O2 solution for 30 min, washed three times in PBS and incubated with 0.1M Tris-HCl buffer containing 0.025% trypsin and 0.1% CaCl2 for 10 min followed by 2N HCl solution at 37 °C for 30 min. Blocking solution containing 2% normal goat serum and 0.25% Triton X-100 was added to the sections for 1 h at room temperature. Next, the sections were incubated for 24 h with blocking solution containing rat monoclonal antibody to BrdU (1:300, RF04-2, Bio-Rad) at 4 °C, followed by incubation in biotinylated rabbit anti-rat antibody (1:400, affinity purified, Lot R1121, Vector Laboratories, Newark, CA, USA) for 1 h. BrdU-positive cells were visualized using the peroxidase method (ABC system and diaminobenzidine as chromogen, Vector Laboratories). The sections were dried, cleared with xylol and cover-slipped with mounting medium (Vector Laboratories, Newark, CA, USA). Ki67 and BrdU-positive cells were counted in every sixth section within the dentate gyrus. All counts were performed using an Olympus BX-51 microscope. To estimate the total number of Ki67 and BrdU-positive cells in a given region, every sixth section was analyzed to obtain the sum of cell counts from each animal and then multiplied by six.

For doublecortin immunohistochemistry, sections were washed three times in 0.1 M Tris-Buffered Saline (TBS) and quenched in 3% H2O2 and 10% MetOH solution for 10 min. Sections were washed again in TBS and were blocked in 5% normal goat serum and 0.25% Triton X-100 in TBS for 1 h at room temperature. Next, sections were incubated with primary antibody (1:500, ab18723, Abcam, Cambridge, UK) in blocking solution for 48 h at 4 °C. The sections were rinsed twice with TBS and incubated with secondary antibody biotinylated goat anti-rabbit antibody (1:1000, Ref no BA-1000, Vector Laboratories) diluted in blocking buffer for 2 h. Doublecortin-positive cells were visualized using the standard immunoperoxidase method (ABC system, Vectastain ABC kit PK-6100, Vector Laboratories), with diaminobenzidine (DAB) as the chromogen (DAB Peroxidase Substrate SK-4100, Vector Laboratories). Total number of doublecortin-positive cells in a given region was obtained from every 24th section and the sum of cell counts was acquired and then multiplied by the 24.

For the determination of the phenotype of the newly generated cells, two sections from each animal which survived three weeks after the BrdU injection were analyzed for co-expression of BrdU and neuronal (calbindin, a marker for mature neurons) or glial (glial fibrillary acidic protein, GFAP, a marker for astrocytes) markers. For immunofluorescent double-labelling, sections were incubated with a mixture of anti-BrdU monoclonal antibody (1:200, RF04-2, Bio-Rad, Hercules, CA, USA) and rabbit anti-calbindin antibody (1:800, AB1778, Chemicon International Inc, Temecula, CA, USA) or rabbit anti-GFAP (1:800, Z0334, Dako, Glostrup, Denmark). Secondary antibodies such as goat anti-rat Alexa-594 antibody (1:800, A11007, Invitrogen, Thermo Fisher Scientific, Waltham, MA, USA) or goat anti-rabbit Alexa-488 (1:700, A11034, Invitrogen, Thermo Fisher Scientific) were used. Confocal microscope (LSM 710 Duo, Carl Zeiss Microscopy GmbH, Oberkochen, Germany) equipped with an argon laser was used to visualize fluorescent signals. 3D images were constructed from series of scans taken at 1.5μm intervals from the dentate gyri, using $40\times$ objective and $2\times$ digital zoom. For illustrative images, $100\times$ objectives were used. Data are expressed as a percentage of BrdU-positive cells found in the granule cell layer and hilus of the dentate gyrus that expressed either calbindin or GFAP.

2.11. Isolation of Brain Cells

Brain cells were isolated as previously described in [40]. Tissues were mechanically homogenized and passed through a 70 μm nylon cell strainer (352350, BD Bioscience) with approximately 10–15 mL of 1X DPBS supplemented with 0.2% glucose into a 50 mL conical tube. Isotonic Percoll dilutions were made by diluting stock Percoll (GE-healthcare, 17-0891-01, Chicago, IL, USA) at a 9:1 ratio with 10X PBS to make stock isotonic Percoll (SIP), which is considered 100% SIP. The layers of Percoll were created by diluting the 100% SIP with 1X Dulbecco's Phosphate Buffered Saline (DPBS) to make 70% SIP, 50% SIP and 35% SIP. Obtained homogenate was then centrifuged at $600\times g$ for 6 min at room temperature. Supernatant was decanted and the pellet resuspended in 6 mL of 70% SIP. The resuspended homogenate was transferred to a 15 mL tube and 3 mL of 50% SIP was carefully layered over. Another 3 mL of 35% SIP was carefully layered on top of the 50% SIP layer, and 2 mL of 1X DPBS was layered on top of the 35% layer. The prepared 15mL tubes were then centrifuged at $2000\times g$ for 20 min at room temperature without brake. Three discrete layers were established after centrifugation. Microglial cells were collected from the interface between 70–50% SIP and astroglial cells were taken from 50–35% SIP; meanwhile, the remaining top layer consisted of myelin, and other cells were collected and subjected to characterization of neuronal marker. All isolated cells were resuspended in sterile 1X DPBS and centrifuged at $600\times g$ for 6 min at room temperature to remove any remaining Percoll. Washed cells were subjected to purity check using qPCR and flow cytometry methods. GLAST was used to check purity of astrocytes, while Cx3cr1 and Slc17a6 were used to check purity of microglial and neuronal cells, followed by quantification of miRNA expression using Taqman miRNA assay.

2.12. RNA Extraction and RT-qPCR

Total RNAs were extracted from brain tissues (hippocampus) by using TRI Reagent® (TR 118) (Molecular Research Center, Inc., Cincinati, OH, USA). To measure mRNA expression, cDNA was synthesized using RevertAid First Strand cDNA Synthesis Kit (Thermo Fisher Scientific) followed by qPCR using $5 \times$ HOT FIREPol® EvaGreen® qPCR Supermix (Solis BioDyne, Tartu, Estonia) on a QuantStudio 12KFlex instrument (Thermo Fisher Scientific) according to the instructions of the respective manufacturers. Primer sequences for target genes were given in the Table S1.

Quantification of miRNA expression was carried out using TaqMan® MicroRNA Assays hsa-miR-146a (Assay ID: 000468, Life technologies) and TaqMan® MicroRNA Assays hsa-miR-146b (Assay ID: 001097, Life technologies, Carlsbad, CA, USA) according to the manufacturer's instructions. For cRNA synthesis, TaqMan® MicroRNA reverse transcription kit (4366596, Thermo Scientific) and for qPCR, 5× HOT FIREPol® Probe qPCR Mix Plus (ROX) (Solis BioDyne) were used, respectively. U6 snRNA (Assay ID: 001973, Life Technologies) was used for the normalization of RT-qPCR. To measure cell-specific expression of miR-146a and miR-146b, the respective cell populations were isolated as described above and then miRNA expression in each cellular population was measured using Taqman miRNA assay.

2.13. Statistical Analysis

GraphPad 8.0.1 was used for statistical analyses and graphical presentations. Student's *t*-test, one-way and two-way ANOVA with Tukey's test were used for post hoc multiple comparisons for statistical analyses, and statistical significance was set at $p < 0.05$. In all figures, data are shown as mean ± standard error of the mean (SEM). For target search, we used TargetScanHuman Release 8.0 (https://www.targetscan.org/vert_80/, accessed on 1 June 2022) [41,42]. Of the transcripts, a total of 299 were conserved sites and 126 were poorly conserved sites. Pathway analysis was performed with g:GOSt tool available in g:Profiler platform, which estimates significance of overlap between functional groups and list of studied genes by calculating enrichment *p*-value using Fisher's one-tailed test [43]. Only significantly ($p < 0.05$) overrepresented pathways associated with neuron development and function are shown. The associations between behavior in NORT and *Gdnf* mRNA expression were analyzed using Pearson's correlation.

3. Results

3.1. miR-146b Is Highly Expressed in Neuronal Cells in the Mouse Brain

First, we evaluated relative expression of miR-146a/b in the hippocampus of *WT* mice and found that both miR-146a and miR-146b were expressed at similar levels in the hippocampal region of mice brains ($p = 0.8472$; Student's *t*-test; Figure 1A). Next, to measure expression of miR-146a/b in different cell types, we isolated three cell populations (microglia, astroglia and remaining cell fraction, i.e., other cells) from adult WT brains. As expected, RT-qPCR and flow cytometry analysis of purified cell types showed that microglial marker *Cx3cr1* mRNA was enriched in microglial fraction (F (2, 6) = 10.83, $p = 0.0102$; one-way ANOVA; Figure 1B). Staining with astroglial marker GLAST by flow cytometry showed enriched number of GLAST+ cells in astroglial fraction (F (2, 6) = 70.69, $p = < 0.0001$; one-way ANOVA; Figure 1C). The other cell fraction defined as neuronal cells expressed high levels of neuronal *Slc17a6* mRNA (F (2, 6) = 65711, $p = <0.0001$; one-way ANOVA; Figure 1D). Interestingly, miR-146a was highly expressed in microglial cells and less in astroglial cells, and was much lower in the fraction containing mainly neuronal cells (F (2, 6) = 23.51, $p = 0.0014$; one-way ANOVA; Figure 1E). In contrast, miR-146b was more highly expressed in the neuronal fraction as compared to microglial and astroglial fractions (F (2, 6) = 18.48, $p = 0.0027$; one-way ANOVA; Figure 1F).

In general, *Mir146b-/-* mice did not present any differences regarding their development, body weight, food and water consumption, and premature mortalities. In addition, no visible abnormalities in brain structures were seen in adult *Mir146b-/-* mice. Volumetric assessment of whole brain ($p = 0.5936$; Student's *t*-test; Figure S1A) and hippocampus ($p = 0.1356$; Student's *t*-test; Figure S1B) did not reveal any changes as compared with *WT* mice. Together, these data demonstrate that despite miR-146b being highly expressed in neurons, lack of miR-146b does not cause major phenotypic impairments.

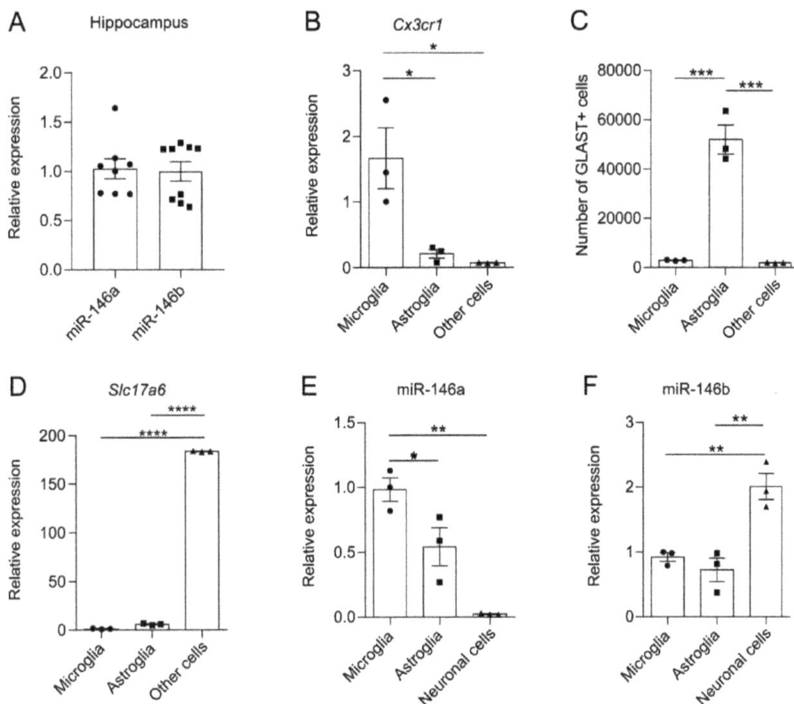

Figure 1. Expression of miR-146a and miR-146b in hippocampus and cellular fractions isolated from WT mice. (**A**) Quantification of miR-146a/b expression in the hippocampus of WT mice. Number of animals in (**A**) = 8–9, Student's *t*-test. (**B–D**) Characteristics of the cellular fractions isolated from the *WT* mouse brain. (**B**) Relative mRNA expression of microglial *Cx3cr1*. (**C**) Number of astroglial marker GLAST+ cells (determined by flow cytometry). (**D**) Relative expression of neuronal marker *Slc17a6* mRNA levels. As fraction of "other cells" abundantly expressed neuronal marker *Slc17a6* mRNA, this fraction was named as neuronal cells. (**E**) Quantification of miR-146a and (**F**) miR-146b in microglia, astroglia and neuronal cells of *WT* mice. Number of cell batches in experiments (**B–F**) = 3, one-way ANOVA followed by Tukey's multiple comparison test. Data represented as mean ± SEM; * $p < 0.05$, ** $p < 0.01$, *** $p < 0.001$, **** $p < 0.0001$.

3.2. Mir146b-/- Mice Display Improved Recognition and Associative Memory

Several recent studies indicate that miR-146a might influence cognitive functions of mice [44–46]; however, no such studies had been performed for miR-146b. Therefore, and because miR-146b expression was higher in neuronal brain cells, we next performed a series of experiments to assess cognitive abilities and behavior of miR-146b-deficient mice. First, we performed NORT test, which takes advantage of the natural preference of mice for novel objects and is widely used to evaluate cognition, and in particular, recognition memory [47]. In this test, during the training session there were no significant differences in the time spent by mice of both genotypes exploring the two familiar objects (Figure S2), indicating that both genotypes *Mir146b-/-* and *WT* mice have the same motivation to explore new objects. However, during the test phase, when one of the familiar objects was replaced by another novel object, *Mir146b-/-* mice showed a significant preference for the novel object as compared to *WT* mice, both at 2-h ($p = 0.0027$; Student's *t*-test; Figure 2A) and 24-h ($p = <0.0001$; Student's *t*-test; Figure 2B) time-points following the training session.

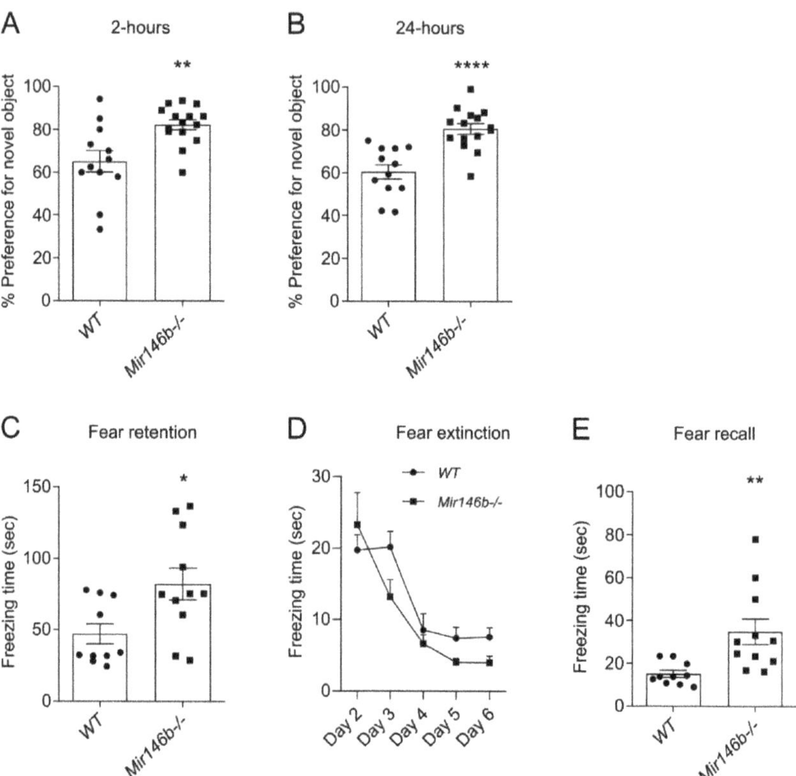

Figure 2. Enhanced recognition and associative memory in *Mir146b-/-* mice. (**A**) Novel object preference of mice at 2-h (short-term memory) and (**B**) 24-h (long-term memory) in novel object recognition test. (**C**) Contextual fear retention. (**D**) Contextual fear extinction and (**E**) tone fear recall of *WT* and *Mir146b-/-* mice in contextual fear conditioning and tone fear recall. Number of animals = 11–15, Student's *t*-test and two-way ANOVA followed by Tukey's multiple comparison test. Data represented as mean ± SEM; * *p* < 0.05, ** *p* < 0.01, **** *p* <0.0001.

Next, we assessed the associative memory of *Mir146b-/-* and *WT* mice by measuring contextual fear retention and extinction and tone fear recall. There were no differences detected in baseline freezing time between *WT* and *Mir146b-/-* mice (Figure S3). The mice were next exposed to the tone paired with foot-shock, and after 24 h of exposure, fear retention was measured in the same context. As expected, both *WT* and *Mir146b-/-* mice showed robust freezing response; however, *Mir146b-/-* mice demonstrated significantly longer freezing time compared to *WT* mice (*p* = 0.0171; Student's *t*-test; Figure 2C). Fear extinction was assessed in absence of tone and foot shock from day 2 to day 6 in the same context; however, no differences in genotypes were detected between *WT* and *Mir146b-/-* mice (F (1, 95) = 3.117, *p* = 0.0807; two-way ANOVA; Figure 2D). Nevertheless, while tone fear recall was measured on day 7, *Mir146b-/-* mice demonstrated significantly longer freezing time compared to the controls (*p* = 0.0063; Student's *t*-test; Figure 2E). These results indicate that miR-146b deficient mice have better ability to recognize novel objects and have better fear memory acquisition and recall as compared to *WT* mice.

3.3. Mir146b-/- Mice Showed Anxiety-like Behaviors but Had No Differences in Depression-like and Social-Dominant Behaviors

As anxiety is often associated with changes in the cognitive domain of the brain [48], we next evaluated anxiety-like behaviors in *Mir146b-/-* mice using OFT and EPM tests. In OFT, *Mir-146b-/-* mice showed no difference in the total distance travelled in the open field arena ($p = 0.9592$; Student's *t*-test; Figure 3A) compared to their *WT* controls. However, the time spent in the central area of the open field was significantly lower in *Mir146b-/-* mice, indicating that these mice are more anxious than control animals ($p = 0.0002$; Student's *t*-test; Figure 3B). However, no changes in the number of total entries ($p = 0.4684$; Student's *t*-test; Figure 3C) and percentage entries onto open arms of the plus maze ($p = 0.4470$; Student's *t*-test; Figure 3D) in *Mir146b-/-* mice were observed. In line with OFT test results, *Mir146b-/-* mice showed significant decrease in the percentage of time spent on the open arms of EPM as compared to the *WT* mice ($p = 0.0027$; Student's *t*-test; Figure 3E).

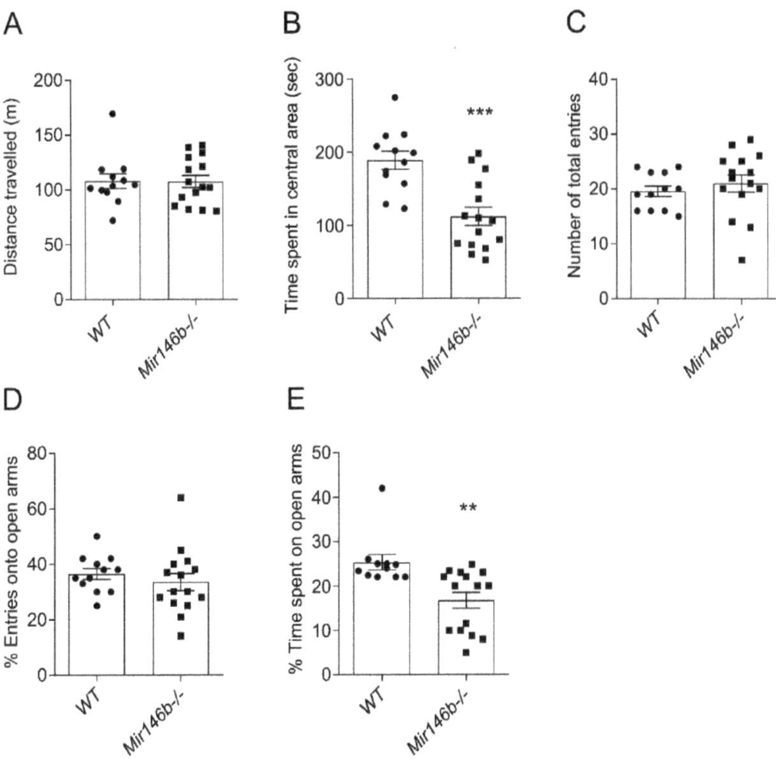

Figure 3. miR-146b deficiency causes anxiety-related behaviors. (**A**) Total distance travelled (locomotor activity). (**B**) Time spent in the central area of the open field test. (**C**) Total number of entries. (**D**) Percentage of entries onto the open arms. (**E**) Percentage of time spent on the open arms of elevated plus maze of *Mir146b-/-* mice and their *WT* littermates. Number of animals = 11–15, Student's *t*-test. Data represented as mean ± SEM; ** $p < 0.01$, *** $p < 0.001$.

Previously, it has been reported that the expression levels of miR-146a/b are inversely correlated with severity of depression in patients with major depressive disorder [49,50]. Therefore, and because the depression is often associated with changes in anxiety and cognitive domain [51], we next evaluated whether miR-146b deficiency may contribute to the depression-like and social dominant behavior using TST and SDT tests. No significant differences between *Mir146b-/-* and *WT* mice were found in TST ($p = 0.2243$; Student's *t*-test; Figure S4A) or SDT tests ($p = 0.2476$; Student's *t*-test; Figure S4B). Together, these results indicate that loss of miR-146b causes anxiety-like behaviors, but does not induce depression-like behaviors or affect social behavior in mice.

3.4. Brain Cell Abundancy Is Altered in the Brain of Mir146b-/- Mice

Cellular composition is considered as another informative characteristic to understand the brain functions [52]. Thus, we next used flow cytometry to quantify major cell types, such as astrocytes, microglia, oligodendrocyte precursor cells (OPC) and neuronal cells in hippocampus (HP) and frontal cortex (FC) of *Mir146b-/-* and *WT* mice.

The gating strategy of brain cells is depicted in the representative dot plots (Figure 4A). Isotype antibody staining as negative controls is shown in (Figure S5). Analysis of the hippocampal cells of *Mir146b-/-* mice demonstrated significantly reduced numbers of astrocytes ($p = 0.0010$; Student's *t*-test; Figure 4B), and no differences were found in the number of microglia ($p = 0.6168$; Student's *t*-test; Figure 4C) and OPCs ($p = 0.0658$; Student's *t*-test; Figure 4D) compared to *WT* mice. Meanwhile, with negative selection we also observed an increased number of neuronal cells ($p = 0.0033$; Student's *t*-test; Figure 4E), while with VGLUT2 staining we found increased number of VGLUT2+ glutamatergic neurons ($p = 0.0019$; Student's *t*-test; Figure 4F) in *Mir146b-/-* mice. The changes in cellular content in FC of *Mir146b-/-* mice were similar to HP, showing decreased numbers of astroglia ($p = 0.0211$; Student's *t*-test; Figure 4G) and no changes in microglial ($p = 0.6372$; Student's *t*-test; Figure 4H) and OPCs number ($p = 0.1786$; Student's *t*-test; Figure 4I), whereas increased numbers of neurons ($p = 0.0339$; Student's *t*-test; Figure 4J) and VGLUT2+ glutamatergic neurons ($p = < 0.0001$; Student's *t*-test; Figure 4K) were observed. To confirm flow cytometry results, we performed immunohistochemistry analysis using neuronal-specific marker NeuN followed by cell counting using tissue sections from FC. Similarly, with flow cytometry results, we detected increased neuronal density of NeuN+ cells ($p = 0.0084$; Student's *t*-test; Figure 4L,M) in FC of *Mir146b-/-* as compared to *WT* mice. Altogether, these results suggest that miR-146b deficiency affects the abundance of astroglial and neuronal cells, with no significant changes of microglial and OPC cells in HP and FC.

3.5. Increased Hippocampal Neurogenesis in Adult Mir146b-/- Mice

miR-146a has been shown to modulate the cell proliferation and differentiation of various cell types in vitro [17,23]. Therefore, and because *Mir146b-/-* mice had increased numbers of neuronal cells, we next assessed adult hippocampal neurogenesis in miR-146b-deficient mice. For the analysis of proliferative activity, we used immunohistochemical detection of proliferation marker Ki67 in the dentate gyrus of the adult hippocampus. We found that the number of Ki67+ cells in the proliferative zone of dentate gyrus was significantly higher in *Mir146b-/-* mice compared to their *WT* littermates ($p = 0.0075$; Student's *t*-test; Figure 5A,E). To track cell survival, BrdU was administered (i.p) to the *Mir146b-/-* and *WT* mice, and three weeks later the brains were processed for immunohistochemistry to visualize BrdU+ cells in the dentate gyrus. A significantly higher fraction of survived BrdU+ cells were detected in *Mir146b-/-* mice compared to *WT* mice ($p = 0.0281$; Student's *t*-test; Figure 5B,F). To assess whether loss of miR-146b affects newly generated cells at earlier stages of differentiation, we used doublecortin to label young post-mitotic neurons and found significantly higher number of doublecortin positive cells in the *Mir146b-/-* mice compared to *WT* mice ($p = 0.0252$; Student's *t*-test; Figure 5C,G).

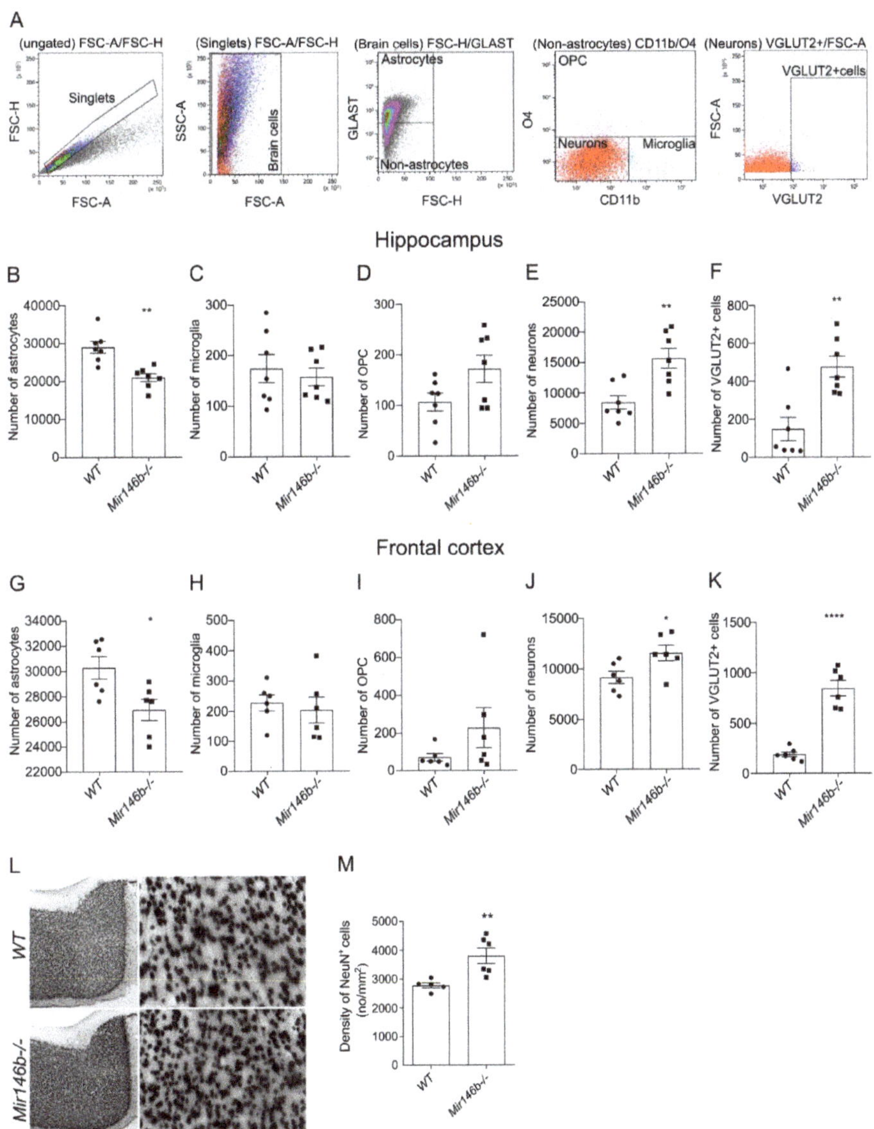

Figure 4. miR-146b deficiency causes altered brain cell abundancy in the hippocampus and frontal cortex of the mouse brain determined by flow cytometry and immunohistochemistry. (**A**) Representative graphs showing flow cytometry gating strategy for astrocytes, microglia, oligodendrocyte precursor cells (OPC) and neurons. (**B**) Number of astrocytes among total brain cells. (**C**) Number of microglia, (**D**) OPC and (**E**) neurons among non-astrocytes. (**F**) Number of VGLUT2+ neurons in the hippocampus and (**G–K**) frontal cortex of *WT* and *Mir146b-/-* mice. (**L**) Representative immunohistochemistry microphotographs of the frontal cortex NeuN+ sections at $40\times$ and $200\times$ magnification and (**M**) quantification of NeuN-positive neuronal density in the frontal cortex of *WT* and *Mir146b-/-* mice. Number of animals = 5–7, Student's *t*-test. Data represented as mean ± SEM; * $p < 0.05$, ** $p < 0.01$, **** $p < 0.0001$.

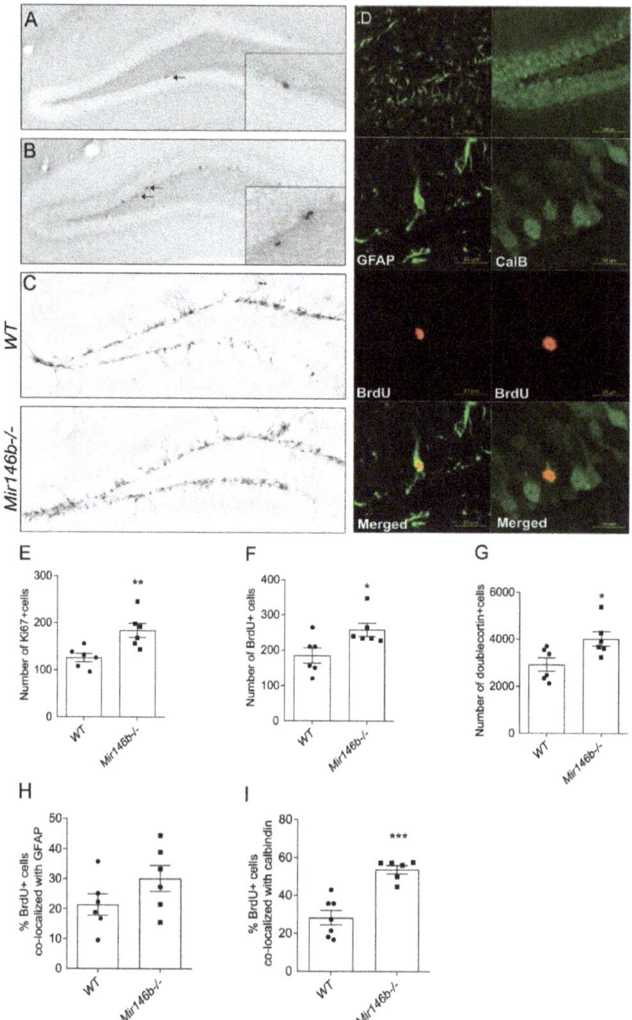

Figure 5. Increased adult hippocampal neurogenesis in *Mir146b-/-* mice. (**A**) Representative immunohistochemistry microphotographs of the hippocampal Ki67+ cells at 100× magnification and inserted microphotographs at 1000× magnification. (**B**) Illustrative microphotographs of the hippocampal BrdU+ cells at 100× magnification and inserted microphotographs at 1000× magnification. (**C**) Represented microphotographs of doublecortin positive cells taken at 100× magnification. (**D**) Illustrative images of BrdU, GFAP and calbindin signal and their co-localization in the hippocampus. (**E**) Quantitative graph showing increased number of Ki67 positive cells. (**F**) Increased number of BrdU+ cells. (**G**) Quantitative graphs showing number of doublecortin positive cells in the hippocampus of *WT* and *Mir146b-/-* mice. (**H**) Percentage of BrdU+ cells with GFAP and (**I**) calbindin in the dentate gyrus of *WT* and *Mir146b-/-* mice. Scale bar = 20 μm. Number of animals = 6, Student's *t*-test. Data represented as mean ± SEM; * $p < 0.05$, ** $p < 0.01$, *** $p < 0.001$.

In order to determine the phenotype of the BrdU+ cells, we performed immunofluorescence analysis with the antibodies against BrdU, neuronal (calbindin) or astroglial (GFAP) markers. We found no differences in percentage of BrdU+ cells that co-localized with astroglial marker, GFAP ($p = 0.1545$; Student's *t*-test; Figure 5D,H), whereas percentage of BrdU+ cells that co-localized with calbindin was higher in *Mir146b-/-* mice compared to the *WT* mice (p = 0.0002; Student's *t*-test; Figure 5D,I). These results indicate that miR-146b deficiency results in increased proliferation, survival and differentiation of progenitors to neuronal lineage but not astroglial lineage.

3.6. miR-146b Deficiency Does Not Cause Microglial Activation in the Hippocampus

As numerous previous studies have demonstrated the immunomodulatory role of miR-146b [25,27], we further explored in more detail microglial morphology, microglial polarization and the expression of *Il1b*, *Tnf*, *Il18* mRNA. No changes were observed in the density of Iba1-positive microglia (p = 0.8618; Student's *t*-test; Figure 6A,B). We analyzed microglial morphological parameters such as microglial cell size (p = 0.7216; Student's *t*-test; Figure 6D), cell body size (p = 0.4740; Student's *t*-test; Figure 6E) and size of dendritic processes (p = 0.6132; Student's *t*-test; Figure 6F), and found no difference between *WT* and *Mir146b-/-* mice. Next, we evaluated the expression of cytokines *Il1b* (p = 0.3853; Student's *t*-test; Figure 6G), *Tnf* (p = 0.3853; Student's *t*-test; Figure 6H), *Il18* (p = 0.2043; Student's *t*-test; Figure 6I) mRNA levels and found no differences between *Mir146b-/-* and the *WT* mice. Similarly, flow cytometry analysis did not reveal changes in the M1 microglial polarization in *Mir146b-/-* mice (p = 0.1780; Student's *t*-test; Figure 6J). It is established that microglia are involved in the regulation of neurogenesis [53] via interaction of microglial fractalkine receptor Cx3cr1 with neuronal Cx3cl1 [54]. We next assessed whether observed enhancement of hippocampal neurogenesis in miR-146b deficient mice could possibly be mediated via fractalkine signaling and measured the expression of *Cx3cr1* mRNA levels in the hippocampus of miR-146b deficient mice and their *WT* littermates. No differences in *Cx3cr1* mRNA expression were observed between the groups (p = 0.6114; Student's *t*-test; Figure 6K). These results indicate that loss of miR-146b neither influences microglial activation nor affects neurogenesis through fractalkine signaling in HP of *Mir146b-/-* mice.

3.7. Glial Cell Line-Derived Neurotrophic Factor (Gdnf) mRNA Is Upregulated in the Hippocampus of Mir146b-/- Mice

To explore which putative miR-146b targets may influence cognition, increased neuronal density in the FC and enhanced hippocampal neurogenesis, we performed a TargetScan [41] search in combination with pathway analysis with g:Profiler [43] as well as a literature search for verified targets. We selected 283 genes containing a total of 299 conserved and 126 poorly conserved sites with TargetScanHuman and subjected this list of genes to pathway and gene ontology group analysis. Interestingly, we detected multiple gene ontology groups associated with neuron development and function to be overrepresented among conserved miR-146a/b targets. Among all selected genes shown in (Table S2), we detected glia-derived neurotrophic factor (*Gdnf*) as a putative target for miR-146a/b. *Gdnf* is known to be expressed in neurons (https://www.brainrnaseq.org/ accessed on 1 June 2022) [55] and previously miR-146a had been shown to be to negatively regulate Gdnf expression [56]. Based on the literature, we also selected for analysis brain-derived neurotrophic factor (*Bdnf*) as a gene. This gene is expressed in astrocytes (https://www.brainrnaseq.org/ accessed on 1 June 2022) [55] and may alter miR-146b expression as Bdnf mutation Val66Met has been associated with altered expression of miR-146b and its downstream targets [57]. In addition, we choose for analysis miR-146a/b target *Irak1*, which is highly expressed in the glial cells (https://www.brainrnaseq.org/ accessed on 1 June 2022) [55]. The expression analysis of the selected targets by RT-qPCR demonstrated that *Gdnf* mRNA expression levels were significantly higher in the hippocampus of *Mir146b-/-* mice compared to WT littermates' levels (p = 0.0477; Student's *t*-test; Figure 7A), whereas there were no changes in *Irak1* levels (p = 0.6547; Student's *t*-test; Figure 7B) and *Bdnf* mRNA levels levels (p = 0.9968; Student's *t*-test; Figure 7C). These data together suggest that miR-146b has capacity to modulate neuronal development due to its influence on expression of *Gdnf* and possible other factors involved in neuronal development.

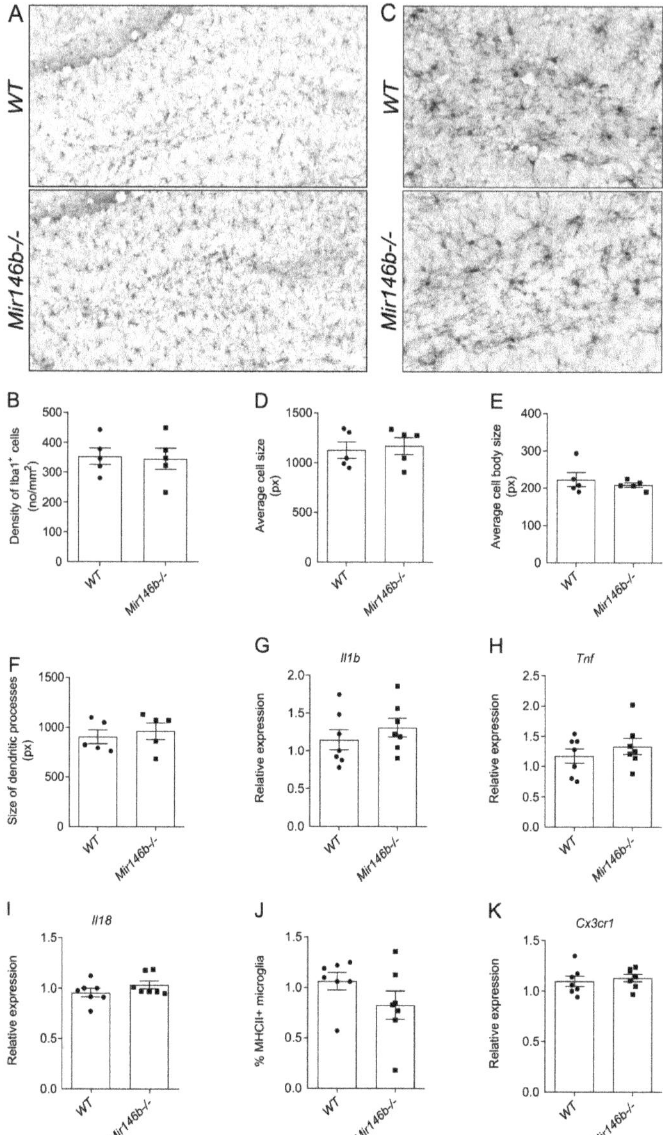

Figure 6. miR-146b deficiency does not cause microglial activation and signs of neuroinflammation in the hippocampus. (**A**) Representative images showing Iba1-positive cells at $100\times$ magnification. (**B**) Quantitative graphs of Iba1 counting in *WT* and *Mir146b-/-* mice. (**C**) Representative images of Iba1 immunohistochemistry for morphological analysis at $400\times$ magnification in the hippocampus of *WT* and *Mir146b-/-* mice. (**D**) Average cell size in pixels. (**E**) Cell body size in pixels. (**F**) Size of dendritic processes in pixels of microglial cells in the hippocampus. Number of animals = 5, Student's *t*-test. (**G**) Relative mRNA expression of cytokines *IL1b*, (**H**) *Tnf* and (**I**) *IL18*. (**J**) Flow cytometry quantification of percentage of MHCII+ M1 type of microglia. (**K**) Relative mRNA expression of *Cx3cr1* in the hippocampus of *WT* and *Mir146b-/-* mice. Number of animals = 7, Student's *t*-test. Data represented as mean ± SEM, respectively.

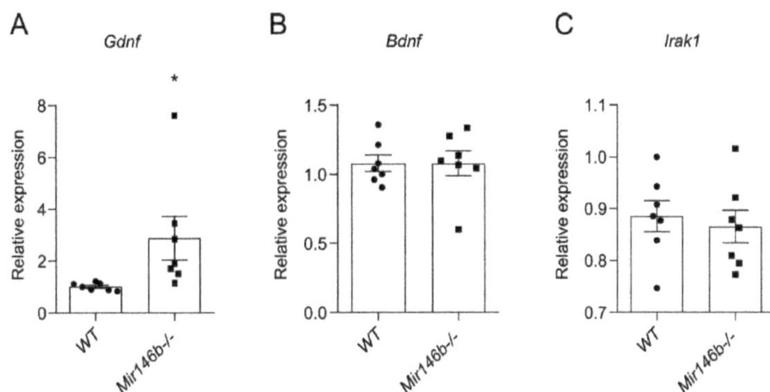

Figure 7. *Gdnf* mRNA is upregulated in *Mir146b-/-* mice. (**A**) Relative mRNA expression of miR-146b targets *Gdnf*. (**B**) *Bdnf* and (**C**) *Irak1* of *WT* and *Mir146b-/-* mice. Number of animals = 7, Student's *t*-test. Data represented as mean ± SEM respectively; * $p < 0.05$.

3.8. Association between Enhanced Cognition in NORT and Gdnf mRNA Expression of Mir146b-/- Mice

To find out whether the enhanced cognition in NORT has any correlation with enhanced expression of *Gdnf* in the hippocampus, we then employed Pearson's analysis. We found that percentage of preference for the novel object in NORT was positively correlated with the levels of *Gdnf* mRNA expression at 2-h ($r = 0.7617$, $p = 0.0466$; Figure 8A) and 24-h ($r = 0.7722$, $p = 0.0419$; Figure 8B) time points in *Mir146b-/-* mice, while no significant correlation was observed in *WT* mice.

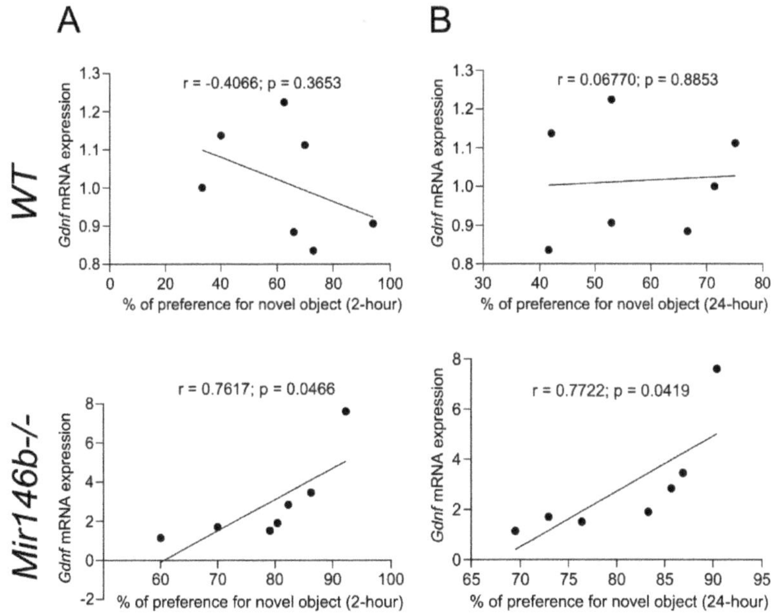

Figure 8. Pearson correlation of percentage of preference for the novel object in NORT with *Gdnf* mRNA expression in the hippocampus of *WT* and miR-146b deficient mice. Correlation analysis was performed at 2 h (**A**) and 24 h (**B**) in NORT.

4. Discussion

The results of this study show that despite sequence similarities between miR-146a/b, their cellular distribution is remarkably different in the mouse brain tissue. While miR-146a is abundantly expressed in the microglial cells, miR-146b was highly expressed in neurons and less in microglia and astroglia. Because the role of the neuronal miR-146b is completely unknown, we next performed a detailed assessment of behavior as well as cellular organization of the brain in *Mir146b-/-* mice.

One of the major findings of our study is that *Mir146b-/-* mice demonstrated enhanced episodic recognition memory. Furthermore, better memory acquisition and recall were shown in the contextual fear conditioning and tone recall tests. Interestingly, *Mir146b-/-* mice also demonstrated slightly increased anxiety in the OFT and EPM tests. In general, anxiety sensitizes sensory cortical systems to innocuous environmental stimuli and might thereby facilitate cognition. Thus, anxiety might play an important adaptive role in the process of cognition [58]. There is a consensus theory that anxiety is associated with better attention control [59], because of improvement in the selectivity of attention and probable better acquisition of the negative emotional stimuli [60]. Thus, the observed enhancement in cognition functions might be at least in part attributed to the increased anxiety levels and arousal. Despite the observed increased anxiety, the immobility time in TST was not changed, suggesting that miR-146b deficient mice do not demonstrate depression-like behavior.

Although we did not observe any changes in the volume of the brain and hippocampus, the experiments using flow cytometry showed abnormalities in the cellular composition in the brain of *Mir146b-/-* mice. We found that loss of miR-146b led to the increased number of neurons, decreased astrocytes and increased VGLUT2+ glutamatergic neurons in HP and FC of *Mir146b-/-* mice. Further immunohistochemistry revealed increased density of the neurons in FC sections of miR-146b deficient mice. Therefore, it might be proposed that the miR-146b deficiency results in the loss of control in the growth of the neuronal population, which results in the higher numbers of neuronal cells. In line with our results, it has previously been shown that miR-146b overexpression by lentivirus vector could inhibit the proliferation of primary hippocampal neural stem cells after transfection [26]. Since the neuronal proliferation in the FC is restricted to the prenatal period [61,62], we speculated that miR-146b has influence on neuronal generation during the early stages of the brain development; however, further studies are needed to confirm this hypothesis. In addition, further experiments are needed to explore whether the observed decrease in the astroglia is also developmentally regulated. During development, the cross-regulatory interactions between elements of different pathways affect the process of cell fate assignment during neural and astroglial tissue patterning [63]. Both neurons and astroglia are produced from the radial glial progenitor cells and cross talk between important signaling pathways such as JAK-STAT signaling, Wnt signaling responsible for this switch from neurogenesis to gliogenesis phase and for differentiation of astrocytes [64]. As miR-146b has been shown to regulate the JAK-STAT [16] and Wnt signaling [65], the absence of miR-146b might cause perturbation in Wnt signaling, which in turn results in the reduction in astroglia during the early stages of development.

We also speculated that capacity of miR-146b to affect neuronal development might persist in the neurogenic region of the adult dentate gyrus. Indeed, mice deficient in miR-146b had enhanced neurogenesis in dentate gyrus, as demonstrated by the higher numbers of neuronal precursors, proliferative cells and their better survival. When assessing the phenotype of newly generated cells surviving, we found that a larger proportion expressed a mature neuronal marker, Calbindin, in miR-146b deficient mice, showing increased differentiation into neurons. As previous studies have demonstrated important roles of the adult dentate gyrus neurogenesis in the memory processing [66,67], we speculate that increased neurogenesis might be responsible for the observed enhanced learning and memory capability in *Mir146b-/-* mice.

As previously miR-146b had been shown to regulate immune responses in variety of cells and tissues [15,17,27], and because microglia can control adult neurogenesis through secretion of various soluble factors including cytokines [68,69], we studied in more detail the morphology of microglia cells. We did not observe signs of microglial activation in *Mir146b-/-* mice. Similarly, no change was observed in the mRNA levels of *Cx3cr1*, a fractalkine receptor essential in the regulation of the interactions between microglia and neurons [70]. These results suggest that microglia or microglia-derived factors are not involved in the promotion of the neuronal phenotype observed in *Mir146b-/-* mice.

In line with the neuronal phenotype of *Mir146b-/-* mice, the pathway analysis of conserved miR-146b targets revealed that neurogenesis-related genes are overrepresented among miR-146b target genes, indicating that miR-146b might influence neuronal development through several different genes. From putative targets, we selected *Gdnf* and explored its expression in the miR-146b deficient mouse brain. Indeed, we observed an increased expression of *Gdnf* mRNA in the hippocampus of *Mir146b-/-* mice. As *Gdnf* participates in proliferation, migration and differentiation of the neural cells [71], and might direct newly generated neurons to the specific neuronal phenotypes [72], we propose that upregulation of *Gdnf* in *Mir146b-/-* mice might contribute to increased neuronal proliferation and survival these mice. Further *Gdnf* mRNA expression was positively correlated with the observed cognitive behaviors in *Mir146b-/-* mice, thus upregulation of *Gdnf* mRNA expression might contribute to the promotion of enhanced cognition observed in miR-146b deficient mice. As previous studies have shown that *Gdnf* improves spatial learning in aged rats [73,74]. Interestingly, no difference was observed in mRNA levels of *Bdnf*, which is a factor potentially regulating miR-146b levels as well as *Irak1*, which is well-known target of miR-146b [19]. However, additional experiments are needed to assess whether there is effect on additional miR-146b targets. Further research is required to provide insight into causal relationships between behavioral and morphological consequences of miR-146b deletion.

In addition, previous studies have shown that electro acupuncture-stimulated hippocampal neurogenesis in the rat model of focal cerebral ischemia and reperfusion was associated with an increased miR-146b expression, while inhibition of miR-146b reduced stimulatory effect of acupuncture on the hippocampal neurogenesis [75]. It should be noted that, aside from the novel neuronal functions of miR-146b described in our study, miR-146b is expressed in microglial cells and is involved in the negative regulation of neuroinflammation [25,27]. Previous studies have demonstrated that focal cerebral ischemia and reperfusion can induce inflammation and that acupuncture is able to reduce neuroinflammation [76]. It is possible that acupuncture might improve neurogenesis via expression of anti-inflammatory miR-146b and that inhibition of neurogenesis by the miR-146b inhibition might involve an inflammatory component. This possibility should be studied in more detail further.

It is also important that further study is conducted around the possible roles of miR-146b in the regulation of apoptosis. There is compelling evidence that miR-146a is an important negative modulator of apoptosis via TAF9b/P53 [77] and SMAD3 pathways [78]. Recent studies demonstrated that miR-146b is also involved in the regulation of apoptosis and that over-expression of miR-146b promoted cell death [79]. Furthermore, miR-146b-5p inhibits tumorigenesis and metastasis of gall bladder cancer by targeting toll-like receptor 4 via the NF-κB pathway [80]. Thus, it is not ruled out that deficiency in miR-146b will result in the suppression of apoptosis, which might also have an impact on the increased neuronal density in the miR-146b deficient mice. Further research is required to provide insight into possible roles of miR-146b in the regulation of neuronal apoptosis as well as causal relationships between behavioral and morphological consequences of miR-146b deletion.

5. Conclusions

In conclusion, our data provide new evidence that miR-146b has important roles in the control of the proliferation and differentiation of neuronal precursors during the development as well as in adulthood. In addition, we show that miR-146b is able to control adult hippocampal neurogenesis, which might be relevant for the observed enhanced cognition and fear. Our data indicate that miR-146b probably exerts its actions via regulation of *Gdnf* expression, which explains why deficiency in miR-146b leads to the activation of neurogenesis. Our data also open new avenues for the regulation of hippocampal neurogenesis via modulation of miR-146b. In addition, miR-146b might be used as a biomarker for tumors of neuronal origin.

Supplementary Materials: The following supporting information can be downloaded at: https://www.mdpi.com/article/10.3390/cells11132002/s1, Figure S1: *Mir146b-/-* mice have no differences in the volume of brain and hippocampus; Figure S2: *Mir146b-/-* and *WT* mice same motivation to explore the objects in NORT; Figure S3: Baseline freezing of *Mir146b-/-* and *WT* mice in CFC. Figure S4: *Mir146b-/-* mice have no depression-like behavior and social dominant behavior; Figure S5: Representative dot plots of isotype controls used in flow cytometric analysis for gating; Table S1: RT-qPCR Primers used in this study; Table S2: Gene ontology groups with neurological function and targeted by miR-146a/b.

Author Contributions: Conceptualization, K.C. and A.Z.; methodology, K.C., K.S., M.J., T.Ž., K.P., L.Y., N.M. and M.P.B.; software, K.C.; validation, K.C. and K.P.; formal analysis, K.C.; investigation, K.C., K.S. and M.J.; resources, A.Z., L.T. and A.R.; data curation, K.C. and K.P.; writing—original draft preparation, K.C.; writing—review and editing, all authors; visualization, K.C.; supervision, A.Z., A.R. and L.T.; project administration, A.Z.; funding acquisition, A.Z., A.R. and L.T. All authors have read and agreed to the published version of the manuscript.

Funding: A.Z. is supported by the Estonian Research Council personal research funding team grant project No. PRG878. L.T. is supported by the Estonian Research Council-European Union Regional Developmental Fund Mobilitas Pluss Program No. MOBTT77 and A.R. is supported by European Regional Development Fund (Project No. 2014-2020.4.01.15-0012) and personal grant from Estonian Research Council, PRG1259.

Institutional Review Board Statement: The use of mice was conducted in accordance with the regulations and guidelines approved by the Laboratory Animal Centre at the Institute of Biomedicine and Translational Medicine, University of Tartu, Estonia. All animal procedures were conducted in accordance with the European Communities Directive (2010/63/EU) with permit (No. 183, 2021) from the Estonian National Board of Animal Experiments.

Informed Consent Statement: Not applicable.

Data Availability Statement: The data presented in this study are available on request from the corresponding author.

Acknowledgments: The authors would like to thank Anu Remm for helping with genotyping of *Mir146b-/-* mice and Mario Plaas for brain volumetric measurements.

Conflicts of Interest: The authors declare no conflict of interest.

References

1. Bartel, D.P. Metazoan Micrornas. *Cell* **2018**, *173*, 20–51. [CrossRef] [PubMed]
2. Brennan, G.P.; Henshall, D.C. MicroRNAs as Regulators of Brain Function and Targets for Treatment of Epilepsy. *Nat. Rev. Neurol.* **2020**, *16*, 506–519. [CrossRef] [PubMed]
3. Su, Z.-F.; Sun, Z.-W.; Zhang, Y.; Wang, S.; Yu, Q.-G.; Wu, Z.-B. Regulatory Effects of MiR-146a/b on the Function of Endothelial Progenitor Cells in Acute Ischemic Stroke in Mice. *Kaohsiung J. Med. Sci.* **2017**, *33*, 369–378. [CrossRef] [PubMed]
4. Adlakha, Y.K.; Saini, N. Brain MicroRNAs and Insights into Biological Functions and Therapeutic Potential of Brain Enriched MiRNA-128. *Mol. Cancer* **2014**, *13*, 33. [CrossRef]
5. Bak, M.; Silahtaroglu, A.; Møller, M.; Christensen, M.; Rath, M.F.; Skryabin, B.; Tommerup, N.; Kauppinen, S. MicroRNA Expression in the Adult Mouse Central Nervous System. *Rna* **2008**, *14*, 432–444. [CrossRef] [PubMed]
6. He, M.; Liu, Y.; Wang, X.; Zhang, M.Q.; Hannon, G.J.; Huang, Z.J. Cell-Type-Based Analysis of MicroRNA Profiles in the Mouse Brain. *Neuron* **2012**, *73*, 35–48. [CrossRef]
7. Kuwabara, T.; Hsieh, J.; Nakashima, K.; Taira, K.; Gage, F.H. A Small Modulatory DsRNA Specifies the Fate of Adult Neural Stem Cells. *Cell* **2004**, *116*, 779–793. [CrossRef]

8. Schratt, G.M.; Tuebing, F.; Nigh, E.A.; Kane, C.G.; Sabatini, M.E.; Kiebler, M.; Greenberg, M.E. A Brain-Specific MicroRNA Regulates Dendritic Spine Development. *Nature* **2006**, *439*, 283–289. [CrossRef]
9. Aten, S.; Hansen, K.F.; Hoyt, K.R.; Obrietan, K. The MiR-132/212 Locus: A Complex Regulator of Neuronal Plasticity, Gene Expression and Cognition. *RNA Dis.* **2016**, *3*, e1375.
10. Yan, H.-L.; Sun, X.-W.; Wang, Z.-M.; Liu, P.-P.; Mi, T.-W.; Liu, C.; Wang, Y.-Y.; He, X.-C.; Du, H.-Z.; Liu, C.-M.; et al. MiR-137 Deficiency Causes Anxiety-like Behaviors in Mice. *Front. Mol. Neurosci.* **2019**, *12*, 260. [CrossRef]
11. Jauhari, A.; Singh, T.; Mishra, S.; Shankar, J.; Yadav, S. Coordinated Action of MiR-146a and Parkin Gene Regulate Rotenone-Induced Neurodegeneration. *Toxicol. Sci.* **2020**, *176*, 433–445. [CrossRef] [PubMed]
12. Launay, J.; Mouillet-Richard, S.; Baudry, A.; Pietri, M.; Kellermann, O. Raphe-Mediated Signals Control the Hippocampal Response to SRI Antidepressants via MiR-16. *Transl. Psychiatry* **2011**, *1*, e56. [CrossRef] [PubMed]
13. Lopez, J.P.; Fiori, L.M.; Cruceanu, C.; Lin, R.; Labonte, B.; Cates, H.M.; Heller, E.A.; Vialou, V.; Ku, S.M.; Gerald, C.; et al. MicroRNAs 146a/b-5 and 425-3p and 24-3p Are Markers of Antidepressant Response and Regulate MAPK/Wnt-System Genes. *Nat. Commun.* **2017**, *8*, 15497. [CrossRef] [PubMed]
14. So, A.Y.; Zhao, J.L.; Baltimore, D. The Yin and Yang of Micro RNA s: Leukemia and Immunity. *Immunol. Rev.* **2013**, *253*, 129–145. [CrossRef] [PubMed]
15. Curtale, G.; Mirolo, M.; Renzi, T.A.; Rossato, M.; Bazzoni, F.; Locati, M. Negative Regulation of Toll-like Receptor 4 Signaling by IL-10–Dependent MicroRNA-146b. *Proc. Natl. Acad. Sci. USA* **2013**, *110*, 11499–11504. [CrossRef]
16. Xiang, M.; Birkbak, N.J.; Vafaizadeh, V.; Walker, S.R.; Yeh, J.E.; Liu, S.; Kroll, Y.; Boldin, M.; Taganov, K.; Groner, B.; et al. STAT3 Induction of MiR-146b Forms a Feedback Loop to Inhibit the NF-KB to IL-6 Signaling Axis and STAT3-Driven Cancer Phenotypes. *Sci. Signal.* **2014**, *7*, ra11. [CrossRef]
17. Hermann, H.; Runnel, T.; Aab, A.; Baurecht, H.; Rodriguez, E.; Magilnick, N.; Urgard, E.; Šahmatova, L.; Prans, E.; Maslovskaja, J.; et al. MiR-146b Probably Assists MiRNA-146a in the Suppression of Keratinocyte Proliferation and Inflammatory Responses in Psoriasis. *J. Invest. Dermatol.* **2017**, *137*, 1945–1954. [CrossRef]
18. Paterson, M.R.; Kriegel, A.J. MiR-146a/b: A Family with Shared Seeds and Different Roots. *Physiol. Genom.* **2017**, *49*, 243–252. [CrossRef]
19. Taganov, K.D.; Boldin, M.P.; Chang, K.-J.; Baltimore, D. NF-KB-Dependent Induction of MicroRNA MiR-146, an Inhibitor Targeted to Signaling Proteins of Innate Immune Responses. *Proc. Natl. Acad. Sci. USA* **2006**, *103*, 12481–12486. [CrossRef]
20. Lee, H.-M.; Kim, T.S.; Jo, E.-K. MiR-146 and MiR-125 in the Regulation of Innate Immunity and Inflammation. *BMB Rep.* **2016**, *49*, 311. [CrossRef]
21. Deng, M.; Du, G.; Zhao, J.; Du, X. MiR-146a Negatively Regulates the Induction of Proinflammatory Cytokines in Response to Japanese Encephalitis Virus Infection in Microglial Cells. *Arch. Virol.* **2017**, *162*, 1495–1505. [CrossRef] [PubMed]
22. Saba, R.; Gushue, S.; Huzarewich, R.L.; Manguiat, K.; Medina, S.; Robertson, C.; Booth, S.A. MicroRNA 146a (MiR-146a) Is over-Expressed during Prion Disease and Modulates the Innate Immune Response and the Microglial Activation State. *PLoS ONE* **2012**, *7*, e30832. [CrossRef] [PubMed]
23. Nguyen, L.S.; Fregeac, J.; Bole-Feysot, C.; Cagnard, N.; Iyer, A.; Anink, J.; Aronica, E.; Alibeu, O.; Nitschke, P.; Colleaux, L. Role of MiR-146a in Neural Stem Cell Differentiation and Neural Lineage Determination: Relevance for Neurodevelopmental Disorders. *Mol. Autism* **2018**, *9*, 38. [CrossRef]
24. Fregeac, J.; Moriceau, S.; Poli, A.; Nguyen, L.S.; Oury, F.; Colleaux, L. Loss of the Neurodevelopmental Disease-Associated Gene MiR-146a Impairs Neural Progenitor Differentiation and Causes Learning and Memory Deficits. *Mol. Autism* **2020**, *11*, 22. [CrossRef] [PubMed]
25. Yang, G.; Zhao, Y. Overexpression of MiR-146b-5p Ameliorates Neonatal Hypoxic Ischemic Encephalopathy by Inhibiting IRAK1/TRAF6/TAK1/NF-AB Signaling. *Yonsei Med. J.* **2020**, *61*, 660. [CrossRef] [PubMed]
26. Dai, Y.; Chen, L.; He, X.; Lin, H.; Jia, W.; Chen, L.; Tao, J.; Liu, W. Construction of MiR-146b Overexpression Lentiviral Vector and the Effect on the Proliferation of Hippocampal Neural Stem Cells. *Chin. J. Tissue Eng. Res.* **2021**, *25*, 3024.
27. Bokobza, C.; Joshi, P.; Schang, A.; Csaba, Z.; Faivre, V.; Montané, A.; Galland, A.; Benmamar-Badel, A.; Bosher, E.; Lebon, S.; et al. MiR-146b Protects the Perinatal Brain against Microglia-Induced Hypomyelination. *Ann. Neurol.* **2022**, *91*, 48–65. [CrossRef]
28. Katakowski, M.; Zheng, X.; Jiang, F.; Rogers, T.; Szalad, A.; Chopp, M. MiR-146b-5p Suppresses EGFR Expression and Reduces in Vitro Migration and Invasion of Glioma. *Cancer Investig.* **2010**, *28*, 1024–1030. [CrossRef]
29. Liu, J.; Xu, J.; Li, H.; Sun, C.; Yu, L.; Li, Y.; Shi, C.; Zhou, X.; Bian, X.; Ping, Y.; et al. MiR-146b-5p Functions as a Tumor Suppressor by Targeting TRAF6 and Predicts the Prognosis of Human Gliomas. *Oncotarget* **2015**, *6*, 29129. [CrossRef]
30. Xia, H.; Qi, Y.; Ng, S.S.; Chen, X.; Li, D.; Chen, S.; Ge, R.; Jiang, S.; Li, G.; Chen, Y.; et al. MicroRNA-146b Inhibits Glioma Cell Migration and Invasion by Targeting MMPs. *Brain Res.* **2009**, *1269*, 158–165. [CrossRef]
31. Zhang, L.; Wang, J.; Fu, Z.; Ai, Y.; Li, Y.; Wang, Y.; Wang, Y. Sevoflurane Suppresses Migration and Invasion of Glioma Cells by Regulating MiR-146b-5p and MMP16. *Artif. Cells Nanomed. Biotechnol.* **2019**, *47*, 3306–3314. [CrossRef] [PubMed]
32. Jürgenson, M.; Aonurm-Helm, A.; Zharkovsky, A. Behavioral Profile of Mice with Impaired Cognition in the Elevated Plus-Maze Due to a Deficiency in Neural Cell Adhesion Molecule. *Pharmacol. Biochem. Behav.* **2010**, *96*, 461–468. [CrossRef] [PubMed]
33. Tronson, N.C.; Schrick, C.; Fischer, A.; Sananbenesi, F.; Pouysségur, J.; Radulovic, J. Regulatory Mechanisms of Fear Extinction and Depression-like Behavior. *Neuropsychopharmacology* **2008**, *33*, 1570–1583. [CrossRef] [PubMed]

34. Yan, L.; Jayaram, M.; Chithanathan, K.; Zharkovsky, A.; Tian, L. Sex-Specific Microglial Activation and SARS-CoV-2 Receptor Expression Induced by Chronic Unpredictable Stress. *Front. Cell. Neurosci.* **2021**, *15*, 750373. [CrossRef] [PubMed]
35. Jürgenson, M.; Aonurm-Helm, A.; Zharkovsky, A. Partial Reduction in Neural Cell Adhesion Molecule (NCAM) in Heterozygous Mice Induces Depression-Related Behaviour without Cognitive Impairment. *Brain Res.* **2012**, *1447*, 106–118. [CrossRef]
36. Piirainen, S.; Chithanathan, K.; Bisht, K.; Piirsalu, M.; Savage, J.C.; Tremblay, M.; Tian, L. Microglia Contribute to Social Behavioral Adaptation to Chronic Stress. *Glia* **2021**, *69*, 2459–2473. [CrossRef]
37. Hovens, I.B.; Nyakas, C.; Schoemaker, R.G. A Novel Method for Evaluating Microglial Activation Using Ionized Calcium-Binding Adaptor Protein-1 Staining: Cell Body to Cell Size Ratio. *Neuroimmunol. Neuroinflamm.* **2014**, *1*, 82–88. [CrossRef]
38. Jürgenson, M.; Zharkovskaja, T.; Noortoots, A.; Morozova, M.; Beniashvili, A.; Zapolski, M.; Zharkovsky, A. Effects of the Drug Combination Memantine and Melatonin on Impaired Memory and Brain Neuronal Deficits in an Amyloid-Predominant Mouse Model of Alzheimer's Disease. *J. Pharm. Pharmacol.* **2019**, *71*, 1695–1705. [CrossRef]
39. Somelar, K.; Jürgenson, M.; Jaako, K.; Anier, K.; Aonurm-Helm, A.; Zvejniece, L.; Zharkovsky, A. Development of Depression-like Behavior and Altered Hippocampal Neurogenesis in a Mouse Model of Chronic Neuropathic Pain. *Brain Res.* **2021**, *1758*, 147329. [CrossRef]
40. Agalave, N.M.; Lane, B.T.; Mody, P.H.; Szabo-Pardi, T.A.; Burton, M.D. Isolation, Culture, and Downstream Characterization of Primary Microglia and Astrocytes from Adult Rodent Brain and Spinal Cord. *J. Neurosci. Methods* **2020**, *340*, 108742. [CrossRef]
41. Agarwal, V.; Bell, G.W.; Nam, J.-W.; Bartel, D.P. Predicting Effective MicroRNA Target Sites in Mammalian MRNAs. *eLife* **2015**, *4*, e05005. [CrossRef] [PubMed]
42. McGeary, S.E.; Lin, K.S.; Shi, C.Y.; Pham, T.M.; Bisaria, N.; Kelley, G.M.; Bartel, D.P. The Biochemical Basis of MicroRNA Targeting Efficacy. *Science* **2019**, *366*, eaav1741. [CrossRef] [PubMed]
43. Reimand, J.; Kull, M.; Peterson, H.; Hansen, J.; Vilo, J. g:Profiler—A Web-Based Toolset for Functional Profiling of Gene Lists from Large-Scale Experiments. *Nucleic Acids Res.* **2007**, *35* (Suppl. 2), W193–W200. [CrossRef] [PubMed]
44. Ansari, A.; Maffioletti, E.; Milanesi, E.; Marizzoni, M.; Frisoni, G.B.; Blin, O.; Richardson, J.C.; Bordet, R.; Forloni, G.; Gennarelli, M.; et al. MiR-146a and MiR-181a Are Involved in the Progression of Mild Cognitive Impairment to Alzheimer's Disease. *Neurobiol. Aging* **2019**, *82*, 102–109. [CrossRef] [PubMed]
45. Cui, L.; Li, Y.; Ma, G.; Wang, Y.; Cai, Y.; Liu, S.; Chen, Y.; Li, J.; Xie, Y.; Liu, G. A Functional Polymorphism in the Promoter Region of MicroRNA-146a Is Associated with the Risk of Alzheimer Disease and the Rate of Cognitive Decline in Patients. *PLoS ONE* **2014**, *9*, e89019. [CrossRef]
46. Zhan-Qiang, H.; Hai-Hua, Q.; Chi, Z.; Miao, W.; Cui, Z.; Zi-Yin, L.; Jing, H.; Yi-Wei, W. MiR-146a Aggravates Cognitive Impairment and Alzheimer Disease-like Pathology by Triggering Oxidative Stress through MAPK Signaling. *Neurología* **2021**. [CrossRef]
47. Ennaceur, A.; Delacour, J. A New One-Trial Test for Neurobiological Studies of Memory in Rats. 1: Behavioral Data. *Behav. Brain Res.* **1988**, *31*, 47–59. [CrossRef]
48. Ma, L. Depression, Anxiety, and Apathy in Mild Cognitive Impairment: Current Perspectives. *Front. Aging Neurosci.* **2020**, *12*, 9. [CrossRef]
49. Enatescu, V.R.; Papava, I.; Enatescu, I.; Antonescu, M.; Anghel, A.; Seclaman, E.; Sirbu, I.O.; Marian, C. Circulating Plasma Micro RNAs in Patients with Major Depressive Disorder Treated with Antidepressants: A Pilot Study. *Psychiatry Investig.* **2016**, *13*, 549. [CrossRef]
50. Hung, Y.-Y.; Wu, M.-K.; Tsai, M.-C.; Huang, Y.-L.; Kang, H.-Y. Aberrant Expression of Intracellular Let-7e, MiR-146a, and MiR-155 Correlates with Severity of Depression in Patients with Major Depressive Disorder and Is Ameliorated after Antidepressant Treatment. *Cells* **2019**, *8*, 647. [CrossRef]
51. Lindert, J.; Paul, K.C.; Lachman, M.E.; Ritz, B.; Seeman, T.E. Depression-, Anxiety-, and Anger and Cognitive Functions: Findings From a Longitudinal Prospective Study. *Front. Psychiatry* **2021**, *12*, 665742. [CrossRef] [PubMed]
52. Von Bartheld, C.S.; Wouters, F.S. Quantitative Techniques for Imaging Cells and Tissues. *Cell Tissue Res.* **2015**, *360*, 1–4. [CrossRef] [PubMed]
53. De Lucia, C.; Rinchon, A.; Olmos-Alonso, A.; Riecken, K.; Fehse, B.; Boche, D.; Perry, V.H.; Gomez-Nicola, D. Microglia Regulate Hippocampal Neurogenesis during Chronic Neurodegeneration. *Brain Behav. Immun.* **2016**, *55*, 179–190. [CrossRef]
54. Bachstetter, A.D.; Morganti, J.M.; Jernberg, J.; Schlunk, A.; Mitchell, S.H.; Brewster, K.W.; Hudson, C.E.; Cole, M.J.; Harrison, J.K.; Bickford, P.C.; et al. Fractalkine and CX3CR1 Regulate Hippocampal Neurogenesis in Adult and Aged Rats. *Neurobiol. Aging* **2011**, *32*, 2030–2044. [CrossRef] [PubMed]
55. Zhang, Y.; Chen, K.; Sloan, S.A.; Bennett, M.L.; Scholze, A.R.; O'Keeffe, S.; Phatnani, H.P.; Guarnieri, P.; Caneda, C.; Ruderisch, N.; et al. An RNA-Sequencing Transcriptome and Splicing Database of Glia, Neurons, and Vascular Cells of the Cerebral Cortex. *J. Neurosci.* **2014**, *34*, 11929–11947. [CrossRef] [PubMed]
56. Kumar, A.; Kopra, J.; Varendi, K.; Porokuokka, L.L.; Panhelainen, A.; Kuure, S.; Marshall, P.; Karalija, N.; Härma, M.-A.; Vilenius, C.; et al. GDNF Overexpression from the Native Locus Reveals Its Role in the Nigrostriatal Dopaminergic System Function. *PLoS Genet.* **2015**, *11*, e1005710. [CrossRef]
57. Hsu, P.-K.; Xu, B.; Mukai, J.; Karayiorgou, M.; Gogos, J.A. The BDNF Val66Met Variant Affects Gene Expression through MiR-146b. *Neurobiol. Dis.* **2015**, *77*, 228–237. [CrossRef]

58. Robinson, O.J.; Vytal, K.; Cornwell, B.R.; Grillon, C. The Impact of Anxiety upon Cognition: Perspectives from Human Threat of Shock Studies. *Front. Hum. Neurosci.* **2013**, *7*, 203. [CrossRef]
59. Hu, K.; Bauer, A.; Padmala, S.; Pessoa, L. Threat of Bodily Harm Has Opposing Effects on Cognition. *Emotion* **2012**, *12*, 28. [CrossRef]
60. Roozendaal, B.; Okuda, S.; Van der Zee, E.A.; McGaugh, J.L. Glucocorticoid Enhancement of Memory Requires Arousal-Induced Noradrenergic Activation in the Basolateral Amygdala. *Proc. Natl. Acad. Sci. USA* **2006**, *103*, 6741–6746. [CrossRef]
61. Kolk, S.M.; Rakic, P. Development of Prefrontal Cortex. *Neuropsychopharmacology* **2022**, *47*, 41–57. [CrossRef] [PubMed]
62. Kroon, T.; van Hugte, E.; van Linge, L.; Mansvelder, H.D.; Meredith, R.M. Early Postnatal Development of Pyramidal Neurons across Layers of the Mouse Medial Prefrontal Cortex. *Sci. Rep.* **2019**, *9*, 5037. [CrossRef] [PubMed]
63. Bejoy, J.; Bijonowski, B.; Marzano, M.; Jeske, R.; Ma, T.; Li, Y. Wnt-Notch Signaling Interactions during Neural and Astroglial Patterning of Human Stem Cells. *Tissue Eng. Part A* **2020**, *26*, 419–431. [CrossRef] [PubMed]
64. Lee, H.-C.; Tan, K.-L.; Cheah, P.-S.; Ling, K.-H. Potential Role of JAK-STAT Signaling Pathway in the Neurogenic-to-Gliogenic Shift in down Syndrome Brain. *Neural Plast.* **2016**, *2016*, 7434191. [CrossRef] [PubMed]
65. Deng, X.; Wu, B.; Xiao, K.; Kang, J.; Xie, J.; Zhang, X.; Fan, Y. MiR-146b-5p Promotes Metastasis and Induces Epithelial-Mesenchymal Transition in Thyroid Cancer by Targeting ZNRF3. *Cell. Physiol. Biochem.* **2015**, *35*, 71–82. [CrossRef] [PubMed]
66. Kee, N.; Teixeira, C.M.; Wang, A.H.; Frankland, P.W. Preferential Incorporation of Adult-Generated Granule Cells into Spatial Memory Networks in the Dentate Gyrus. *Nat. Neurosci.* **2007**, *10*, 355–362. [CrossRef]
67. Vivar, C.; Van Praag, H. Functional Circuits of New Neurons in the Dentate Gyrus. *Front. Neural Circuits* **2013**, *7*, 15. [CrossRef]
68. Araki, T.; Ikegaya, Y.; Koyama, R. The Effects of Microglia-and Astrocyte-derived Factors on Neurogenesis in Health and Disease. *Eur. J. Neurosci.* **2021**, *54*, 5880–5901. [CrossRef]
69. Kim, S.U.; de Vellis, J. Microglia in Health and Disease. *J. Neurosci. Res.* **2005**, *81*, 302–313. [CrossRef]
70. Paolicelli, R.C.; Bolasco, G.; Pagani, F.; Maggi, L.; Scianni, M.; Panzanelli, P.; Giustetto, M.; Ferreira, T.A.; Guiducci, E.; Dumas, L.; et al. Synaptic Pruning by Microglia Is Necessary for Normal Brain Development. *Science* **2011**, *333*, 1456–1458. [CrossRef]
71. Gianino, S.; Grider, J.R.; Cresswell, J.; Enomoto, H.; Heuckeroth, R.O. GDNF Availability Determines Enteric Neuron Number by Controlling Precursor Proliferation. *Development* **2003**, *130*, 2187–2198. [CrossRef] [PubMed]
72. Cortés, D.; Carballo-Molina, O.A.; Castellanos-Montiel, M.J.; Velasco, I. The Non-Survival Effects of Glial Cell Line-Derived Neurotrophic Factor on Neural Cells. *Front. Mol. Neurosci.* **2017**, *10*, 258. [CrossRef] [PubMed]
73. Pelleymounter, M.A.; Cullen, M.J.; Baker, M.B.; Healy, D. Glial Cell Line-Derived Neurotrophic Factor (GDNF) Improves Spatial Learning in Aged Fischer 344 Rats. *Psychobiology* **1999**, *27*, 397–401. [CrossRef]
74. Pertusa, M.; Garcia-Matas, S.; Mammeri, H.; Adell, A.; Rodrigo, T.; Mallet, J.; Cristofol, R.; Sarkis, C.; Sanfeliu, C. Expression of GDNF Transgene in Astrocytes Improves Cognitive Deficits in Aged Rats. *Neurobiol. Aging* **2008**, *29*, 1366–1379. [CrossRef]
75. Zhang, S.; Jin, T.; Wang, L.; Liu, W.; Zhang, Y.; Zheng, Y.; Lin, Y.; Yang, M.; He, X.; Lin, H.; et al. Electro-Acupuncture Promotes the Differentiation of Endogenous Neural Stem Cells via Exosomal MicroRNA 146b after Ischemic Stroke. *Front. Cell. Neurosci.* **2020**, *14*, 223. [CrossRef]
76. Li, N.; Guo, Y.; Gong, Y.; Zhang, Y.; Fan, W.; Yao, K.; Chen, Z.; Dou, B.; Lin, X.; Chen, B.; et al. The Anti-Inflammatory Actions and Mechanisms of Acupuncture from Acupoint to Target Organs via Neuro-Immune Regulation. *J. Inflamm. Res.* **2021**, *14*, 7191. [CrossRef]
77. Pan, J.-A.; Tang, Y.; Yu, J.-Y.; Zhang, H.; Zhang, J.-F.; Wang, C.-Q.; Gu, J. MiR-146a Attenuates Apoptosis and Modulates Autophagy by Targeting TAF9b/P53 Pathway in Doxorubicin-Induced Cardiotoxicity. *Cell Death Dis.* **2019**, *10*, 668. [CrossRef]
78. Qiu, M.; Li, T.; Wang, B.; Gong, H.; Huang, T. MiR-146a-5p Regulated Cell Proliferation and Apoptosis by Targeting SMAD3 and SMAD4. *Protein Pept. Lett.* **2020**, *27*, 411–418. [CrossRef]
79. Xu, J.; Zhang, Z.; Chen, Q.; Yang, L.; Yin, J. MiR-146b Regulates Cell Proliferation and Apoptosis in Gastric Cancer by Targeting PTP1B. *Dig. Dis. Sci.* **2020**, *65*, 457–463. [CrossRef]
80. Ouyang, B.; Pan, N.; Zhang, H.; Xing, C.; Ji, W. MiR-146b-5p Inhibits Tumorigenesis and Metastasis of Gallbladder Cancer by Targeting Toll-like Receptor 4 via the Nuclear Factor-κB Pathway. *Oncol. Rep.* **2021**, *45*, 15. [CrossRef]

Article

TRIM32 Deficiency Impairs the Generation of Pyramidal Neurons in Developing Cerebral Cortex

Yan-Yun Sun [1,2,†], Wen-Jin Chen [3,†], Ze-Ping Huang [1,2], Gang Yang [4], Ming-Lei Wu [1,2], De-En Xu [5], Wu-Lin Yang [6,7], Yong-Chun Luo [8], Zhi-Cheng Xiao [9], Ru-Xiang Xu [3,*,‡] and Quan-Hong Ma [1,2,*,‡]

[1] Department of Neurology and Clinical Research Center of Neurological Disease, The Second Affiliated Hospital of Soochow University, Suzhou 215123, China; yysun@suda.edu.cn (Y.-Y.S.); 20194254018@stu.suda.edu.cn (Z.-P.H.); 20204250100@stu.suda.edu.cn (M.-L.W.)

[2] Jiangsu Key Laboratory of Neuropsychiatric Diseases, Institute of Neuroscience, Soochow University, Suzhou 215123, China

[3] Department of Neurosurgery, Sichuan Academy of Medical Sciences and Sichuan Provincial People's Hospital, University of Electronic Science and Technology of China, Chengdu 610072, China; uscyxychenwj@126.com

[4] Lab Center, Medical College of Soochow University, Suzhou 215123, China; yanggang@suda.edu.cn

[5] Wuxi No. 2 People's Hospital, Wuxi 214001, China; xudeen@njmu.edu.cn

[6] Anhui Province Key Laboratory of Medical Physics and Technology, Institute of Health and Medical Technology, Hefei Institutes of Physical Science, Chinese Academy of Sciences, Hefei 230031, China; yangw@cmpt.ac.cn

[7] Hefei Cancer Hospital, Chinese Academy of Sciences, Hefei 230031, China

[8] Department of Neurosurgery, First Medical Center of Chinese PLA General Hospital, Beijing 100028, China; luoyong4581@plagh.cn

[9] Department of Anatomy and Developmental Biology, Monash University, Clayton 3800, Australia; zhicheng.xiao@monash.edu

* Correspondence: maquanhong@suda.edu.cn (Q.-H.M.); xuruxiang1123@uestc.edu.cn (R.-X.X.)

† These authors contributed equally to this work.

‡ These authors contributed equally to this work.

Citation: Sun, Y.-Y.; Chen, W.-J.; Huang, Z.-P.; Yang, G.; Wu, M.-L.; Xu, D.-E.; Yang, W.-L.; Luo, Y.-C.; Xiao, Z.-C.; Xu, R.-X.; et al. TRIM32 Deficiency Impairs the Generation of Pyramidal Neurons in Developing Cerebral Cortex. Cells 2022, 11, 449. https://doi.org/10.3390/cells 11030449

Academic Editors: FengRu Tang and Cord Brakebusch

Received: 8 December 2021
Accepted: 25 January 2022
Published: 28 January 2022

Publisher's Note: MDPI stays neutral with regard to jurisdictional claims in published maps and institutional affiliations.

Abstract: Excitatory-inhibitory imbalance (E/I) is a fundamental mechanism underlying autism spectrum disorders (ASD). TRIM32 is a risk gene genetically associated with ASD. The absence of TRIM32 causes impaired generation of inhibitory GABAergic interneurons, neural network hyperexcitability, and autism-like behavior in mice, emphasizing the role of TRIM32 in maintaining E/I balance, but despite the description of TRIM32 in regulating proliferation and differentiation of cultured mouse neural progenitor cells (NPCs), the role of TRIM32 in cerebral cortical development, particularly in the production of excitatory pyramidal neurons, remains unknown. The present study observed that TRIM32 deficiency resulted in decreased numbers of distinct layer-specific cortical neurons and decreased radial glial cell (RGC) and intermediate progenitor cell (IPC) pool size. We further demonstrated that TRIM32 deficiency impairs self-renewal of RGCs and IPCs as indicated by decreased proliferation and mitosis. A TRIM32 deficiency also affects or influences the formation of cortical neurons. As a result, TRIM32-deficient mice showed smaller brain size. At the molecular level, RNAseq analysis indicated reduced Notch signalling in TRIM32-deficient mice. Therefore, the present study indicates a role for TRIM32 in pyramidal neuron generation. Impaired generation of excitatory pyramidal neurons may explain the hyperexcitability observed in TRIM32-deficient mice.

Keywords: TRIM32; excitatory-inhibitory imbalance; cortex development; ASD; NPCs

1. Introduction

The imbalance between excitability and inhibitory activity in brain circuits is one of the key mechanisms underlying neurodevelopmental disorders, such as autism spectrum disorders (ASD), which are characterized by impaired social behaviours and repetitive and stereotypic behaviours [1]. The E/I balance in the brain is coordinated by glutamatergic pyramidal neurons and GABAergic interneurons. In cerebral cortex, the pyramidal neurons are arranged in six layers, while the inhibitory interneurons are scattered. There are definite temporal and spatial orders in which excitatory pyramidal neurons and inhibitory interneurons are generated [2–4]. For example, the cortical pyramidal neurons are generated following an "inside-out" order, while deep-layer cortical neurons are generated following an "out-side-in" order [5,6]. The later born cortical neurons will move across the earlier-born neurons which reside in the deep laminar of cortex [6]. Any disturbances during these sequential processes will lead to developmental disorders such as ASD.

In the developing dorsal ventricular zone, RGCs produce both upper layer and deep layer cortical pyramidal neurons. RGCs can either directly generate cortical pyramidal neurons or generate them via intermediate progenitor cells (IPCs). RGCs generate both neurons and glia while IPCs only generate neurons. Intrinsic and extrinsic signals coordinate the proliferation and differentiation of RGCs and IPCs, thus regulating the development of cerebral cortex. Exploring the molecular mechanisms underlying such regulation would help to understand the pathogenesis of developmental neurological disorders.

Tripartite motif (TRIM) 32 belongs to the TRIM family, has a RING finger-like structure as its major feature and contains one or two zinc ions called "B-Boxes" and a related coiled-coil region [7]. TRIM32 has E3 ubiquitin ligase activity [8]. It plays an important role in the ubiquitin-protease degradation of proteins. TRIM32 is expressed in a variety of systems. In the nervous system, TRIM32 is primarily expressed on neural progenitor cells (NPCs). TRIM32 is ubiquitously found in the cytoplasm of NPCs, although it translocates to the nuclei once the NPCs differentiate into neurons [9]. In cultured NPCs, knocking-down of TRIM32 increases proliferation of NPCs, while decreasing neuronal differentiation [10,11]. Rare copy number variation analysis has shown that the loss of *TRIM32* gene is strongly associated with autism and attention deficit hyperactivity disorder [12,13]. Consistent with this finding, TRIM32 knockout mice exhibit ASD-like behaviors and hyperexcitability, accompanied with decreased numbers of interneurons in the telencephalon [14]. These studies indicate that TRIM32, as an essential modulator in NPCs, play import roles in maintenance of cortical development. In this paper, we describe TRIM32's role in the production of pyramidal neurons in developing cortex. Absence of TRIM32 leads to a smaller size of brain. TRIM32-deficient mice exhibit decreased numbers of both upper- and deep-layer cortical neurons, accompanied with a reduced proliferation of both RGCs and IPCs. According to the present study, loss of TRIM32 impairs the generation of cortical pyramidal neurons by reducing the size of the pool of NPCs.

2. Materials and Methods

2.1. Mice

TRIM32 KO mice (TRIM32$^{-/-}$) were kindly provided by Professor Jens C. Schwamborn from the University of Luxembourg. The mouse uses the BGA355 embryonic parent cell line to capture and insert a 5 kb genomic fragment into the second exon of TRIM32 with a SA-IRES-bgeopA expression cassette. This mouse background is a 129SvEvBrd X C57 BL/6 heterozygote and then reverts to C57 BL back/6 precursors for more than 8 generations [15]. Adult mice were reared on an adequate supply of food and water for 12 h. Day-night light cycle at 25 °C. Mice used in this study were from heterozygous breeding pairs (TRIM32$^{+/-}$) and TRIM32$^{+/+}$ littermates were used as controls. All animal experiments were carried out following the Institutional Animal Care and Use Committee of Soochow University.

2.2. Genotyping Detection of TRIM32 Mice

Before using the animals in different experiments, TRIM32 mice were genotyped. We used the following primer sequences to detect genotyping: TRIM32 WT1 (5′-3′): GGAGA-GACACTATTTCCTAAGTCA; TRIM32 WT2 (5′-3′): GTTCAGGTGAGAAGCTGCTGCA; TRIM32 Mu (5′-3′): GGGACAGGATAAGTATGACATCA. The primer pairs WT1 and WT2 and WT1 and Mu were used to designate WT and knockout mice, respectively. The PCR reaction conditions were set at 94 °C (5 min) for enzyme activation, 35 cycles of denaturation at 94 °C for 30 s, annealing at 60 °C for 30 s, extension at 72 °C for 30 s, then 72 °C for 7 min and finally at 4 ∘C infinity. The amplified DNA was separated on a 2% agarose gel by electrophoresis using a current of 120 mA for 60 min. WT and knockout bands were detected at 250 and 300 bp, respectively, compared to the standard DNA ladder.

2.3. Calculate the Embryonic Age of Mice

Mice were mated the previous day and examined for vaginal plugs the next morning. Mouse embryos with vaginal embolism were counted as E0.

2.4. Antibodies

The primary antibodies were rabbit TRIM32 antibody (sc-99011, Santa Cruz, Dallas, TX, USA), mouse TRIM32 antibody (SAB1407164, Sigma, St. Louis, MO, USA), rabbit anti-TBR1 (ab3190, Abcam, Cambridge, UK), rabbit anti- BCL11B (ab28448, Abcam, Cambridge, UK), rabbit anti-CUX1 (sc-13024, Santa Cruz, Dallas, TX, USA), chicken anti-PAX6 (AB_528427, Developmental Studies Hybridoma Bank, Lowa, IA, USA), rabbit anti-TBR2 (ab23345, Abcam, Cambridge, UK), Click-iT™ EdU imaging kit (C10086, Thermo Fisher, Rockford, IL, USA), rabbit anti-PH3 (9713P, Cell Signaling Technology, Boston, MA, USA), rabbit anti-active caspase-3 (C8487, Sigma, St. Louis, MO, USA). The corresponding secondary antibodies conjugated with Alexa fluorophores 488/555/647 (A21202, A31570, A28181, A21206, A31572, A32795, A11039) were from Invitrogen.

2.5. Immunofluorescence Staining and Image Analysis

Mice were perfused with ice-cold PBS followed by 4% paraformaldehyde and post-fixed in 4% paraformaldehyde for 4 h followed by dehydration in 10%, 20%, and 30% sucrose, respectively. TRIM32$^{-/-}$ and TRIM32$^{+/+}$ mouse cortex or subventricular zone slices with the same anatomical position were taken for staining. Cryosections were washed three times with PBS containing 0.3% Triton X-100, 10 min each, and then non-specific binding sites were blocked with 10% BSA and goat (or donkey) serum at 10% for 1 h, then overnight incubated with primary antibody at 4 °C. Sections were washed three times in PBS and incubated with appropriate fluorescent secondary antibodies for 2 h at room temperature, then washed three times with PBS and mounted in mounting medium containing DAPI FluoromountG® (010020, Southern Biotech, Birmingham, AL, USA). The stained sections were examined with a confocal laser scanning microscope LSM 700 (Zeiss, Oberkochen, Germany). The same site and area (50 μm × 300 μm) were taken in the stained section and the number of marker cells was calculated. Images were taken with a confocal microscope. For each mouse genotype, we selected anatomical sections from at least three mice for serial analysis. The area of DAPI$^+$, TBR1$^+$, BCL11B$^+$, CUX1$^+$, PAX6$^+$, TBR2$^+$, Edu$^+$, PH3$^+$, active Caspase3$^+$, or Edu$^+$/TBR1$^+$ cells in each image was quantified using Image J software as described [14].

2.6. Edu Pulse Chase

2.6.1. The Ability of NPCs Proliferation

Pregnant mice were injected intraperitoneally with Edu (50 mg/kg body weight) at E14.5 and E16.5 and the offspring were sacrificed after 30 min. Coronal brain sections were taken for Edu staining to analyse cell proliferation.

2.6.2. The Ability of NPCs to Differentiate into TBR1 Positive Neurons

The pregnant mice were injected intraperitoneally with Edu (50 mg/kg body weight) at E13.5, and the offspring were sacrificed at E18.5. Coronal brain sections were stained for Edu and TBR1. The numbers of Edu$^+$TBR1$^+$ cells were quantified.

2.7. Sample Preparation and RNA-seq Analysis

Under anaesthesia with sodium pentobarbital (80 mg/kg) administered by intraperitoneal injection, WT and TRIM32 KO mice were sacrificed at E18.5 and the brain removed.

Isolated mice were immediately stored in RNAlater® solution (Ambion, Rockford, IL, USA) at 4 °C. Tissue samples were transferred to −80 °C for storage until analysis after 24 h. RNA was extracted with TRIzol® reagent (Life Technologies, Rockford, IL, USA) and RNeasy kit (Qiagen, Dusseldorf, Germany) for RNA sequencing. The R DESeq2 software package was used to analyse differential expression in RNAseq, and the built-in algorithm in DESeq2 was used for normalization. Pairwise comparisons between the brains of two groups are performed on all genes, and further analysis was performed with a fold change >2 to extract the differentially expressed genes ($p < 0.05$). A total of 39 upregulated genes and 36 down-regulated genes were obtained. KEGG enrichment was performed by the DA-VID webtool (https://david.ncifcrf.gov/home.jsp) on 28 December 2021. The protein-protein interaction network was done by the GeneMANIA webtool (http://genemania.org/) on 14 January 2022.

2.8. Statistical Analysis

Values are presented as mean $\pm$ SEM. Data were analysed using Student's *t*-test using SPSS 18.0 software (SPSS, Chicaco, IL, USA). Significance in differences was accepted at $p < 0.05$. * $p < 0.05$; ** $p < 0.01$; *** $p < 0.001$.

3. Results

3.1. The Expression Pattern of TRIM32 in the Cortex during Embryo

According to previous studies, TRIM32 is expressed by NPCs, including interneuron progenitors within the developing ventricular zone [10,11,14]. In order to determine which types of NPCs express TRIM32, we immunostained for the markers Sox2 (a marker for NPCs), Pax6 (an indicator of radial glial cells) or Tbr2 (an indicator of IPCs) in the dorsal subventricular zone (dSVZ) of E14. 5. It was found that both Sox2$^+$ and Pax6$^+$ cells expressed TRIM32. In contrast, only a few Tbr2$^+$ IPCs expressed TRIM32 (Figure 1A). TRIM32 is primarily expressed in RGCs rather than IPCs in the dSVZ, as shown by these results. In dSVZ cells, TRIM32 was detected in both the cytoplasm and nucleus of RGCs and IPCs. Whereas TRIM32 exhibited nuclear location in the cortical neurons labelled with TBR1, a marker of layer VI neurons, in the cortical plate (CP) at E13.5 and 18.5. However, TRIM32 was transiently expressed in a population of cells below the layer VI cortex at E16.5, where it was located in the cytoplasm (Figure 1B). The dynamic expression of TRIM32 in developing brains suggests it may be an essential regulator of brain development.

3.2. TRIM32 Deficiency Mice Exhibit Reduced Size of Brain

At E18.5 (Figure 2A), when neurons are complete, TRIM32$^{-/-}$ mice displayed a smaller brain [16]. To measure the width of the neocortex and cortical plate in the cortex, we used DAPI immunofluorescence staining and Image J to measure the width. According to the results, the width of the neocortex (Figure 2B,C) and the cortical plate (Figure 2D,E) was decreased in the cortex of E18.5 TRIM32$^{-/-}$ mice. These results indicate that TRIM32 regulated the size of developing brain.

3.3. TRIM32 Deficiency Results in Reduced Generation of Cortical Neurons in Developing Cortex

Further analysis examined whether deficiency of TRIM32 causes abnormal numbers of pyramidal neurons at different developmental stages using these neuronal markers. TBR1 is expressed in layer VI neurons, BCL11B is expressed in layer V neurons, CUX1 is expressed in layer II-IV neurons [2]. The results showed that layer VI Tbr1$^+$ cells were reduced in the cortex of TRIM32$^{-/-}$ mice at E14.5, E16.5 and E18.5 (Figure 3A,B). Layer V BCL11b$^+$ cells were downregulated in the cortex of TRIM32$^{-/-}$ mice at E16.5 and E18.5 (Figure 3C,D), while layer II-IV CUX1$^+$ neurons in the cortex of TRIM32$^{-/-}$ mice were downregulated at E18.5 and P30 (Figure 3E,F). To further confirm that the decreased number of layer-specific cortical neurons is due to impaired cortical neuron generation, we performed a pulse tracking experiment to label newborn layer V cortical neurons. Pregnant mice were injected intraperitoneally with Edu at E13.5 and mice were scarified at E18.5 and immunostained for Edu and TBR1. Quantitative analysis showed that Edu$^+$ had lower numbers of TBR1$^+$ in TRIM32$^{-/-}$ VZ/SVZ compared to TRIM32$^{+/+}$ mice (Figure 4A,B). Thus, these results suggest that the absence of TRIM32 leads to reduced formation of pyramidal neurons during brain development. Pyramidal neurons are decreased until adulthood, when brain development is completed, excluding the possibility that TRIM32 deficiency delays the generation of cortical neurons without altering the overall number.

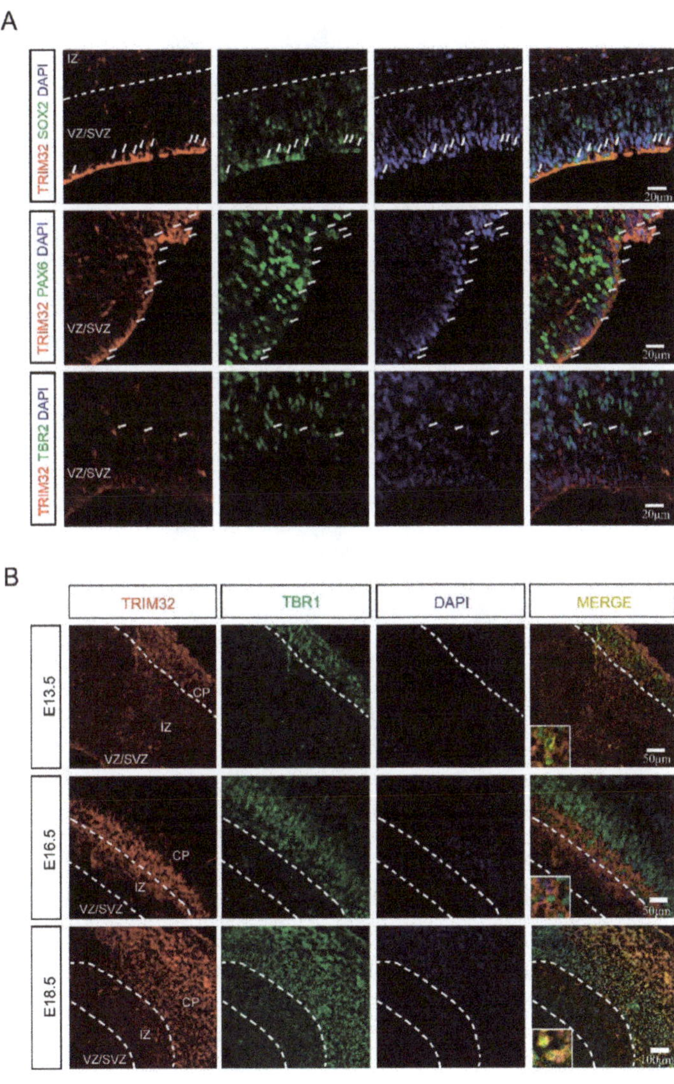

Figure 1. The expression pattern of TRIM32 in the developing cortex. (**A**) The coronal sections of telencephalon from E14.5 mice were immunostained for TRIM32, SOX2/PAX6/TBR2 and DAPI. Scale bars = 20 μm. (**B**) The coronal sections of telencephalon from E13.5, E16.5 and E18.5 mice were immunostained for TRIM32, TBR1 and DAPI. Scale bars = 50 μm (upper and middle panel); Scale bar = 100 μm (bottom panel).

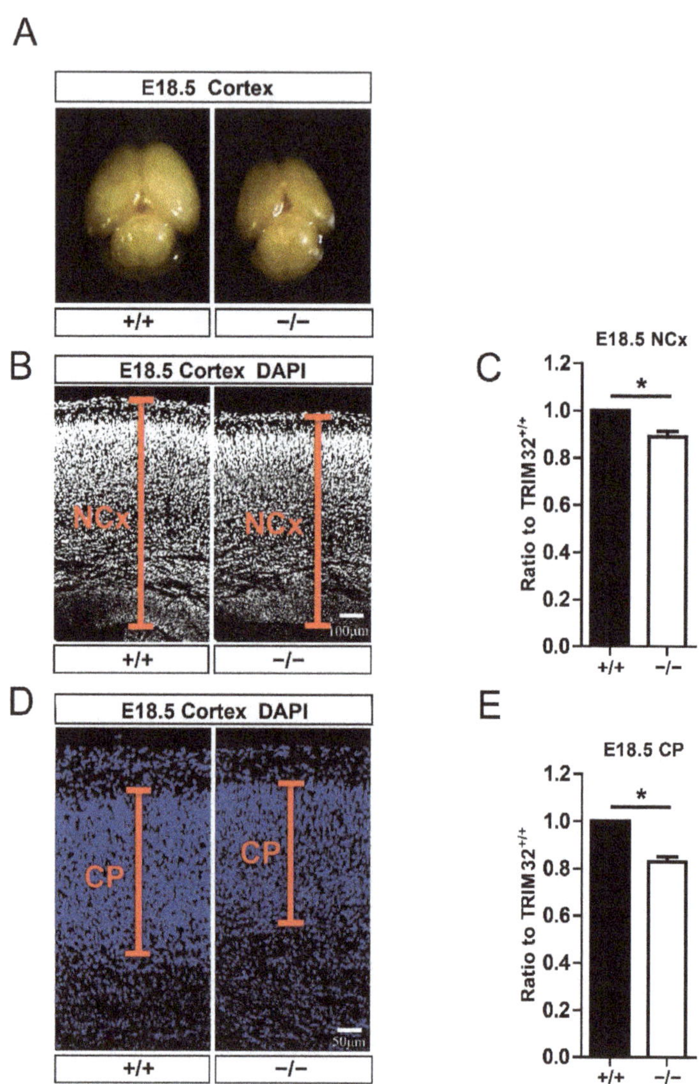

Figure 2. Brain size and cortical width of E18.5 TRIM32$^{-/-}$ mice and TRIM32$^{+/+}$ littermates. (**A**) Representative images of brain size. (**B,D**) The dorsal telencephalon was stained for DAPI. The width of neocortex (NCx) and cortical plate (CP) was indicated. Scale bars = 100 μm. (**C,E**) The relative width of TRIM32$^{-/-}$ neocortex and cortical plate. The width of TRIM32$^{+/+}$ was normalized to 1.0. n = 15 slices for 3 mice/genotype. Scale bars = 50 μm. * $p < 0.05$.

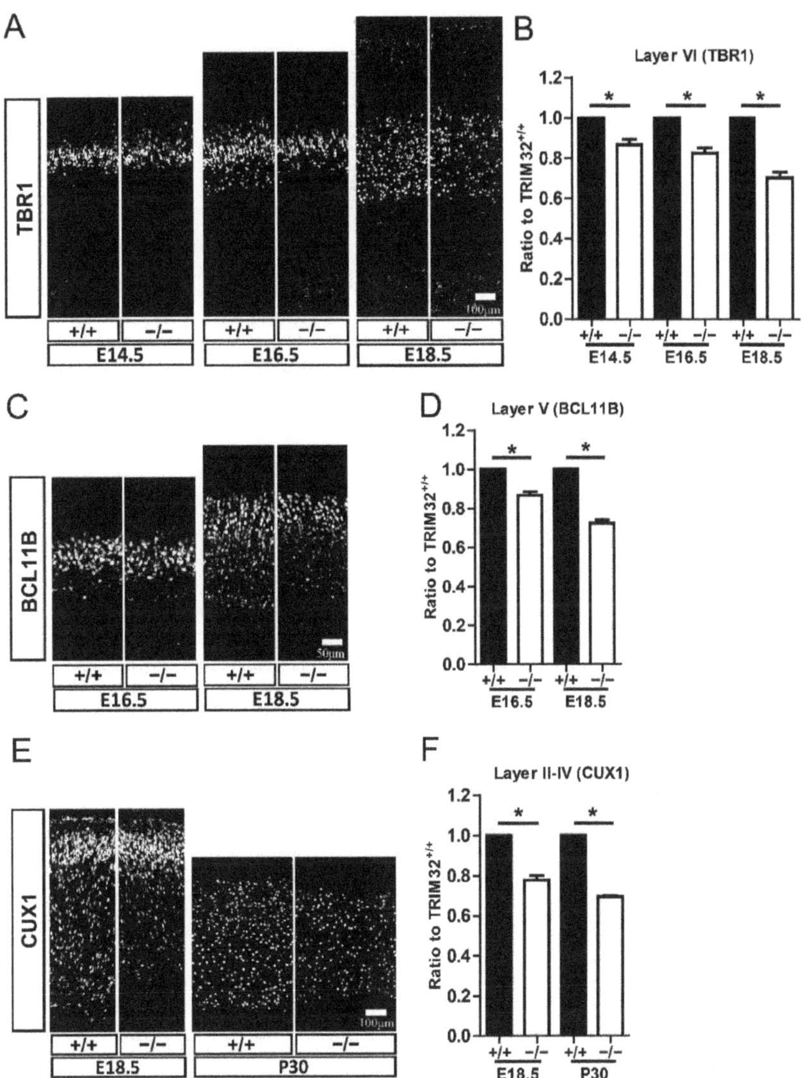

Figure 3. TRIM32$^{-/-}$ mice exhibited decreased numbers of both deep-and upper-layer cortical neurons. The coronal sections of cerebral cortex at distinct developmental stages as indicated were immunostained for TBR1 (**A**), BCL11B (**C**) and CUX1 (**E**). Relative density of TBR1$^+$ (**B**), BCL11b$^+$ (**D**) and CUX1$^+$ (**F**) cells in TRIM32$^{-/-}$ brains. The density of above cells in TRIM32$^{+/+}$ brains were normalized to 1.0. n = 15 slices for 3 mice/genotype. Scale bars = 100 μm (**A**,**E**); Scale bars = 50 μm (**C**). $^*p < 0.05$.

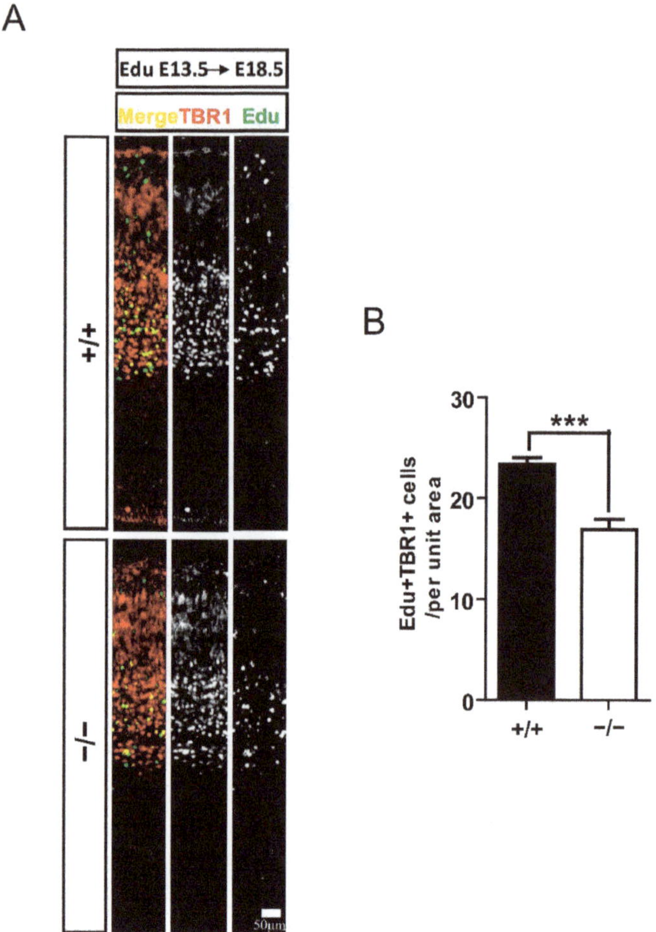

Figure 4. Impaired generation of layer VI cortical neurons in TRIM32$^{-/-}$ mice. Edu were injected intraperitoneally at E13.5. The mice were scarified at E18.5. The coronal sections were immunostained for Edu and TBR1 (**A**). Density of EdU$^+$TBR1$^+$ cells (**B**). n = 15 slices for 3 mice/genotype. Scale bars= 50 μm. *** $p < 0.001$.

3.4. TRIM32 Deficiency Causes Smaller Size of Neural Progenitor Pool

Since TRIM32 deficiency resulted in a decreased number of cortical neurons, we wonder if such an observation is caused by an altered number of progenitor cells. IPC in the dorsal subventricular zone (dSVZ) [17,18]. The latter can also directly generate cortical pyramidal neurons in all layers [19–21].

Therefore, we examined whether the number of RGCs and IPCs was altered by the absence of TRIM32 by immunostaining for PAX6 and TBR2, which label RGCs and IPCs, respectively. Results showed that TRIM32$^{-/-}$ mice had decreased numbers of PAX6$^+$ RGC (Figure 5A,B) and TBR2$^+$ IPC (Figure 5C,D) in VZ/SVZ at E14.5 and E16.5 compared to TRIM32$^{+/+}$ littermates, suggesting that TRIM32 deficiency results in a smaller neural progenitor pool size.

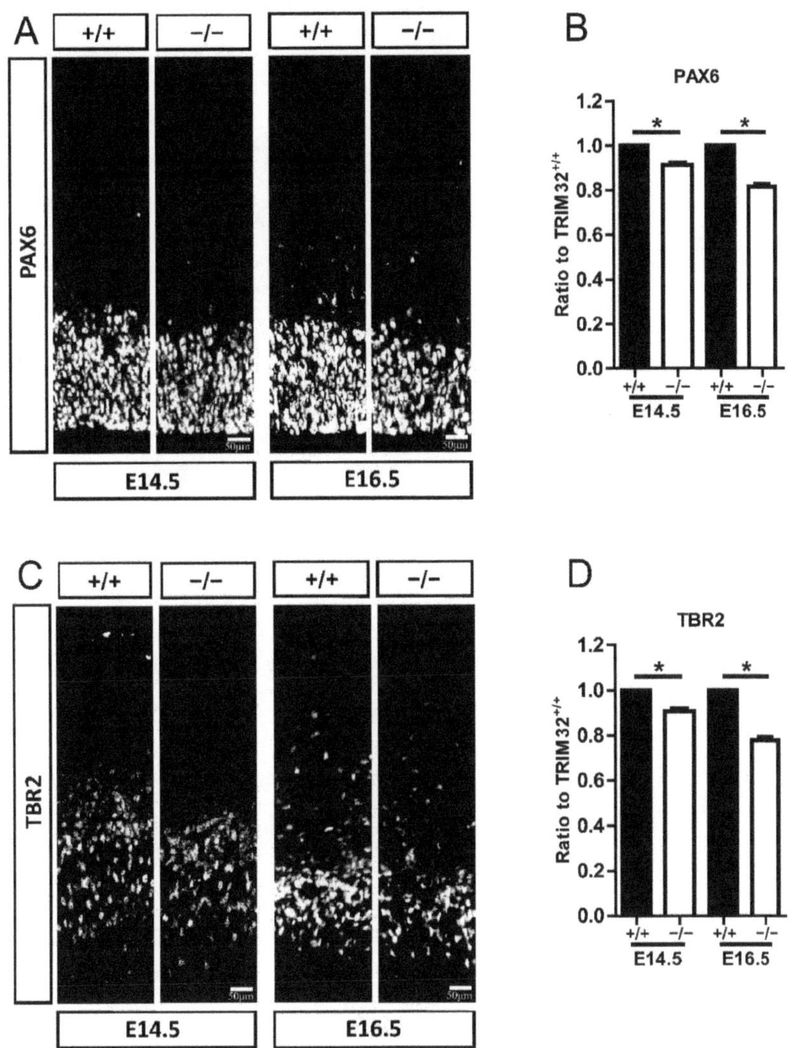

Figure 5. TRIM32$^{-/-}$ mice exhibited decreased numbers of RGCs and IPCs. The coronal sections of cerebral cortex at distinct E14.5 and E16.5 were immunostained for either PAX6 (**A**) or TBR2 (**C**). Relative density of PAX6^{+} (**B**) and TBR2^{+} cells (**D**) in TRIM32$^{-/-}$ brains. The density of above cells in TRIM32$^{+/+}$ brains were normalized to 1.0. n = 15 slices for 3 mice/genotype. Scale bars = 50 µm. * $p < 0.05$.

3.5. TRIM32 Deficiency Decreases Proliferation and Mitosis of Both RGCs and IPCs

To further investigate the mechanisms underlying the smaller RGC and IPC cluster size caused by the absence of TRIM32, we used Edu to target NPCs in the VZ/SVZ zone at E14.5 and E16.5, respectively to mark. After 30 min, the offspring were sacrificed. Coronal brain sections were taken for Edu staining to analyse NPC proliferation. We found that Edu^{+} cells were reduced in TRIM32$^{-/-}$ VZ/SVZ at E14.5 and E16.5, respectively, compared to TRIM32$^{+/+}$ mice (Figure 6A,B). We also compare the mitotic NPCs in the basal and apical VZ, corresponding to RGC and IPC, respectively. PH3 (mitosis-specific marker) positive cells in both basal VZ and apical VZ showed decreased numbers at E14.5 (Figure 6C,D).

In contrast, only PH3$^+$ cells in the apical VZ show reduced numbers in E16.5 TRIM32$^{-/-}$ mice compared to those in TRIM32$^{+/+}$ littermates (Figure 6C,E), indicating reduced mitosis of both RGCs and of IPC in developing TRIM32$^{-/-}$ brains. These results indicate that TRIM32 deficiency leads to reduced self-renewal in NPCs.

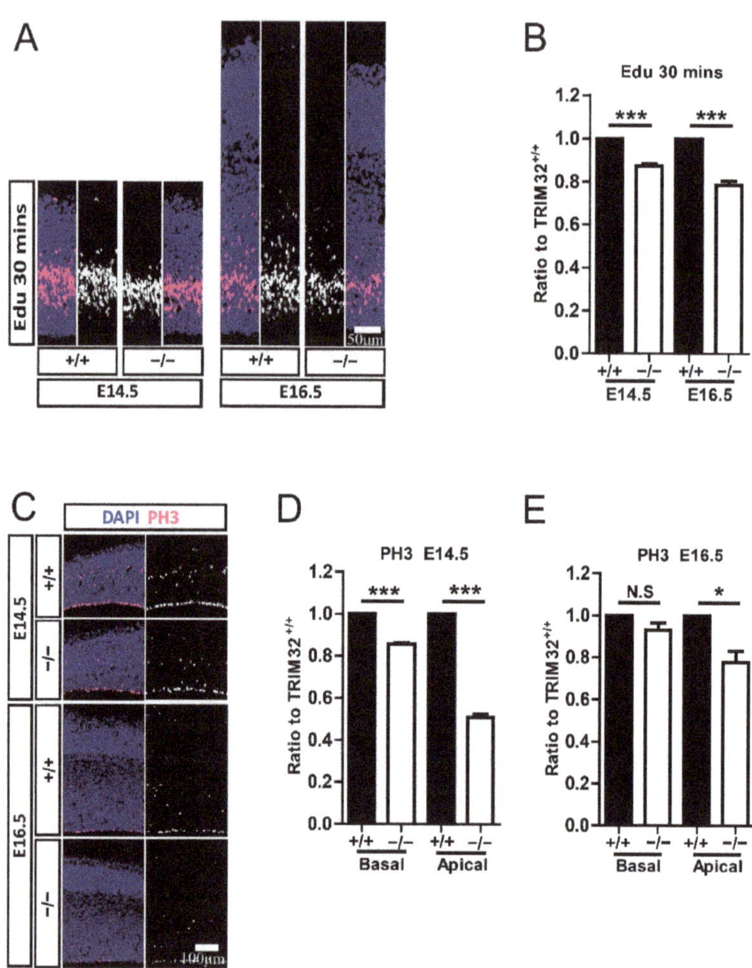

Figure 6. TRIM32$^{-/-}$ mice exhibited decreased proliferation and mitosis. The mice were scarified 30 min after being injected intraperitoneally with Edu at either E14.5 or E16.5. The coronal cortical sections were immunostained for either Edu (**A**) or PH3 (**C**) and DAPI. Relative density of Edu$^+$ (**B**) or PH3$^+$ cells (**D,E**) in TRIM32$^{-/-}$ brains. The density of above cells in TRIM32$^{+/+}$ brains were normalized to 1.0. n = 15 slices for 3 mice/genotype. Scale bars = 50 μm (**A**); Scale bars = 100 μm (**C**). * $p < 0.05$; *** $p < 0.001$; N.S: no significance.

3.6. TRIM32 Deficiency Does Not Affect Apoptosis in Developing Cortex

Apoptosis is an essential mechanism in regulating the number of cortical neurons and NPCs. The brain tends to produce excessive numbers of neurons during development, and some neurons undergo apoptosis at a later stage, ultimately determining the number of neurons in the brain. It has also been reported that the increase in neurons caused by decreased apoptosis causes neuropsychiatric diseases [22,23]. To investigate whether apoptosis is responsible for the reduced numbers of cortical neurons and NPCs, we immunostained Caspase3 to label apoptotic cells in the developing cerebrum. TRIM32$^{-/-}$ mice and their wild-type littermates displayed identical numbers of caspase3$^+$ cells in the dorsal telencephalon (Figure 7). Thus, TRIM32 deficiency does not affect the apoptosis of cortical neurons and NPCs in the developing cortex. The decreased number of cortical neurons and NPCs observed in TRIM32/brains was not due to increased apoptosis.

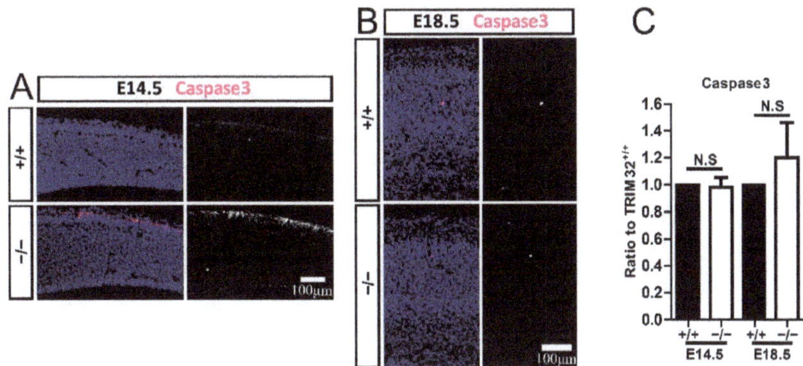

Figure 7. TRIM32 deficiency does not affect apoptosis. The coronal cortical sections were immunostained for Caspase3 and DAPI at E14.5 and E18.5 (**A,B**). Relative density of Caspase3$^+$ cells (**C**) in TRIM32$^{-/-}$ brains. The density of above cells in TRIM32$^{+/+}$ brains were normalized to 1.0. $n = 15$ slices for 3 mice/genotype. Scale bars = 100 μm. N.S: no significance.

3.7. Downstream Signalling Regulated by TRIM32

The mammalian target of rapamycin (mTOR) signalling pathway plays an essential role in brain development [14,24]. mTOR, when overactivated, is one of the central pathways in the etiology of ASD [25,26]. TRIM32 maintained mTOR activity by promoting proteosomal degradation of G protein signalling protein 10 (RGS10). The absence of TRIM32 leads to suppression of mTOR signalling, which in turn leads to increased autophagy, followed by accelerated c-myc degradation [14]. TRIM32 regulates the formation of GABAergic interneurons via this signalling pathway. To further identify additional signalling pathways regulated by the absence of TRIM32, we performed RNAseq analysis on the brains of WT and TRIM32$^{-/-}$ mice. The results showed that Notch signalling was downregulated in TRIM32$^{-/-}$ mice (Figure 8A). Stem cells and neuronal migration in embryonic and adult brain neurogenesis [27]. mTOR is a positive regulator of Notch signalling in human and mouse cells that acts through induction of the STAT3/p63/Jagged signalling cascade [28]. In the downregulated protein interaction diagram, the genes directly related to the Notch pathway are deltx E3 ubiquitin ligase (DTX2) and mastermind-like transcription coactivator 3 (MAML3) (Figure 8B). We were more interested in genes like homeobox A1 (HOXA1) and MAML3 (Figure 8B,C). HOXA1 has been reported to play a crucial role in multiple biological processes and can mediate gene expression and cell differentiation [29,30]. In addition, HOXA1 could be regulated by the AKT/mTOR signalling pathway and the Notch1 signalling pathway [31,32]. MAML3 was part of the Notch1-containing ternary complex in vivo [33], and has been reported to act as a transcriptional coactivator of NOTCH2. Heynen et al. reported that the results indicate an important mechanistic role for MAML3 in retinoic acid-mediated proliferation and differentiation [34]. DTX2 is a homologue of DTX1. The latter regulates Notch signalling in a Su(H)/RBPJ-independent signalling pathway [35]. DTX2 activity is required for neural crest formation, indicating its role in neural development. In addition to Notch signalling, DTX2 also regulates BMP signalling, which plays an essential role in cortical development [35,36]. In conclusion, HOXA1, DTX2 and MAML3 may be involved in mTOR or Notch signalling to affect pyramidal neuron generation. Considering that the interacting proteins can be direct or indirect, this is not clear yet, so we will check these issues in the next investigations.

A

Term	PValue	Fold Enrichment	FDR
mmu04330:Notch signaling pathway	0.049869	34.87982	0.448818
mmu04740:Olfactory transduction	0.312444	2.373765	1
mmu04120:Ubiquitin mediated proteolysis	1	6.017997	1
mmu05200:Pathways in cancer	1	2.163432	1
mmu04614:Renin-angiotensin system	1	24.41587	1
mmu05143:African trypanosomiasis	1	24.41587	1
mmu05220:Chronic myeloid leukemia	1	11.86883	1
mmu00360:Phenylalanine metabolism	1	37.15459	1
mmu04010:MAPK signaling pathway	1	3.404604	1

B

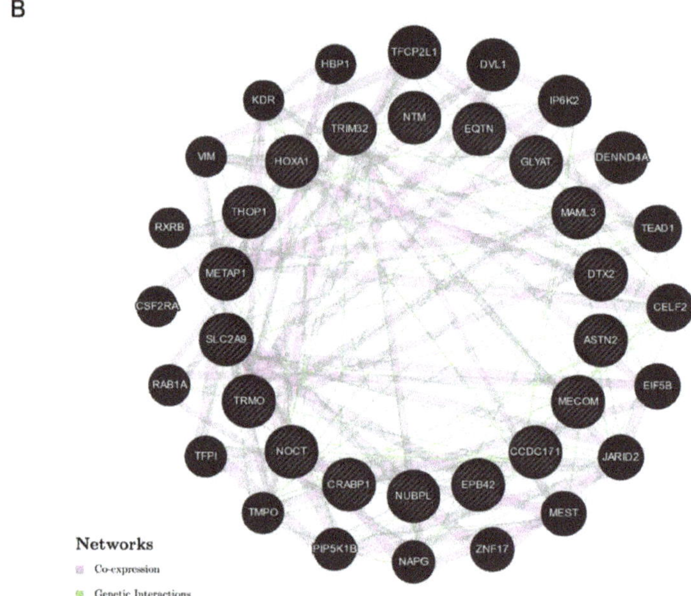

C

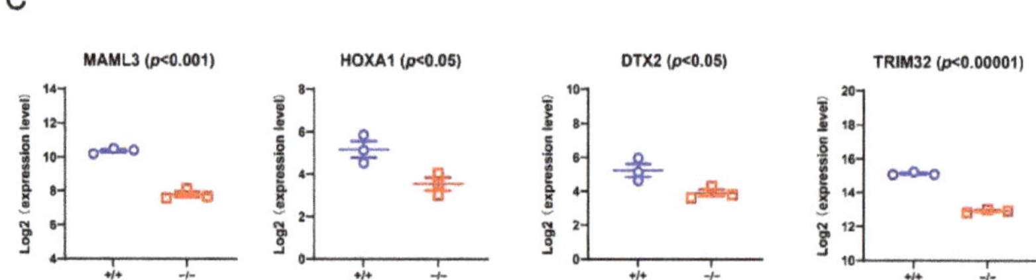

Figure 8. TRIM32 deficiency impairs Notch signalling pathway. (**A**) The enriched pathways for downregulated genes in TRIM32$^{-/-}$ mice. (**B**) The downregulated protein interaction diagram. The inner circle represents all identifiable down-regulated genes, and the outer circle represents the genes that might be predicted to interact with these down-regulated genes. The types of interactions between genes are illustrated. (**C**) Expression levels of MAML3, HOXA1, DTX2, and TRIM32 were compared in TRIM32$^{+/+}$ mice (+/+) and TRIM32$^{-/-}$ mice (−/−).

4. Discussion

The imbalance between excitability and inhibitory neural activity (E/I) of brain circuits is a fundamental mechanism for the pathogenesis of ASD. The E/I imbalance could be caused by abnormal generation of GABAergic or pyramidal neurons or their dysfunction. The present study shows that *TRIM32*, an ASD risk gene [12], is an important regulator in the formation of cortical pyramidal neurons. This observation helps explain the E/I imbalance observed in TRIM32$^{-/-}$ mice [37]. MGE-derived GABAergic interneurons begin to produce at day E9.5 and peak at day E13.5 [38]. Pyramidal neurons are mainly produced during the period from E11 to E17 [16]. Interestingly, TRIM32 is expressed in both RGCs located in the dorsal VZ and in interneuron progenitors in L/MGE. Consistent with our previous observation that TRIM32 deficiency causes the loss of multiple subtypes of GABAergic interneurons in the adult brain, the present study observed the loss of pyramidal neurons in the cortex of 1-month-old TRIM32$^{-/-}$ mice as the neuron excitatory pyramidal cells begin their generation and development. Cortical pyramidal neurons are obviously not enough to compensate for the loss of GABAergic interneurons. In addition, the density of neurons can affect the development of neurons, for example, the density of dendritic spines, the arrangement of synapses. In line with this point, indeed, note the dysfunction of TRIM32-deficient synapses, which also exhibited hyperexcitability [37]. Therefore, all these abnormal processes can eventually lead to an E/I imbalance, which in turn causes autism-like behaviours in TRIM32$^{-/-}$ mice [39].

In terms of the cellular mechanism, we have further observed that the absence of TRIM32 affects the self-renewal of both RGCs and IPCs, consistent with the findings that TRIM32 deficiency results in reduced neural proliferation both in vivo and in vitro progenitor cells, including L/MGE parents. Research has shown that some children with ASD have abnormal brain size [40]. Interestingly, in the current study, poor self-renewal in NPCs, including RGC, IPC and L/MGE parents, causes smaller NPC pool size, ultimately resulting in smaller brain size. Using the pulse chase experiment, we also confirmed decreased neuronal differentiation caused by TRIM32 deficiency. Thus, impaired self-renewal coupled with decreased neuronal differentiation contributes to the decreased number of pyramidal neurons in TRIM32-deficient mice.

In terms of the molecular mechanism, our previous results suggest that TRIM32 maintains mTOR activity by promoting RGS10 degradation by proteasomes. Increased mTOR-mediated autophagy promotes c-myc degradation and impairs proliferation of TRIM32-deficient L/MGE precursors [14]. Since TRIM32-deficient RGCs share a similar cellular property to TRIM32-deficient L/MGE progenitors, we propose that these two different progenitors share a similar molecular mechanism. To identify additional downstream signalling pathways caused by TRIM32 deficiency, we performed RNAseq analysis embryonic cortex with TRIM32 deficiency. Notch signalling plays a critical role in neural stem cell maintenance and neurogenesis in both the embryonic and adult brain [27]. Notch is also a substrate for autophagy. Its autophagic degradation is required in stem cell development and neurogenesis [41]. In the down-regulated protein interaction diagram, we identified HOXA1 and MAML3 proteins. HOXA1 acted as a DNA-binding transcription factor that mediated cell differentiation [29,30]. It is worth noting that the Hoxa1 gene is genetically connected to ASD [42–44], and Hoxa1 is involved in the proliferation of MIR99 cells by the AKT/mTOR signalling [31]. MAML3 was recognized in vivo as part of the Notch1-containing ternary complex [33], and acted as a transcriptional coactivator for NOTCH2. MAML3 showed an important role in the proliferation and differentiation mechanism [34]. The downregulated protein Rab1A, a conserved small guanosine triphosphatase (GTPase), predominantly regulates the transport of vesicular proteins from the endoplasmic reticulum (ER) to the Golgi apparatus and is involved in mediating Notch signalling, adhesion cell and cell migration [45,46]. Therefore, HOXA1 and MAML3 may be involved in mTOR or Notch signalling to affect pyramidal neuron generation, which will be explored further in the future.

5. Conclusions

In summary, this study demonstrates the effects of TRIM32 on the generation of pyramidal neurons in embryonic development and explores their mechanism. In TRIM32$^{-/-}$ mice, NCx and CP were decreased in the cortex, cortex size was decreased, and the number of pyramidal neurons was also decreased. TRIM32 deficiency decreased RGC and IPC cells, proliferating and mitotic neural progenitor cells in VZ/SVZ, decreased cell migration and differentiation in TBR1 positive neuron in cortex, but does not affect TRIM32 apoptosis regulates the generation of cortical pyramidal neurons cells influencing the proliferation, migration and differentiation of RGCs or IPCs and may pass through the mTOR or Notch signalling pathway. The completion of this research provides a theoretical basis for elucidating the mechanism of generation of pyramidal neurons in the embryonic period and provides strong data support for exploring the onset and development of related neurodevelopmental diseases.

Author Contributions: Conceptualization, Q.-H.M. and R.-X.X.; methodology, Y.-Y.S. and W.-J.C.; validation, Z.-P.H., G.Y. and M.-L.W.; formal analysis, D.-E.X.; investigation, Y.-Y.S., W.-J.C., Z.-P.H., G.Y., M.-L.W., D.-E.X. and W.-L.Y.; resources, Q.-H.M. and R.-X.X.; data curation, Y.-Y.S. and W.-J.C.; writing—original draft preparation, Y.-Y.S.; writing—review and editing, D.-E.X., W.-L.Y., Y.-C.L., Z.-C.X., R.-X.X. and Q.-H.M.; supervision, Q.-H.M. and R.-X.X.; funding acquisition, Q.-H.M. All authors have read and agreed to the published version of the manuscript.

Funding: This work was supported by National Natural Science Foundation of China (92049120, 81870897, 81901296, 81971295, 81671189), Natural Science Foundation of Jiangsu Province (BK20181436), Team Innovation Funding Program of Second Affiliated Hospital of Soochow University (XKTJ-TD202003), Guangdong Key Project in "Development of new tools for diagnosis and treatment of Autism" (2018B030335001), translational medicine fund of WuXi municipal health commission (ZM010), Priority Academic Program Development of Jiangsu Higher Education Institution (PAPD), Jiangsu Provincial Medical Key Discipline Project (ZDXKB2016022) and Jiangsu Key Laboratory of Translational Research and Therapy for Neuro-Psycho-Diseases (BM2013003), Military Medical Science Research Project (16CXZ001).

Institutional Review Board Statement: The study was conducted according to the guidelines of the Declaration of Helsinki and approved by the Institutional Animal Care and Use Committee of Soochow University (No. 201809A100).

Informed Consent Statement: Not applicable.

Data Availability Statement: All data generated or analysed during this study are available from the corresponding author upon reasonable request.

Acknowledgments: We thank Jens C. Schwamborn from University of Luxembourg for generously providing TRIM32 deficient mice, and Diji Kuriakose from Monash University for suggestions on the manuscript.

Conflicts of Interest: The authors declare no conflict of interest.

References

1. Lee, E.; Lee, J.; Kim, E. Excitation/Inhibition Imbalance in Animal Models of Autism Spectrum Disorders. *Biol. Psychiatry* **2017**, *81*, 838–847. [CrossRef] [PubMed]
2. Wu, Z.Q.; Li, D.; Huang, Y.; Chen, X.P.; Huang, W.; Liu, C.F.; Zhao, H.Q.; Xu, R.X.; Cheng, M.; Schachner, M.; et al. Caspr Controls the Temporal Specification of Neural Progenitor Cells through Notch Signaling in the Developing Mouse Cerebral Cortex. *Cereb. Cortex* **2017**, *27*, 1369–1385. [CrossRef] [PubMed]
3. Seuntjens, E.; Nityanandam, A.; Miquelajauregui, A.; Debruyn, J.; Stryjewska, A.; Goebbels, S.; Nave, K.A.; Huylebroeck, D.; Tarabykin, V. Sip1 regulates sequential fate decisions by feedback signaling from postmitotic neurons to progenitors. *Nat. Neurosci.* **2009**, *12*, 1373–1380. [CrossRef] [PubMed]
4. Chen, J.L.; Nedivi, E. Highly specific structural plasticity of inhibitory circuits in the adult neocortex. *Neuroscientist* **2013**, *19*, 384–393. [CrossRef]
5. Borrell, V.; Gotz, M. Role of radial glial cells in cerebral cortex folding. *Curr. Opin. Neurobiol.* **2014**, *27*, 39–46. [CrossRef]
6. Nadarajah, B.; Parnavelas, J.G. Modes of neuronal migration in the developing cerebral cortex. *Nat. Rev. Neurosci.* **2002**, *3*, 423–432. [CrossRef]

7. Reymond, A.; Meroni, G.; Fantozzi, A.; Merla, G.; Cairo, S.; Luzi, L.; Riganelli, D.; Zanaria, E.; Messali, S.; Cainarca, S.; et al. The tripartite motif family identifies cell compartments. *EMBO J.* **2001**, *20*, 2140–2151. [CrossRef]

8. Kudryashova, E.; Kudryashov, D.; Kramerova, I.; Spencer, M.J. Trim32 is a ubiquitin ligase mutated in limb girdle muscular dystrophy type 2H that binds to skeletal muscle myosin and ubiquitinates actin. *J. Mol. Biol.* **2005**, *354*, 413–424. [CrossRef]

9. Hillje, A.L.; Beckmann, E.; Pavlou, M.A.S.; Jaeger, C.; Pacheco, M.P.; Sauter, T.; Schwamborn, J.C.; Lewejohann, L. The neural stem cell fate determinant TRIM32 regulates complex behavioral traits. *Front. Cell. Neurosci.* **2015**, *9*, 75. [CrossRef]

10. Hillje, A.L.; Worlitzer, M.M.A.; Palm, T.; Schwamborn, J.C. Neural Stem Cells Maintain Their Stemness through Protein Kinase C zeta-Mediated Inhibition of TRIM32. *Stem Cells* **2011**, *29*, 1437–1447. [CrossRef]

11. Schwamborn, J.C.; Berezikov, E.; Knoblich, J.A. The TRIM-NHL Protein TRIM32 Activates MicroRNAs and Prevents Self-Renewal in Mouse Neural Progenitors. *Cell* **2009**, *136*, 913–925. [CrossRef] [PubMed]

12. Lionel, A.C.; Tammimies, K.; Vaags, A.K.; Rosenfeld, J.A.; Ahn, J.W.; Merico, D.; Noor, A.; Runke, C.K.; Pillalamarri, V.K.; Carter, M.T.; et al. Disruption of the ASTN2/TRIM32 locus at 9q33.1 is a risk factor in males for autism spectrum disorders, ADHD and other neurodevelopmental phenotypes. *Hum. Mol. Genet.* **2014**, *23*, 2752–2768. [CrossRef] [PubMed]

13. Lionel, A.C.; Crosbie, J.; Barbosa, N.; Goodale, T.; Thiruvahindrapuram, B.; Rickaby, J.; Gazzellone, M.; Carson, A.R.; Howe, J.L.; Wang, Z.; et al. Rare copy number variation discovery and cross-disorder comparisons identify risk genes for ADHD. *Sci. Transl. Med.* **2011**, *3*, 95ra75. [CrossRef] [PubMed]

14. Zhu, J.W.; Zou, M.M.; Li, Y.F.; Chen, W.J.; Liu, J.C.; Chen, H.; Fang, L.P.; Zhang, Y.; Wang, Z.T.; Chen, J.B.; et al. Absence of TRIM32 Leads to Reduced GABAergic Interneuron Generation and Autism-like Behaviors in Mice via Suppressing mTOR Signaling. *Cereb. Cortex* **2020**, *30*, 3240–3258. [CrossRef] [PubMed]

15. Nicklas, S.; Otto, A.; Wu, X.L.; Miller, P.; Stelzer, S.; Wen, Y.F.; Kuang, S.H.; Wrogemann, K.; Patel, K.; Ding, H.; et al. TRIM32 Regulates Skeletal Muscle Stem Cell Differentiation and Is Necessary for Normal Adult Muscle Regeneration. *PLoS ONE* **2012**, *7*, e30445. [CrossRef]

16. Mitsuhashi, T.; Takahashi, T. Genetic regulation of proliferation/differentiation characteristics of neural progenitor cells in the developing neocortex. *Brain Dev.* **2009**, *31*, 553–557. [CrossRef]

17. Tarabykin, V.; Stoykova, A.; Usman, N.; Gruss, P. Cortical upper layer neurons derive from the subventricular zone as indicated by Svet1 gene expression. *Development* **2001**, *128*, 1983–1993. [CrossRef]

18. Cubelos, B.; Sebastian-Serrano, A.; Kim, S.; Moreno-Ortiz, C.; Redondo, J.M.; Walsh, C.A.; Nieto, M. Cux-2 controls the proliferation of neuronal intermediate precursors of the cortical subventricular zone. *Cereb. Cortex* **2008**, *18*, 1758–1770. [CrossRef]

19. McConnell, S.K.; Kaznowski, C.E. Cell cycle dependence of laminar determination in developing neocortex. *Science* **1991**, *254*, 282–285. [CrossRef]

20. Price, J.; Thurlow, L. Cell lineage in the rat cerebral cortex: A study using retroviral-mediated gene transfer. *Development* **1988**, *104*, 473–482. [CrossRef]

21. Reid, C.B.; Liang, I.; Walsh, C. Systematic widespread clonal organization in cerebral cortex. *Neuron* **1995**, *15*, 299–310. [CrossRef]

22. Buss, R.R.; Sun, W.; Oppenheim, R.W. Adaptive roles of programmed cell death during nervous system development. *Annu. Rev. Neurosci.* **2006**, *29*, 1–35. [CrossRef] [PubMed]

23. Fang, W.Q.; Chen, W.W.; Jiang, L.; Liu, K.; Yung, W.H.; Fu AK, Y.; Ip, N.Y. Overproduction of upper-layer neurons in the neocortex leads to autism-like features in mice. *Cell Rep.* **2014**, *9*, 1635–1643. [CrossRef] [PubMed]

24. Lipton, J.O.; Sahin, M. The neurology of mTOR. *Neuron* **2014**, *84*, 275–291. [CrossRef]

25. Crino, P.B. The mTOR signalling cascade: Paving new roads to cure neurological disease. *Nat. Rev. Neurol.* **2016**, *12*, 379–392. [CrossRef]

26. Costa-Mattioli, M.; Monteggia, L.M. mTOR complexes in neurodevelopmental and neuropsychiatric disorders. *Nat. Neurosci.* **2013**, *16*, 1537–1543. [CrossRef]

27. Imayoshi, I.; Sakamoto, M.; Yamaguchi, M.; Mori, K.; Kageyama, R. Essential roles of Notch signaling in maintenance of neural stem cells in developing and adult brains. *J. Neurosci.* **2010**, *30*, 3489–3498. [CrossRef]

28. Ma, J.; Meng, Y.; Kwiatkowski, D.J.; Chen, X.; Peng, H.; Sun, Q.; Zha, X.; Wang, F.; Wang, Y.; Jing, Y.; et al. Mammalian target of rapamycin regulates murine and human cell differentiation through STAT3/p63/Jagged/Notch cascade. *J. Clin. Investig.* **2010**, *120*, 103–114. [CrossRef]

29. Maamar, H.; Cabili, M.N.; Rinn, J.; Raj, A. linc-HOXA1 is a noncoding RNA that represses Hoxa1 transcription in cis. *Gene Dev.* **2013**, *27*, 1260–1271. [CrossRef]

30. De Kumar, B.; Parker, H.J.; Paulson, A.; Parrish, M.E.; Zeitlinger, J.; Krumlauf, R. Hoxa1 targets signaling pathways during neural differentiation of ES cells and mouse embryogenesis. *Dev. Biol.* **2017**, *432*, 151–164. [CrossRef]

31. Zhang, L.; Liu, X.L.; Yuan, Z.; Cui, J.; Zhang, H. MiR-99a suppressed cell proliferation and invasion by directly targeting HOXA1 through regulation of the AKT/mTOR signaling pathway and EMT in ovarian cancer. *Eur. Rev. Med. Pharmacol.* **2019**, *23*, 4663–4672.

32. Wu, Q.X.; Lu, S.T.; Zhang, L.; Zhao, L.J. LncRNA HOXA-AS2 Activates the Notch Pathway to Promote Cervical Cancer Cell Proliferation and Migration. *Reprod. Sci.* **2021**, *28*, 3000–3009. [CrossRef] [PubMed]

33. Kitagawa, M. Notch signalling in the nucleus: Roles of Mastermind-like (MAML) transcriptional coactivators. *J. Biochem.* **2016**, *159*, 287–294. [CrossRef] [PubMed]

34. Heynen, G.J.J.E.; Nevedomskaya, E.; Palit, S.; Basheer, N.J.; Lieftink, C.; Schlicker, A.; Zwart, W.; Bernards, R.; Bajpe, P.K. Mastermind-Like 3 Controls Proliferation and Differentiation in Neuroblastoma. *Mol. Cancer Res.* **2016**, *14*, 411–422. [CrossRef] [PubMed]

35. Endo, Y.; Osumi, N.; Wakamatsu, Y. Deltex/Dtx mediates NOTCH signaling in regulation of Bmp4 expression in cranial neural crest formation during avian development. *Dev. Growth Differ.* **2003**, *45*, 241–248. [CrossRef]

36. Gamez, B.; Rodriguez-Carballo, E.; Ventura, F. BMP signaling in telencephalic neural cell specification and maturation. *Front. Cell. Neurosci.* **2013**, *7*, 87. [CrossRef] [PubMed]

37. Ntim, M.; Li, Q.F.; Zhang, Y.; Liu, X.D.; Li, N.; Sun, H.L.; Zhang, X.; Khan, B.; Wang, B.; Wu, Q.; et al. TRIM32 Deficiency Impairs Synaptic Plasticity by Excitatory-Inhibitory Imbalance via Notch Pathway. *Cereb. Cortex* **2020**, *30*, 4617–4632. [CrossRef] [PubMed]

38. Miyoshi, G.; Butt, S.J.B.; Takebayashi, H.; Fishell, G. Physiologically distinct temporal cohorts of cortical interneurons arise from telencephalic olig2-expressing precursors. *J. Neurosci.* **2007**, *27*, 7786–7798. [CrossRef]

39. Sohal, V.S.; Rubenstein JL, R. Excitation-inhibition balance as a framework for investigating mechanisms in neuropsychiatric disorders. *Mol. Psychiatry* **2019**, *24*, 1248–1257. [CrossRef]

40. Fombonne, E.; Roge, B.; Claverie, J.; Courty, S.; Fremolle, J. Microcephaly and macrocephaly in autism. *J. Autism Dev. Disord.* **1999**, *29*, 113–119. [CrossRef]

41. Wu, X.; Fleming, A.; Ricketts, T.; Pavel, M.; Virgin, H.; Menzies, F.M.; Rubinsztein, D.C. Autophagy regulates Notch degradation and modulates stem cell development and neurogenesis. *Nat. Commun.* **2016**, *7*, 10533. [CrossRef] [PubMed]

42. Conciatori, M.; Stodgell, C.J.; Hyman, S.L.; O'Bara, M.; Militerni, R.; Bravaccio, C.; Trillo, S.; Montecchi, F.; Schneider, C.; Melmed, R.; et al. Association between the HOXA1 A218G polymorphism and increased head circumference in patients with autism. *Biol. Psychiatry* **2004**, *55*, 413–419. [CrossRef] [PubMed]

43. Raznahan, A.; Lee, Y.; Vaituzis, C.; Tran, L.; Mackie, S.; Tiemeier, H.; Clasen, L.; Lalonde, F.; Greenstein, D.; Pierson, R.; et al. Allelic Variation Within the Putative Autism Spectrum Disorder Risk Gene Homeobox A1 and Cerebellar Maturation in Typically Developing Children and Adolescents. *Autism Res.* **2012**, *5*, 93–100. [CrossRef] [PubMed]

44. Muscarella, L.A.; Guarnieri, V.; Sacco, R.; Curatolo, P.; Manzi, B.; Alessandrelli, R.; Giana, G.; Militerni, R.; Bravaccio, C.; Lenti, C.; et al. Candidate gene study of HOXB1 in autism spectrum disorder. *Mol. Autism* **2010**, *1*, 9. [CrossRef]

45. Mukhopadhyay, A.; Nieves, E.; Che, F.Y.; Wang, J.; Jin, L.J.; Murray, J.W.; Gordon, K.; Angeletti, R.H.; Wolkoff, A.W. Proteomic analysis of endocytic vesicles: Rab1a regulates motility of early endocytic vesicles. *J. Cell Sci.* **2011**, *124*, 765–775. [CrossRef]

46. Hutagalung, A.H.; Novick, P.J. Role of Rab GTPases in Membrane Traffic and Cell Physiology. *Physiol. Rev.* **2011**, *91*, 119–149. [CrossRef]

Article

NXN Gene Epigenetic Changes in an Adult Neurogenesis Model of Alzheimer's Disease

Idoia Blanco-Luquin [1,*], Blanca Acha [1], Amaya Urdánoz-Casado [1], Eva Gómez-Orte [2], Miren Roldan [1], Diego R. Pérez-Rodríguez [3], Juan Cabello [2] and Maite Mendioroz [1,4]

[1] Neuroepigenetics Laboratory-Navarrabiomed, Hospital Universitario de Navarra (HUN), Universidad Pública de Navarra (UPNA), IdiSNA (Navarra Institute for Health Research), 31008 Pamplona, Spain; blanca.acha.santamaria@navarra.es (B.A.); amaya.urdanoz.casado@navarra.es (A.U.-C.); mroldana@navarra.es (M.R.); maitemendilab@gmail.com (M.M.)

[2] CIBIR (Center for Biomedical Research of La Rioja), 26006 Logroño, Spain; emgomez@riojasalud.es (E.G.-O.); juan.cabello@riojasalud.es (J.C.)

[3] Neurophysiology Department, Hospital Universitario de Navarra (HUN), IdiSNA (Navarra Institute for Health Research), 31008 Pamplona, Spain; drpr25@gmail.com

[4] Department of Neurology, Hospital Universitario de Navarra (HUN), IdiSNA (Navarra Institute for Health Research), 31008 Pamplona, Spain

* Correspondence: iblancol@navarra.es; Tel.: +34-848425739

Abstract: In view of the proven link between adult hippocampal neurogenesis (AHN) and learning and memory impairment, we generated a straightforward adult neurogenesis *in vitro* model to recapitulate DNA methylation marks in the context of Alzheimer's disease (AD). Neural progenitor cells (NPCs) were differentiated for 29 days and Aβ peptide 1–42 was added. mRNA expression of Neuronal Differentiation 1 (*NEUROD1*), Neural Cell Adhesion Molecule 1 (*NCAM1*), Tubulin Beta 3 Class III (*TUBB3*), RNA Binding Fox-1 Homolog 3 (*RBFOX3*), Calbindin 1 (*CALB1*), and Glial Fibrillary Acidic Protein (*GFAP*) was determined by RT-qPCR to characterize the culture and framed within the multistep process of AHN. Hippocampal DNA methylation marks previously identified in Contactin-Associated Protein 1 (*CNTNAP1*), SEPT5-GP1BB Readthrough (*SEPT5-GP1BB*), T-Box Transcription Factor 5 (*TBX5*), and Nucleoredoxin (*NXN*) genes were profiled by bisulfite pyrosequencing or bisulfite cloning sequencing; mRNA expression was also measured. *NXN* outlined a peak of DNA methylation overlapping type 3 neuroblasts. Aβ-treated NPCs showed transient decreases of mRNA expression for *SEPT5-GP1BB* and *NXN* on day 9 or 19 and an increase in DNA methylation on day 29 for *NXN*. *NXN* and *SEPT5-GP1BB* may reflect alterations detected in the brain of AD human patients, broadening our understanding of this disease.

Keywords: adult hippocampal neurogenesis; NPCs; Alzheimer's disease; Aβ peptide; DNA methylation; gene expression; *NXN*; *CNTNAP1*; *SEPT5-GP1BB*; *TBX5*

Citation: Blanco-Luquin, I.; Acha, B.; Urdánoz-Casado, A.; Gómez-Orte, E.; Roldan, M.; Pérez-Rodríguez, D.R.; Cabello, J.; Mendioroz, M. NXN Gene Epigenetic Changes in an Adult Neurogenesis Model of Alzheimer's Disease. *Cells* **2022**, *11*, 1069. https://doi.org/10.3390/cells11071069

Academic Editors: FengRu Tang and Luisa Alexandra Meireles Pinto

Received: 27 January 2022
Accepted: 20 March 2022
Published: 22 March 2022

Publisher's Note: MDPI stays neutral with regard to jurisdictional claims in published maps and institutional affiliations.

1. Introduction

Adult neurogenesis (AN) is the process of forming functional neurons *de novo*. In the adult mammalian brain, neurogenesis occurs predominantly in specific brain niches: the subgranular zone (SGZ) of the dentate gyrus (DG) of the hippocampus and the subventricular zone (SVZ) lining the lateral ventricles [1,2]. During the process of adult hippocampal neurogenesis (AHN), neural stem cells (NSCs) self-renew and differentiate, giving rise to transient amplifying progenitors (TAPs), neuroblasts, and eventually mature neurons, astrocytes, and oligodendrocytes.

AHN regulators can be divided into intrinsic or extrinsic factors, that is, transcription factors (TFs) synthesized by the developing neural precursors and neurons, and growth factors and neurotrophins secreted from the surrounding niche, respectively [3]. Epigenetic mechanisms tightly regulate extrinsic and intrinsic factors [4], controlling both temporal and spatial gene expression. Sequential steps of AN are regulated directly or indirectly by *de novo* methylation and maintenance of methylation marks [5]. Each distinct human brain region (cerebral cortex, cerebellum, and pons) has a characteristic DNA methylation signature [6], and even within brain regions such as the hippocampus, global methylation varies between neuronal subtypes [7].

During both physiological and pathological aging in humans, AHN clearly emerges as a robust phenomenon [8]. AHN is involved with the normal functionality of hippocampal circuits, which demonstrates an important link between AN and cognitive processes [9]. Thus, impaired neurogenesis may negatively impact the survival of adult-born neurons and contribute to learning and memory failure, as occurs with aging and neurological disorders, e.g., Alzheimer's disease (AD) [8,10,11].

AD is the most common neurodegenerative disorder, characterized by progressive memory loss and cognitive decline caused by widespread loss of neurons and synaptic connections in the cortex, hippocampus, amygdala, and basal forebrain, and by a gradually significant loss of brain mass. The amyloid precursor protein (APP) plays a key role in normal brain development by influencing NSC proliferation, cell fate specification, and neuronal maturation [10]. However, its derivative, the amyloid β (Aβ) peptide, a cleavage product of the APP enzymatic processing, is the major component of amyloid plaques, one of the hallmark pathologies found in brains of AD patients. Monomeric Aβ can self-aggregate to form oligomers, protofibrils, and amyloid fibrils, which deposit as amyloid plaques. Although the impact of Aβ on neurogenesis is still controversial, it is well known that Aβ plaques can cause severe damage to neurons and astrocytes, which results in the gradual loss of neurons associated with AD symptoms [11].

Remarkable alterations in AHN have been detected at early stages of AD, even before the onset of hallmark lesions or neuronal loss [8,12]. Impairments in epigenetic mechanisms lead to the generation of damaged neurons from NSCs, exacerbating the loss of neurons and deficits in learning and memory that characterize AD pathology [11]. Indeed, we and others have described epigenetic changes in DNA methylation in the hippocampus of AD patients at the genome-wide level [6,13]. In a previous study, we reported altered DNA methylation in the AD hippocampus occurring at specific regulatory regions crucial for neuronal differentiation; moreover, a set of neurogenesis-related genes were identified in the damaged tissue [6]. Hence, a better understanding of AHN impairment observed at the initial and later stages of AD by noninvasive methods may reveal insights into the pathogenesis of AD. What is more, restoration of normal levels of AHN may provide a potential therapeutic strategy to delay or halt AD-linked cognitive decline [8,12].

Here, we propose an intuitive *in vitro* approach to assess a stepwise lineage progression, as occurs during *in vivo* neurogenesis, by using human neural progenitor cells (NPCs) derived from an induced pluripotent stem cell (iPSC) line as the starting source material. In order to infer whether the differentiation of human NPCs into mature neurons is disrupted in the AD microenvironment, we designed an observational descriptive study by generating an *in vitro* model triggered by prolonged exposure to nanomolar concentrations of Aβ peptide 1–42. Next, we evaluated DNA methylation levels and mRNA expression changes of specific neurogenesis-related candidate genes.

2. Materials and Methods

2.1. NPCs Culture, Neuronal Differentiation and Aβ Peptide Administration

NPCs Derived from XCL1 DCXpGFP (ACS5005™, American Type Culture Collection, ATCC, Manassas, VA, USA) were cultured following manufacturer recommendations. Briefly, 0.30×10^6 NPCs were seeded onto a CellMatrix Basement Membrane Gel (ATCC® ACS3035™) coated 12-well plate and incubated in NPC expansion medium: complete growth medium including DMEM/F-12 (Gibco, Fisher Scientific, Waltham, MA, USA), supplemented with the Growth Kit for Neural Progenitor Cell Expansion (ATCC® ACS3003) and then maintained in a humidified incubator (5% CO_2, 37 °C).

Neuronal differentiation experiments were carried out for 9, 19, and 29 days by plating NPCs at a seeding density of 80,000 viable cells/cm^2 in 6-well coated culture plates. First, NPCs were incubated in an expansion medium (day 0). From day 1 (post-seeding), half of the medium was changed for differentiation medium every 2–3 days throughout the duration of the culture period. Complete Differentiation Medium consisted of serum-free neuronal basal BrainPhys™ Neuronal Medium, formulated to improve the electrophysiological and synaptic properties of the neurons [14], NeuroCult™ SM1 Neuronal Supplement (1:50), N2 Supplement-A (1:100), Recombinant Human Brain-Derived Neurotrophic Factor (BDNF, 20 ng/mL), Recombinant Human Glial-Derived Neurotrophic Factor (GDNF, 20 ng/mL), Dibutyryl-cAMP (1 mM) and ascorbic acid (200 nM) (STEMCELL Technologies, Vancouver, BC, Canada). Half-fresh medium containing Aβ protein fragment 1–42 (50 nM; Sigma-Aldrich, St. Louis, MO, USA) or DMSO (Sigma-Aldrich) as a vehicle was added once a week.

NPCs were harvested on day 0 and 9, 19, and 29 days of differentiation for both conditions by detaching them with Accutase (Innovative Cell Technologies, San Diego, CA, USA), then washed with Dulbecco's phosphate-buffered saline (DPBS, Sigma-Aldrich), centrifuged at 13,000 rpm and frozen at −80 °C. All experiments were performed in triplicate.

2.2. Selection of Candidate Epigenetic Marks in AD

A set of differentially methylated positions (DMPs) in AD was produced from a methylome dataset generated in a previous study described elsewhere [6]. In brief, the Infinium HumanMethylation450 BeadChip array (Illumina, Inc., San Diego, CA, USA) was performed at the Roswell Park Cancer Institute Genomics Shared Resource (Buffalo, NY, USA) to measure DNA methylation levels in CpG sites (also named *positions*) in a cohort of 26 pure AD cases and 12 controls. A total of 118 AD-related DMPs were identified in the hippocampus of AD cases compared to controls. Here, we selected four of the above-identified DMPs in AD patients compared to controls (absolute β-difference $\geq$ 0.085 and *p*-value $\leq$ 0.05) and analyzed them due to their relationship with neurogenesis (Table 1 and Supplementary Figure S1).

Table 1. Selected differentially methylated positions (DMPs) in AD hippocampus measured by 450 K Illumina BeadChip array. The table shows four DMPs prioritized by beta difference (delta) criteria. Each CpG site was annotated by UCSC hg19 build.

DMPs	Genomic Coordinates		Beta Difference	*p*-Value	Genes
cg16308533	17	40838983	0.118	0.004	*CNTNAP1*
cg04533276	22	19709548	0.117	0.007	*SEPT-GP1BB*
cg18689332	12	114837666	0.106	0.000	*TBX5*
cg19987768	17	750306	−0.162	0.043	*NXN*

2.3. DNA Methylation Levels Assessed by Bisulfite Pyrosequencing

Genomic DNA was isolated from frozen cell pellets of basal NPCs and control or Aβ peptide treated NPCs incubated in differentiation media for 9, 19, or 29 days by using the FlexiGene DNA Kit (Qiagen, Redwood City, CA, USA). Next, 500 ng of genomic DNA was bisulfite converted using the EpiTect Bisulfite Kit (Qiagen) according to the manufacturer's protocol. Primer pairs to amplify and sequence the chosen CpG genomic positions were designed with PyroMark Assay Design version 2.0.1.15 (Qiagen) (Supplementary Table S1) and bisulfite PCR reactions were carried out on a VeritiTM Thermal Cycler (Applied Biosystems, Foster City, CA, USA). Next, 20 µL of the biotinylated PCR product was immobilized using streptavidin-coated Sepharose beads (GE Healthcare Life Sciences, Piscataway, NJ, USA) and 0.4 µM of sequencing primer annealed to purified DNA strands. Pyrosequencing was performed using PyroMark Gold Q96 reagents (Qiagen) on a PyroMark™ Q96 ID System (Qiagen). For each particular CpG, DNA methylation levels were expressed as the percentage of methylated cytosines over the sum of total cytosines. Unmethylated and methylated DNA samples (EpiTect PCR Control DNA Set, Qiagen) were used as controls for the pyrosequencing reaction.

2.4. Extension of NXN Gene Methylation Mapping by Bisulfite Cloning Sequencing

Previously bisulfite-converted genomic DNA was used to validate pyrosequencing results. Primer pair sequences were designed by MethPrimer [15] (Supplementary Table S1). PCR products were cloned using the TopoTA Cloning System (Invitrogen, Carlsbad, CA, USA); a minimum of 10–12 independent clones were sequenced for each triplicate, cell condition, and region (Sanger sequencing) [16]. Methylation graphs were obtained with the QUMA software [17].

2.5. Neurogenesis Markers mRNA Expression: Analysis by Real-Time Quantitative PCR (RT-qPCR)

Total RNA was extracted from frozen pellets of basal NPCs and the control or Aβ peptide treated NPCs incubated in differentiation media for 9, 19, or 29 days using the RNeasy Mini kit (QIAGEN, Redwood City, CA, USA) following the manufacturer's instructions. Genomic DNA was digested with DNase I (RNase-Free DNase Set, Qiagen). RNA concentration and purity were determined using a NanoDrop spectrophotometer. Complementary DNA (cDNA) was reversely transcribed from 1000 ng total RNA with SuperScript® III First-Strand Synthesis Reverse Transcriptase (Invitrogen) after priming with oligo-d (T) and random primers. RT-qPCR reactions were performed in duplicate with Power SYBR Green PCR Master Mix (Invitrogen) in a QuantStudio 12 K Flex Real-Time PCR System (Applied Biosystems, Foster City, CA, USA). Sequences of primer pairs were designed using a real-time PCR tool (IDT, Coralville, IA, USA) (listed in Supplementary Table S1). Relative mRNA expression levels of lineage-specific genes in a particular sample were calculated as previously described [18] and the geometric mean of the *ACTB* and *GAPDH* genes used as reference to normalize the expression values.

2.6. Immunofluorescence Staining

NPCs were seeded on Nunc™ Lab-Tek™ II chamber slides (Thermo Fisher Scientific, Waltham, MA, USA), coated with CellMatrix Basement Membrane Gel. Cells were either left untreated or treated with Aβ protein fragment 1–42 (50 nM) in differentiation media, as described above. After 9, 19, or 29 days of incubation, cells were fixed with 4% formalin (OPPAC, Noain, Spain) for 15 min; next, they were permeabilized using 0.5% TWEEN® 20 (Sigma-Aldrich) in DPBS and blocked with 10% fetal bovine serum (Sigma-Aldrich) containing 0.5% Tween in DPBS for 30 min at room temperature. Rabbit monoclonal anti-NeuN [EPR12763] (Cat# ab177487, RRID:AB_2532109; 1:300), anti-GFAP [EP672Y] (Cat# ab33922, RRID:AB_732571; 1:300), anti-Synaptophysin [YE269] (Cat# ab32127, RRID:AB_2286949; 1:200) and anti-Ki67 [SP6] (Cat# ab16667, RRID:AB_302459; 1:500) primary antibodies (Abcam, Cambridge, UK) diluted in blocking buffer were added and incubated overnight at 4 °C. After three washing steps, Alexa Fluor® 647 donkey anti-rabbit secondary antibody (Abcam Cat# ab150075, RRID:AB_2752244; 1:500) was added and incubated for 30 min at room temperature in the dark. Following three washing steps, the slides were mounted with ProLong™ Gold Antifade Mountant with DAPI (Molecular Probes, OR, USA). Immunofluorescence images were obtained using a Cytation 5 Cell Imaging Multi-Mode Reader and analyzed with the Gen5™ software (BioTek, Winooski, VT, USA).

2.7. Statistical Data Analysis

Statistical analyses were performed with the SPSS version 21.0 (IBM, Inc., Armonk, NY, USA) and GraphPad Prism version 6.00 for Windows (GraphPad Software, La Jolla, CA, USA). We first checked that all continuous variables had a normal distribution using the one-sample Shapiro–Wilk test. Significance level was set at p-value < 0.05. Differences between the various time points for mRNA levels of specific genes and percentages of DNA methylation were assessed by one-way analysis of variance (one-way ANOVA) followed by post hoc Tukey's honestly significant difference test. In cases where the Levene test did not show homogeneity of variance, Welch's ANOVA followed by Dunnett's T3 were conducted. Non-parametric data were analyzed using the Kruskal–Wallis test. A paired *t*-test was used to analyze differences in methylation or expression levels of the studied genes between Aβ peptide treated and control groups at each time point. GraphPad Prism version 6.00 for Windows was used to draw the graphs.

3. Results

3.1. Time-Related Changes in Cultured NPCs during Neural Differentiation

To determine whether neural differentiation was effectively induced, we first examined any morphological modifications of the cells over time. As shown in Figure 1A, NPCs exposure to differentiation medium led to an increase in the number and length of neuritic extensions, which even connected with the extensions of neighboring cells in comparison with basal cells grown in proliferation medium at Time 0. These changes in cell morphology, typical of cells undergoing differentiation [19,20], were noticed from the first time point (day 9), becoming more evident over time in response to directed neurogenesis.

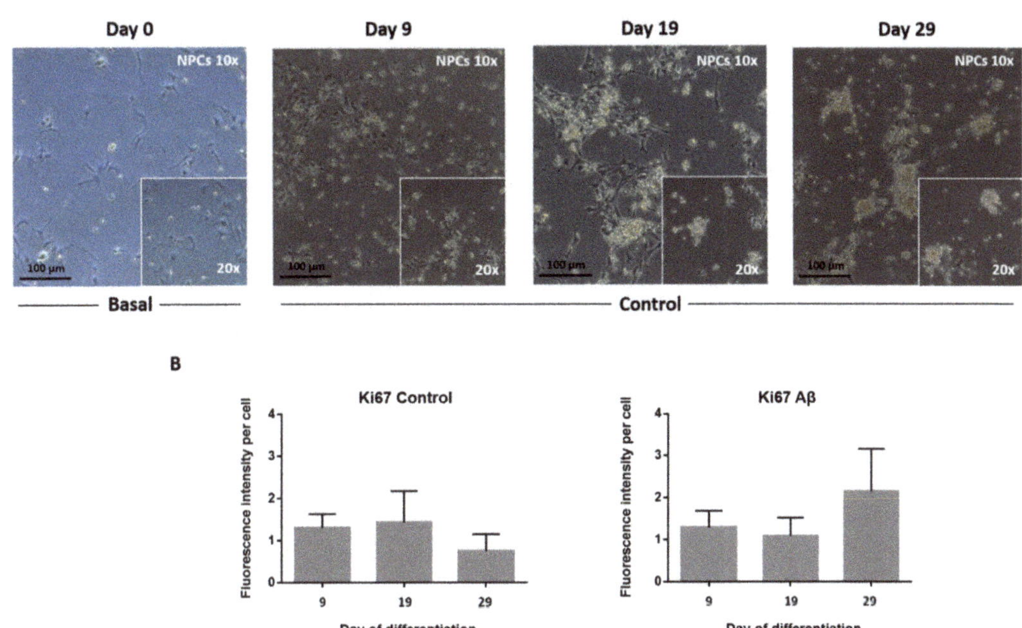

Figure 1. Phenotypic examination of NPCs directed differentiation in culture and Ki67 protein expression. (**A**) Phase-contrast images on days 0, 9, 19, and 29 of basal cells incubated in expansion medium and control cells incubated in differentiation medium (10× magnification with 20× magnification inset lens; the scale bar is 100 µm). (**B**) The graph shows Ki67 proliferation marker expression for control and Aβ-treated NPCs at 9, 19, and 29 days of culture in differentiation medium. Data represent the mean value ± standard error of the mean (SEM).

The total cell number in NPC cultures remained steady because of no proliferation, confirmed by unchanged Ki67 protein marker expression in control or exposed to Aβ peptide cells (Figure 1B), which was associated with a gradual boost of cell differentiation. In fact, immunofluorescence (IF) staining revealed neuronal nuclei (NeuN) and synaptophysin protein expression, which mark neurons and synaptic vesicles in the NPC culture (Figure 2).

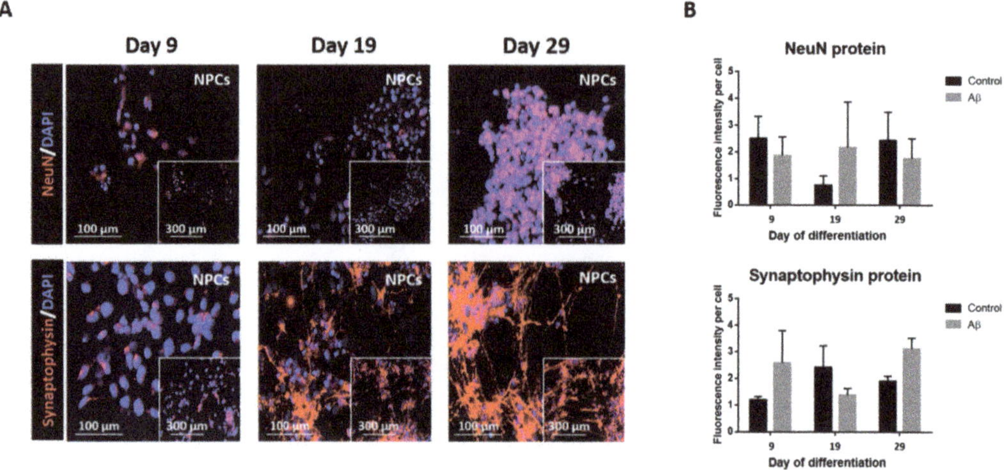

Figure 2. Immunofluorescence staining of NPC differentiation. (**A**) Representative images show NeuN and synaptophysin protein expression on days 9, 19, and 29 in NPCs incubated in differentiation medium (20× magnification (the scale bar is 100 μm) with 10× magnification 4 × 4 montage inset (the scale bar is 300 μm)). (**B**) The graphs show NeuN and synaptophysin markers expression for control and Aβ treated NPCs at 9, 19, and 29 days of culture in differentiation medium. Data represent the mean value ± SEM.

To confirm the above observations, we explored if gene expression profiles of different TFs and molecular markers had changed in our *in vitro* model across consecutive stages of driven neuronal differentiation. For that, we measured mRNA expression levels of the Neuronal Differentiation 1 (*NEUROD1*), Neural Cell Adhesion Molecule 1 (*NCAM1*), Tubulin Beta 3 Class III (*TUBB3*), RNA Binding Fox-1 Homolog 3 (*RBFOX3*), Calbindin 1 (*CALB1*), and Glial Fibrillary Acidic Protein (*GFAP*) genes by RT-qPCR (Figure 3). Expression levels of all genes but *CALB1* changed over time.

NEUROD1 mRNA expression levels of NPCs increased in differentiation medium. Statistically significant increases of mRNA expression for this basic helix-loop-helix (bHLH) TF on days 9 (p-value < 0.05), 19 (p-value < 0.05) and 29 (p-value < 0.001) were observed in comparison to basal cells.

In our *in vitro* model, *NCAM1* mRNA expression overlapped that of *NEUROD1*. We found a statistically significant increase from the addition of differentiation medium to the cell culture ($F_{(3,17)}$ = 31.85, p-value = 3.3634 × 10^{-7}), which was more pronounced on day 19 (p-value < 0.001). Significant differences were also seen between days 9 and 19 (p-value < 0.001), days 9 and 29 (p-value < 0.01) and between basal cells and any of the other time points: from day 0 to day 9 (p-value < 0.01) and from day 0 to day 29 (p-value < 0.001).

Once the proliferation medium was changed for differentiation medium, NPCs began to express *TUBB3* mRNA, a gene marker with a key role for proper axon guidance and maintenance. This increase remained constant over time in comparison to basal cells (p-value < 0.01). However, no changes were observed between the first, second, and third time points.

RBFOX3 encodes the NeuN antigen, which has been widely used as a marker for postmitotic neurons. In our study, *RBFOX3* mRNA expression progressively rises over time, proving the successful achievement of progenitor-to-neuron differentiation. Statistically significant differences in the rise of mRNA expression between day 0 and day 9 (p-value < 0.01), day 9 and day 19 (p-value < 0.01) and day 9 and day 29 (p-value < 0.05) were seen. Likewise, all other differences between any time point with respect to basal cells were also statistically significant: from day 0 to day 19 (p-value < 0.01) and from day 0 to day 29 (p-value < 0.05).

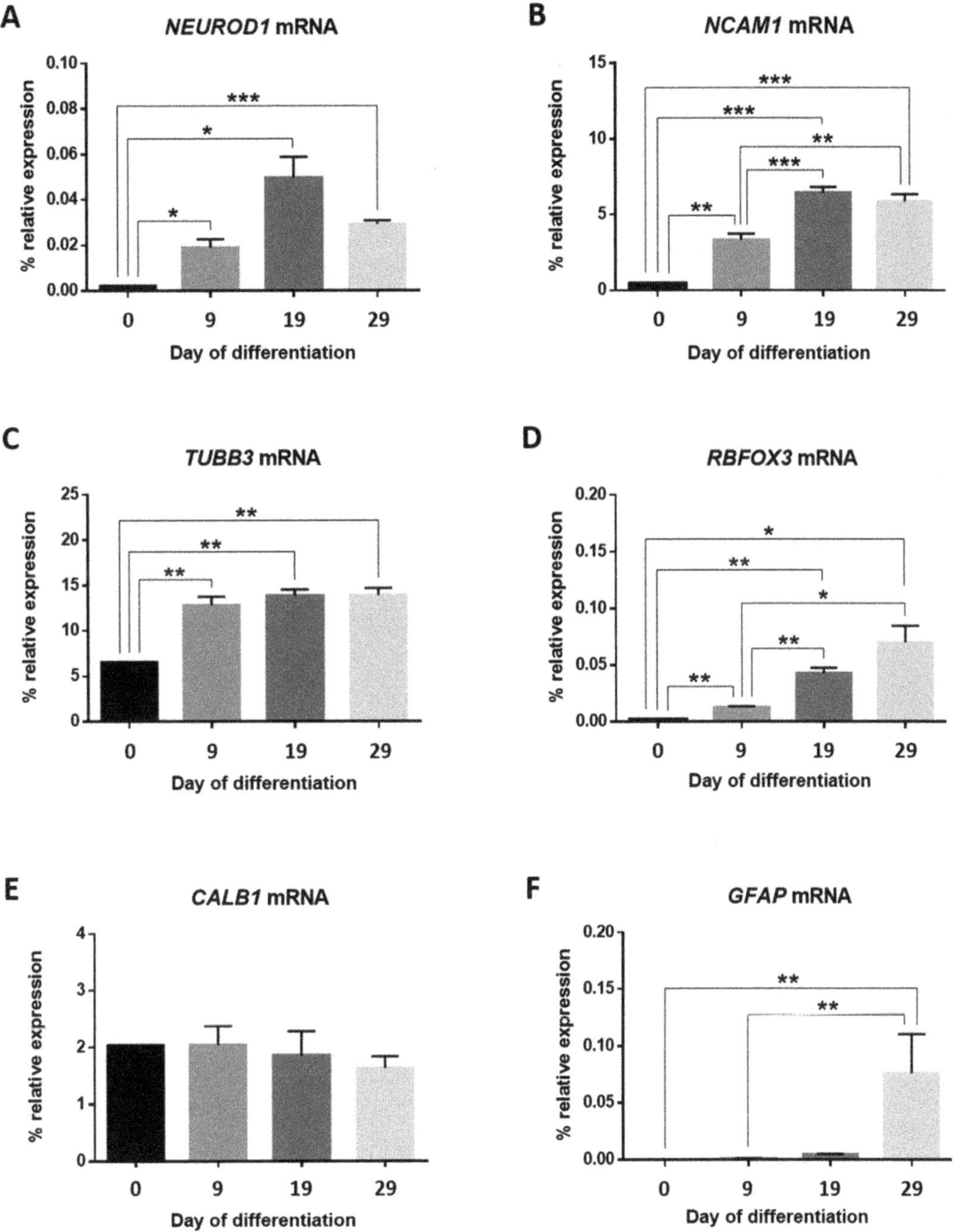

Figure 3. *NEUROD1* (**A**), *NCAM1* (**B**), *TUBB3* (**C**), *RBFOX3* (**D**), *CALB1* (**E**), and *GFAP* (**F**) gene expression profiles. Bar graphs show mRNA percentages of relative expression for each gene relative to the geometric mean of *ACTB* and *GAPDH* housekeeping gene expression for NPCs at each time point of culture. Data represent the mean value ± SEM; * *p*-value < 0.05; ** *p*-value < 0.01; *** *p*-value < 0.001.

Regarding *CALB1* mRNA expression, and given that this gene encodes a protein expressed in mature granule cells, no significant changes were detected.

A statistically significant rise in *GFAP* mRNA expression was observed on day 29 in comparison with basal cells (*p*-value < 0.01) and day 9 of differentiation (*p*-value < 0.01). This suggested the presence of NPCs-derived astrocytes in the culture.

None of the neuronal lineage-specific genes showed significant mRNA expression differences between day 19 and day 29.

3.2. Assessment of Epigenetic Markers Involved in Neurogenesis in Differentiating NPCs

DNA methylation levels of four neurogenesis-related genes previously found to be altered in the AD hippocampus [6] were quantified by bisulfite pyrosequencing. The same genomic loci identified in the human hippocampus were used to assess DNA methylation levels, corresponding to the genes Contactin-Associated Protein 1 (*CNTNAP1*), *SEPT5-GP1BB* Readthrough (*SEPT5-GP1BB*), T-Box Transcription Factor 5 (*TBX5*), and Nucleoredoxin (*NXN*) (Table 1 and Supplementary Figure S1).

No significant differences in DNA methylation levels were observed for *CNTNAP1*, *SEPT5-GP1BB*, and *TBX5* throughout the differentiation process within the time frame of this study (Figure 4A–C).

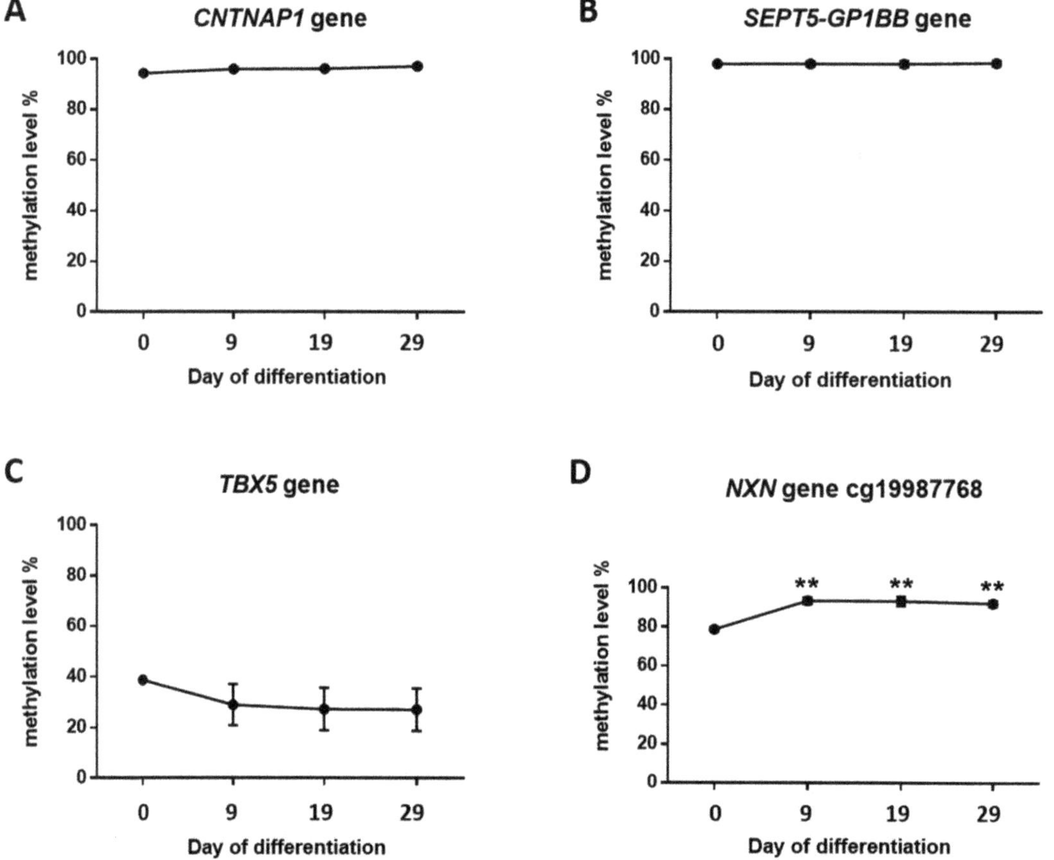

Figure 4. *CNTNAP1* (**A**), *SEPT5-GP1BB* (**B**), *TBX5* (**C**), and *NXN* (**D**) DNA methylation levels in differentiating NPCs. Graphs represent percentages of methylation levels measured by pyrosequencing on days 0, 9, 19, and 29. Vertical lines: SEM. ** *p*-value < 0.01.

Nonetheless, changes in *NXN* methylation levels were observed. Two CpG positions were assessed for the *NXN* gene. For the first one, DNA methylation levels increased on day 9 (p-value < 0.01) and were maintained over time; statistically significant differences were also seen on day 19 (p-value < 0.01) and day 29 (p-value < 0.01) with respect to basal cells (Figure 4D). Regarding the CpG following cg19987768, the pyrogram revealed a similar methylation pattern (day 9 vs. day 0: p-value < 0.05; day 19 vs. day 0: p-value < 0.01; day 29 vs. day 0: p-value < 0.05) (Supplementary Figure S2A). The same differences in methylation levels were observed for both CpGs together (day 9 vs. day 0: p-value < 0.001; day 19 vs. day 0: p-value < 0.001; day 29 vs. day 0: p-value < 0.001) (Supplementary Figure S2B). These findings led us to extend the methylation local mapping for the *NXN* gene using bisulfite cloning sequencing. We confirmed that average DNA methylation levels across all CpG sites for the amplicon were statistically significantly higher at every time point in comparison to day 0 (day 9 vs. day 0: p-value < 0.001; day 19 vs. day 0: p-value < 0.001; day 29 vs. day 0: p-value < 0.05) (Figure 5). Additionally, this approach revealed a decrease in *NXN* DNA methylation levels on day 29, which was statistically significant with respect to day 9 (p-value < 0.01).

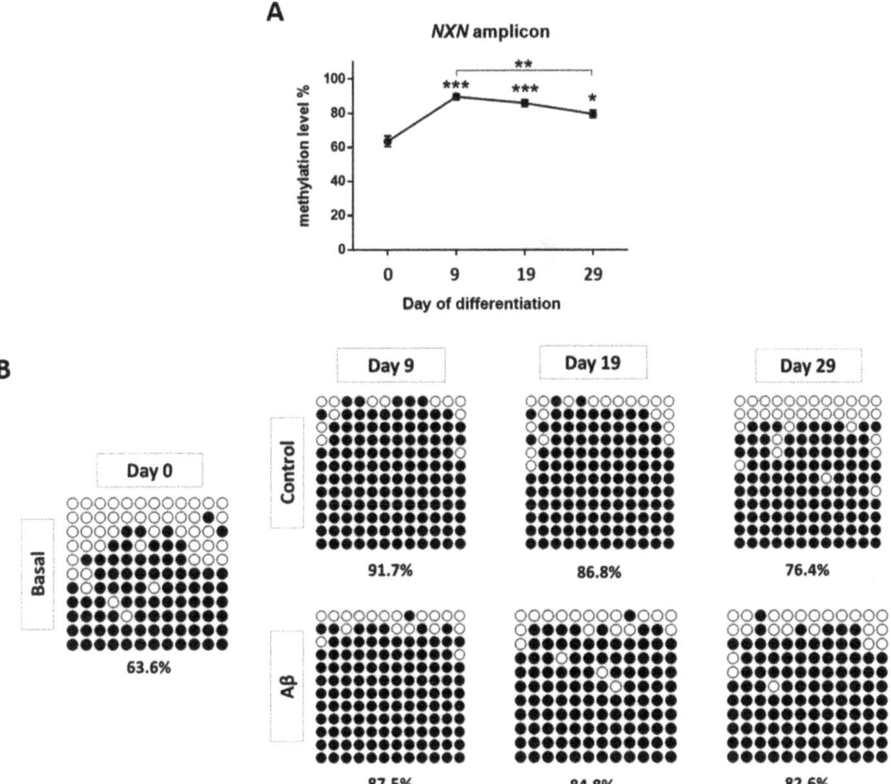

Figure 5. *NXN* DNA methylation levels by bisulfite cloning sequencing. (**A**) Percentages of DNA methylation for *NXN* over time. (**B**) *NXN* extended mapping is illustrated by black/white circle-style figures. Black and white circles denote methylated and unmethylated cytosines, respectively. Each column represents a single CpG site in the examined amplicon, and each line represents an individual DNA clone. Average percentages of methylation for each analyzed sample are indicated at the bottom. * p-value < 0.05; ** p-value < 0.01; *** p-value < 0.001.

We also measured mRNA expression levels of these markers by RT-qPCR (Figure 6).

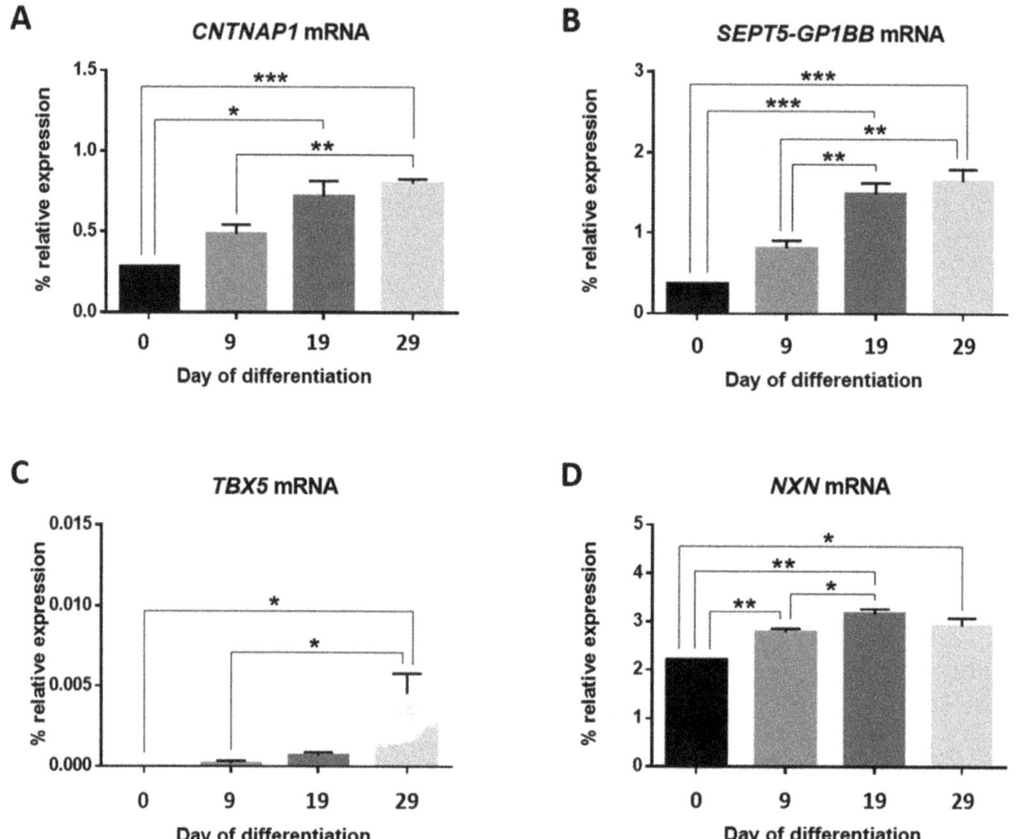

Figure 6. mRNA expression profiles for the *CNTNAP1* (**A**), *SEPT5-GP1BB* (**B**), *TBX5* (**C**), and *NXN* (**D**) genes. Bar graphs represent the percentages of relative mRNA expression for each gene relative to the geometric mean of the *ACTB* and *GAPDH* housekeeping gene expression for NPCs at each time point of culture. Mean values ± SEM. * *p*-value < 0.05; ** *p*-value < 0.01; *** *p*-value < 0.001.

CNTNAP1 mRNA expression levels progressively increased over time with statistically significant differences on day 19 (*p*-value < 0.05) and day 29 (*p*-value < 0.001) in comparison to basal cells. Moreover, significant expression differences were noticed between day 9 and day 29 (*p*-value < 0.01) (Figure 6A).

From day 19, a significant increase in mRNA expression for *SEPT5-GP1BB* was detected (*p*-value < 0.01) and maintained on day 29 (*p*-value < 0.01). Furthermore, mRNA expression on day 19 (*p*-value< 0.001) and day 29 (*p*-value< 0.001) was also significantly higher than for day 0 (Figure 6B).

mRNA levels for the *TBX5* gene increased on day 29 with statistically significant differences in comparison to the cells in culture on day 0 (*p*-value < 0.05) and day 9 (*p*-value < 0.05) (Figure 6C).

Finally, significant differences were observed from the addition of the differentiation medium for the *NXN* gene in terms of gene expression (day 9 vs. day 0: *p*-value < 0.01; day 19 vs. day 0: *p*-value < 0.01; day 29 vs. day 0: *p*-value < 0.05) (Figure 6D). The increase in mRNA expression continued to day 19 (0.384 ± 0.117; *p*-value < 0.05).

Overall, similar transcriptional patterns for the *TBX5* and *GFAP* genes and the *NXN*, *NCAM1* and *RBFOX3* genes during the NPCs culture period, were observed.

3.3. Effect of Aβ Peptide Addition on Cultured NPCs during the Stages of Neurogenesis

To mimic the cell environment in AD, we exposed NPCs to Aβ peptide 1–42 once a week during the differentiation period. First, we assessed whether the expression levels of the genes selected to characterize each stage of neurogenesis in culture were altered due to the addition of the Aβ peptide (Figure 7).

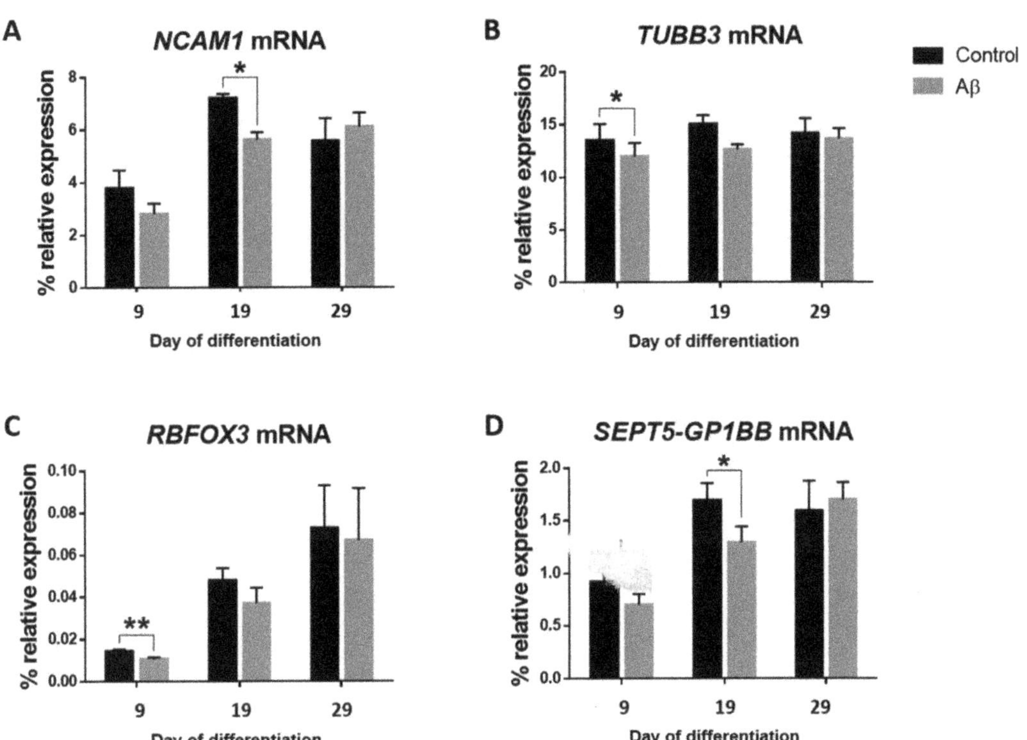

Figure 7. Effect of the addition of Aβ peptide 1–42 during the differentiation period. mRNA expression of the *NCAM1* (**A**), *TUBB3* (**B**), *RBFOX3* (**C**), and *SEPT5-GP1BB* (**D**) genes relative to the geometric mean of *ACTB* and *GAPDH* housekeeping genes expression was determined for the controls and Aβ peptide treated NPCs on days 9, 19, and 29. Vertical lines represent the SEM. * p-value < 0.05; ** p-value < 0.01.

We found transient and mild treatment-specific differences in mRNA expression for some of the studied lineage-specific genes. The Aβ peptide reduced *NCAM1* expression (p-value < 0.05) on day 19 (Figure 7A), and *TUBB3* (p-value < 0.05) and *RBFOX3* (p-value < 0.01) expression on day 9 (Figure 7B,C). Interestingly, such decreases occurred at the beginning or in between the studied time window, but these differences were no longer significant at the end time point (day 29).

Next, we assessed how the addition of Aβ peptide affected mRNA expression of neurogenesis-related genes and if the changes had any relationship with their methylation status.

We observed a statistically significant decrease in *SEPT5-GP1BB* mRNA on day 19 (p-value < 0.05) (Figure 7D) and an increase in the percentage of DNA methylation with a trend towards statistical significance on day 29 (p-value = 0.082) with the addition of the Aβ peptide to the culture.

Finally, the Aβ peptide slightly reduced *NXN* mRNA expression on day 9 (p-value < 0.05) which is maintained until day 19 (p-value < 0.05) (Figure 8A). *NXN* methylation seems to decrease on day 9 but does not reach statistical significance (p-value = 0.11). One possible explanation for this is that the sample size is insufficient to show statistical significance. Interestingly, a rise in the percentage of *NXN* methylation level of Aβ peptide-treated cells was seen on day 29, measuring all amplicon CpG sites (p-value < 0.05) (Figure 8B), when the decrease in *NXN* mRNA expression is no longer observed.

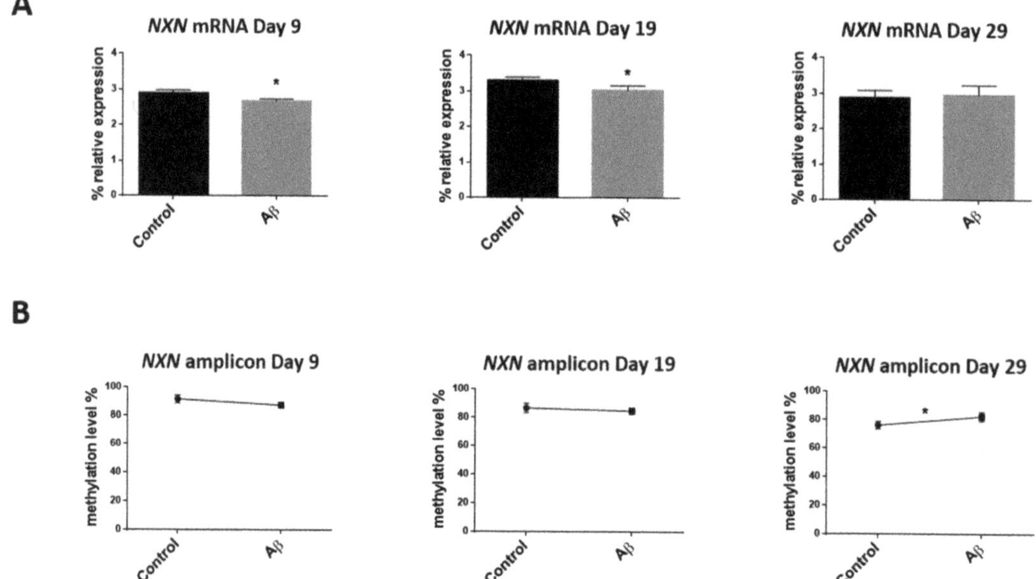

Figure 8. Effect of Aβ peptide 1–42 addition on the *NXN* gene during the differentiation period. mRNA expression relative to the geometric mean of *ACTB* and *GAPDH* housekeeping genes expression (**A**). DNA methylation level in the extended mapping amplicon (**B**) were determined for the controls and Aβ peptide-treated neural progenitor cells on days 9, 19, and 29. Vertical lines represent the SEM. * p-value < 0.05.

4. Discussion

To date, a broad overview of the stages of AHN exists. This complex multistep process can be divided into four phases: a precursor cell phase, an early survival phase, a postmitotic maturation phase, and a late survival phase. Type 1 radial glia-like cells (RGLs) represent the NSC population that can differentiate into TAPs (type 2 cells), which initially have a glial (type 2a) and then a neuronal (type 2b) phenotype. Through a migratory neuroblast-like stage (type 3), lineage-committed cells exit the cell cycle ahead of maturation into dentate granule neurons functionally integrated into the hippocampal circuitry [21,22]. Based on cell morphology TFs expression and a set of marker proteins, distinct milestones have been established [21]. In this study, we examined the expression dynamics of key markers in order to characterize a directed human NPCs differentiation model across distinct differentiation stages (Figure 9) to test new AHN epigenetic and expression markers that might be associated with AD.

During stage 1 (proliferation phase), type 1 RGL cells express GFAP. However, no differences in *GFAP* expression are detected until day 19 after the addition of the differentiation medium. This suggests that our *in vitro* NPCs culture window starts after the proliferative phase, during stage 2, when type-2 cells (differentiation phase) lose the GFAP marker [22]. Thus, in contrast to their *in vivo* counterparts in the SGZ of the brain (some authors describe that the *in vitro* expanded NSCs are less neurogenic and mainly biased towards an astrocytic fate upon differentiation [20]), *GFAP* expression on day 19 would correspond to a subset of astrocytes present in our NPCs culture [23].

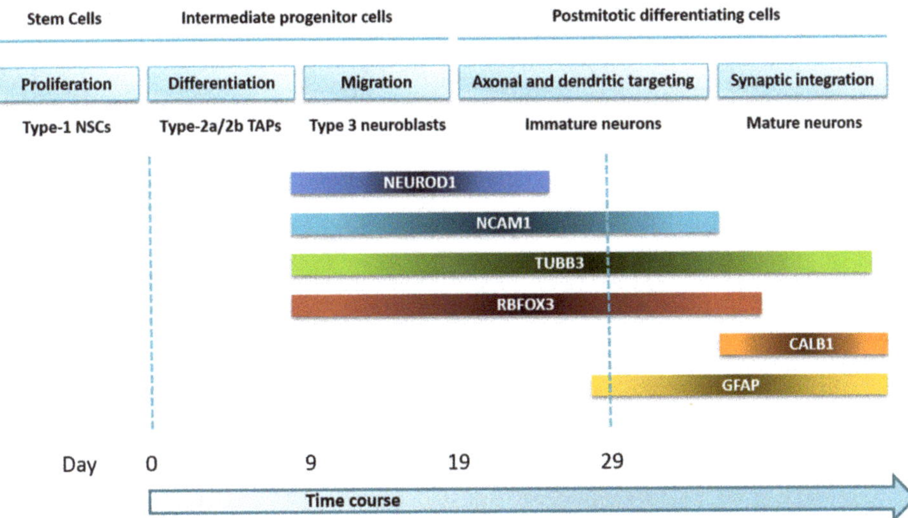

Figure 9. Expression pattern of AHN lineage-specific genes, assessed to characterize our NPCs *in vitro* model. The diagram illustrates *NEUROD1*, *NCAM1*, *TUBB3*, *RBFOX3*, *CALB1*, and *GFAP* gene expression profiles during directed neuronal differentiation for our time window NPCs culture model, based on the developmental stages of AHN within the neurogenic niche of the DG.

In stage 3 (migration phase), migrating neuroblasts display the polysialylated form of NCAM (PSA-NCAM), a marker that appears at the late stage of AN and seems to persist in young postmitotic neurons [24]. Accordingly, our results suggest the presence of a plateau between day 19 and day 29 for *NCAM1* mRNA expression. Most PSA-NCAM-positive cells express NeuroD and NeuN, but not GFAP, which supports the abovementioned findings [24]. bHLH TF *NEUROD1* plays an essential role in the differentiation and survival of neuronal precursors in the SGZ. NeuroD1 deletion leads to new granule neurons depletion and their failure to integrate into the DG [25]. In line with findings by Xuan Yu et al. [26], we observed a rise of *NEUROD1* gene expression during our culture time window. Moreover, expression of NeuroD can also be detected in PSA-NCAM-positive cells, precedes it [24], and reaches the highest point in late-stage type 2b and type 3 cells [2]. Once the newly generated neurons become postmitotic, they begin to express the NeuN marker, which is consistent with an earlier *RBFOX3* mRNA expression in our model. We found that *RBFOX3* expression increases until days 19 and 29 of differentiation, showing an expression profile similar to that of *NCAM1*.

Next, cells become postmitotic entering stage 4 (axonal and dendritic targeting). Immature neurons still express PSA-NCAM and, at the same time, can also be marked by NeuN. *TUBB3*, involved in axon guidance and maintenance, is expressed simultaneously; it encodes a class III member of the beta-tubulin protein family, characteristic of early postmitotic and differentiated neurons and some mitotically active neuronal precursors. This is consistent with the increase in *TUBB3* mRNA detected in our model, prior to its translation into protein. *TUBB3* mRNA expression persists in neurons displaying high complexity and electrophysiological properties, such as very low capacitance, high input resistance, depolarized resting membrane potential, and lack of synaptic activity, which show immunoreactivity for NeuN and thus represent postmitotic neurons [24,27].

Finally, mature granule cells establish their synaptic contacts and become functionally integrated into the hippocampus in stage 5 (synaptic integration), expressing calbindin along with NeuN but without co-expressing PSA-NCAM [24]. We do not find variations in *CALB1* mRNA expression within the analyzed culture time window, which may occur later in time. We indeed detect synaptophysin in the IF study on day 29, which suggests that our time window ends early at the synaptic integration phase.

Hence, by culturing NPCs as a monolayer in a medium that accelerates neuronal differentiation by enhancing synaptic activity [14], we achieve a less time-consuming differentiation strategy that resembles the *in vivo* developmental program of human hippocampal DG, which differs from that of the SVZ [28], as we are able to generate developing neurons potentially expressing relevant features of the AHN process.

Once the first objective was accomplished, we evaluated whether a set of AD-related differentially methylated genes targeted specific AHN milestones. These genes had been identified in a previous study of the human hippocampus and annotated as neurogenesis genes following a curated review of the literature [6]. No differences in DNA methylation for the *CNTNAP1*, *SEPT5-GP1BB*, and *TBX5* genes were identified within the period of this study. Only one or two CpGs were analyzed for each gene, those that had been identified as differentially methylated in the hippocampus of AD patients, so changes in DNA methylation may be present in other regions of the gene and may not have been detected with our approach. Still, changes in DNA methylation may occur before or after our time window.

However, it is worth noting that all the above genes undergo mRNA expression changes, suggesting they could be considered potential molecular markers of different AHN stages (Figure 10). Further studies should be carried out to confirm this.

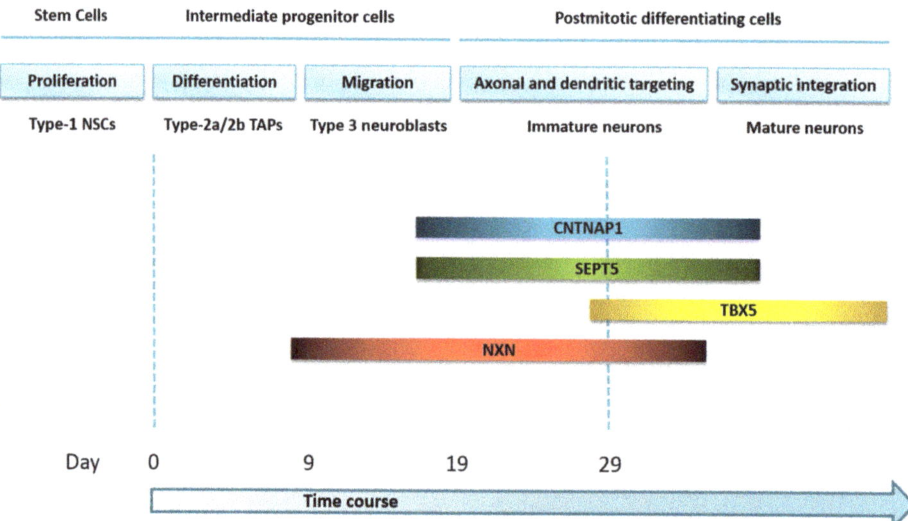

Figure 10. Expression patterns of neurogenesis-related genes evaluated in our *in vitro* model on NPCs. The illustration depicts the expression profiles of the *CNTNAP1*, *SEPT5-GP1BB*, *TBX5*, and *NXN* genes during directed neuronal differentiation of our time window culture model on NPCs, according to the developmental stages of AHN within the neurogenic niche of the DG.

CNTNAP1 and *SEPT5-GP1BB* mRNA expression levels increase on day 19/29, possibly identifying immature neurons, when axonal and dendritic targeting occurs. Indeed, *CNTNAP1* encodes a type I integral membrane protein that regulates the intracellular processing and transport of contactin to the cell surface [29,30], also known as contactin-associated protein (CASPR), which is present in synapses and interacts with AMPA (α-amino-3-hydroxy-5-methyl-4-isoxazolepropionic acid) glutamate receptors that mediate fast excitatory synaptic transmission in the central nervous system (CNS) [31]. CASPR is an adhesion molecule crucial to forming axoglial paranodal junctions surrounding the nodes of Ranvier in myelinated axons [32].

Known to be a negative regulator of neurite outgrowth in CNS neurons [30], CASPR1 plays an essential role in the timing of neuron and astrocyte development in the mouse cerebral cortex by repressing the transcription of the Notch effector Hes1. In radial glial cells, CASPR1 deficiency delays the generation of cortical neurons and induces the early formation of cortical astrocytes without affecting the number of progenitor cells. Thus, during the neurogenic period, CASPR1 is highly expressed, while during the gliogenic period its expression decreases [32,33]. Moreover, CASPR1 has been reported to be under the regulation of the astrocytic methyl-CpG-binding protein 2 (MeCP2) along with key myelin genes and proteins [34].

For its part, *SEPT5-GP1BB* is originated from naturally occurring read-through transcription between the neighboring *SEPT5* (*SEPTIN5*) and *GP1BB* (Glycoprotein Ib Platelet Subunit Beta) genes on chromosome 22. Inefficient use of an imperfect polyA signal in the upstream *SEPT5* gene causes transcription to continue into the *GP1BB* gene. The Genotype Tissue Expression (GTEx) Project established by the National Institutes of Health (NIH) Common Fund shows the highest median expression of this gene in the brain cortex, but to the best of our knowledge, this is the first study describing *SEPT5-GP1BB* as a possible key marker of temporal specification of cell fate in neurogenesis.

The *TBX5* gene displays the highest level of mRNA expression on day 29. It belongs to a phylogenetically conserved family of genes sharing a common DNA-binding domain, the T-box, which encodes TFs involved in the regulation of developmental processes. Accordingly, it is considered pivotal in the establishment of the cardiac lineage [35]. Moreover, *TBX5* regulates the development of the vertebrate eye [36] and limb skeletogenesis [37]. Here, we observe a statistically significant increase in *TBX5* mRNA at the end time point of our culture (day 29), and therefore, we propose it as a transcriptional candidate marker of postmitotic differentiating cells that may exhibit a peak of expression in the transition of immature to mature neurons.

The most relevant findings of our study relate to the *NXN* gene. At CpG site resolution, *NXN* shows differential methylation at every time point in comparison to basal cells. Moreover, when we extend the mapping and further average across all CpG sites of the amplicon, we confirm these findings and show that peak methylation of *NXN* occurs on day 9. Such curve outlined by the percentage of *NXN* DNA methylation would range from type-2a/2b TAPs to immature neurons, peaking at type 3 neuroblasts. This may allow to discriminate the migration stage of neurogenesis.

Interestingly, the increase in *NXN* methylation is associated with higher mRNA expression levels during our culture time window. DNA methylation at gene promoter regions usually represses gene expression through the recruitment of methylated DNA-binding protein family members, such as methyl-CpG-binding protein 1 (MBD1) and MeCP2. Nevertheless, DNA methylation roles in gene regulation appear complex and multi-faceted and genome structure integration becomes of major importance [38]. In the same way that CG-rich and CG-poor regulatory elements may undergo distinct modes of epigenetic regulation [38], DNA methylation has been linked to gene activation within the transcribed regions and the highest levels of gene body methylation may enhance transcription [39]. Indeed, it is precisely in this region where the studied DMPs are located (Supplementary Figure S1). Thus, DNA methylation has been previously correlated with increased expression in human embryonic stem (ES) cells in an *in vitro*-induced differentiation work [40,41]. Furthermore, gene expression is not only regulated by methylation in the same region, but by other epigenetic mechanisms or methylation in other regulatory areas. Several gene regulatory elements seem to communicate on the same or different chromosomes. Enhancers and insulators participate in this higher-order organization of chromatin [42]. In fact, sequential recruitment of lineage-restricted transcription factors leads to enhancers being activated or maintained in a poised state upon stem cell differentiation [43].

NXN is a ubiquitously expressed endogenous antioxidant, member of the thioredoxin antioxidant superfamily [44,45]. In brain sections of mice, there is a predominant neuronal expression of *NXN* in septal nuclei and the hippocampus, in which its deletion results are embryonically lethal, mainly due to cranial defects and deformities [46]. Specifically, immunoreactive signals of *NXN* were found in fibers in the cortex, hippocampus, and cerebellum [46].

In proliferating cells, NXN sequesters dishevelled segment polarity protein 2 (DVL2). Upon the increase in ROS, NXN releases DVL2, relaying the WNT signal to downstream effectors. As a result, cytosolic β-catenin accumulates and shuttles to the nucleus where it drives specific expression of target genes relevant to neuronal differentiation [44,45,47]. NXN also retains a pool of inactive Dvl by preventing the possible interaction of Dvl and kelch-like protein 12 (KLHL12) and its subsequent ubiquitination and degradation, ensuring a prompt activation upon Wnt stimulation [46]. In agreement with this, it has been proved that *NXN* knockdown of SH-SY5Y human neuroblastoma cells increases proliferation and cell cycle reentry [48]. Accordingly, in our *in vitro* model, the increased expression of *NXN* mRNA levels is consistent with the absence of cell proliferation.

The literature points to interactions with further partners that include histone deacetylase 6 (HDAC6), heat shock protein 90 kDa (HSP90), and calcium calmodulin kinase 2a (Camk2a), a postsynaptic kinase crucial for neuronal plasticity [46,48]. Moreover, *NXN* may be implicated in transcriptional regulation, promoting the induction of the TFs CREB (cAMP response element-binding protein), NFκB (nuclear factor kappa B), and AP-1 (activator protein-1) [46].

In the context of AD, it is known that Aβ peptides are generated after the cleavage of APP by γ-secretase in the amyloidogenic pathway [10]. In previous models, the physiological concentration of Aβ peptides in the brain revealed a positive effect on neuroplasticity and learning, showing improved hippocampal long-term potentiation (LTP), while high nanomolar Aβ administration resulted in impaired cognition [49,50], suggesting a hormetic nature [51]. Because low picomolar levels of extracellular concentrations of Aβ in the normal brain have been estimated, in our experiments we chose a concentration of Aβ peptide 1–42 in the nanomolar range (50 nM), added once a week during the 29 days of culture, a single dose determined by the average of concentrations used by Gulisano et al. [52] and Malmsten et al. [53].

It has been reported that the synthetic Aβ peptide 1–42 oligomer decreases human NSC proliferative potential and appears to favor glial differentiation; it reduces neuronal cell fates [10] or suppresses the number of functional human ES cells-derived neurons [54]. Nonetheless, Bernabeu-Zornoza et al. showed that 1 µM monomeric Aβ peptide 1–42 promoted human NSCs proliferation by increasing the pool of glial precursors, without affecting neurogenesis [55]. On the other hand, differentiating neurospheres exposed to fibrillar Aβ decreased neuronal differentiation and induced gliogenesis [54]. The existing controversies may be due to Aβ isoforms, peptide concentrations, aggregation state, administration times, or type of NSCs/NPCs from different species or culture systems used in each experiment [55].

In our work, some of the analyzed genes show a mild decrease in mRNA expression after Aβ 1–42 addition. This transient effect is evident on day 9 or day 19, not occurring on day 29, suggesting that, despite affecting genes involved in the fate of neurogenesis, probably before cells maturation and leading to a decrease in differentiation, the addition of nanomolar concentrations of Aβ is somehow counteracted in the long-term. A time-dependent reversal of the effects of picomolar Aβ on synaptic plasticity and memory had been already seen by Koppensteiner et al., attributable to the enzyme neprilysin, whose levels are reduced with aging and in the brains of AD patients [56]. In fact, a study in which mutant APP was overexpressed to ensure Aβ release exclusively by mature neurons, found neither a positive nor a negative effect in AHN [57]. Hence, our simplistic model may shed light on early AD neurogenesis events, before Aβ deposition cannot be overcome.

A transcriptomic analysis of several human AD profiles demonstrated upregulation of neural progenitor markers expression and downregulation of later neurogenic markers, implying that neurogenesis is reduced in AD due to compromised maturation [58]. Interestingly, the authors showed downregulation of *NCAM1* expression in the hippocampus of early-stage AD, as well as of *NCAM1*, *TUBB*, and *RBFOX3* in late-stage AD, which is in line with our results after the addition of Aβ to the culture. Moreover, Moreno-Jimenez et al. recently provided evidence for substantial maturation impairment underlying AD progression. They identified a decline in doublecortin-expressing cells that co-expressed PSA-NCAM in the DG starting at Braak stage III, followed by a reduction in the expression of NeuN and βIII-tubulin, among others, at some of the subsequent stages of the disease [12].

Our results also show a decrease in *SEPT5-GP1BB* mRNA expression on day 19 when Aβ 1–42 was added to the culture. Again, this suggests that even low levels of Aβ peptide deposit may already have an effect on neuronal fate. For *NXN*, such a decrease in mRNA expression was also observed on day 9 and day 19 cultures. No changes were seen on day 29 when the percentage of methylation levels in the *NXN* amplicon increased in differentiating cells with Aβ 1–42.

Thus, *NXN* emerges as a candidate gene that needs to be further studied to address its ability to determine not only the temporal sequence of neurogenesis but simultaneously the differences in the AD brain due to Aβ peptide deposition.

AHN confers a unique mode of plasticity to the mature mammalian brain. Research in this field requires non-invasive monitoring to understand the lifelong impact [59]. Easier than manipulating NSCs, in part because of the time saving, our NPCs model facilitates studying gene expression levels in an *in vitro* cell culture platform within a human context [60]. Moreover, this straightforward approach may help further understand the alterations affecting specific lineage cell types in presence of the Aβ peptide, including early pathological changes, possibly associated with prodromal phases. On the other hand, other cell types are involved in pathogenesis, particularly microglia, which play a major role, together with neuroinflammation, in the risk of developing AD and its progression. In consequence, co-cultures with other cell types present at neurogenic niches, such as microglia, may be implemented to overcome the limitations presented by the characteristics of an *in vivo* niche environment.

Finally, the development of AHN monitoring methods as biomarkers for cognitive function in live individuals will be crucial to staging AD progress. Moreover, studying the utility of TF reprogramming to preserve endogenous AHN may contribute to cognitive resilience in AD [58]. However, despite the enthusiasm, the prospect of using adult NSCs therapeutically as a regenerative source needs to address neuronal integration and its impact on host mature neural circuits [59]. It will involve strategies to accomplish the NSC pool maintenance, generation of correct neuronal subtypes, suppression of glial fates, and differentiation and survival of immature neurons [2].

5. Conclusions

In this work, we present the transcriptional profiles of a number of genes involved in specific stages of the AHN process for a thorough understanding of the lineage-restricted fate during human neuronal differentiation. The addition of Aβ peptide 1–42 to our human NPCs culture model, generates results that are similar to those obtained in human AD samples regarding the expression of the *NCAM1*, *TUBB3*, and *RBFOX3* genes, offering an *in vitro* opportunity to study AHN impairment in the AD context. Considering this approach, the *NXN* gene shows a rise in DNA methylation, the maximum being coincident in time with type 3 neuroblasts and displays differential DNA methylation in immature neurons in presence of the Aβ peptide. Moreover, *CNTNAP1*, *SEPT5-GP1BB*, *TBX5*, as well as *NXN* were revealed as mRNA expression molecular markers for specific stages of AHN. Finally, differentiating NPCs decrease their *SEPT5-GP1BB* or *NXN* mRNA expression at different neurogenesis time points with the addition of the Aβ peptide to the culture.

Supplementary Materials: The following supporting information can be downloaded at: https://www.mdpi.com/article/10.3390/cells11071069/s1. Supplementary Table S1. Bisulfite pyrosequencing, bisulfite cloning sequencing and RT-qPCR primers. Supplementary Figure S1. Genomic positions of the CpGs analyzed by pyrosequencing. Supplementary Figure S2. *NXN* DNA methylation levels in differentiating NPCs.

Author Contributions: I.B.-L. contributed to the conception and design of the work, running the experiments, analysis and interpretation of data, figure design and drawing and drafting/review of the manuscript for content; B.A. contributed to running the experiments, statistical analysis and review of the manuscript for content; A.U.-C. contributed to figure design and drawing and review of the manuscript for content; E.G.-O. contributed to the design and review of the manuscript for content. M.R. contributed to running the experiments; D.R.P.-R. contributed to the interpretation of data and review of the manuscript for content; J.C. contributed to the design and review of the manuscript for content; M.M. contributed to the conception and design of the work, analysis and interpretation of data, statistical analysis, study supervision, drafting/review of the manuscript for content and funding acquisition. All authors have read and agreed to the published version of the manuscript.

Funding: This research was funded by the Spanish Government through a grant from the Institute of Health Carlos III (FIS PI17/02218), jointly funded by the European Regional Development Fund (ERDF), European Union, A way of shaping Europe; the Trans-Pyrenean Biomedical Research Network (REFBIO II-MOMENEU project) and the Government of Navarra through two grants from the Department of Industry of the Government of Navarra (PI058 iBEAS-Plus and PI055 iBEAS-Plus). AUC received a grant Doctorandos industriales 2018–2020 and a Predoctoral grant (2019) founded by the Department of Industry and Health of the Government of Navarra. MM received a grant Programa de intensificación- (LCF/PR/PR15/51100006) founded by Fundación Bancaria la Caixa and Fundación Caja-Navarra, and Contrato de intensificación from the Institute of Health Carlos III (INT19/00029).

Institutional Review Board Statement: Not applicable.

Informed Consent Statement: Not applicable.

Data Availability Statement: The data presented in this study are available on request from the corresponding author.

Acknowledgments: We want to kindly thank Valle Coca (Navarrabiomed BrainBank, technical support), Paula Aldaz and Imanol Arozarena (Cancer Signalling Research Unit, Navarrabiomed, technical and scientific support), Natalia Ramirez (Haematological Oncology Research Unit, Navarrabiomed, scientific support), Ibai Tamayo, Arkaitz Galbete, Mónica Enguita, Berta Ibañez, and Julián Librero (Methodology Unit, Navarrabiomed, technical support) for their help.

Conflicts of Interest: The authors declare no conflict of interest.

Abbreviations

Aβ: amyloid β; AD: Alzheimer's disease; AHN: adult hippocampal neurogenesis; AMPA: α-amino-3-hydroxy-5-methyl-4-isoxazolepropionic acid; AN: adult neurogenesis; ANOVA: two-way analysis of variance; AP-1: activator protein-1; APP: amyloid precursor protein; BDNF: Brain-Derived Neurotrophic Factor; bHLH: basic helix-loop-helix; *CALB1*: calbindin 1; Camk2a: calcium calmodulin kinase 2a; CASPR: contactin-associated protein; cDNA: complementary DNA; *CNTNAP1*: Contactin-Associated Protein 1; *CREB*: cAMP response element-binding protein; DG: dentate gyrus; CNS: central nervous system; DMPs: differentially methylated positions; DVL2: Dishevelled segment polarity protein 2; ES: embryonic stem; GDNF: Glial-Derived Neurotrophic Factor; *GFAP*: Glial Fibrillary Acidic Protein; *GP1BB*: Glycoprotein Ib Platelet Subunit Beta; HDAC6: histone deacetylase 6; HSP90: heat shock protein 90 kDa; IF: immunofluorescence; iPSCs: induced pluripotent stem cells; KLHL12: kelch-like protein 12; LTP: long-term potentiation; MBD1: methyl-CpG-binding protein 1; MeCP2: methyl-CpG-binding protein 2; *NCAM1*: Neural Cell Adhesion Molecule 1; NeuN: neuronal nuclei; *NEUROD1*: Neuronal Differentiation 1; NFκB: nuclear factor kappa-B; NPCs: neural progenitor cells; NSCs: neural stem cells; *NXN*: Nucleoredoxin; PSA-NCAM: polysialylated form of NCAM; PSCs: pluripotent stem cells; *RBFOX3*: RNA Binding Fox-1 Homolog 3; RGLs: radial glia-like cells;

RT-qPCR: real-time quantitative PCR; SE: standard error; SEM: standard error of the mean; *SEPT5*: *SEPTIN5*; *SEPT5-GP1BB*: *SEPT5-GP1BB* Readthrough; SGZ: subgranular zone; SVZ: subventricular zone; TAPs: transient amplifying progenitors; *TBX5*: T-Box Transcription Factor 5; TFs: transcription factors; *TUBB3*: Tubulin Beta 3 Class III.

References

1. Hollands, C.; Bartolotti, N.; Lazarov, O. Alzheimer's Disease and Hippocampal Adult Neurogenesis; Exploring Shared Mechanisms. *Front. Neurosci.* **2016**, *10*, 178. [CrossRef] [PubMed]
2. Hsieh, J. Orchestrating transcriptional control of adult neurogenesis. *Genes Dev.* **2012**, *26*, 1010–1021. [CrossRef] [PubMed]
3. Covic, M.; Karaca, E.; Lie, D.C. Epigenetic regulation of neurogenesis in the adult hippocampus. *Heredity* **2010**, *105*, 122–134. [CrossRef] [PubMed]
4. Ma, D.K.; Marchetto, M.C.; Guo, J.U.; Ming, G.L.; Gage, F.H.; Song, H. Epigenetic choreographers of neurogenesis in the adult mammalian brain. *Nat. Neurosci.* **2010**, *13*, 1338–1344. [CrossRef]
5. Fitzsimons, C.P.; van Bodegraven, E.; Schouten, M.; Lardenoije, R.; Kompotis, K.; Kenis, G.; van den Hurk, M.; Boks, M.P.; Biojone, C.; Joca, S.; et al. Epigenetic regulation of adult neural stem cells: Implications for Alzheimer's disease. *Mol. Neurodegener.* **2014**, *9*, 25. [CrossRef]
6. Altuna, M.; Urdánoz-Casado, A.; Sánchez-Ruiz de Gordoa, J.; Zelaya, M.V.; Labarga, A.; Lepesant, J.M.J.; Roldán, M.; Blanco-Luquin, I.; Perdones, Á.; Larumbe, R.; et al. DNA methylation signature of human hippocampus in Alzheimer's disease is linked to neurogenesis. *Clin. Epigenet.* **2019**, *11*, 91. [CrossRef]
7. Jobe, E.M.; Zhao, X. DNA Methylation and Adult Neurogenesis. *Brain Plast.* **2017**, *3*, 5–26. [CrossRef]
8. Mu, Y.; Gage, F.H. Adult hippocampal neurogenesis and its role in Alzheimer's disease. *Mol. Neurodegener.* **2011**, *6*, 85. [CrossRef]
9. Baglietto-Vargas, D.; Sanchez-Mejias, E.; Navarro, V.; Jimenez, S.; Trujillo-Estrada, L.; Gomez-Arboledas, A.; Sanchez-Mico, M.; Sanchez-Varo, R.; Vizuete, M.; Davila, J.C.; et al. Dual roles of Abeta in proliferative processes in an amyloidogenic model of Alzheimer's disease. *Sci. Rep.* **2017**, *7*, 10085. [CrossRef]
10. Coronel, R.; Bernabeu-Zornoza, A.; Palmer, C.; Muniz-Moreno, M.; Zambrano, A.; Cano, E.; Liste, I. Role of Amyloid Precursor Protein (APP) and Its Derivatives in the Biology and Cell Fate Specification of Neural Stem Cells. *Mol. Neurobiol.* **2018**, *55*, 7107–7117. [CrossRef]
11. Li, X.; Bao, X.; Wang, R. Neurogenesis-based epigenetic therapeutics for Alzheimer's disease (Review). *Mol. Med. Rep.* **2016**, *14*, 1043–1053. [CrossRef]
12. Moreno-Jimenez, E.P.; Flor-Garcia, M.; Terreros-Roncal, J.; Rabano, A.; Cafini, F.; Pallas-Bazarra, N.; Avila, J.; Llorens-Martin, M. Adult hippocampal neurogenesis is abundant in neurologically healthy subjects and drops sharply in patients with Alzheimer's disease. *Nat. Med.* **2019**, *25*, 554–560. [CrossRef]
13. Ellison, E.M.; Bradley-Whitman, M.A.; Lovell, M.A. Single-Base Resolution Mapping of 5-Hydroxymethylcytosine Modifications in Hippocampus of Alzheimer's Disease Subjects. *J. Mol. Neurosci.* **2017**, *63*, 185–197. [CrossRef]
14. Satir, T.M.; Nazir, F.H.; Vizlin-Hodzic, D.; Hardselius, E.; Blennow, K.; Wray, S.; Zetterberg, H.; Agholme, L.; Bergström, P. Accelerated neuronal and synaptic maturation by BrainPhys medium increases Aβ secretion and alters Aβ peptide ratios from iPSC-derived cortical neurons. *Sci. Rep.* **2020**, *10*, 601. [CrossRef]
15. Li, L.C.; Dahiya, R. MethPrimer: Designing primers for methylation PCRs. *Bioinformatics* **2002**, *18*, 1427–1431. [CrossRef]
16. Snyder, M.W.; Kircher, M.; Hill, A.J.; Daza, R.M.; Shendure, J. Cell-free DNA Comprises an In Vivo Nucleosome Footprint That Informs Its Tissues-of-Origin. *Cell* **2016**, *164*, 57–68. [CrossRef]
17. Kumaki, Y.; Oda, M.; Okano, M. QUMA: Quantification tool for methylation analysis. *Nucleic Acids Res.* **2008**, *36*, W170–W175. [CrossRef]
18. Livak, K.J.; Schmittgen, T.D. Analysis of relative gene expression data using real-time quantitative PCR and the 2(-Delta Delta C(T)) Method. *Methods* **2001**, *25*, 402–408. [CrossRef]
19. Compagnucci, C.; Piemonte, F.; Sferra, A.; Piermarini, E.; Bertini, E. The cytoskeletal arrangements necessary to neurogenesis. *Oncotarget* **2016**, *7*, 19414–19429. [CrossRef]
20. Azari, H. Isolation and enrichment of defined neural cell populations from heterogeneous neural stem cell progeny. *Methods Mol. Biol.* **2013**, *1059*, 95–106. [CrossRef]
21. Kempermann, G.; Song, H.; Gage, F.H. Neurogenesis in the Adult Hippocampus. *Cold Spring Harb. Perspect. Biol.* **2015**, *7*, a018812. [CrossRef]
22. Zhang, J.; Jiao, J. Molecular Biomarkers for Embryonic and Adult Neural Stem Cell and Neurogenesis. *BioMed Res. Int.* **2015**, *2015*, 727542. [CrossRef]
23. Pierret, C.; Morrison, J.A.; Rath, P.; Zigler, R.E.; Engel, L.A.; Fairchild, C.L.; Shi, H.; Maruniak, J.A.; Kirk, M.D. Developmental cues and persistent neurogenic potential within an in vitro neural niche. *BMC Dev. Biol.* **2010**, *10*, 5. [CrossRef]
24. Von Bohlen Und Halbach, O. Immunohistological markers for staging neurogenesis in adult hippocampus. *Cell Tissue Res.* **2007**, *329*, 409–420. [CrossRef]

25. Shohayeb, B.; Diab, M.; Ahmed, M.; Ng, D.C.H. Factors that influence adult neurogenesis as potential therapy. *Transl. Neurodegener.* **2018**, *7*, 4. [CrossRef]

26. Yu, D.X.; Di Giorgio, F.P.; Yao, J.; Marchetto, M.C.; Brennand, K.; Wright, R.; Mei, A.; McHenry, L.; Lisuk, D.; Grasmick, J.M.; et al. Modeling hippocampal neurogenesis using human pluripotent stem cells. *Stem Cell Rep.* **2014**, *2*, 295–310. [CrossRef]

27. Ambrogini, P.; Lattanzi, D.; Ciuffoli, S.; Agostini, D.; Bertini, L.; Stocchi, V.; Santi, S.; Cuppini, R. Morpho-functional characterization of neuronal cells at different stages of maturation in granule cell layer of adult rat dentate gyrus. *Brain Res.* **2004**, *1017*, 21–31. [CrossRef]

28. Ertaylan, G.; Okawa, S.; Schwamborn, J.C.; Del Sol, A. Gene regulatory network analysis reveals differences in site-specific cell fate determination in mammalian brain. *Front. Cell. Neurosci.* **2014**, *8*, 437. [CrossRef]

29. Huang, T.; Nguyen, L.H.; Lin, T.V.; Gong, X.; Zhang, L.; Kim, G.B.; Sarkisian, M.R.; Breunig, J.J.; Bordey, A. In utero electroporation-based translating ribosome affinity purification identifies age-dependent mRNA expression in cortical pyramidal neurons. *Neurosci. Res.* **2019**, *143*, 44–52. [CrossRef]

30. Gollan, L.; Salomon, D.; Salzer, J.L.; Peles, E. Caspr regulates the processing of contactin and inhibits its binding to neurofascin. *J. Cell Biol.* **2003**, *163*, 1213–1218. [CrossRef]

31. Santos, S.D.; Iuliano, O.; Ribeiro, L.; Veran, J.; Ferreira, J.S.; Rio, P.; Mulle, C.; Duarte, C.B.; Carvalho, A.L. Contactin-associated protein 1 (Caspr1) regulates the traffic and synaptic content of α-amino-3-hydroxy-5-methyl-4-isoxazolepropionic acid (AMPA)-type glutamate receptors. *J. Biol. Chem.* **2012**, *287*, 6868–6877. [CrossRef] [PubMed]

32. Zou, Y.; Zhang, W.F.; Liu, H.Y.; Li, X.; Zhang, X.; Ma, X.F.; Sun, Y.; Jiang, S.Y.; Ma, Q.H.; Xu, D.E. Structure and function of the contactin-associated protein family in myelinated axons and their relationship with nerve diseases. *Neural Regen. Res.* **2017**, *12*, 1551–1558. [CrossRef] [PubMed]

33. Wu, Z.Q.; Li, D.; Huang, Y.; Chen, X.P.; Huang, W.; Liu, C.F.; Zhao, H.Q.; Xu, R.X.; Cheng, M.; Schachner, M.; et al. Caspr Controls the Temporal Specification of Neural Progenitor Cells through Notch Signaling in the Developing Mouse Cerebral Cortex. *Cereb. Cortex* **2017**, *27*, 1369–1385. [CrossRef] [PubMed]

34. Buch; Lipi, B.; Langhnoja; Jaldeep, L.; Pillai, P.P.; Prakash, P. Role of astrocytic MeCP2 in regulation of CNS myelination by affecting oligodendrocyte and neuronal physiology and axo-glial interactions. *Exp. Brain Res.* **2018**, *236*, 3015–3027. [CrossRef]

35. Chen, W.; Zhang, L.; Shao, S.X.; Wang, H.P.; Cui, S.J.; Zhang, Y.N.; Kong, X.Z.; Yin, Q.; Zhang, J.P. Transcription factors GATA4 and TBX5 promote cardiomyogenic differentiation of rat bone marrow mesenchymal stromal cells. *Histol. Histopathol.* **2015**, *30*, 1487–1498. [CrossRef]

36. Grajales-Esquivel, E.; Luz-Madrigal, A.; Bierly, J.; Haynes, T.; Reis, E.S.; Han, Z.; Gutierrez, C.; McKinney, Z.; Tzekou, A.; Lambris, J.D.; et al. Complement component C3aR constitutes a novel regulator for chick eye morphogenesis. *Dev. Biol.* **2017**, *428*, 88–100. [CrossRef]

37. You, L.; Zou, J.; Zhao, H.; Bertos, N.R.; Park, M.; Wang, E.; Yang, X.J. Deficiency of the chromatin regulator BRPF1 causes abnormal brain development. *J. Biol. Chem.* **2015**, *290*, 7114–7129. [CrossRef]

38. Murao, N.; Noguchi, H.; Nakashima, K. Epigenetic regulation of neural stem cell property from embryo to adult. *Neuroepigenetics* **2016**, *5*, 1–10. [CrossRef]

39. Jjingo, D.; Conley, A.B.; Yi, S.V.; Lunyak, V.V.; Jordan, I.K. On the presence and role of human gene-body DNA methylation. *Oncotarget* **2012**, *3*, 462–474. [CrossRef]

40. Yu, D.H.; Ware, C.; Waterland, R.A.; Zhang, J.; Chen, M.H.; Gadkari, M.; Kunde-Ramamoorthy, G.; Nosavanh, L.M.; Shen, L. Developmentally programmed 3′ CpG island methylation confers tissue- and cell-type-specific transcriptional activation. *Mol. Cell. Biol.* **2013**, *33*, 1845–1858. [CrossRef]

41. Tirado-Magallanes, R.; Rebbani, K.; Lim, R.; Pradhan, S.; Benoukraf, T. Whole genome DNA methylation: Beyond genes silencing. *Oncotarget* **2017**, *8*, 5629–5637. [CrossRef]

42. Zhao, H.; Dean, A. Organizing the genome: Enhancers and insulators. *Biochem. Cell Biol.* **2005**, *83*, 516–524. [CrossRef]

43. Palstra, R.J.; Grosveld, F. Transcription factor binding at enhancers: Shaping a genomic regulatory landscape in flux. *Front. Genet.* **2012**, *3*, 195. [CrossRef]

44. Rharass, T.; Lantow, M.; Gbankoto, A.; Weiss, D.G.; Panáková, D.; Lucas, S. Ascorbic acid alters cell fate commitment of human neural progenitors in a WNT/β-catenin/ROS signaling dependent manner. *J. Biomed. Sci.* **2017**, *24*, 78. [CrossRef]

45. Rharass, T.; Lemcke, H.; Lantow, M.; Kuznetsov, S.A.; Weiss, D.G.; Panáková, D. Ca²⁺-mediated mitochondrial reactive oxygen species metabolism augments Wnt/β-catenin pathway activation to facilitate cell differentiation. *J. Biol. Chem.* **2014**, *289*, 27937–27951. [CrossRef]

46. Urbainsky, C.; Nölker, R.; Imber, M.; Lübken, A.; Mostertz, J.; Hochgräfe, F.; Godoy, J.R.; Hanschmann, E.M.; Lillig, C.H. Nucleoredoxin-Dependent Targets and Processes in Neuronal Cells. *Oxidative Med. Cell. Longev.* **2018**, *2018*, 4829872. [CrossRef]

47. Funato, Y.; Miki, H. Nucleoredoxin, a novel thioredoxin family member involved in cell growth and differentiation. *Antioxid. Redox Signal.* **2007**, *9*, 1035–1057. [CrossRef]

48. Valek, L.; Tegeder, I. Nucleoredoxin Knockdown in SH-SY5Y Cells Promotes Cell Renewal. *Antioxidants* **2021**, *10*, 449. [CrossRef]

49. Lazarevic, V.; Fienko, S.; Andres-Alonso, M.; Anni, D.; Ivanova, D.; Montenegro-Venegas, C.; Gundelfinger, E.D.; Cousin, M.A.; Fejtova, A. Physiological Concentrations of Amyloid Beta Regulate Recycling of Synaptic Vesicles via Alpha7 Acetylcholine Receptor and CDK5/Calcineurin Signaling. *Front. Mol. Neurosci.* **2017**, *10*, 221. [CrossRef]

50. Garcia-Osta, A.; Alberini, C.M. Amyloid beta mediates memory formation. *Learn. Mem.* **2009**, *16*, 267–272. [CrossRef]

51. Puzzo, D.; Privitera, L.; Palmeri, A. Hormetic effect of amyloid-beta peptide in synaptic plasticity and memory. *Neurobiol. Aging* **2012**, *33*, 1484.e15–1484.e24. [CrossRef]

52. Gulisano, W.; Melone, M.; Li Puma, D.D.; Tropea, M.R.; Palmeri, A.; Arancio, O.; Grassi, C.; Conti, F.; Puzzo, D. The effect of amyloid-β peptide on synaptic plasticity and memory is influenced by different isoforms, concentrations, and aggregation status. *Neurobiol. Aging* **2018**, *71*, 51–60. [CrossRef]

53. Malmsten, L.; Vijayaraghavan, S.; Hovatta, O.; Marutle, A.; Darreh-Shori, T. Fibrillar β-amyloid 1-42 alters cytokine secretion, cholinergic signalling and neuronal differentiation. *J. Cell. Mol. Med.* **2014**, *18*, 1874–1888. [CrossRef]

54. Wicklund, L.; Leao, R.N.; Stromberg, A.M.; Mousavi, M.; Hovatta, O.; Nordberg, A.; Marutle, A. Beta-amyloid 1-42 oligomers impair function of human embryonic stem cell-derived forebrain cholinergic neurons. *PLoS ONE* **2010**, *5*, e15600. [CrossRef] [PubMed]

55. Coronel, R.; Palmer, C.; Bernabeu-Zornoza, A.; Monteagudo, M.; Rosca, A.; Zambrano, A.; Liste, I. Physiological effects of amyloid precursor protein and its derivatives on neural stem cell biology and signaling pathways involved. *Neural Regen. Res.* **2019**, *14*, 1661–1671. [CrossRef] [PubMed]

56. Koppensteiner, P.; Trinchese, F.; Fa, M.; Puzzo, D.; Gulisano, W.; Yan, S.; Poussin, A.; Liu, S.; Orozco, I.; Dale, E.; et al. Time-dependent reversal of synaptic plasticity induced by physiological concentrations of oligomeric Abeta42: An early index of Alzheimer's disease. *Sci. Rep.* **2016**, *6*, 32553. [CrossRef] [PubMed]

57. Tincer, G.; Mashkaryan, V.; Bhattarai, P.; Kizil, C. Neural stem/progenitor cells in Alzheimer's disease. *Yale J. Biol. Med.* **2016**, *89*, 23–35. [PubMed]

58. Gatt, A.; Lee, H.; Williams, G.; Thuret, S.; Ballard, C. Expression of neurogenic markers in Alzheimer's disease: A systematic review and metatranscriptional analysis. *Neurobiol. Aging* **2019**, *76*, 166–180. [CrossRef]

59. Bond, A.M.; Ming, G.L.; Song, H. Adult Mammalian Neural Stem Cells and Neurogenesis: Five Decades Later. *Cell Stem Cell* **2015**, *17*, 385–395. [CrossRef]

60. Efthymiou, A.; Shaltouki, A.; Steiner, J.P.; Jha, B.; Heman-Ackah, S.M.; Swistowski, A.; Zeng, X.; Rao, M.S.; Malik, N. Functional screening assays with neurons generated from pluripotent stem cell-derived neural stem cells. *J. Biomol. Screen.* **2014**, *19*, 32–43. [CrossRef]

Article

Cell Type-Specific Role of RNA Nuclease SMG6 in Neurogenesis

Gabriela Maria Guerra [1,†], Doreen May [1], Torsten Kroll [1], Philipp Koch [1], Marco Groth [1], Zhao-Qi Wang [1,2,*], Tang-Liang Li [1,3] and Paulius Grigaravičius [1,*]

[1] Leibniz Institute on Aging—Fritz Lipmann Institute (FLI), Beutenbergstr 11, 07745 Jena, Germany; gabrielagrra@gmail.com (G.M.G.); Doreen.May@leibniz-fli.de (D.M.); Torsten.Kroll@leibniz-fli.de (T.K.); Philipp.Koch@leibniz-fli.de (P.K.); Marco.Groth@leibniz-fli.de (M.G.); li.tangliang@sdu.edu.cn (T.-L.L.)

[2] Faculty of Biological Sciences, Friedrich-Schiller University of Jena, Beutenbergstr 11, 07745 Jena, Germany

[3] State Key Laboratory of Microbial Technology, Shandong University, No. 72, Binhai Road, Qingdao 266237, China

* Correspondence: zhao-qi.wang@leibniz-fli.de (Z.-Q.W.); paulius.grigaravicius@leibniz-fli.de (P.G.); Tel.: +49-3641-656415 (Z.-Q.W.); +49-3641-656419 (P.G.)

† Present address: Deutsches Rheuma-Forschungszentrum (DRFZ), Virchowweg 12, 10117 Berlin, Germany.

Abstract: SMG6 is an endonuclease, which cleaves mRNAs during nonsense-mediated mRNA decay (NMD), thereby regulating gene expression and controling mRNA quality. SMG6 has been shown as a differentiation license factor of totipotent embryonic stem cells. To investigate whether it controls the differentiation of lineage-specific pluripotent progenitor cells, we inactivated *Smg6* in murine embryonic neural stem cells. Nestin-Cre-mediated deletion of *Smg6* in mouse neuroprogenitor cells (NPCs) caused perinatal lethality. Mutant mice brains showed normal structure at E14.5 but great reduction of the cortical NPCs and late-born cortical neurons during later stages of neurogenesis (i.e., E18.5). *Smg6* inactivation led to dramatic cell death in ganglionic eminence (GE) and a reduction of interneurons at E14.5. Interestingly, neurosphere assays showed self-renewal defects specifically in interneuron progenitors but not in cortical NPCs. RT-qPCR analysis revealed that the interneuron differentiation regulators *Dlx1* and *Dlx2* were reduced after *Smg6* deletion. Intriguingly, when *Smg6* was deleted specifically in cortical and hippocampal progenitors, the mutant mice were viable and showed normal size and architecture of the cortex at E18.5. Thus, SMG6 regulates cell fate in a cell type-specific manner and is more important for neuroprogenitors originating from the GE than for progenitors from the cortex.

Keywords: SMG6; NMD; neurogenesis; neurodevelopmental syndromes

Citation: Guerra, G.M.; May, D.; Kroll, T.; Koch, P.; Groth, M.; Wang, Z.-Q.; Li, T.-L.; Grigaravičius, P. Cell Type-Specific Role of RNA Nuclease SMG6 in Neurogenesis. *Cells* **2021**, *10*, 3365. https://doi.org/10.3390/cells10123365

Academic Editor: FengRu Tang

Received: 27 October 2021
Accepted: 26 November 2021
Published: 30 November 2021

1. Introduction

Cell fate relies on the correct "read" of the genetic code and its translation into functional proteins. Nonsense-mediated mRNA decay (NMD) is a cellular surveillance mechanism that is involved in controlling the quality of mRNA [1–3]. It degrades transcripts that, after a nonsense mutation or alternative splicing events, harbor a premature termination codon (PTC) before (>50–55 nucleotides) an exon–exon junction complex (EJC). The stable interaction of UPF1 with eRFs at the PTC site recruits the NMD factors UPF2, UPF3 and the kinase SMG1. SMG1 phosphorylates UPF1 and UPF2 thereby promoting the recruitment of the endonuclease SMG6 or the SMG5/SMG7-mediated exonuclease for RNA degradation [1]. The branches of SMG6- and SMG5/7-mediated NMD pathways have been shown to overlap, yet with distinct differences in certain populations of target transcripts [4]. Recently, it was demonstrated that SMG6 (Suppressor with morphogenetic effect on genitalia protein 6) endonuclease activity can depend on the SMG5/SMG7 heterodimer [5]. Moreover, non-mutant transcripts can also be targeted by NMD, for example the transcripts with uORF and long 3′UTR, thereby regulating normal gene expression [4,6].

NMD deficiency leads to an accumulation of deleterious mRNA products, which can be translated not only into mutated malfunctional proteins but also into functional ones, which then may cause tissue dysfunction or pathogenesis [3,7–9]. The complete knockout of NMD genes, such as *Upf1* [10], *Upf2* [11] or *Smg1* [12] resulted in embryonic lethality, highlighting the importance of NMD for early development. Previously, we have shown that the complete deletion of *Smg6* in mouse germline blocks differentiation of embryonic stem (ES) cells into germ layers and thereby results in early embryonic lethality [13]. Furthermore, we demonstrated that SMG6 is required for the production of induced pluripotent stem cells (iPS) from somatic cells [13]. This identified SMG6-mediated NMD as a fundamental regulator of cell fate change, thereby controlling differentiation and development. In that study, the telomeric function of SMG6, originally identified as *Est1a* (Ever Shorter Telomere 1a) in yeast, was proven to be negligible.

In humans, mutations or deletion of NMD factors have been associated with several neurological disorders [14]. Human UPF3B, encoded by the X-linked gene *UPF3B*, was the first NMD factor linked with human neurodevelopmental syndromes such as X-linked intellectual disability (ID) with and without autism, childhood onset schizophrenia (COS) and attention deficit hyperactivity disorder (ADHD) [15,16]. Mutations of *UPF3B* led to the loss or truncation of the protein and caused ID in males of several families [15,16]. Genetic studies have identified mutations or copy number variants in *UPF2*, *UPF3*, *SMG9* and the exon-junction complex component *RBM8A* in human patients with neurological symptoms [17–19]. Furthermore, NMD proteins (UPF3b, UPF1, UPF2, SMG1) have been shown to participate in axon outgrowth, synapse formation and, thus, in various behavioral processes [20–26]. All these observations suggest that, in addition to its general role in the early embryonic development, NMD has specific functions in self-renewal and differentiation of neuroprogenitor cells as well as in neuronal functionality. However, the precise molecular mechanism underlying these processes remains largely unknown.

In order to elucidate the function of SMG6 in neurogenesis during brain development, we applied genetic and cellular studies to inactivate *Smg6* in various neural progenitor cells. We found that SMG6 is more critical for the cell fate determination of interneuron progenitors in the ganglionic eminence (GE) compared to cortical neuroprogenitor cells (NPCs).

2. Material and Methods

2.1. Mice

Smg6^{flox}(*Smg6$^{tm1.1Zqw}$*), Smg6$^{+/\Delta}$(*Smg6$^{tm1.2Zqw}$*) [13], Nes-Cre (*Tg(Nes-cre)1Kln*) [27], Emx1-Cre (*Emx1$^{tm1(cre)Krj}$*) [28], Rosa26-CreER^{T2} (Gt(ROSA)26Sor$^{tm1(cre/ERT2)Tyj}$), *Smg6*-CNSΔ and *Smg6*-CoHiΔ mice were bred and housed in the mouse facility of Fritz Lipmann Institute (FLI, Jena, Germany). Mice were fed ad libitum with standard laboratory chow and water in ventilated cages under a 12 h light/dark cycle. All animal work was conducted according to the German animal welfare legislation and approved by the Thüringer Landesamt für Verbraucherschutz (TLV). The genotyping of mice was performed by PCR on DNA extracted from tail tissue as previously described [13].

2.2. Immunoblot Analysis

Proteins were extracted using RIPA buffer supplemented with 1 mM PMSF and 20–50 µg of cell lysates were separated with SDS–PAGE as described [29]. The primary antibodies rabbit anti-*Smg6*/Est1A (1:1500; Abcam, Berlin, Germany), mouse anti-GAPDH (1:20,000; Sigma-Aldrich, Taufkirchen, Germany), mouse anti-Actin (1:20,000; Sigma-Aldrich), and secondary antibodies HRP-conjugated goat anti-rabbit IgG or goat anti-mouse IgG (1:10,000; DAKO, Frankfurt am Main, Germany), were used.

2.3. RT-qPCR

Total RNA was isolated from neural stem cells using the RNeasy Mini Kit (Qiagen, Hilden, Germany) following the manufacture's recommendations. After genomic DNA removal by DNAse I, the cDNA library was generated using SuperScript™ III Reverse Transcriptase (Invitrogen, Thermo Fisher Scientific, Waltham, MA, USA). The quantitative real-time PCR (qPCR) was performed in triplicates for each sample using Platinum™ SYBR™ Green qPCR SuperMix-UDG (Invitrogen) and a LightCycler® 480 Instrument (Roche). The sequences of the used primers are summarized in Table 1.

Table 1. Primers used for qPCR analysis.

Transcript	Primer Sequence (5′ to 3′)	Reference
Smg6	AGGAATTGGACAGCCAACAG TCTCGGTTTTATCCGGTTTG	Li et al., 2015 [13]
Gas5	TTTCCGGTCCTTCATTCTGA TCTTCTATTTGAGCCTCCATCCA	Weischenfeldt et al., 2008 [11]
Ifrd1	ATCGGACTGTTCAACCTTTCAG GCACTCTTATCAAGGGTTAGGTC	Park et al., 2017 [30]
Gad67	CACAGGTCACCCTCGATTTTT ACCATCCAACGATCTCTCTCATC	MGH primerbank [31]
Dlx1	GGCTGTGTTTATGGAGTTTGGG CCTGGGTTTACGGATCTTTTTC	Wang et al., 2016 [32]
Dlx2	GTGGCTGATATGCACTCGACC GCTGGTTGGTGTAGTAGCTGC	MGH primerbank [31]
Mash1	TCTCCTGGGAATGGACTTTG GGTTGGCTGTCTGGTTTGTT	Kraus et al., 2013 [33]
Actin	AGAGGGAAATCGTGCGTGAC CAATAGTGATGACCTGGCCGT	Li et al., 2015 [13]
Gapdh	GTGTTCCTACCCCCAATGTGT ATTGTCATACCAGGAAATGAGCTT	Li et al., 2015 [13]

2.4. Neurosphere Formation Assay and In Vitro Neuronal Stem Cell Differentiation

Neuroprogenitor cells were isolated and used for neurosphere formation assays from E13.5 cortices and ganglionic eminences as previously described [29]. After isolation, neuroprogenitor cells were plated for neurosphere formation in DMEM/F12 medium (Gibco, Thermo Fisher Scientific, Waltham, MA, USA) supplemented with 2% B-27 supplement (Invitrogen), 1X penicillin and streptomycin (Thermo Fischer Scientific, Waltham, MA, USA), 20 ng/mL EGF and 10 ng/mL bFGF (PeproTech, East Windsor, NJ, USA). Formed neurosphere numbers and cell numbers were counted after 7 days in culture. For the in vitro differentiation, neurospheres were trypsinised and cells were plated on poly-L-lysine (10 µg/mL overnight at room temperature, P5899, Sigma-Aldrich) and then laminin (10 µg/mL, 30 min at 37 °C, L2020, Sigma-Aldrich)-coated flat bottom 96-well plates (CellCarrier 96 Ultra, Cat# 6005550). After 2 days, the differentiation was initiated by changing the neural stem cell medium by differentiation medium: DMEM/F12 (Gibco) supplemented with 1% FSC (Thermo Fischer Scientific), 2% B27 (Invitrogen), 1X penicillin and streptomycin (Thermo Fischer Scientific). Cells were fixed, permeabilized and applied for immunofluorescent staining as described previously [34]. β-Tubulin III (TUJ1) antibody (1:200, Covance, MMS-435P, Princeton, NJ, USA) was used to detect neurons, and GFAP antibody (1:400, Dako, Z0334, Frankfurt am Main, Germany) was used to detect astrocytes.

2.5. Histology, TUNEL Reaction and Immunofluorescent Staining

Brains from E14.5 and E18.5 embryos were fixed overnight with 4% paraformaldehyde (PFA) (pH 7.2) and cryopreserved in 30% sucrose overnight. Neg-50 (Thermo Fischer Scientific) frozen section medium was used to embed the brains followed by cryosectioning (Microm™ HM 550 Cryostat, Thermo Fischer Scientific, Waltham, MA, USA) of 12 µm thick slices. After antigen retrieval in citrate buffer for 40 min at 95 °C, immunostaining with the following primary antibodies was performed: rabbit anti-SOX2 (1:200, Abcam, Ab97959); rabbit anti-TBR2/Eomes (1:200, Abcam, Ab23345); rabbit anti-TBR1 (1:200, Abcam, Ab31940); rat anti-CTIP2 (1:200, Abcam, Ab18465); rabbit anti-CUX1/CDP (1:100, Santa Cruz, Sc-13024, Heidelberg, Germany); rabbit anti-A-calbindin D-28k (1:1000, Swant, CB38, Burgdorf, Switzerland); anti-phospho-histone H3 (Ser10) (1:400, Cell Signaling, 9071, Danvers, MA, USA); and anti-Cleaved Caspase-3 (1:200, Cell Signaling, 9661S).

Immunoreactivity was visualized using secondary antibodies anti-rabbit IgG conjugated with Cy3 (1:200, Sigma-Aldrich), Cy2 or Cy5 (1:200, Jackson ImmunoResearch), donkey anti-mouse IgG conjugated with Cy3 (1:200; Sigma-Aldrich), goat anti-rabbit Biotin conjugated (1:400, Vector Laboratories, Burlingame, CA, USA) and donkey anti-rat IgG conjugated with Alexa-488 (1:200, Sigma-Aldrich), streptavidin-Cy3 (1:800, Sigma-Aldrich). Overall cell death was detected using TUNEL reaction, as described previously [35]. In all cases, the nuclei were counterstained using DAPI (1:10,000, Sigma-Aldrich) and mounted in ProLong Gold Antifade reagent (P36930, Invitrogen).

2.6. Microscopy and Image Analysis

The images of whole brain sections were acquired using BX61VS Olympus Virtual microscope and processed with Olympus Olyvia 2.9 computer program. The immunofluorescence images of the in vitro differentiation of neural stem cells were acquired using the microscope ImageXpress Micro Confocal (short IXMC) from Molecular Device (MD). In each well, a z-stack of seven images with 1 μm distance was recorded using $10\times$ Plan Apo objective with confocal mode (50 μm slit disc) at four sites. For subsequent image analysis, a custom module in the MetaXpress software from MD (version 6.2.3.) was created and applied on maximum projections of each z-stack. Each cell was defined by its nucleus using a mask derived from the DAPI channel. Depending on the intensity within this mask in the other channels, each nucleus was classified as belonging to a neuron, astrocyte or unclassified cell. This led to a summarized number of each cell type in each well.

2.7. RNA-Seq and Bioinformatic Analysis

NPCs were isolated from E13.5 embryo brains after crossing $Smg6^{flox}$ and Rosa26-CreER^{T2} mice, and kept in neurosphere cultures for 2 days followed by 4-Hydroxytamoxifen (4-OHT, Sigma) treatment to induce the $Smg6$ deletion. Total RNA was isolated 6 days after 4-OHT treatment from cells with genotypes $Smg6^{flox/flox}$; Rosa26-CreER^{T2} (without 4-OHT) (ctr), $Smg6^{flox/flox}$ treated with 4-OHT (ctr + 4-OHT) and $Smg6^{flox/flox}$; Rosa26-CreER^{T2} treated with 4-OHT ($Smg6$-iKO)) using the RNeasy Mini Kit (Qiagen) and following the manufacturer's manual. The RNA integrity was checked using an Agilent Bioanalyzer 2100 (Agilent Technologies). All samples showed a RIN (RNA integrity number) higher than 9. Approximately 800 ng of total RNA was used for library preparation using a TruSeq Stranded Total RNA (RiboZero Gold) according to the manufacturer's protocol. The libraries were pooled into one and sequenced in three lanes using HiSeq2500 (Illumina) in single-read high-output mode, which created reads with a length of 50 bp. Sequencing reads were extracted using bcl2FastQ v1.8.4. On average, 73 million reads per sample were obtained. For expression analysis, the raw reads were mapped with STAR (version 2.5.4b, parameters: –alignIntronMax 100000–outSJfilterReads –outSAMmultNmax Unique –outFilterMismatchNoverLmax 0.04) [36] to the *Mus musculus* genome (GRCm38) with the Ensembl genome annotation (Release 92). For each Ensembl gene, reads that mapped uniquely to one genomic position were counted with FeatureCounts (version 1.5.0, multi-mapping or multi-overlapping reads were not counted, stranded mode was set to "–s 2", Ensembl release 92 gene annotation) [37]. The table of raw counts per gene per sample was analysed with R (version 3.5.0) using the package DESeq2 (version 1.20.0) [38]. The sample group $Smg6$-iKO ($n = 4$) was contrasted with the sample group Ctr + 4-OHT ($n = 4$), with $Smg6$-iKO being the reference level. The sample group Ctr ($n = 3$) was also contrasted with the sample group $Smg6$-iKO ($n = 4$) and the Ctr + 4-OHT ($n = 4$). For each gene of the comparison, the p-value was calculated using the Wald significance test. Resulting p-values were adjusted for multiple testing with Benjamini & Hochberg correction. Genes with an adjusted p-value < 0.05 were considered differentially expressed (DEGs). The log2 fold changes (LFCs) were shrunk with lfcShrink to control for variance of LFC estimates for genes with low read counts. The changes in molecular pathways as well as their possible upstream regulators were identified by analyzing abovementioned three pairwise comparisons using Ingenuity Pathway Analysis (IPA) program (Qiagen).

2.8. Statistical Analysis

Depending on the distribution of the data points, unpaired two-tailed Student's *t*-test or Mann–Whitney U (MWU) test were used to calculate significance. Data sets underwent Shapiro–Wilk test for the normal distribution. If the data set passed the Shapiro–Wilk test (p value > 0.05), Student's *t*-test was used, if it did not pass (p value < 0.05), Mann–Whitney U test was applied. Statistical analyses were performed using GraphPad Prism 8 (GraphPad Software, San Diego, CA, USA). The type of the test performed is indicated in each figure legend. Indication for significance was used as follows: n.s. > 0.05, * < 0.05, ** < 0.01, *** < 0.001, **** < 0.0001.

3. Results

3.1. Smg6 Deletion in CNS Compromises Embryonic Neurogenesis and Newborn Viability

In order to understand the biological function of SMG6 specifically in the differentiation program of committed lineage NPCs in the central nervous system (CNS), we generated a conditional knockout mouse model in which *Smg6* was deleted in NPCs from embryonic day E10.5 by intercrossing $Smg6^{F/\Delta}$ [13] and *Nestin-Cre* transgenic mice [27] to generate CNS-deleted mice (*Smg6*-CNSΔ). We obtained an expected number of *Smg6*-CNSΔ embryos at E14.5 and a slightly reduced number at E18.5 (Figure S1A) according to the Mendelian ratio. However, all *Smg6*-CNSΔ newborns died within 1–2 days after birth (Figure S1B).

To investigate the role of SMG6 in neurogenesis, we first analysed brains at E14.5 and found that an efficient *Smg6* deletion in the CNS (Figure S1C) yielded normal embryo body and brain weight as well as cortex (CTX) thickness (Figure S1D–G). Furthermore, cortical cellularity and populations of SOX2+ neural progenitor cells and TBR2+ intermediate progenitors (IPs), as well as populations of early newborn neurons positive for TBR1 and CTIP2, were the same as in controls at this stage of development (Figure 1A–C). Next, we examined the brains just before birth at E18.5 and found that *Smg6*-CNSΔ fetuses had normal body and brain weights (Figure S1H–J). However, we detected a significantly smaller CTX in E18.5 *Smg6*-CNSΔ brains (Figure 1D), indicating mild microcephaly and defects in embryonic neurogenesis. Concomitantly, the *Smg6*-CNSΔ cortices presented a significant reduction of cortical cellularity at this stage (Figure 1E), likely responsible for the reduction of the CTX.

Mouse CTX contain well defined cellular layers composed of neural precursors in the ventricular and subventricular proliferative zones (VZ and SVZ, respectively), early born neurons in the middle part (layers VI and V) and late born neurons in the upper part of the cortical plate (layers II/III). Next, we histologically analysed the structure of the E18.5 *Smg6*-CNSΔ brains. It revealed that all the neuronal layers in *Smg6*-CNSΔ cortices were formed, but with a significant decrease of SOX2+ NPCs in the VZ and TBR2+ IPs in the SVZ, indicating an exhaustion of NPC pools during the late embryonic brain development (Figure 1F,G). The numbers of early born neurons in layers VI and V positive for TBR1 and CTIP2 were normal, in contrast to a significant reduction of the late born neurons in the layers II/III judged by the CUX1+ population (Figure 1F,G). These findings indicate that SMG6 is dispensable for early cortical neurogenesis, but its deletion prematurely depleted NPC pools and compromised cortical neurogenic production, affecting the cellularity of the CTX during later development.

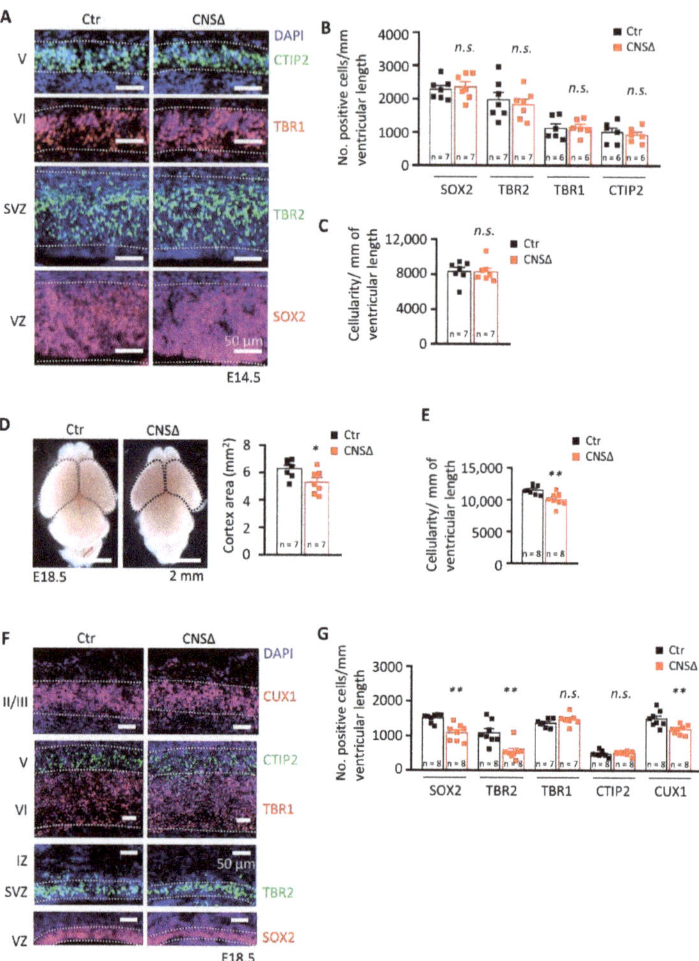

Figure 1. SMG6 deficiency in the central nervous system causes perinatal lethality. (**A**) Representative immunofluorescent staining of E14.5 cortices, coronal sections. SOX2 and TBR2 stained the progenitors of the VZ and SVZ, respectively. TBR1 and CTIP2 stained the early born neurons of layers VI and V, respectively. DAPI was used to stain the cell nucleus. VZ: ventricular zone, SVZ: subventricular zone. (**B**) Quantification of the neuroprogenitor cells (SOX2 and TBR2) and neurons (TBR1 and CTIP2) in E14.5 cortices in A, Student's *t*-test was applied for all markers. (**C**) Quantification of the total cellularity in the E14.5 cortices, Student's *t*-test. (**D**) Smaller CTX area of the E18.5 *Smg6*-CNSΔ brains, Student's *t*-test. (**E**) Quantification of the total cellularity in the E18.5 cortices, Student's *t*-test. (**F**) Immunofluorescent staining of E18.5 coronal sections showing only cortical region. SOX2 and TBR2 labels NPCs and intermediate progenitor cells in the VZ and SVZ, respectively. TBR1 and CTIP2 stained the early born neurons in layers VI and V. CUX1 stained the later born neurons in layers IV, III and II. DAPI was used to counterstain the cell nucleus. VZ: ventricular zone, SVZ: subventricular zone. (**G**) Quantification of the NPC cells (SOX2 and TBR2) and neurons (TBR1, CTIP2 and CUX1) in E18.5 cortices shown in D (MWU test for SOX2 and TBR2, Student's *t*-test for the rest). For all graphs: *n*—number of embryos analysed. Error bars represent SEM, statistic comparison as indicated in each graph description—n.s. > 0.05, * $p < 0.05$, ** $p < 0.01$.

3.2. SMG6 Is Essential for Survival of Neural Cells

In order to investigate the cause for the reduction of NPCs and late born neurons of E18.5 *Smg6*-CNSΔ brains, we analysed cell death by TUNEL and Active-Caspase 3 (Act-Cas3) staining. At both E14.5 and E18.5 stages *Smg6*-CNSΔ brains exhibited a significant increase of TUNEL and Act-Cas3 signals in the areas of CTX and GE (Figures 2A–H and S2A–H). The cell death at E14.5 was mainly found in the proliferative VZ and SVZ areas, where double TBR2 + TUNEL+ staining confirmed the death of IPs (Figure 2A,B). Intriguingly, TUNEL staining in GE (Figure 2C,D) detected clearly higher (Figure 2D versus Figure 2Bi) cell death in GE than in CTX at E14.5 brain sections. At E18.5, TUNEL positive cells in CTX were additionally detected in the intermediate zone (IZ) as well as in the cortical plate (CP) (Figure 2E,F), suggesting neuronal death at a late stage of brain development. However, it stayed at comparable levels to E14.5 (Figure 2Bi versus Figure 2Fi) whereas in GE (Figure 2G,H) we detected less TUNEL signals compared to younger embryos (Figure 2H versus Figure 2D). Immunofluorescence staining for Act-Cas3 confirmed elevated Caspase 3-dependent cell death in the CTX as well as in the GE (Figure S2A–H). Counting of phospho- Histone H3 (Ser10)+ cells did not reveal any difference for the number of mitotic cells in the CTX and GE between mutant and control animals at E14.5 (Figure S3A,B). Interestingly, SMG6 seems to be more crucial for the survival of NPCs of GABAergic interneurons (IN) that are generated in the GE. To confirm this, we stained E14.5 cortices with calbindin, an interneuron marker [39–41], and detected significantly less calbindin+ IN in *Smg6*-CNSΔ GE (Figure 2I,J) compared to controls. These observations indicate that at early neurogenesis, SMG6 absence affects the survival of NPCs prominently in the GE, and to a lesser extend in the CTX.

3.3. Smg6 Deletion Compromises the Self-Renewal and Differentiation Capacity of Neuroprogenitors

To investigate the renewal capacity of SMG6-deficient NPCs, we performed the in vitro neurosphere formation assay using neural stem cells isolated from the CTX or the GE at E13.5 (Figure S4A). The control and *Smg6*-CNSΔ cortical NPCs formed the same number of neurospheres after 7 days (Figure 3A,Ai), containing a comparable number of cells (Figure 3Aii), indicating a dispensable role of SMG6 in the renewal capacity of cortical NPCs. In contrast, mutant GE NPCs gave rise to a similar number of neurospheres (Figure 3B,Bi), but these neurospheres contained much fewer cells (Figure 3Bii). These results indicate that SMG6 is specifically critical for the renewal capacity of the GE NPCs, but less so for those from the CTX. Concomitantly, the mRNA expression levels of the transcription factors *Dlx1*, *Dlx2* and *Mash1*, known to drive interneuron differentiation [42–47], were dramatically reduced in the neurospheres originating from SMG6-deficient GE (Figure 3C). These findings indicate a specific role of SMG6 in interneuron progenitors.

Because SMG6 is essential for ES cell differentiation [13], we next studied the differentiation potential of *Smg6*-deleted neuroprogenitors. To this end, we used neurospheres derived originally from NPCs of the CTX or GE (Figure S4A) and induced their differentiation in vitro. We found significantly less neurons, judged by TUJ1 staining, at 6 and 8 days post differentiation (dpd) of NPCs originated from both the CTX and GE (Figure 3D,G,H,K). The percentage of GFAP-positive cells, representing astrocytes, was modestly but significantly increased in *Smg6* mutant cultures (Figure 3F,J). In addition, we observed overall reduced cellularity at 8 dpd compared to controls. It is of note that the reduction of cell numbers at 8 dpd was more prominent in differentiation cultures from the GE (76%) than in cultures from the CTX (29%) (Figure 3E,I). In summary, SMG6 plays a role in the differentiation program in vitro and is required for the survival of differentiating cells particularly of GE derived NPCs.

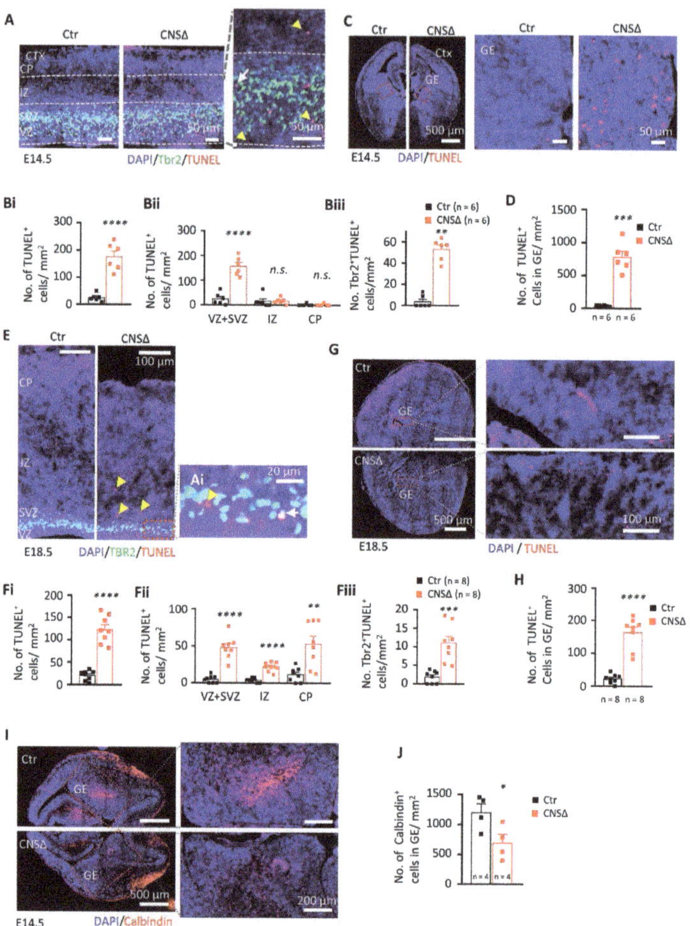

Figure 2. SMG6 deficiency causing death of neuronal cells. (**A**) Co-staining of dying cells using TUNEL and intermediate progenitors by labelling TBR2 on coronal sections of E14.5 embryo brains counterstained by DAPI for cell nucleus. Yellow head arrow marks TUNEL-positive cells, white arrow—TUNEL- and TBR2-positive cells. Quantification of the total TUNEL-positive cells (**Bi**) their distribution in different cortical layers (**Bii**) and dying TBR2-positive (TUNEL$^+$TBR2$^+$) cells (**Biii**) at E14.5 indicate neural cell death mainly in area of progenitors. Statistics MWU for IZ and CP in Bii and Biii, Welch *t*-test in Bi and Student's *t*-test for the rest. (**C**) TUNEL staining of GE at E14.5 with quantification respectively in (**D**) using Student's *t*-test. (**E**) Co-staining of dying cells using TUNEL and intermediate progenitors by TBR2 on coronal sections of E18.5 embryo brains counterstained by DAPI for cell nucleus. Yellow head arrow marks TUNEL-positive cells, white arrow—TUNEL- and TBR2-positive cells. Quantification performed in the same manner at E14.5 (**Fi**–**Fiii**) shows increased cell death in all cortical layers. Statistics by Student's *t*-test. (**G**) TUNEL staining of GE at E18.5 quantified in (**H**) using Welch *t*-test. (**I**) Immuno-fluorescent staining and quantification (**J**) of calbindin-positive cells in coronal sections of E14.5 embryo brains. DAPI counterstains the cell nucleus. Statistics by Student's *t*-test. For all graphs: VZ—ventricular zone; SVZ—subventricular zone; IZ—Intermediate zone; CP—cortical plate; *n*—number of embryos analysed. Error bars represent SEM. Significance—n.s. > 0.05, * *p* < 0.05, ** *p* < 0.01, *** *p* < 0.001, **** *p* < 0.0001.

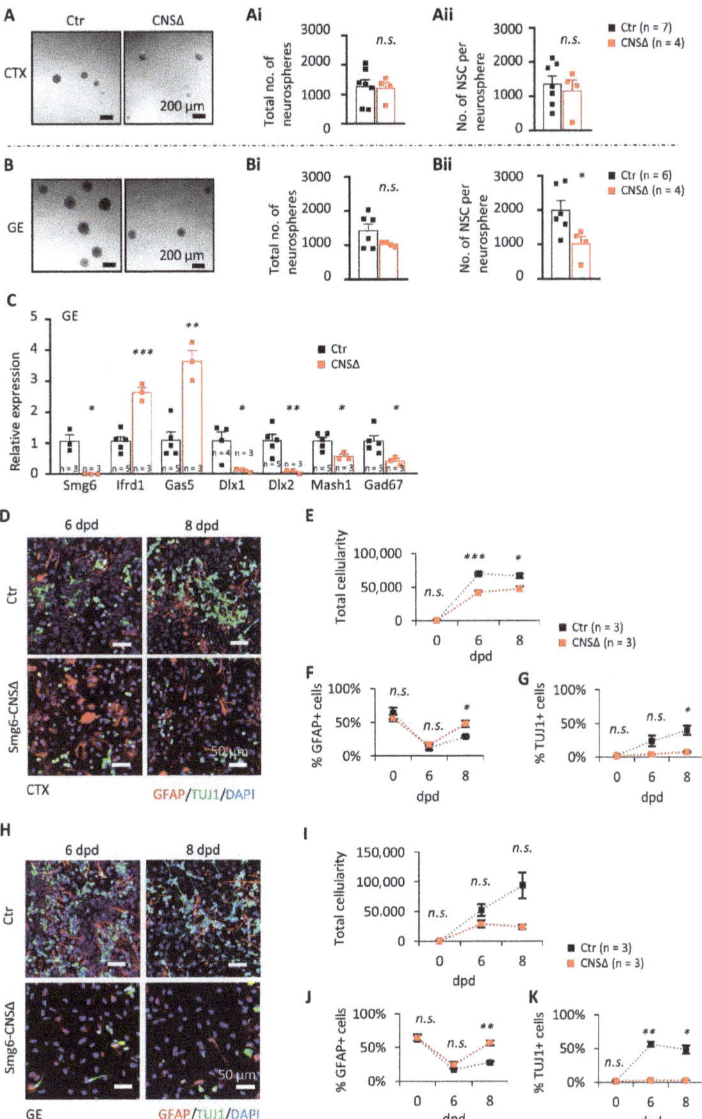

Figure 3. Neuroprogenitor renewal and differentiation are impaired without SMG6, especially in the NPCs from GE. (**A**) In vitro neurosphere assay on NPCs from CTX shows no significant difference in the number of neurospheres (**Ai**) as well as in the cell number per neurosphere (**Aii**) after 7 days in culture. The numbers of neurospheres formed from GE cells (**B**) also did not differ (**Bi**), but had fewer cells per neurosphere (**Bii**). (**C**) qPCR analysis on GE neurospheres shows relative expression changes of indicated gene mRNAs after *Smg6* deletion. (**D–G**) Differentiation capacity of progenitors from CTX and (**H–K**) GE at 6 and 8 days after differentiation induction (dpd). (**D,H**) Co-stainings of in vitro differentiated cultures from CTX and GE, respectively, at the indicated time points with quantifications of total cellularity in (**E,I**), GFAP-positive cells (**E,J**) and neurons labelled with TUJ1 staining (**G,K**). For all graphs: *n*—number of embryos used for cell isolations. Error bars represent SEM. Statistics by unpaired Student's *t*-test in (**A–C**) and Welch *t*-test in (**E–G,I–K**) significance— n.s. > 0.05, * $p < 0.05$, ** $p < 0.01$, *** $p < 0.001$.

3.4. SMG6 Deficiency Exclusively in the Cortex Does Not Inhibit Corticogenesis

Nestin-Cre drives deletion of the *Smg6* gene in the whole brain; thus, CTX and GE both are affected in *Smg6*-CNSΔ mice. To further dissect the SMG6 function in different populations of NPCs, we deleted *Smg6* only in the cortical/hippocampal progenitor cells at day E9.5 by crossing *Smg6*-floxed mouse with the Emx1-Cre mouse model (designated as *Smg6*-CoHiΔ). Intriguingly, in contrast to *Smg6*-CNSΔ mutants, the *Smg6*-CoHiΔ mice were born at expected ratios (Figure S4B) and were viable during the observation period of 20 months. We confirmed specific deletion of the SMG6 protein in the CTX, but not in other parts of the brain such as the GE and the hind/mid brain (Figure S4F). Notably, the SMG6 levels seem to be reduced less efficiently than in *Smg6*-CNSΔ embryos because the CTX of *Smg6*-CoHiΔ embryos contain a high number of cells, i.e., IN that are not affected by the deletion and thus have normal levels of SMG6. The body and brain weight of *Smg6*-CoHiΔ embryos as well as the CTX were normal at E18.5 (Figures S4C–E and 4A,B). Moreover, the cellularity and thickness of *Smg6*-CoHiΔ cortices were the same as controls (Figure 4C–E). Microscopic analysis of immunofluorescent staining revealed a similar number of TBR1+ and CTIP2+ neurons as well as SOX2+ and TBR2+ neuroprogenitor populations between mutant and control littermates (Figure 4D,F). In contrast to *Smg6*-CNSΔ, TUNEL assay did not detect obvious cell death in the CTX (Figure 5A,B) nor in the GE (Figure 5C,D). However, we detected a significant increase of cellular death only in the retrosplenial cortex of E18.5 *Smg6*-CoHiΔ embryos (Figure 5E,F), indicating that SMG6 is vital specifically and only for this small population of cortical cells.

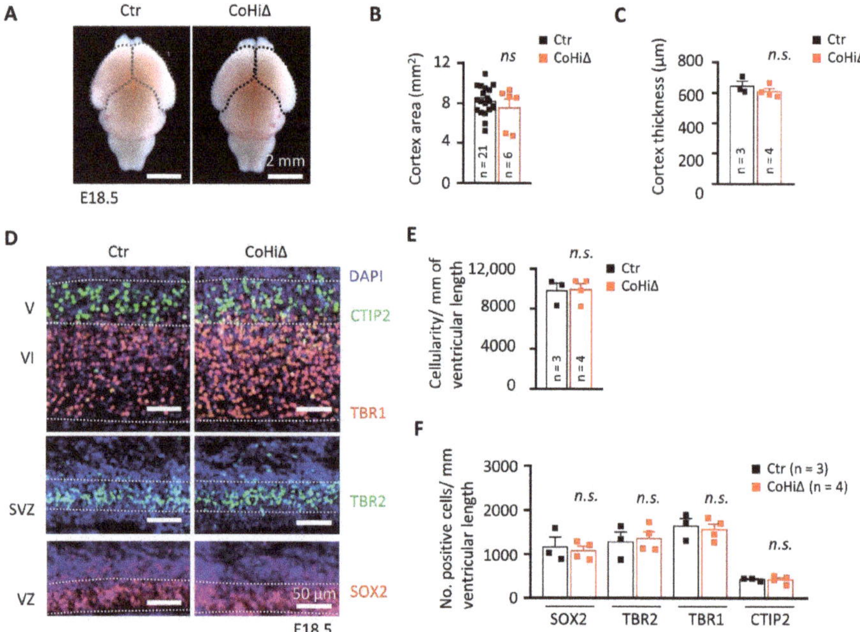

Figure 4. SMG6 deficiency solely in cortical and hippocampal neuroprogenitors does not inhibit corticogenesis. (**A**) Comparison of the E18.5 brains and quantification of the CTX area (**B**). (**C**) Quantification of the CTX thickness. (**D**) Immunofluorescent staining of E18.5 coronal sections showing only cortical regions. SOX2 and TBR2 labels NPCs and intermediate progenitors in the VZ and SVZ, respectively. TBR1 and CTIP2 stains early born neurons in layers VI and V. DAPI counterstains the cell nucleus. VZ: ventricular zone, SVZ: subventricular zone. (**E**) Quantification of the total cellularity in the E18.5 cortices. (**F**) Quantification of the neuroprogenitor cells (SOX2 and TBR2) and neurons (TBR1 and CTIP2) in E18.5 cortices shown in D. For all graphs: *n*—number of embryos analysed. Error bars represent SEM. Statistics by unpaired Student's *t*-test, except in (**B**) the MWU was used, significance—n.s. > 0.05.

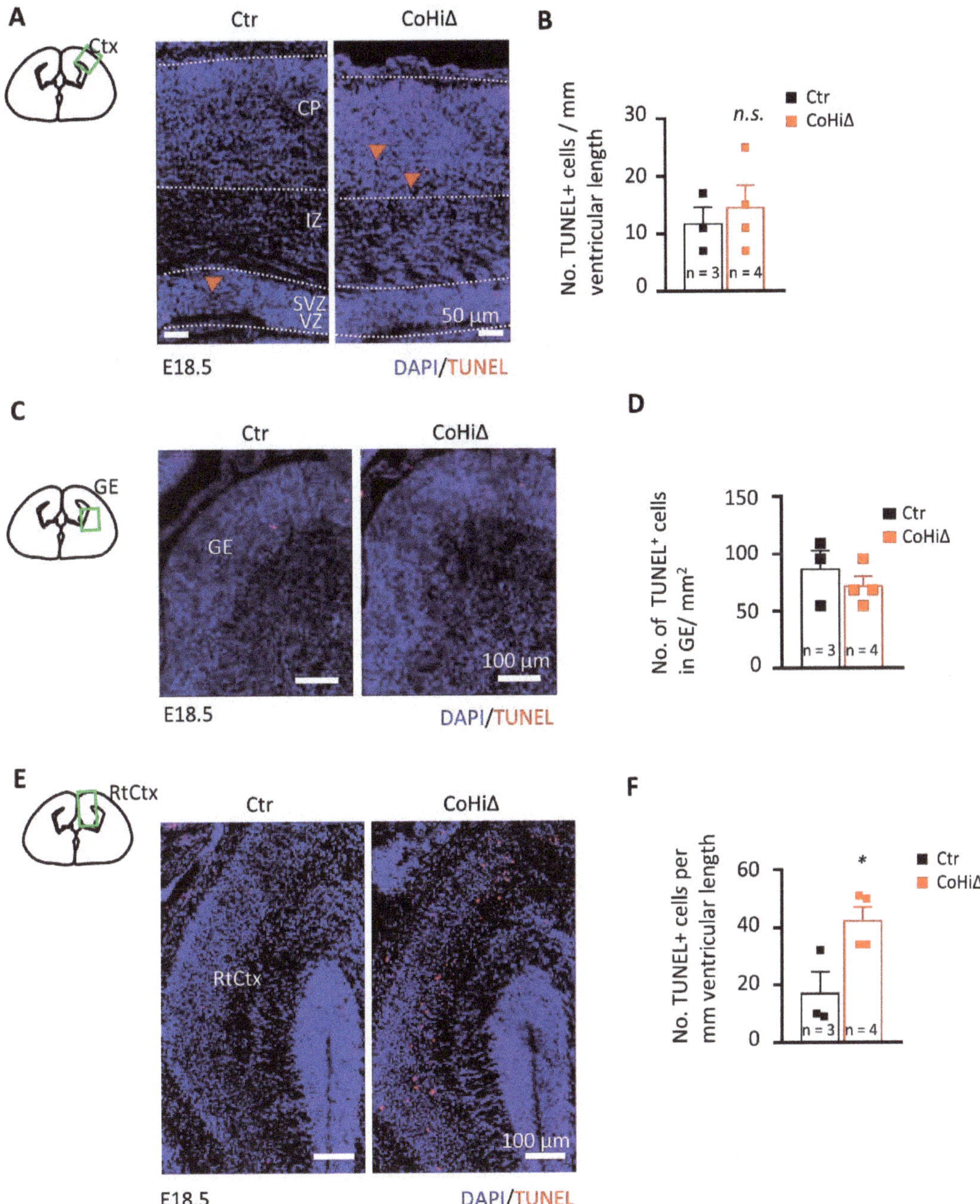

Figure 5. SMG6 deficiency in cortical and hippocampal neuroprogenitors cause neuronal cell death in the retrosplenial cortex. TUNEL staining comparison of E18.5 brain sections in CTX (**A**), GE (**C**) and retrosplenial cortex (RtCtx) (**D**). (**B**,**E**,**F**) Quantifications of total TUNEL positive cells in the respective brain parts. For all graphs: *n*—number of embryos analysed. Error bars represent SEM. Statistics by unpaired Student's *t*-test, except in (**F**) where the MWU was used, significance—n.s. > 0.05, * *p* < 0.05.

Taken together, using two different neural specific Cre mouse models, we demonstrate that GE NPCs and interneurons are particularly vulnerable to SMG6 loss. We show that cortical NPCs and neurons are affected mainly if *Smg6* is deleted simultaneously in NPCs from both brain parts GE and CTX. We conclude that the defects of cortical neurogenesis in the *Smg6*-CNSΔ model are likely sensitized by *Smg6* deletion in the IN progenitors derived from the GE.

3.5. SMG6 Null Mutation Activates DNA Repair and p53 Pathways Causing Cell Cycle Dysregulation

The finding that *Smg6* deletion has a stronger effect on IN progenitors prompted us to investigate the transcriptional programs initiated by the loss of SMG6. We analysed the total transcriptome profile of SMG6-deficient NPCs isolated from E13.5 *Smg6*-CER (Smg6^{flox} crossed with Rosa26-CreER^{T2}) mice brains and cultured for 6 days in the presence of 4-OHT that induces *Smg6* deletion (*Smg6*-iKO). RNA-seq data comparison of controls (Ctr + 4-OHT) with *Smg6*-iKO cells revealed 859 differentially expressed genes (DEGs) (cutoff adjusted $p < 0.05$), containing 385 upregulated and 474 downregulated DEGs (Figure 6A, Supplementary Table S1). Confirming its function in NMD, *Smg6* knockout resulted in the presence of the prominent NMD target genes within the upregulated DEGs (underlined in Figure 6A). The analysis using IPA (Ingenuity Pathway Analysis) showed that the majority of the DEGs are involved in DNA repair and cell cycle pathways (Figure 6B). Furthermore, we used the IPA upstream regulator tool to find which possible regulator was upstream, and whether the activation or silencing of it could explain the observed gene expression changes. The upstream analysis of all DEGs predicted the activation of *Trp53*, *Cdkn2a* and *Cdkn1a* genes (positive Z-score) that might have caused the detected changes (Figure 6C), indicating activation of DNA repair pathways that would lead to cell cycle arrest and eventually cell death. Interestingly, we found that *FoxM1* is among the top 30 strongest upstream regulators of DEGs, although with a negative Z-score, indicating that an inhibition of the FOXM1 function could also be a reason for the detected expression changes. FOXM1 is known to promote *Dlx1* gene expression [48] and its silencing could cause the down regulation of *Dlx1*, which we observed in neurospheres after *Smg6* deletion (Figure 3C).

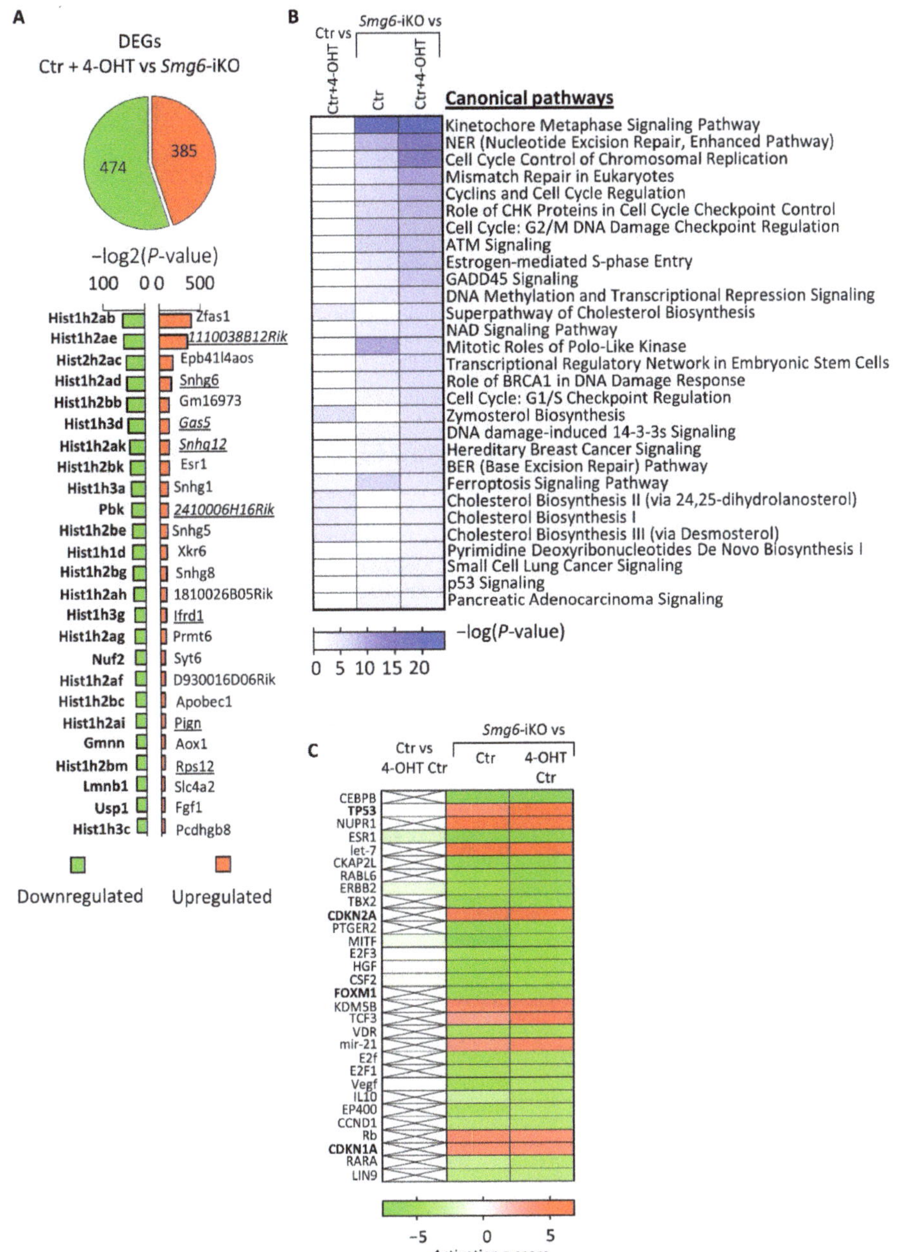

Figure 6. RNAseq analysis of neuroprogenitors after 4-OHT induced deletion of *Smg6*. (**A**) Total amount of identified differentially expressed genes (DEGs) with the 25 strongest up- and down-regulated genes. (**B**) Top 30 altered molecular pathways identified using IPA software. Control comparison of control cells treated and not treated with 4-OHT is shown in the left column. (**C**) Top 30 predicted upstream regulators of the detected changes in the transcriptome. Control comparison of control cells treated and not treated with 4-OHT is shown in the left column. For Ctr condition cells from 3 embryos and for each of Ctr + 4OHT and *Smg6*-iKO conditions cells from four embryos were used. DEGs for analysis by IPA were defined by adjusted *p*-value $\leq$ 0.05 and log2FoldChange $\leq$ −0.2 and $\geq$0.2.

4. Discussion

A complete deletion of key NMD genes such as *Smg1*, *Upf1* or *Upf2* results in early embryonic lethality, demonstrating an essential function of RNA metabolism/quality control during development [10–12]. We previously showed that *Smg6* deletion in ES cells had no impact on their viability but blocked their ability to differentiate into germ layers during mouse development. We concluded that SMG6-mediated NMD is a license factor for the cell fate determination of pluripotent stem cells [13]. Interestingly, in the present study, we show that if *Smg6* is deleted in committed neural stem cells during development at E10.5 (Nestin-Cre) or at E9.5 (Emx1-Cre), these NPCs are viable, can differentiate into various cell types and thus are able to support development of the entire brain. However, later embryonic neurogenesis and thus production of late born neurons is compromised in the Nestin-Cre mediated *Smg6* deletion model. This is in immense difference to the pluripotent ES cells, where the deletion of *Smg6* blocked differentiation through overexpression of the pluripotency gene *c-Myc* [13]. In contrast to ES cells, the overexpression of *c-Myc* during neurogenesis is known rather to support not only the self-renewal but also the neuronal differentiation of the NPCs [49,50].

Nestin-Cre mediated deletion of *Smg6* resulted in defects in both CTX and GE. In the CTX we detected a reduction of the NPC pool in VZ and SVZ and less production of late born neurons at E18.5 (Figure 1F,G), indicating that defects appear rather late in corticogenesis. Interestingly, we found that *Smg6* inactivation affects GE-derived NPCs earlier, i.e., at E14.5, judged by higher cellular death in histology and a decreased self-renewal of IN progenitors in the neurosphere assay compared to results obtained from CTX. Furthermore, we detected a reduced generation of calbindin-positive GABAergic neurons at E14.5, indicating dysregulation of the interneuron production. The striking finding is that Emx1-Cre mediated *Smg6* inactivation exclusively in cortical and hippocampal progenitors (*Smg6*-CoHiΔ mice), which spared IN progenitors, caused no comparable cortical defects. However, Emx1-Cre was reported to be more efficient and mediating stronger cortical defects than Nestin-Cre in other models [34,51,52]. *Smg6*-CoHiΔ mice were viable during the observation period of 20 months. This cell type specificity of SMG6 is supported by in vitro neurosphere assays, where we found that SMG6 null compromises only progenitors from GE but not from CTX. Thus, we conclude that SMG6 plays a less important role in the cortical progenitor renewal and differentiation, but is specifically required for IN neuroprogenitors and their derived cell lineages during brain development. Given these observations and the fact that GE-derived INs migrate tangentially to the CTX and regulate cortical neurogenesis via the GABA release [53,54], it is plausible that *Smg6* deletion caused malfunction of INs, which affected cortical neurogenesis (Figure 7).

The role of SMG6 on interneuron progenitors is reminiscent of a recent study showing that exon-junction complex factor RBM8A is critical for interneuron development [55]. Consistent with the defects of SMG6 deficient interneurons, key transcription factors that control the cell fate of interneuron progenitors were found dysregulated. *Dlx1* and *Dlx2* expression was dramatically reduced in SMG6-deficient neurospheres. DLX1 and DLX2 transcription factors promote or repress the expression of other transcription factors that are responsible for triggering IN differentiation, migration and maturation [42–47]. The downregulation of *Dlx1* gene correlates well with the IPA upstream regulator analysis of the RNA-seq data (Figure 6C), which predicted the inactivation of FOXM1, a known positive regulator of *Dlx1* [48]. Remarkably, along the same line to *Smg6*-CNSΔ mice, perinatal lethality has also been reported in mice with *Dlx1/2* double knock out (KO) [42,45,56] due to impaired differentiation and migration of GABAergic INs in the neocortex [45]. Disruptions of other genes such as *Gad1*, *Gad2*, *Nkx2.1* and *Sox6*, known to regulate GABAergic neuron development, were also often linked to perinatal lethality [57–62]. Unfortunately, the precise reason for the perinatal lethality of the above mentioned and *Smg6*-CNSΔ mice is unclear.

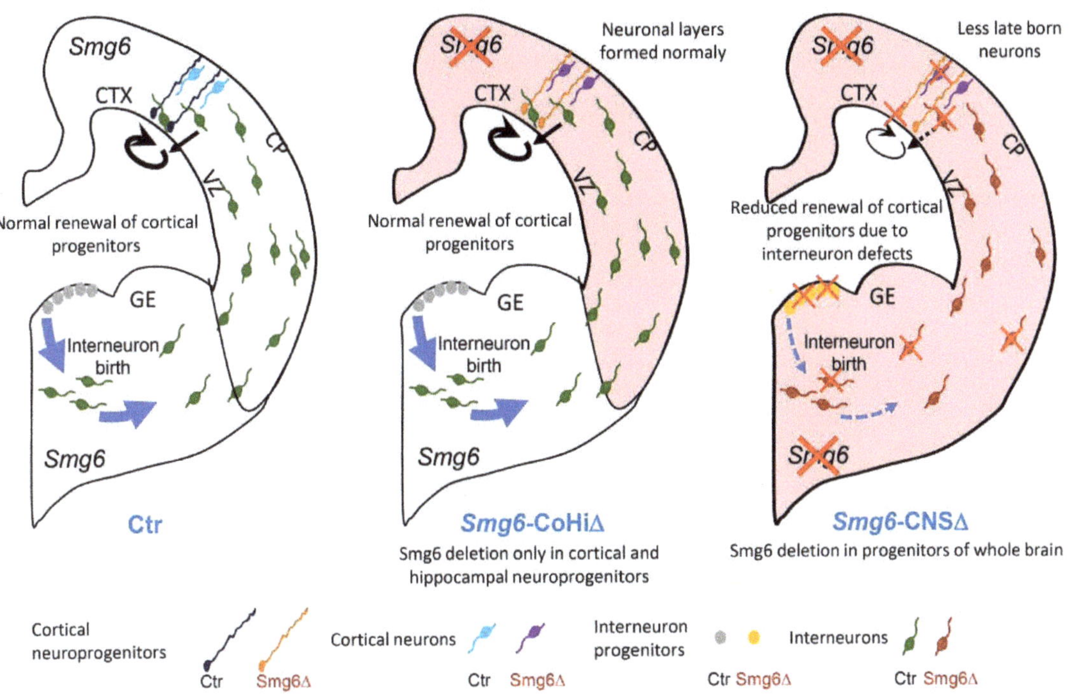

Figure 7. The role of SMG6 in embryonic neurogenesis. *Smg6* deletion in cortical and hippocampal NPCs (*Smg6*-CoHiΔ) does not inhibit corticogenesis. However, if SMG6 is absent in the whole nervous system (*Smg6*-CNSΔ) it causes cell death and also renewal defects of GE neural stem cells leading to defective interneurons tangentially migrating to the CTX. Consequently, the self-renewal of cortical NPCs and the production of late born neurons are impaired possibly because of environmental changes caused by defects in interneurons.

The transcriptional changes in the SMG6-deficient NPCs confirmed the defective NMD by changes in known NMD target transcripts. The majority of the top 30 dysregulated pathways belong to DNA repair, cell cycle and p53 related pathways, which can be modulated also by the NMD [63–66].

Since the transcriptome and thus the expression of NMD targets highly depends on the cell type, it is plausible that deletion of *Smg6* may have cell type-dependent consequences in cell fate. In this regard, the conditional knock-out of *Upf2* in the adult hematopoietic system preferentially compromised the viability of hematopoietic stem cells and progenitors, but not that of terminally differentiated T cells [11]. In addition, the *Upf2* null mutation did not affect the proliferation of fetal hepatocytes, but compromised their maturation process [67]. Furthermore, UPF3B was demonstrated to be very important for a subset of olfactory sensory neurons [68]. Taking these studies together, SMG6, or in general NMD, regulates cell fate programs highly depending on the cell type and developmental stage.

In conclusion, although SMG6 is essential for the differentiation of pluripotent ES cells, it is less important for the differentiation of committed cortical NPCs. Using various mouse models, we demonstrate that SMG6, as a general endonuclease of NMD for aberrant RNA, plays distinct roles in different cell types: CTX versus GE neuroprogenitor cells (Figure 7) versus ES cells. These genetic results allow predicting that the importance of NMD varies dramatically, depending on the transcriptional program of a specific cell type.

Supplementary Materials: The following are available online at https://www.mdpi.com/article/10 .3390/cells10123365/s1, Supplementary Figure S1: Smg6 deletion in all neuroprogenitors; Figure S2: Smg6 deletion in the central nervous system cause Caspase3 dependent cell death in the cortex and ganglionic eminence; Figure S3: Proliferation of neuroprogenitors in the cortex and ganglionic eminence of E14.5 brains of indicated genotypes; Figure S4: Smg6 deletion only in cortex and hippocampus does not affect embryo development. Supplementary Table S1: The list of identified DEGs.

Author Contributions: Conceptualization, G.M.G., Z.-Q.W., T.-L.L. and P.G.; methodology, G.M.G., D.M., T.K., P.K., M.G., T.-L.L. and P.G.; software, T.K.; validation, G.M.G., T.K., P.K., M.G. and D.M.; formal analysis, G.M.G., T.K., P.K., M.G. and P.G.; investigation, G.M.G., T.-L.L., D.M., T.K., P.K., M.G. and P.G.; resources, Z.-Q.W. and P.G.; data curation, G.M.G., T.K., P.K., M.G. and P.G.; writing—draft preparation, review and editing, G.M.G., Z.-Q.W. and P.G.; visualization, G.M.G., Z.-Q.W. and P.G.; supervision, Z.-Q.W. and P.G.; project administration, Z.-Q.W. and P.G.; funding acquisition, Z.-Q.W. All authors have read and agreed to the published version of the manuscript.

Funding: G.M.G. was a member of the Leibniz Gradual School on Aging (LGSA) at the FLI. This work was supported by the Deutsche Forschungsgemeinschaft (DFG) (WA2627/8-1) and the DFG-funded RTG1715. TL was supported by grants from National Natural Science Foundation of China (Nos. 31770871 and 81571380) and a grant (No. KF2020005) from NHC Key Laboratory of Birth Defect for Research and Prevention (Hunan Provincial Maternal and Child Health Care Hospital, Changsha, China). The article is published via an IOAP funding of Friedrich-Schiller-University Jena and Thüringer University and State Library Jena (ThULB).

Institutional Review Board Statement: All animal work was conducted according to the German animal welfare legislation and approved by the Thüringer Landesamt für Verbraucherschutz (TLV) (License number: 03-042/16).

Informed Consent Statement: Not applicable.

Data Availability Statement: The RNAseq dataset presented in this study is available on GEO (GSE186964, https://www.ncbi.nlm.nih.gov/geo/query/acc.cgi?acc=GSE186964, accessed on 26 November 2021) and other datasets upon request from the corresponding author.

Acknowledgments: We are grateful to Patrick Elsner for excellent assistance in the maintenance of the mouse colonies as well as to the imaging and histology facilities for their support. We thank all members of the Wang lab for critical discussion of the project.

Conflicts of Interest: The authors declare no competing interest.

References

1. Lykke-Andersen, S.; Jensen, T.H. Nonsense-mediated mRNA decay: An intricate machinery that shapes transcriptomes. *Nat. Rev. Mol. Cell Biol.* **2015**, *16*, 665–677. [CrossRef] [PubMed]
2. Palacios, I.M. Nonsense-mediated mRNA decay: From mechanistic insights to impacts on human health. *Brief. Funct. Genom.* **2013**, *12*, 25–36. [CrossRef]
3. Han, X.; Wei, Y.L.; Wang, H.; Wang, F.L.; Ju, Z.Y.; Li, T.L. Nonsense-mediated mRNA decay: A 'nonsense' pathway makes sense in stem cell biology. *Nucleic Acids Res.* **2018**, *46*, 1038–1051. [CrossRef]
4. Colombo, M.; Karousis, E.D.; Bourquin, J.; Bruggmann, R.; Muhlemann, O. Transcriptome-wide identification of NMD-targeted human mRNAs reveals extensive redundancy between SMG6-and SMG7-mediated degradation pathways. *RNA* **2017**, *23*, 189–201. [CrossRef]
5. Boehm, V.; Kueckelmann, S.; Gerbracht, J.V.; Kallabis, S.; Britto-Borges, T.; Altmuller, J.; Kruger, M.; Dieterich, C.; Gehring, N.H. SMG5-SMG7 authorize nonsense-mediated mRNA decay by enabling SMG6 endonucleolytic activity. *Nat. Commun.* **2021**, *12*, 3965. [CrossRef] [PubMed]
6. Barbosa, C.; Peixeiro, I.; Romao, L. Gene Expression Regulation by Upstream Open Reading Frames and Human Disease. *PLoS Genet.* **2013**, *9*, e1003529. [CrossRef] [PubMed]
7. Frischmeyer, P.A.; Dietz, H.C. Nonsense-mediated mRNA decay in health and disease. *Hum. Mol. Genet.* **1999**, *8*, 1893–1900. [CrossRef] [PubMed]
8. Nickless, A.; Bailis, J.M.; You, Z.S. Control of gene expression through the nonsense-mediated RNA decay pathway. *Cell Biosci.* **2017**, *7*, 26. [CrossRef]
9. Pawlicka, K.; Kalathiya, U.; Alfaro, J. Nonsense-Mediated mRNA Decay: Pathologies and the Potential for Novel Therapeutics. *Cancers* **2020**, *12*, 765. [CrossRef]
10. Medghalchi, S.M.; Frischmeyer, P.A.; Mendell, J.T.; Kelly, A.G.; Lawler, A.M.; Dietz, H.C. Rent1, a trans-effector of nonsense-mediated mRNA decay, is essential for mammalian embryonic viability. *Hum. Mol. Genet.* **2001**, *10*, 99–105. [CrossRef]

11. Weischenfeldt, J.; Damgaard, I.; Bryder, D.; Theilgaard-Monch, K.; Thoren, L.A.; Nielsen, F.C.; Jacobsen, S.E.W.; Nerlov, C.; Porse, B.T. NMD is essential for hematopoietic stem and progenitor cells and for eliminating by-products of programmed DNA rearrangements. *Genes Dev.* **2008**, *22*, 1381–1396. [CrossRef]

12. McIlwain, D.R.; Pan, Q.; Reilly, P.T.; Elia, A.J.; McCracken, S.; Wakeham, A.C.; Itie-Youten, A.; Blencowe, B.J.; Mak, T.W. Smg1 is required for embryogenesis and regulates diverse genes via alternative splicing coupled to nonsense-mediated mRNA decay. *Proc. Natl. Acad. Sci. USA* **2010**, *107*, 12186–12191. [CrossRef]

13. Li, T.L.; Shi, Y.; Wang, P.; Guachalla, L.M.; Sun, B.F.; Joerss, T.; Chen, Y.S.; Groth, M.; Krueger, A.; Platzer, M.; et al. Smg6/Est1 licenses embryonic stem cell differentiation via nonsense-mediated mRNA decay. *EMBO J.* **2015**, *34*, 1630–1647. [CrossRef] [PubMed]

14. Jaffrey, S.R.; Wilkinson, M.F. Nonsense-mediated RNA decay in the brain: Emerging modulator of neural development and disease. *Nat. Rev. Neurosci.* **2018**, *19*, 715–728. [CrossRef] [PubMed]

15. Addington, A.M.; Gauthier, J.; Piton, A.; Hamdan, F.F.; Raymond, A.; Gogtay, N.; Miller, R.; Tossell, J.; Bakalar, J.; Germain, G.; et al. A novel frameshift mutation in UPF3B identified in brothers affected with childhood onset schizophrenia and autism spectrum disorders. *Mol. Psychiatry* **2011**, *16*, 238–239. [CrossRef] [PubMed]

16. Tarpey, P.S.; Raymond, F.L.; Nguyen, L.S.; Rodriguez, J.; Hackett, A.; Vandeleur, L.; Smith, R.; Shoubridge, C.; Edkins, S.; Stevens, C.; et al. Mutations in UPF3B, a member of the nonsense-mediated mRNA decay complex, cause syndromic and nonsyndromic mental retardation. *Nat. Genet.* **2007**, *39*, 1127–1133. [CrossRef]

17. Nguyen, L.S.; Kim, H.G.; Rosenfeld, J.A.; Shen, Y.P.; Gusella, J.F.; Lacassie, Y.; Layman, L.C.; Shaffer, L.G.; Gecz, J. Contribution of copy number variants involving nonsense-mediated mRNA decay pathway genes to neuro-developmental disorders. *Hum. Mol. Genet.* **2013**, *22*, 1816–1825. [CrossRef]

18. Sartor, F.; Anderson, J.; McCaig, C.; Miedzybrodzka, Z.; Muller, B. Mutation of genes controlling mRNA metabolism and protein synthesis predisposes to neurodevelopmental disorders. *Biochem. Soc. Trans.* **2015**, *43*, 1259–1265. [CrossRef]

19. Shaheen, R.; Anazi, S.; Ben-Omran, T.; Seidahmed, M.Z.; Caddle, L.B.; Palmer, K.; Ali, R.; Alshidi, T.; Hagos, S.; Goodwin, L.; et al. Mutations in SMG9, Encoding an Essential Component of Nonsense-Mediated Decay Machinery, Cause a Multiple Congenital Anomaly Syndrome in Humans and Mice. *Am. J. Hum. Genet.* **2016**, *98*, 643–652. [CrossRef]

20. Colak, D.; Ji, S.J.; Porse, B.T.; Jaffrey, S.R. Regulation of Axon Guidance by Compartmentalized Nonsense-Mediated mRNA Decay. *Cell* **2013**, *153*, 1252–1265. [CrossRef]

21. Huang, L.; Shum, E.Y.; Jones, S.H.; Lou, C.H.; Dumdie, J.; Kim, H.; Roberts, A.J.; Jolly, L.A.; Espinoza, J.L.; Skarbrevik, D.M.; et al. A Upf3b-mutant mouse model with behavioral and neurogenesis defects. *Mol. Psychiatry* **2018**, *23*, 1773–1786. [CrossRef] [PubMed]

22. Johnson, J.L.; Stoica, L.; Liu, Y.W.; Zhu, P.J.; Bhattacharya, A.; Buffington, S.A.; Huq, R.; Eissa, N.T.; Larsson, O.; Porse, B.T.; et al. Inhibition of Upf2-Dependent Nonsense-Mediated Decay Leads to Behavioral and Neurophysiological Abnormalities by Activating the Immune Response. *Neuron* **2019**, *104*, 665–679.e8. [CrossRef] [PubMed]

23. Jolly, L.A.; Homan, C.C.; Jacob, R.; Barry, S.; Gecz, J. The UPF3B gene, implicated in intellectual disability, autism, ADHD and childhood onset schizophrenia regulates neural progenitor cell behaviour and neuronal outgrowth. *Hum. Mol. Genet.* **2013**, *22*, 4673–4687. [CrossRef]

24. Long, A.A.; Mahapatra, C.T.; Woodruff, E.A.; Rohrbough, J.; Leung, H.T.; Shino, S.; An, L.L.; Doerge, R.W.; Metzstein, M.M.; Pak, W.L.; et al. The nonsense-mediated decay pathway maintains synapse architecture and synaptic vesicle cycle efficacy. *J. Cell Sci.* **2010**, *123*, 3303–3315. [CrossRef]

25. Mooney, C.M.; Jimenez-Mateos, E.M.; Engel, T.; Mooney, C.; Diviney, M.; Veno, M.T.; Kjems, J.; Farrell, M.A.; O'Brien, D.F.; Delanty, N.; et al. RNA sequencing of synaptic and cytoplasmic Upf1-bound transcripts supports contribution of nonsense-mediated decay to epileptogenesis. *Sci. Rep.* **2017**, *7*, 41517. [CrossRef]

26. Notaras, M.; Allen, M.; Longo, F.; Volk, N.; Toth, M.; Jeon, N.L.; Klann, E.; Colak, D. UPF2 leads to degradation of dendritically targeted mRNAs to regulate synaptic plasticity and cognitive function. *Mol. Psychiatry* **2020**, *25*, 3360–3379. [CrossRef]

27. Tronche, F.; Kellendonk, C.; Kretz, O.; Gass, P.; Anlag, K.; Orban, P.C.; Bock, R.; Klein, R.; Schutz, G. Disruption of the glucocorticoid receptor gene in the nervous system results in reduced anxiety. *Nat. Genet.* **1999**, *23*, 99–103. [CrossRef]

28. Gorski, J.A.; Talley, T.; Qiu, M.S.; Puelles, L.; Rubenstein, J.L.R.; Jones, K.R. Cortical excitatory neurons and glia, but not GABAergic neurons, are produced in the Emx1-expressing lineage. *J. Neurosci.* **2002**, *22*, 6309–6314. [CrossRef]

29. Frappart, P.O.; Tong, W.M.; Demuth, I.; Radovanovic, I.; Herceg, Z.; Aguzzi, A.; Digweed, M.; Wang, Z.Q. An essential function for NBS1 in the prevention of ataxia and cerebellar defects. *Nat. Med.* **2005**, *11*, 538–544. [CrossRef]

30. Park, G.; Horie, T.; Kanayama, T.; Fukasawa, K.; Iezaki, T.; Onishi, Y.; Ozaki, K.; Nakamura, Y.; Yoneda, Y.; Takarada, T.; et al. The transcriptional modulator Ifrd1 controls PGC-1alpha expression under short-term adrenergic stimulation in brown adipocytes. *FEBS J.* **2017**, *284*, 784–795. [CrossRef] [PubMed]

31. Spandidos, A.; Wang, X.W.; Wang, H.J.; Seed, B. PrimerBank: A resource of human and mouse PCR primer pairs for gene expression detection and quantification. *Nucleic Acids Res.* **2010**, *38*, D792–D799. [CrossRef]

32. Wang, Y.; Wu, Q.; Yang, P.; Wang, C.; Liu, J.; Ding, W.; Liu, W.; Bai, Y.; Yang, Y.; Wang, H.; et al. LSD1 co-repressor Rcor2 orchestrates neurogenesis in the developing mouse brain. *Nat. Commun.* **2016**, *7*, 10481. [CrossRef] [PubMed]

33. Kraus, P.; Sivakamasundari, V.; Lim, S.L.; Xing, X.; Lipovich, L.; Lufkin, T. Making sense of Dlx1 antisense RNA. *Dev. Biol.* **2013**, *376*, 224–235. [CrossRef]

34. Grigaravicius, P.; Kaminska, E.; Hubner, C.A.; McKinnon, P.J.; von Deimling, A.; Frappart, P.O. Rint1 inactivation triggers genomic instability, ER stress and autophagy inhibition in the brain. *Cell Death Differ.* **2016**, *23*, 454–468. [CrossRef] [PubMed]
35. Gruber, R.; Zhou, Z.; Sukchev, M.; Joerss, T.; Frappart, P.O.; Wang, Z.Q. MCPH1 regulates the neuroprogenitor division mode by coupling the centrosomal cycle with mitotic entry through the Chk1-Cdc25 pathway. *Nat. Cell Biol.* **2011**, *13*, 1325–1334. [CrossRef]
36. Dobin, A.; Davis, C.A.; Schlesinger, F.; Drenkow, J.; Zaleski, C.; Jha, S.; Batut, P.; Chaisson, M.; Gingeras, T.R. STAR: Ultrafast universal RNA-seq aligner. *Bioinformatics* **2013**, *29*, 15–21. [CrossRef]
37. Liao, Y.; Smyth, G.K.; Shi, W. featureCounts: An efficient general purpose program for assigning sequence reads to genomic features. *Bioinformatics* **2014**, *30*, 923–930. [CrossRef]
38. Love, M.I.; Huber, W.; Anders, S. Moderated estimation of fold change and dispersion for RNA-seq data with DESeq2. *Genome Biol.* **2014**, *15*, 550. [CrossRef] [PubMed]
39. Cauli, B.; Audinat, E.; Lambolez, B.; Angulo, M.C.; Ropert, N.; Tsuzuki, K.; Hestrin, S.; Rossier, J. Molecular and physiological diversity of cortical nonpyramidal cells. *J. Neurosci.* **1997**, *17*, 3894–3906. [CrossRef] [PubMed]
40. Gonchar, Y.; Burkhalter, A. Three distinct families of GABAergic neurons in rat visual cortex. *Cereb. Cortex* **1997**, *7*, 347–358. [CrossRef]
41. Kubota, Y.; Kawaguchi, Y. 3 Classes of Gabaergic Interneurons in Neocortex and Neostriatum. *Jpn. J. Physiol.* **1994**, *44*, S145–S148. [PubMed]
42. Anderson, S.A.; Qiu, M.S.; Bulfone, A.; Eisenstat, D.D.; Meneses, J.; Pedersen, R.; Rubenstein, J.L.R. Mutations of the homeobox genes Dlx-1 and Dlx-2 disrupt the striatal subventricular zone and differentiation of late born striatal neurons. *Neuron* **1997**, *19*, 27–37. [CrossRef]
43. Cobos, I.; Borello, U.; Rubenstein, J.L.R. Dlx transcription factors promote migration through repression of axon and dendrite growth. *Neuron* **2007**, *54*, 873–888. [CrossRef] [PubMed]
44. Dai, X.J.; Iwasaki, H.; Watanabe, M.; Okabe, S. Dlx1 transcription factor regulates dendritic growth and postsynaptic differentiation through inhibition of neuropilin-2 and PAK3 expression. *Eur. J. Neurosci.* **2014**, *39*, 531–547. [CrossRef] [PubMed]
45. Le, T.N.; Du, G.Y.; Fonseca, M.; Zhou, Q.P.; Wigle, J.T.; Eisenstat, D.D. Dlx homeobox genes promote cortical interneuron migration from the basal forebrain by direct repression of the semaphorin receptor Neuropilin-2. *J. Biol. Chem.* **2007**, *282*, 19071–19081. [CrossRef]
46. Long, J.E.; Garel, S.; Alvarez-Dolado, M.; Yoshikawa, K.; Osumi, N.; Alvarez-Buylla, A.; Rubenstein, J.L.R. Dlx-dependent and -independent regulation of olfactory bulb interneuron differentiation. *J. Neurosci.* **2007**, *27*, 3230–3243. [CrossRef]
47. Stuhmer, T.; Puelles, L.; Ekker, M.; Rubenstein, J.L. Expression from a Dlx gene enhancer marks adult mouse cortical GABAergic neurons. *Cereb. Cortex* **2002**, *12*, 75–85. [CrossRef] [PubMed]
48. Chan, D.W.; Hui, W.W.Y.; Wang, J.J.; Yung, M.M.H.; Hui, L.M.N.; Qin, Y.; Liang, R.R.; Leung, T.H.Y.; Xu, D.; Chan, K.K.L.; et al. DLX1 acts as a crucial target of FOXM1 to promote ovarian cancer aggressiveness by enhancing TGF-beta/SMAD4 signaling. *Oncogene* **2017**, *36*, 1404–1416. [CrossRef]
49. Nagao, M.; Campbell, K.; Burns, K.; Kuan, C.Y.; Trumpp, A.; Nakafuku, M. Coordinated control of self-renewal and differentiation of neural stem cells by Myc and the p19(ARF)-p53 pathway. *J. Cell Biol.* **2008**, *183*, 1243–1257. [CrossRef]
50. Zinin, N.; Adameyko, I.; Wilhelm, M.; Fritz, N.; Uhlen, P.; Ernfors, P.; Henriksson, M.A. MYC proteins promote neuronal differentiation by controlling the mode of progenitor cell division. *EMBO Rep.* **2014**, *15*, 383–391. [CrossRef]
51. Liang, H.X.; Hippenmeyer, S.; Ghashghaei, H.T. A Nestin-cre transgenic mouse is insufficient for recombination in early embryonic neural progenitors. *Biol. Open* **2012**, *1*, 1200–1203. [CrossRef]
52. Lee, Y.; Shull, E.R.P.; Frappart, P.O.; Katyal, S.; Enriquez-Rios, V.; Zhao, J.F.; Russell, H.R.; Brown, E.J.; McKinnon, P.J. ATR maintains select progenitors during nervous system development. *EMBO J.* **2012**, *31*, 1177–1189. [CrossRef] [PubMed]
53. Haydar, T.F.; Wang, F.; Schwartz, M.L.; Rakic, P. Differential modulation of proliferation in the neocortical ventricular and subventricular zones. *J. Neurosci.* **2000**, *20*, 5764–5774. [CrossRef] [PubMed]
54. LoTurco, J.J.; Owens, D.F.; Heath, M.J.S.; Davis, M.B.E.; Kriegstein, A.R. GABA and glutamate depolarize cortical progenitor cells and inhibit DNA synthesis. *Neuron* **1995**, *15*, 1287–1298. [CrossRef]
55. McSweeney, C.; Dong, F.P.; Chen, M.; Vitale, J.; Xu, L.; Crowley, N.; Luscher, B.; Zou, D.H.; Mao, Y.W. Full function of exon junction complex factor, Rbm8a, is critical for interneuron development. *Transl. Psychiatry* **2020**, *10*, 379. [CrossRef] [PubMed]
56. Qiu, M.S.; Bulfone, A.; Ghattas, I.; Meneses, J.J.; Christensen, L.; Sharpe, P.T.; Presley, R.; Pedersen, R.A.; Rubenstein, J.L.R. Role of the Dlx homeobox genes in proximodistal patterning of the branchial arches: Mutations of Dlx-1, Dlx-2, and Dlx-1 and -2 alter morphogenesis of proximal skeletal and soft tissue structures derived from the first and second arches. *Dev. Biol.* **1997**, *185*, 165–184. [CrossRef]
57. Asada, H.; Kawamura, Y.; Maruyama, K.; Kume, H.; Ding, R.G.; Kanbara, N.; Kuzume, H.; Sanbo, M.; Yagi, T.; Obata, K. Cleft palate and decreased brain gamma-aminobutyric acid in mice lacking the 67-kDa isoform of glutamic acid decarboxylase. *Proc. Natl. Acad. Sci. USA* **1997**, *94*, 6496–6499. [CrossRef]
58. Kash, S.F.; Tecott, L.H.; Hodge, C.; Baekkeskov, S. Increased anxiety and altered responses to anxiolytics in mice deficient in the 65-kDa isoform of glutamic acid decarboxylase. *Proc. Natl. Acad. Sci. USA* **1999**, *96*, 1698–1703. [CrossRef]

59. Sussel, L.; Marin, O.; Kimura, S.; Rubenstein, J.L.R. Loss of Nkx2.1 homeobox gene function results in a ventral to dorsal molecular respecification within the basal telencephalon: Evidence for a transformation of the pallidum into the striatum. *Development* **1999**, *126*, 3359–3370. [CrossRef]
60. Butt, S.J.B.; Sousa, V.H.; Fuccillo, M.V.; Hjerling-Leffler, J.; Miyoshi, G.; Kimura, S.; Fishell, G. The requirement of Nkx2-1 in the temporal specification of cortical interneuron subtypes. *Neuron* **2008**, *59*, 722–732. [CrossRef]
61. Azim, E.; Jabaudon, D.; Fame, R.M.; Macklis, J.D. SOX6 controls dorsal progenitor identity and interneuron diversity during neocortical development. *Nat. Neurosci.* **2009**, *12*, 1238–1247. [CrossRef] [PubMed]
62. Smits, P.; Li, P.; Mandel, J.; Zhang, Z.P.; Deng, J.M.; Behringer, R.R.; de Crombrugghe, B.; Lefebvre, V. The transcription factors L-sox5 and sox6 are essential for cartilage formation. *Dev. Cell* **2001**, *1*, 277–290. [CrossRef]
63. Cowen, L.E.; Tang, Y. Identification of nonsense-mediated mRNA decay pathway as a critical regulator of p53 isoform beta. *Sci. Rep.* **2017**, *7*, 17535. [CrossRef] [PubMed]
64. Gonzalez-Huici, V.; Wang, B.; Gartner, A. A Role for the Nonsense-Mediated mRNA Decay Pathway in Maintaining Genome Stability in Caenorhabditis elegans. *Genetics* **2017**, *206*, 1853–1864. [CrossRef]
65. Gudikote, J.P.; Cascone, T.; Poteete, A.; Sitthideatphaiboon, P.; Wu, Q.; Morikawa, N.; Zhang, F.; Peng, S.; Tong, P.; Li, L.; et al. Inhibition of nonsense-mediated decay rescues p53beta/gamma isoform expression and activates the p53 pathway in MDM2-overexpressing and select p53-mutant cancers. *J. Biol. Chem.* **2021**, *297*, 101163. [CrossRef] [PubMed]
66. Janke, R.; Kong, J.; Braberg, H.; Cantin, G.; Yates, J.R.; Krogan, N.J.; Heyer, W.D. Nonsense-mediated decay regulates key components of homologous recombination. *Nucleic Acids Res.* **2016**, *44*, 5218–5230. [CrossRef]
67. Thoren, L.A.; Norgaard, G.A.; Weischenfeldt, J.; Waage, J.; Jakobsen, J.S.; Damgaard, I.; Bergstrom, F.C.; Blom, A.M.; Borup, R.; Bisgaard, H.C.; et al. UPF2 Is a Critical Regulator of Liver Development, Function and Regeneration. *PLoS ONE* **2010**, *5*, e11650. [CrossRef] [PubMed]
68. Tan, K.; Jones, S.H.; Lake, B.B.; Dumdie, J.N.; Shum, E.Y.; Zhang, L.J.; Chen, S.; Sohni, A.; Pandya, S.; Gallo, R.L.; et al. The role of the NMD factor UPF3B in olfactory sensory neurons. *eLife* **2020**, *9*, e57525. [CrossRef]

Review

Emerging Roles of N6-Methyladenosine Modification in Neurodevelopment and Neurodegeneration

Liqi Shu [1,†], **Xiaoli Huang** [2,3,†], **Xuejun Cheng** [2,3] and **Xuekun Li** [2,3,4,5,*]

[1] Department of Neurology, The Warren Alpert Medical School of Brown University, Providence, RI 02908, USA; Shuliqi@gmail.com
[2] The Children's Hospital, School of Medicine, Zhejiang University, Hangzhou 310052, China; xiaolihuang@zju.edu.cn (X.H.); xuejun_cheng@zju.edu.cn (X.C.)
[3] National Clinical Research Center for Child Health, Hangzhou 310052, China
[4] The Institute of Translational Medicine, School of Medicine, Zhejiang University, Hangzhou 310029, China
[5] Zhejiang University Cancer Center, Zhejiang University, Hangzhou 310029, China
* Correspondence: xuekun_li@zju.edu.cn
† These authors contributed equally.

Abstract: N6-methyladenosine (m^{6}A), the most abundant modification in messenger RNAs (mRNAs), is deposited by methyltransferases ("writers") Mettl3 and Mettl14 and erased by demethylases ("erasers") Fto and Alkbh5. m^{6}A can be recognized by m^{6}A-binding proteins ("readers"), such as Yth domain family proteins (Ythdfs) and Yth domain-containing protein 1 (Ythdc1). Previous studies have indicated that m^{6}A plays an essential function in various fundamental biological processes, including neurogenesis and neuronal development. Dysregulated m^{6}A modification contributes to neurological disorders, including neurodegenerative diseases. In this review, we summarize the current knowledge about the roles of m^{6}A machinery, including writers, erasers, and readers, in regulating gene expression and the function of m^{6}A in neurodevelopment and neurodegeneration. We also discuss the perspectives for studying m^{6}A methylation.

Keywords: N6-methyladenosine; Mettl3; Mettl14; Fto; Ythdf1; neurodevelopment; neurodegeneration

Citation: Shu, L.; Huang, X.; Cheng, X.; Li, X. Emerging Roles of N6-Methyladenosine Modification in Neurodevelopment and Neurodegeneration. *Cells* **2021**, *10*, 2694. https://doi.org/10.3390/cells10102694

Academic Editors: FengRu Tang and Cord Brakebusch

Received: 18 August 2021
Accepted: 29 September 2021
Published: 9 October 2021

Publisher's Note: MDPI stays neutral with regard to jurisdictional claims in published maps and institutional affiliations.

1. Introduction

Epigenetics refers to the heritable changes in gene expression and cell state caused by some specific mechanisms, aside from the occurrence of potential genetic sequences. More than 170 types of RNA modifications, including N6-methyladenosine (m^{6}A), 5-methylcytidine (m^{5}C), N1-methyladenosine (m^{1}A), and N7-methylguanosine (m^{7}G), have been identified in mammalian transcripts, and the most abundant internal RNA modification is N6-methyladenosine (m^{6}A) [1,2]. m^{6}A is installed by methyltransferases (writers), removed by demethylases (erasers), and recognized by m^{6}A binding proteins (readers). Methyltransferase-like 3 (Mettl3) and methyltransferase-like 14 (Mettl14) form the core of the methyltransferase complex; AlkB homolog 5 protein (Alkbh5) and Fat mass and obesity-associated protein (Fto) are identified as demethylases; YTH domain family proteins (Ythdf1, Ythdf2, Ythdf3) and YTH domain-containing family protein 1 (Ythdc1) are essential reader proteins.

m^{6}A modification is precisely catalyzed by a multi-subunit methyltransferase enzyme complex containing Mettl3, Mettl14, and other accessory components such as Wilms tumor 1-associated protein (Wtap), a mammalian splicing factor [3]. Mettl3 has catalytic activity, while Mettl14 acts as the RNA-binding platform and facilitates the recognition of Mettl3 [4]. Mettl3 and Mettl14 form heterodimers, which interact with Wtap. Wtap does not possess any methylation activity but interacts with Mettl3 and Mettl14 and promotes the recruitment of the Mettl3–Mettl14 complex to target transcripts [5]. The presence of m^{6}A modification induces the preferential binding of certain proteins, i.e., m^{6}A readers,

Ythdf family proteins, and Ythdc1. In addition, m⁶A modification is reversible and can be removed by demethylases, including Fto and Alkbh5. Therefore, m⁶A machinery consists of multiple components that have diverse functions and make the field colorful (Figure 1).

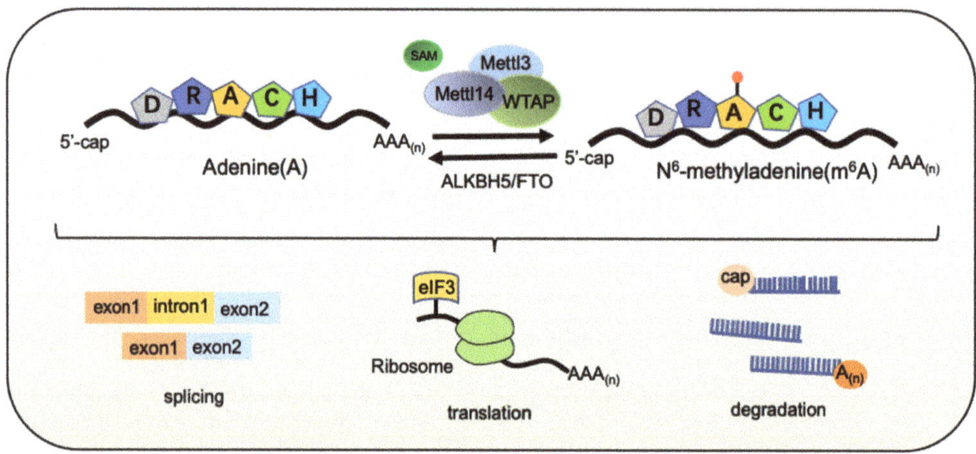

Figure 1. Schematic illustration of m⁶A modification. m⁶A methylation is catalyzed by the methyltransferase complex containing Mettl3, Mettl14, and an adaptor protein, such as WTAP. Fto and Alkbh5 can function as demethylases, and Yth family proteins can recognize m⁶A sites. m⁶A modification in mammals is presented on the consensus sequence DRACH (D = A/G/U, R = A/G, H = A/C/U). Reversible m⁶A modification plays important roles in regulating RNA metabolism, including RNA splicing, nuclear export, translation, and degradation in the specific context. Mettl3, methyltransferase-like 3; Mettl14, methyltransferase-like 14; WTAP, Wilms tumor 1-associating protein; Fto, fat mass and obesity-associated protein; ALKBH5, AlkB homolog 5.

m⁶A-specific methylated RNA immunoprecipitation (MeRIP) with next-generation sequencing data has revealed that m⁶A is non-randomly distributed in mRNAs but is especially enriched at the 5′ and 3′ UTRs [6,7]. m⁶A has been shown to impact RNA metabolism, including mRNA stability, translation, splicing, and localization; consequently, m⁶A regulates gene expression and involves diverse biological processes [2,8]. Present findings show that m⁶A modulates brain function [9,10] and regulates neurogenesis [11–18], brain development [7,17–19], axon regeneration [20], and learning and memory [13,15]. The dysregulation of m⁶A has been found in a set of neurological disorders, such as Alzheimer's disease, Fragile X syndrome, attention-deficit/hyperactivity disorder (ADHD), and intellectual disability [19,21–24]. In this review, we summarize the recent findings regarding the function and biological consequences of m⁶A modification in the neural system, from neural development to brain function and neurological disorders.

2. m⁶A and Neurogenesis

2.1. Writers

During embryonic neurogenesis, Mettl14 displays the highest expression in radial glia cells, and *Mettl14* knockout (KO) in embryonic mouse brains extends the cell cycle of radial glia cells and induces aberrant cortical neurogenesis. Similar defects were induced by Mettl3 knockdown [11]. Mettl14 also regulates the cell cycle of human cortical neuronal progenitor cells [11]. The deletion of *Mettl14* in embryonic neural stem cells (eNSCs) led to a remarkable decrease in proliferation and immature differentiation in vitro and in vivo [16]. In addition, *Mettl3* knockdown reduced the proliferation and skewed the differentiation of adult neural stem cells (aNSCs) towards neuronal lineage, while the newborn neurons displayed immature morphology [12]. Transcriptome analysis revealed that the deficiency of either *Mettl3* or *Mettl14* affected the expression of transcripts related to neurogenesis, the cell cycle, and neuronal development [11,12,16]. *Mettl3* conditional-knockout mice showed severe developmental defects of the cerebellum and cell death [17]. These results suggest an essential and conserved function of m^{6}A in maintaining normal neurogenesis in the mammalian brain (Figure 2A).

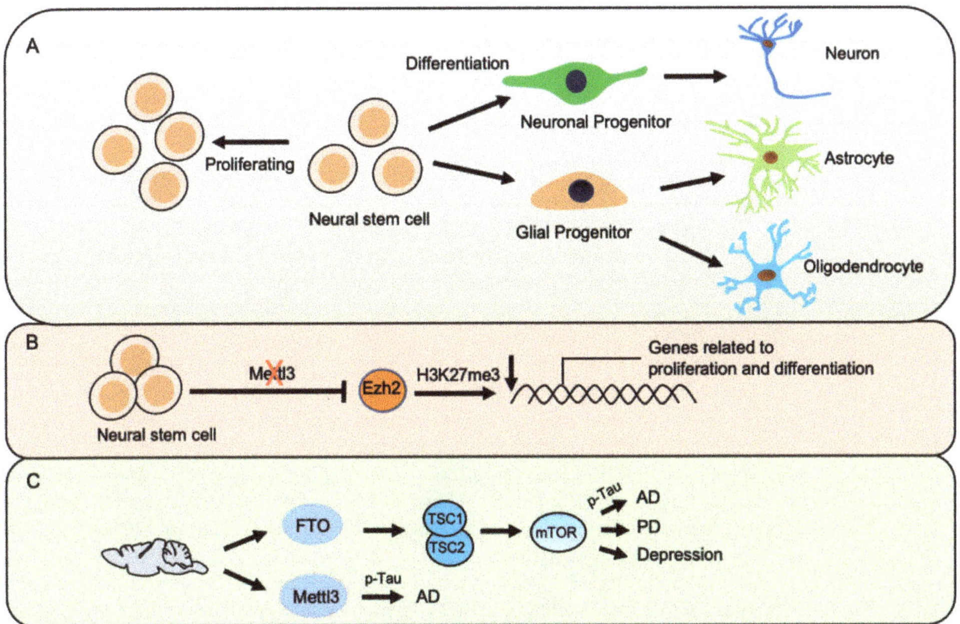

Figure 2. m^{6}A modification in neural development and neurological disorders. (**A**). Schematic representation of neurogenesis. Neural stem cells have the capability to self-renew and differentiate into neural cells, such as neurons, astrocytes, and oligodendrocytes. (**B**). Loss of m^{6}A modification affects histone modifications, including H3K27me3 and H3K27ac, which regulate the expression of genes related to the proliferation and differentiation of neural stem cells. (**C**). The modulation of m^{6}A modification machinery contributes to neurodegenerative diseases, including Alzheimer's disease and Parkinson's disease, through the regulation of multiple pathways, such as mTOR. AD, Alzheimer's disease; PD, Parkinson's disease; TSC1, tuberous sclerosis 1; TSC2, tuberous sclerosis 2.

m^{6}A regulates gene expression not only through regulating RNA metabolism but also via modulating mRNAs encoding histone modifiers and transcription factors [25]. In mouse eNSCs, transcripts for histone acetyltransferases CBP (CREB binding protein) and p300 are m^{6}A-modified [16]. In addition, transcripts for histone methyltransferase Ezh2 are also m^{6}A-modified, and *Mettl3* knockdown reduces the level of Ezh2 and consequent histone H3 trimethylation at lysine 27 (H3K27me3) in aNSCs [12]. Ectopic Ezh2 could rescue *Mettl3*-knockdown-induced deficits in aNSCs [12]. These findings suggest a crosstalk between RNA modification and transcriptional regulation and reveal a new layer of the mechanism regulating neurogenesis (Figure 2C).

2.2. Erasers

The fat mass and obesity-associated (Fto) gene was originally referred to as an obesity-risk gene and is the first identified m^{6}A demethylase [26]. The loss-of-function mutation of the *Fto* gene caused growth retardation and severe neurodevelopmental disorders, including microcephaly, functional brain defects, and delayed psychomotor activity in humans [27–29]. *Fto*-deficient mice showed increased postnatal mortality, significant loss of adipose tissue and body mass, and disordered energy homeostasis [27,30]. The constitutive loss of *Fto* decreased brain size and body weight, impaired the pool of adult neural stem cells (aNSCs), and impaired the learning and memory of mice [15]. Specific ablation of *Fto* in aNSCs also inhibited neurogenesis and neuronal development [13]. In addition, specific deletion of *Fto* in lipids led to decreased neurogenesis and increased apoptosis [14]. These findings indicate that Fto regulates neurogenesis through diverse pathways, including affecting brain-derived neurotrophic factor (BDNF) signaling, the expression of platelet-derived growth factor receptor (Pdgfra) and suppressor of cytokine signaling 5 (Socs5), and adenosine levels [13–15].

Another m^6A demethylase, Alkbh5, is primarily localized in the nuclear speckles. Alkbh5-mediated demethylation activity affects nuclear RNA export and RNA metabolism and, consequently, regulates gene expression. The cerebellum of *Alkbh5*-deficient mice did not show detectable changes in weight and morphology, but *Alkbh5*-KO mice were more sensitive to hypoxia and showed a significantly reduced size of whole brain and cerebellum compared to control littermates [18]. In addition, the number of proliferating cells was significantly increased, but mature neurons were reduced in the cerebellum of *Alkbh5*-deficient mice [18], which suggests that *Alkbh5* deficiency affects the proliferation and differentiation of neuronal progenitor cells.

2.3. Readers

Ythdf1 is preferentially expressed in the hippocampus of mouse brains. Genetic deletion of *Ythdf1* impaired the learning and memory of mice, whereas it did not affect gross hippocampal and cortical histology, neurogenesis, and motor abilities [31]. Electrophysiological data showed that *Ythdf1*-deficient neurons had reduced spine density and decreased amplitude and frequency of miniature excitatory postsynaptic currents, which could be rescued by ectopic Ythdf1 [31]. This study further showed that Ythdf1 facilitates learning and memory by promoting the translation of target transcripts, including Gria1, Grin1, and Camk2a induced by neuronal stimulation.

m^6A reader Ythdf2 is critical for embryonic development and has a lethal effect in mice [32]. *Ythdf2*-deficient mice embryos were alive at embryonic day 12.5, 14.5, and 18.5 but displayed abnormal brain development, including reduced cortical thickness and decreased proliferation of neural stem/progenitor cells (NSPCs) [32]. In addition, *Ythdf2* deficiency skewed the differentiation of NSCs towards neuronal lineage, but newborn neurons had fewer and shorter neurites [32].

Fragile X mental retardation protein FMRP can bind mRNAs, and FMRP target mRNAs are significantly enriched for m^6A modification [22]. The loss of the FMRP coding gene *Fmr1* altered the m^6A landscape and reduced the expression of FMRP-targeted long mRNAs in the cerebral cortex of adult mice. In addition, FMRP can interact with Ythdf2 [22]. This study provides a new layer of mechanism that specifies how FMRP regulates neuronal development and brain function.

3. m^6A and Neural Development

m^6A is abundant in the mammalian brain transcriptome, relative to other organs, and more than 25% of human transcripts are m^6A-modified [6,7,33]. During embryonic and postnatal brain development, m^6A displays temporal and spatial features, and specific m^6A modification sites are present in transcripts across brain regions [6,11,21], which suggests an important role of m^6A in neural development. Conditional deletion of *Mettl14* led to smaller sizes of newborn pups, and all died before postnatal day 25 (P25) [11]. Mettl14-cKO pups showed enlarged ventricles, delayed depletion of PAX6$^+$ radial glial cells, a type of neural stem cells, and prolonged cell-cycle progression [11]. Similar phenotypes were also observed in the brains of embryonic mice with *Mettl3* knockdown [11]. m^6A sequencing showed that transcripts with m^6A modification were related to the cell cycle and neuronal differentiation [11]. In addition, during the postnatal cerebellum development, the global level of m^6A decreases from P7 to P60, and m^6A is developmentally/temporally modulated [18]. Specific m^6A peaks at P7 were close to stop codon regions, whereas P60-specific m^6A peaks were near start codons [18]. *Mettl3* deficiency induces embryonic lethal effects, and the acute knockdown and specific ablation of *Mettl3* both induced remarkable cortical and cerebellar defects, including a reduced number of Purkinje cells and the increased apoptosis of cerebellar granule cells [17,18].

Fto-deficient mice showed a decreased body weight compared to control mice, and the sizes of whole and distinct brain regions were also decreased remarkably [15]. In contrast to control mice, which exhibited locomotor activity induced by cocaine, *Fto*-deficient mice significantly lost their response to cocaine [34]. Mechanistically, Fto can also demethylase mRNAs involved in dopamine signaling, including Ped1b, Girk2, and Syn1; consequently, Fto can alter dopamine midbrain circuitry [34]. *Alkbh5*-knockout mice also showed drastically smaller cerebella and reduced mature neurons [18]. Collectively, these findings highlight the critical function of m^6A in neural development.

4. m^6A in Axonal and Synaptic Development

Acute knockdown of *Mettl3* led to remarkable decreases of newborn neurons upon the differentiation of aNSCs, which displayed an immature morphology, with a reduced number of intersections and decreased total dendritic length [12]. In addition, *Mettl3* knockdown also inhibited the morphological development of cultured hippocampal neurons [12]. Fto was enriched in the dendrites and synapses of neurons and can be locally translated into axons [35]. Treatment with a Fto activity inhibitor promoted m^6A signals but inhibited axon elongation by regulating the axonal translation of Gap-43 [36]. In addition, transcripts for Roundabout (Robo) family member Robo3.1, an axon guidance receptor, were m^6A-modified, and m^6A reader Ythdf1 regulated axon guidance via the promotion of the translation of Robo3.1 [37]. Beyond affecting axon growth, m^6A also regulates axon regeneration. Peripheral nerve injury induces a dynamic m^6A landscape and enhances the expression of mRNAs modified by m^6A, including Sox11, Atf3, and Gadd45a [20]. *Mettl14* ablation in mature neurons promoted the translation in the adult dorsal root ganglion (DRG) and reduced the length of the longest neuronal process [20]. Similar effects were also observed in adult DRGs of *Ythdf1*-KO mice.

In addition, m^6A modification that was identified in the synaptic transcriptome and in transcripts with m^6A peaks in the stop codon but not in the start codon are associated with neurological dysfunction, including intellectual disability, microcephaly, and seizures [38]. m^6A level was negatively correlated with transcript abundance in synaptosomal RNAs, suggesting the local degradation of m^6A mRNA [38]. Interestingly, m^6A peaks in the stop codon did not show a strong effect on the synaptic location of transcripts [38]. Furthermore, in contrast to hypomethylated transcripts, hypermethylated transcripts were highly related to synaptic development and neurological disorders, including intellectual disability, autism, and schizophrenia. [38].

5. m^6A and Gliogenesis

Astrocytes and oligodendrocytes are two major macroglia cells in the brain that account for at least 50% of brain cells and are involved in diverse biological processes and brain function. In addition, to induce abnormal neurogenesis, acute knockdown of *Mettl3* induces precocious astrocytes upon the differentiation of NSCs [12]. Constitutive deletion of *Mettl14* can significantly reduce astrogenesis in embryonic mice brains [11]. Furthermore, *Ythdf2*-deficient NSCs only generate neuronal cells but not glial cells upon the differentiation [32]. Genetic ablation of *Ythdf2* also increased the sensitivity of newborn neurons to reactive oxygen species stress [32]. Mechanistically, the expression of some transcripts related to neural development and differentiation, axon guidance, and synapse development (i.e., Nrp2, Nrxn3, Flrt2, Ptprd, Ddr2) was remarkably upregulated in *Ythdf2*-deficient NSCs [32]. One identified mechanism is that *Ythdf2* deficiency represses m^6A-modified mRNA clearance [32]. These findings indicate that m^6A writers and reader(s) are essential for the proper temporal progression of neurogenesis and gliogenesis.

In addition to its important roles in astrocytes, differential m^6A peaks were detected in transcripts during the differentiation of oligodendrocyte precursor cells (OPCs) to mature oligodendrocytes. Specific inactivation of *Mettl14* in oligodendrocytes reduces the number of mature oligodendrocytes and, consequently, leads to hypomyelination [39]. Furthermore, *Mettl14* deficiency inhibits oligodendrocyte differentiation, including morphological development, but does not affect OPCs. One potential mechanism is that the loss of *Mettl14* induces the abnormal splicing of myriad RNA transcripts, including neurofascin 155 [39]. Proline-rich coiled-coil 2A (Prrc2a) is a novel m^{6}A reader and is highly expressed in OPCs. *Prrc2a* deficiency reduces the proliferation of OPCs and decreases the expression of oligodendroglial lineage-related transcripts via the direct modulation of the half-life of Olig2 mRNA [40]. Consequently, *Prrc2a*-deficient mice exhibited hypomyelination and impaired locomotive and cognitive abilities [40].

6. m^{6}A and Brain Function

Specific deletion of *Mettl3* in CaMKIIα-expressing neurons impairs long-term potentiation, which enhances long-term memory consolidation via the modulation of the translation of immediate-early genes, such as Arc, Egr1, and c-Fos [41]. Genetic ablation of *Mettl14* in dopamine D1 receptor (D1R)-expressing striatonigral neurons or dopamine D2 receptor (D2R)-expressing striatopallidal neurons also decreased the expression of neuron- and synapse-specific proteins, decreased the number of striatal cells double-labeled for mature neuronal marker NeuN and Mettl14, and increased neuronal excitability [42]. Behavioral tests show that *Mettl14* deficiency in these two types of neurons impairs sensorimotor learning and reversal learning [42].

The constitutive or NSC-specific deletion of *Fto* not only causes aberrant neurogenesis, it also impairs the learning and memory abilities of mice [13,15]. In addition, fear condition training induced dynamic m^{6}A modification, and the majority peaks were present in mRNAs. *Fto*-specific knockdown in the mouse medial prefrontal cortex (mPFC) enhanced the cued fear memory [43]. *Ythdf1*-KO mice exhibit deficits in spatial learning and memory and contextual learning [31]. *Ythdf1* deficiency also impaired basal synaptic transmission and long-term potentiation of mice, which can be rescued by ectopic Ythdf1 [31]. Ythdf1 modulates learning and memory formation mainly by promoting the translation of neuronal-stimulation-related transcripts. Heat shock stress can specifically increase m^{6}A modification in 5′UTR and can alter the cellular localization and expression of Ythdf2, but not Fto, Mettl3, Mettl14, and Wtap [44]. The level of m^{6}A modification in 5′UTR was correlated with the expression of a set of transcripts, especially the Hsp70 gene *Hspa1a* [44].

7. m^{6}A and Neurological Disorders

Consistent with important functions in neural development [18,32], neurogenesis [11,12,15,16], learning and memory [12,13,15,42] and stress response [44,45], the present evidence also indicates that m^{6}A modification is involved in several neurological disorders, including Alzheimer's disease (AD), and Parkinson's disease (PD), schizophrenia, and attention-deficit/hyperactivity disorder (ADHD) via the regulation of gene expression and RNA metabolism [10,11,46–50]. Next, we discuss the function of m^{6}A modification in neurodegenerative diseases, including Alzheimer's disease and Parkinson's disease.

A temporal feature of m^{6}A modification has been revealed during postnatal brain development and aging [6,12]. In the brain of amyloid precursor protein (APP)/presenilin-1 (PS1) (APP/PS1) transgenic AD mouse models, m^{6}A levels increased in the cortex and hippocampus, and the expressions of Mettl3 and Fto increased and decreased, respectively, compared with control mice [48]. Very recently, Shafik et al. found that m^{6}A peaks decreased during the maturation stage of postnatal brain development (postnatal 2 weeks to 6 weeks), whereas these peaks increased during the process of aging (26 weeks and 52 weeks) [21]. In addition, this study also showed increased Fto expression and decreased Mettl3 expression. The differentially methylated transcripts were enriched in the signaling pathways related to Alzheimer's disease, and differential m^{6}A methylation is associated with decreased protein expression in an AD mouse model, which was further validated in a Drosophila transgenic AD model [21]. In agreement with this study, the Fto protein level increased in the brain tissues of transgenic AD mice, and Fto depletion did not affect the level of amyloid β 42 (Aβ42) but significantly increased the level of phosphorylated Tau in the neurons from an AD mice model [51]. They further found that Fto regulates Tau phosphorylation by activating mTOR signaling. Yoon et al. performed MeRIP, followed by next-generation sequencing with forebrain organoids, and the ontology analysis of human-specific m^{6}A-targeted transcripts showed an enrichment in neurodegenerative disorders, including Alzheimer's disease [11]. Taken together, these findings suggest that m^{6}A modification could play a pivotal function in the progression of AD (Figure 2C).

Acute knockdown of *Mettl14* in substantia nigra reduced m^{6}A levels and impaired motor function and locomotor activity [52]. Nuclear receptor-related protein 1 (Nurr1), pituitary homeobox 3 (Pitx3) and engrailed1 (En1) are related to tyrosine hydroxylase expression and dopaminergic function, and their expression was remarkably reduced by *Mettl14* depletion [52]. The specific knockout of *Fto* in dopaminergic neurons impairs the dopamine neuron-dependent behavioral response by regulating dopamine transmission, which implies the important role of Fto-mediated m^{6}A demethylation in regulating dopaminergic midbrain circuitry [34]. In a Parkinson's disease (PD) rat model, the overall level of m^{6}A in the striatum decreased, and the Fto level significantly increased [53]. Either ectopic Fto or treatment with m^{6}A inhibitors reduces m^{6}A levels and induces oxidative stress and apoptosis of dopamine neurons, partially by promoting the expression of N-methyl-D-aspartate (NMDA) receptor 1 [53]. Consistently, *Fto* knockdown increases m^{6}A levels and reduces apoptosis in vitro [53]. In addition, a large cohort study with 1647 Han Chinese individuals with Parkinson's disease (PD) has identified 214 rare variants in 10 genes with m^{6}A modification; however, no significant association was observed between these variants and the risk for PD according to their analysis [54]. Therefore, the roles of m^{6}A modification still need more comprehensive investigation (Figure 2C).

8. Conclusions and Perspectives

As the most abundant modification in mRNAs, previous studies have revealed the dynamic features of m^{6}A modification and have uncovered its important function in a variety of biological processes and diseases. It seems that the more we explore m6A modification, the more complicated it becomes. First, m^{6}A modification is reversible and includes multiple key "players": writers, erasers, and readers. The interaction between these key players and other epigenetic modifications, such as histone modifiers, makes the field more complicated. Second, the complexity of m^{6}A modification also lies in the fact that it is hard to define a promoting or repressing function of m^{6}A modification in a set of diseases. The deficiency of m^{6}A writers and erasers could show similar effects on the diseases but could not exhibit contrary effects as routinely thought. Third, m^{6}A modification can regulate a defined biological process, i.e., the maintenance, renewal, and differentiation of neural stem cells by modulating diverse gene expression and signaling pathways. In addition, multiple players of m^{6}A modification exhibit effects on the same biological process, such as neurogenesis. It is hard to distinguish whether the effect is independent of each other, and it remains unclear whether they crosstalk. Therefore, how m^{6}A writers, erasers, and readers cooperate to regulate adult neurogenesis still needs more investigation.

Although dramatic progress has been made in understanding the function of m^{6}A modification, future studies should devote more effort to uncovering the multi-faceted nature of the associated mechanisms. The interaction between m^{6}A modification and histone modifiers suggests a colorful landscape wherein m^{6}A modification interacts with other epigenetic machinery, i.e., DNA modifications and non-coding RNAs. In addition, considering a substantial enrichment of m^{6}A in the 5′ and 3′ UTRs of transcripts, do multiple writers, erasers, and readers have binding specificity for distinct regions? Finally, establishing a more precise spatiotemporal landscape of m^{6}A in the pathological context could be of clinical significance. With the technical advances of sequencing, we anticipate the identification of key m^{6}A site(s) that can contribute to the diagnosis and treatment of specific diseases.

Author Contributions: All authors have read and agreed to the published version of the manuscript.

Funding: This work was supported by the National Key Research and Development Program of China (2017YFE0196600 to X.L.) and the National Natural Science Foundation of China (grants 92049108 to X.L.).

Conflicts of Interest: The authors declare no conflict of interest.

References

1. Li, S.; Mason, C.E. The pivotal regulatory landscape of RNA modifications. *Annu. Rev. Genom. Hum. Genet.* **2014**, *15*, 127–150. [CrossRef] [PubMed]
2. Wiener, D.; Schwartz, S. The epitranscriptome beyond m6A. *Nat. Rev. Genet.* **2021**, *22*, 119–131. [CrossRef]
3. Meng, T.-G.; Lu, X.; Guo, L.; Hou, G.-M.; Ma, X.-S.; Li, Q.-N.; Huang, L.; Fan, L.-H.; Zhao, Z.-H.; Ou, X.-H.; et al. Mettl14 is required for mouse postimplantation development by facilitating epiblast maturation. *FASEB J.* **2019**, *33*, 1179–1187. [CrossRef] [PubMed]
4. Yu, J.; Shen, L.; Liu, Y.; Ming, H.; Zhu, X.; Chu, M.; Lin, J. The m6A methyltransferase METTL3 cooperates with demethylase ALKBH5 to regulate osteogenic differentiation through NF-κB signaling. *Mol. Cell. Biochem.* **2020**, *463*, 203–210. [CrossRef]
5. Ping, X.-L.; Sun, B.-F.; Wang, L.; Xiao, W.; Yang, X.; Wang, W.-J.; Adhikari, S.; Shi, Y.; Lv, Y.; Chen, Y.-S.; et al. Mammalian WTAP is a regulatory subunit of the RNA N6-methyladenosine methyltransferase. *Cell Res.* **2014**, *24*, 177–189. [CrossRef] [PubMed]
6. Meyer, K.; Saletore, Y.; Zumbo, P.; Elemento, O.; Mason, C.E.; Jaffrey, S.R. Comprehensive analysis of mRNA methylation reveals enrichment in 3′ UTRs and near stop codons. *Cell* **2012**, *149*, 1635–1646. [CrossRef] [PubMed]
7. Dominissini, D.; Moshitch-Moshkovitz, S.; Schwartz, S.; Salmon-Divon, M.; Ungar, L.; Osenberg, S.; Cesarkas, K.; Jacob-Hirsch, J.; Amariglio, N.; Kupiec, M.; et al. Topology of the human and mouse m6A RNA methylomes revealed by m6A-seq. *Nature* **2012**, *485*, 201–206. [CrossRef]
8. Gilbert, W.V.; Bell, T.A.; Schaening, C. Messenger RNA modifications: Form, distribution, and function. *Science* **2016**, *352*, 1408–1412. [CrossRef]
9. Nainar, S.; Marshall, P.; Tyler, C.R.; Spitale, R.C.; Bredy, T.W. Evolving insights into RNA modifications and their functional diversity in the brain. *Nat. Neurosci.* **2016**, *19*, 1292–1298. [CrossRef]
10. Livneh, I.; Moshitch-Moshkovitz, S.; Amariglio, N.; Rechavi, G.; Dominissini, D. The m6A epitranscriptome: Transcriptome plasticity in brain development and function. *Nat. Rev. Neurosci.* **2020**, *21*, 36–51. [CrossRef]
11. Yoon, K.-J.; Ringeling, F.R.; Vissers, C.; Jacob, F.; Pokrass, M.; Jimenez-Cyrus, D.; Su, Y.; Kim, N.-S.; Zhu, Y.; Zheng, L.; et al. Temporal control of mammalian cortical neurogenesis by m6A methylation. *Cell* **2017**, *171*, 877–889.e17. [CrossRef]
12. Chen, J.; Zhang, Y.-C.; Huang, C.; Shen, H.; Sun, B.; Cheng, X.; Zhang, Y.-J.; Yang, Y.-G.; Shu, Q.; Yang, Y.; et al. m6A regulates neurogenesis and neuronal development by modulating histone methyltransferase Ezh2. *Genom. Proteom. Bioinform.* **2019**, *17*, 154–168. [CrossRef]
13. Cao, Y.; Zhuang, Y.; Chen, J.; Xu, W.; Shou, Y.; Huang, X.; Shu, Q.; Li, X. Dynamic effects of Fto in regulating the proliferation and differentiation of adult neural stem cells of mice. *Hum. Mol. Genet.* **2020**, *29*, 727–735. [CrossRef]
14. Gao, H.; Cheng, X.; Chen, J.; Ji, C.; Guo, H.; Qu, W.; Dong, X.; Chen, Y.; Ma, L.; Shu, Q.; et al. Fto-modulated lipid niche regulates adult neurogenesis through modulating adenosine metabolism. *Hum. Mol. Genet.* **2020**, *29*, 2775–2787. [CrossRef]
15. Li, L.; Zang, L.; Zhang, F.; Chen, J.; Shen, H.; Shu, L.; Liang, F.; Feng, C.; Chen, D.; Tao, H.; et al. Fat mass and obesity-associated (FTO) protein regulates adult neurogenesis. *Hum. Mol. Genet.* **2017**, *26*, 2398–2411. [CrossRef] [PubMed]
16. Wang, Y.; Li, Y.; Yue, M.; Wang, J.; Kumar, S.; Wechsler-Reya, R.J.; Zhang, Z.; Ogawa, Y.; Kellis, M.; Duester, G.; et al. N(6)-methyladenosine RNA modification regulates embryonic neural stem cell self-renewal through histone modifications. *Nat. Neurosci.* **2018**, *21*, 195–206. [CrossRef] [PubMed]
17. Wang, C.-X.; Cui, G.-S.; Liu, X.; Xu, K.; Wang, M.; Zhang, X.-X.; Jiang, L.-Y.; Li, A.; Yang, Y.; Lai, W.-Y.; et al. METTL3-mediated m6A modification is required for cerebellar development. *PLoS Biol.* **2018**, *16*, e2004880. [CrossRef] [PubMed]
18. Ma, C.; Chang, M.; Lv, H.; Zhang, Z.-W.; Zhang, W.; He, X.; Wu, G.; Zhao, S.; Zhang, Y.; Wang, D.; et al. RNA m6A methylation participates in regulation of postnatal development of the mouse cerebellum. *Genome Biol.* **2018**, *19*, 68. [CrossRef]
19. Chang, M.; Lv, H.; Zhang, W.; Ma, C.; He, X.; Zhao, S.; Zhang, Z.-W.; Zeng, Y.-X.; Song, S.; Niu, Y.; et al. Region-specific RNA m6A methylation represents a new layer of control in the gene regulatory network in the mouse brain. *Open Biol.* **2017**, *7*, 170166. [CrossRef] [PubMed]
20. Weng, Y.-L.; Wang, X.; An, R.; Cassin, J.; Vissers, C.; Liu, Y.; Liu, Y.; Xu, T.; Wang, X.; Wong, S.Z.H.; et al. Epitranscriptomic m6A regulation of axon regeneration in the adult mammalian nervous system. *Neuron* **2018**, *97*, 313–325.e6. [CrossRef]
21. Shafik, A.M.; Zhang, F.; Guo, Z.; Dai, Q.; Pajdzik, K.; Li, Y.; Kang, Y.; Yao, B.; Wu, H.; He, C.; et al. N6-methyladenosine dynamics in neurodevelopment and aging, and its potential role in Alzheimer's disease. *Genome Biol.* **2021**, *22*, 17. [CrossRef]
22. Zhang, F.; Kang, Y.; Wang, M.; Li, Y.; Xu, T.; Yang, W.; Song, H.; Wu, H.; Shu, Q.; Jin, P. Fragile X mental retardation protein modulates the stability of its m6A-marked messenger RNA targets. *Hum. Mol. Genet.* **2018**, *27*, 3936–3950. [CrossRef] [PubMed]

23. Yang, C.; Hu, Y.; Zhou, B.; Bao, Y.; Li, Z.; Gong, C.; Wang, S.; Xiao, Y. The role of m6A modification in physiology and disease. *Cell Death Dis.* **2020**, *11*, 960. [CrossRef] [PubMed]
24. Choudhry, Z.; Sengupta, S.M.; Grizenko, N.; Thakur, G.A.; Fortier, M.-E.; Schmitz, N.; Joober, R. Association between obesity-related gene FTO and ADHD. *Obesity* **2013**, *21*, E738–E744. [CrossRef] [PubMed]
25. Wei, J.; He, C. Chromatin and transcriptional regulation by reversible RNA methylation. *Curr. Opin. Cell Biol.* **2021**, *70*, 109–115. [CrossRef] [PubMed]
26. Yang, S.; Wei, J.; Cui, Y.-H.; Park, G.; Shah, P.; Deng, Y.; Aplin, A.E.; Lu, Z.; Hwang, S.; He, C.; et al. m6A mRNA demethylase FTO regulates melanoma tumorigenicity and response to anti-PD-1 blockade. *Nat. Commun.* **2019**, *10*, 2782. [CrossRef] [PubMed]
27. Fischer, J.; Koch, L.; Emmerling, C.; Vierkotten, J.; Peters, T.; Brüning, J.C.; Rüther, U. Inactivation of the Fto gene protects from obesity. *Nature* **2009**, *458*, 894–898. [CrossRef]
28. Boissel, S.; Reish, O.; Proulx, K.; Kawagoe-Takaki, H.; Sedgwick, B.; Yeo, G.S.H.; Meyre, D.; Golzio, C.; Molinari, F.; Kadhom, N.; et al. Loss-of-function mutation in the dioxygenase-encoding FTO gene causes severe growth retardation and multiple malformations. *Am. J. Hum. Genet.* **2009**, *85*, 106–111. [CrossRef]
29. Daoud, H.; Zhang, D.; McMurray, F.; Yu, A.; Luco, S.M.; Vanstone, J.; Jarinova, O.; Carson, N.; Wickens, J.; Shishodia, S.; et al. Identification of a pathogenic FTO mutation by next-generation sequencing in a newborn with growth retardation and developmental delay. *J. Med. Genet.* **2016**, *53*, 200–207. [CrossRef]
30. Fu, Y.; Jia, G.; Pang, X.; Wang, R.N.; Wang, X.; Li, C.J.; Smemo, S.; Dai, Q.; Bailey, K.A.; Nobrega, M.A.; et al. FTO-mediated formation of N6-hydroxymethyladenosine and N6-formyladenosine in mammalian RNA. *Nat. Commun.* **2013**, *4*, 1798. [CrossRef]
31. Shi, H.; Zhang, X.; Weng, Y.-L.; Lu, Z.; Liu, Y.; Lu, Z.; Li, J.; Hao, P.; Zhang, Y.; Zhang, F.; et al. m6A facilitates hippocampus-dependent learning and memory through YTHDF1. *Nature* **2018**, *563*, 249–253. [CrossRef]
32. Li, M.; Zhao, X.; Wang, W.; Shi, H.; Pan, Q.; Lu, Z.; Perez, S.P.; Suganthan, R.; He, C.; Bjørås, M.; et al. Ythdf2-mediated m6A mRNA clearance modulates neural development in mice. *Genome Biol.* **2018**, *19*, 69. [CrossRef]
33. Yoon, K.-J.; Ming, G.-L.; Song, H. Epitranscriptomes in the adult mammalian brain: Dynamic Changes regulate behavior. *Neuron* **2018**, *99*, 243–245. [CrossRef] [PubMed]
34. Hess, M.; Hess, S.; Meyer, K.; Verhagen, L.A.W.; Koch, L.; Brönneke, H.S.; Dietrich, M.; Jordan, S.D.; Saletore, Y.; Elemento, O.; et al. The fat mass and obesity associated gene (Fto) regulates activity of the dopaminergic midbrain circuitry. *Nat. Neurosci.* **2013**, *16*, 1042–1048. [CrossRef] [PubMed]
35. Walters, B.; Mercaldo, V.; Gillon, C.; Yip, M.; Neve, R.L.; Boyce, F.M.; Frankland, P.W.; Josselyn, S. The role of the RNA demethylase FTO (fat mass and obesity-associated) and mRNA methylation in hippocampal memory formation. *Neuropsychopharmacology* **2017**, *42*, 1502–1510. [CrossRef] [PubMed]
36. Yu, J.; Chen, M.; Huang, H.; Zhu, J.; Song, H.; Zhu, J.; Park, J.; Ji, S.-J. Dynamic m6A modification regulates local translation of mRNA in axons. *Nucleic Acids Res.* **2018**, *46*, 1412–1423. [CrossRef] [PubMed]
37. Zhuang, M.; Li, X.; Zhu, J.; Zhang, J.; Niu, F.; Liang, F.; Chen, M.; Li, D.; Han, P.; Ji, S.-J. The m6A reader YTHDF1 regulates axon guidance through translational control of Robo3.1 expression. *Nucleic Acids Res.* **2019**, *47*, 4765–4777. [CrossRef] [PubMed]
38. Merkurjev, D.; Hong, W.-T.; Iida, K.; Oomoto, I.; Goldie, B.J.; Yamaguti, H.; Ohara, T.; Kawaguchi, S.-Y.; Hirano, T.; Martin, K.C.; et al. Synaptic N6-methyladenosine (m6A) epitranscriptome reveals functional partitioning of localized transcripts. *Nat. Neurosci.* **2018**, *21*, 1004–1014. [CrossRef] [PubMed]
39. Xu, H.; Dzhashiashvili, Y.; Shah, A.; Kunjamma, R.B.; Weng, Y.-L.; Elbaz, B.; Fei, Q.; Jones, J.S.; Li, Y.I.; Zhuang, X.; et al. m6A mRNA methylation is essential for oligodendrocyte maturation and CNS myelination. *Neuron* **2019**, *105*, 293–309.e5. [CrossRef]
40. Wu, R.; Li, A.; Sun, B.; Sun, J.-G.; Zhang, J.; Zhang, T.; Chen, Y.; Xiao, Y.; Gao, Y.; Zhang, Q.; et al. A novel m6A reader Prrc2a controls oligodendroglial specification and myelination. *Cell Res.* **2018**, *29*, 23–41. [CrossRef] [PubMed]
41. Zhang, Z.; Wang, M.; Xie, D.; Huang, Z.; Zhang, L.; Yang, Y.; Ma, D.; Li, W.; Zhou, Q.; Yang, Y.-G.; et al. METTL3-mediated N6-methyladenosine mRNA modification enhances long-term memory consolidation. *Cell Res.* **2018**, *28*, 1050–1061. [PubMed]
42. Koranda, J.L.; Dore, L.; Shi, H.; Patel, M.; Vaasjo, L.O.; Rao, M.N.; Chen, K.; Lu, Z.; Yi, Y.; Chi, W.; et al. Mettl14 is essential for epitranscriptomic regulation of striatal function and learning. *Neuron* **2018**, *99*, 283–292.e5. [CrossRef] [PubMed]
43. Widagdo, J.; Zhao, Q.-Y.; Kempen, M.-J.; Tan, M.C.; Ratnu, V.S.; Wei, W.; Leighton, L.; Spadaro, P.A.; Edson, J.; Anggono, V.; et al. Experience-dependent accumulation of N 6 -methyladenosine in the prefrontal cortex is associated with memory processes in mice. *J. Neurosci.* **2016**, *36*, 6771–6777. [CrossRef] [PubMed]
44. Zhou, J.; Wan, J.; Gao, X.; Zhang, X.; Jaffrey, S.; Qian, S.-B. Dynamic m6A mRNA methylation directs translational control of heat shock response. *Nature* **2015**, *526*, 591–594. [CrossRef]
45. Engel, M.; Eggert, C.; Kaplick, P.M.; Eder, M.; Röh, S.; Tietze, L.; Namendorf, C.; Arloth, J.; Weber, P.; Rex-Haffner, M.; et al. The role of m6A/m-RNA methylation in stress response regulation. *Neuron* **2018**, *99*, 389–403.e9. [CrossRef]
46. Zhao, B.; Roundtree, I.A.; He, C. Post-transcriptional gene regulation by mRNA modifications. *Nat. Rev. Mol. Cell Biol.* **2016**, *18*, 31–42. [CrossRef]
47. Du, K.; Zhang, L.; Lee, T.; Sun, T. m6A RNA methylation controls neural development and is involved in human diseases. *Mol. Neurobiol.* **2018**, *56*, 1596–1606. [CrossRef]
48. Han, M.; Liu, Z.; Xu, Y.; Liu, X.; Wang, D.; Li, F.; Wang, Y.; Bi, J. Abnormality of m6A mRNA methylation is involved in Alzheimer's disease. *Front. Neurosci.* **2020**, *14*, 98. [CrossRef]

49. Engel, M.; Chen, A. The emerging role of mRNA methylation in normal and pathological behavior. *Genes Brain Behav.* **2017**, *17*, e12428. [CrossRef]
50. Yue, Y.; Liu, J.; He, C. RNA N6-methyladenosine methylation in post-transcriptional gene expression regulation. *Genes Dev.* **2015**, *29*, 1343–1355. [CrossRef]
51. Li, H.; Ren, Y.; Mao, K.; Hua, F.; Yang, Y.; Wei, N.; Yue, C.; Li, D.; Zhang, H. FTO is involved in Alzheimer's disease by targeting TSC1-mTOR-Tau signaling. *Biochem. Biophys. Res. Commun.* **2018**, *498*, 234–239. [CrossRef] [PubMed]
52. Teng, Y.; Liu, Z.; Chen, X.; Liu, Y.; Geng, F.; Le, W.; Jiang, H.; Yang, L. Conditional deficiency of m6A methyltransferase Mettl14 in substantia nigra alters dopaminergic neuron function. *J. Cell. Mol. Med.* **2021**, *25*, 8567–8572. [CrossRef] [PubMed]
53. Chen, X.; Yu, C.; Guo, M.; Zheng, X.; Ali, S.; Huang, H.; Zhang, L.; Wang, S.; Huang, Y.; Qie, S.; et al. Down-regulation of m6A mRNA methylation is involved in dopaminergic neuronal death. *ACS Chem. Neurosci.* **2019**, *10*, 2355–2363. [CrossRef]
54. Qin, L.; Min, S.; Shu, L.; Pan, H.; Zhong, J.; Guo, J.; Sun, Q.; Yan, X.; Chen, C.; Tang, B.; et al. Genetic analysis of N6-methyladenosine modification genes in Parkinson's disease. *Neurobiol. Aging* **2020**, *93*, 143.e9–143.e13. [CrossRef] [PubMed]

Article

Ghrelin Regulates Expression of the Transcription Factor Pax6 in Hypoxic Brain Progenitor Cells and Neurons

Irina I. Stoyanova [1,*], Andrii Klymenko [2], Jeannette Willms [2], Thorsten R. Doeppner [3,4], Anton B. Tonchev [1] and David Lutz [2,*]

1   Department of Anatomy and Cell Biology, Faculty of Medicine, Research Institute of the Medical University, 9002 Varna, Bulgaria; anton.tonchev@mu-varna.bg
2   Department of Neuroanatomy and Molecular Brain Research, Ruhr University Bochum, 44801 Bochum, Germany; andrii.klymenko@rub.de (A.K.); jeannette.willms@ruhr-uni-bochum.de (J.W.)
3   Department of Neurology, University Medical Center Goettingen, 37075 Goettingen, Germany; thorsten.doeppner@med.uni-goettingen.de
4   Research Institute for Health Sciences and Technologies (SABITA), Medipol University, Istanbul 34810, Turkey
*   Correspondence: stoyanovai@yahoo.co.uk (I.I.S.); david.lutz@rub.de (D.L.)

Abstract: The nature of brain impairment after hypoxia is complex and recovery harnesses different mechanisms, including neuroprotection and neurogenesis. Experimental evidence suggests that hypoxia may trigger neurogenesis postnatally by influencing the expression of a variety of transcription factors. However, the existing data are controversial. As a proof-of-principle, we subjected cultured cerebral cortex neurons, cerebellar granule neurons and organotypic cerebral cortex slices from rat brains to hypoxia and treated these cultures with the hormone ghrelin, which is well-known for its neuroprotective functions. We found that hypoxia elevated the expression levels and stimulated nuclear translocation of ghrelin's receptor GHSR1 in the cultured neurons and the acute organotypic slices, whereas ghrelin treatment reduced the receptor expression to normoxic levels. GHSR1 expression was also increased in cerebral cortex neurons of mice with induced experimental stroke. Additional quantitative analyses of immunostainings for neuronal proliferation and differentiation markers revealed that hypoxia stimulated the proliferation of neuronal progenitors, whereas ghrelin application during the phase of recovery from hypoxia counteracted these effects. At the mechanistic level, we provide a link between the described post-ischemic phenomena and the expression of the transcription factor Pax6, an important regulator of neural progenitor cell fate. In contrast to the neurogenic niches in the brain where hypoxia is known to increase Pax6 expression, the levels of the transcription factor in cultured hypoxic cerebral cortex cells were downregulated. Moreover, the application of ghrelin to hypoxic neurons normalised the expression levels of these factors. Our findings suggest that ghrelin stimulates neurogenic factors for the protection of neurons in a GHSR1-dependent manner in non-neurogenic brain areas such as the cerebral cortex after exposure to hypoxia.

Keywords: GHSR1; ghrelin; hypoxia; neurogenesis; transcription factors; progenitor cells

Citation: Stoyanova, I.I.; Klymenko, A.; Willms, J.; Doeppner, T.R.; Tonchev, A.B.; Lutz, D. Ghrelin Regulates Expression of the Transcription Factor Pax6 in Hypoxic Brain Progenitor Cells and Neurons. *Cells* **2022**, *11*, 782. https://doi.org/10.3390/cells11050782

Academic Editor: FengRu Tang

Received: 27 December 2021
Accepted: 22 February 2022
Published: 23 February 2022

Publisher's Note: MDPI stays neutral with regard to jurisdictional claims in published maps and institutional affiliations.

1. Introduction

Stroke is a medical condition of impeded blood supply to the brain and oxygen shortage (hypoxia). The recovery from hypoxia requires intra- and extracellular processes recapitulating nervous system development such as extracellular matrix reorganisation, neurogenesis, and stimulation of neuronal plasticity. A prolonged hypoxia state may result in ischemia followed by neuronal death and reduction of neuronal density [1]. Interestingly, hypoxia can also trigger neurogenesis within the surrounding tissue [2,3] prenatally as well as postpartum [4,5]. Neurogenesis in the adult brain is essential and occurs during the entire postnatal life. It is believed to be restricted to the subventricular zone and the subgranular zone of the dentate gyrus, though it has been observed in some other areas of the mammalian brain, e.g., the subcallosal zone [6], the striatum [7], the amygdala [8], and also the neocortex [9], albeit at considerably lower levels. The adult neurogenesis goes through the same consecutive stages as the embryonic one does and yields post-mitotic functionally integrated new neurons [10].

Exposure to hypoxia during the early postnatal period activates generation of Tbr1-positive spiny pyramidal neurons in vitro as well as ex vivo [11]. In vitro, some of the early neuronal stem cells can acquire multipotency and undergo long-term self-renewal upon exposure to hormones and growth- and transcription factors [12].

Pax6 is a member of the paired-box and homeobox-containing gene family (PAX) of transcription factors and its early expressed protein Pax6 is a key transcription factor in the generation of neuronal lineages during the development of the central nervous system [13–15]. Moreover, *Pax6* is highly conserved between species: there is no difference between the amino acid sequence of the human and mouse Pax6, thus pointing to a pivotal role in brain development [16]. Indeed, *Pax6* regulates corticogenesis, numbers and arrangement of cortical cells in layers as well as the ratio of excitatory and inhibitory neurons [5]; these effects are dose-dependent [17]. *Pax6* is considered a neurogenic fate determinant directing astrocyte-to-neuron conversion during adult neurogenesis.

Unlike the neurogenic niches, where hypoxia increases Pax6 expression, the levels in the neocortex are downregulated [2]. Does this decreased neocortical expression of Pax6 act neuroprotectively on cortical neurons? Could Pax6 upregulation influence neurogenesis, and thus, benefit recovery from hypoxia? As a proof-of-principle, we subjected dissociated cerebral cortex neurons, cerebellar granule neurons and acute cerebral cortex slices to hypoxia and treated them with the hormone ghrelin, which is well-known to act neuroprotectively against oxidative stress in vivo [18] and in vitro [19]. We then examined the effect of hypoxia on the expression levels of ghrelin's receptor GHSR1 (growth hormone secretagogue receptor 1) in cultured neurons and acute organotypic slices. We also confirmed that the expression of GHSR1 is upregulated in the cerebral cortex of mice with transitional middle cerebral artery occlusion used as an in vivo stroke model. Furthermore, we subjected hypoxic cultures of mature cerebral cortex neurons that had been treated with ghrelin to immunostaining for Pax6, Ki67 (neuronal proliferation and differentiation marker), or NeuN (neuronal nuclear antigen) aiming at analysing the changes in the ratio of cells positive for these markers upon hypoxia vs. normoxic conditions.

2. Materials and Methods

2.1. Animals Used for Primary Cell Cultures

All animal experiments were conducted according to German law and approved by the corresponding committee on animal use, being conformed to the guidelines set by the European Union. Wistar rats were bred and maintained at 22 °C on a 12 h light/dark cycle with *ad libitum* food and water access in the Animal Facility of the Medical Faculty at the Ruhr University Bochum, Germany. For in vitro experiments, a mixed population of female and male offspring was used. The manuscript was prepared following the ARRIVE guidelines for animal research [20].

2.2. Animals Used for Experimental Ischemia In Vivo

Permission for the animal experiments was received by the Lower Saxony State Office for Consumer Protection and Food Safety of the City Oldenburg (Niedersächsisches Landesamt für Verbraucherschutz und Lebensmittelsicherheit (LAVES)/Oldenburg, contract No 33.9-42502-04-11/0622). Six-month-old C57Bl6 mice were subjected to transient focal cerebral ischemia as described [21]. Briefly, animals were anaesthetised by inhalation of 0.8–1.5% isoflurane, 30% O_2, and remainder N_2O. The rectal temperature was maintained at 36.5–37.0 °C by employing a feedback-controlled heating system under continuous blood flow monitoring using a laser Doppler flow (LDF) system (Perimed, Järfälla, Sweden). Occlusion of the middle cerebral artery was achieved with a 7–0 silicon coated nylon monofilament (tip diameter of 180 μm; Doccol, Sharon, MA, USA) that was withdrawn after 45 min to entail transient cerebral ischemia. LDF recordings were continued for additional 15 min to monitor and verify the appropriate reperfusion of the brain. Animals were sacrificed on day 28 after the induction of stroke and subjected to immunostaining for detection of the ghrelin receptor GHSR1 as described in Section 2.6.

2.3. Dissociated Cell Cultures

Cerebral cortex neurons were obtained from newborn Wistar rats, from six plating procedures, five pups (from the same mother) per plating. In brief, neonates were anaesthetised by isoflurane inhalation and decapitated. Brains were isolated, the cerebral cortices were dissected and placed in Neurobasal medium (Thermo Fisher Scientific, Waltham, MA, USA). After removal of the meninges, the cerebral cortices were collected in a chemically defined R12 culture medium [22] with trypsin for chemical dissociation. Thereafter, 150 µL of soybean trypsin inhibitor and 125 µL of DNAse I (20.000 units, Thermo Fisher Scientific) were added, followed by trituration for mechanical dissociation of the cells. The suspension was centrifuged at 1200 rpm at 4 °C for 5 min. Cells were plated on glass coverslips at a density of approximately 3000 cells/mm^2. The glass coverslips were pre-coated with 20 mg/mL poly-L-lysine (Merck, Darmstadt, Germany) to enhance cell adhesion. Cells were allowed to settle at 37 °C and 5% CO_2 for 2 h (incubation in R12 medium, optimised with 50 ng/mL nerve growth factor (Thermo Fisher Scientific)). The medium was serum-free to suppress glial cell proliferation, thus keeping glial cell numbers lower than 5% [22]. The medium was renewed twice a week. The cultures were stored in an incubator under standard conditions of 37 °C, 100% humidity and 5% CO_2 for a period of one or three weeks prior to hypoxia.

Cerebellar granule neurons were obtained from a mixed population of 4-day-old female and male Wistar rats. Briefly, the animals were anaesthetised by isoflurane inhalation and decapitated. Brains were isolated, the cerebella were dissected and placed in Neurobasal medium (Thermo Fisher Scientific). After removal of the meninges, the cerebella were incubated in 0.025% trypsin (Merck) in Hank's balanced salt solution (HBSS, Thermo Fisher Scientific) at 37 °C for 30 min. The tissue was then incubated in HBSS containing 1% BSA (Merck) and 1% w/v trypsin inhibitor (cat.# T-6522, Merck) at 37 °C for 5 min. After washing in HBSS, the tissue was triturated with a pipette tip, and the dissociated neurons were cultured in neurobasal medium supplemented with 2% B-27 (Thermo Fisher Scientific), 0.5 mM L-glutamine (Thermo Fisher Scientific), 100 units/mL penicillin (Thermo Fisher Scientific), and 100 µg/mL streptomycin (Thermo Fisher Scientific) at a density of 1.7×10^4 cells per well of a 24-well plate coated with poly-L-lysine (Merck). Cerebellar granule neurons were cultured for 24 h prior to hypoxia induction.

2.4. Acute Cerebral Cortex Slices

Acute cerebral cortex slices were obtained from a mixed population of 4-day-old female and male Wistar rats. The animals were anaesthetised by isoflurane inhalation and decapitated. Brains were isolated, the brain hemispheres were dissected in DMEM medium (Merck) and sliced at a thickness of 400 µm perpendicularly to their longitudinal axis using the McIlwain tissue chopper (Plano, Wetzlar, Germany). Slices were placed onto Millicell membrane inserts (Merck) and transferred into six-well plates with 1 mL of nutrition medium per well (25% heat-inactivated horse serum, 25% HBSS, 50% DMEM, 2 mM glutamine, pH 7.2). Slices were maintained under standard conditions of 37 °C, 100% humidity and 5% CO_2 for a period of 24 h prior to hypoxia.

2.5. Induction of Hypoxia and Treatment with Ghrelin

One-third of the cultured cells/acute slices were kept under normoxic conditions (37 °C, 100% humidity and 5% CO_2) and the rest of the cultures were exposed to hypoxia for 6 h. Hypoxia was achieved by subjecting the cultures and slices to air evacuation [23], which took ~3–4 s. The induction of hypoxia begins with lowering the partial oxygen pressure (pO_2) from ~160 mmHg to less than 25 mmHg within 30 min. This process was visually inspected with an oxygen indicator. The specimens were air-evacuated and sealed in a plastic bag, and then incubated at 37 °C and in a humidified atmosphere for 6 h (for further details see also [23]). After six hours of hypoxia, the cultures and slices were maintained under normoxic conditions (37 °C, 100% humidity and 5% CO_2) for 24 h. Thereby, half of these cultures/slices were supplemented with human ghrelin peptide (cat.# ab199421, Abcam, Cambridge, UK) for the duration of 24 h (final concentration of 0.5 μM as described elsewhere [24–27]). The other half of the cultures were kept in a plain medium, also for 24 h, and used as a control.

2.6. Immunostaining

Immunostaining was performed on 36 independent cultures, with equal cell density for each experimental condition (normoxia, hypoxia, hypoxia + ghrelin). Primary cell cultures and acute slices were fixed in 4% paraformaldehyde/phosphate buffered saline pH 7.4 (PBS, Merck) at room temperature for 20 min, washed in PBS for 5 min, immersed in 1% bovine serum albumin (in PBS containing 0.01% Triton-X) for 20 min and immunostained for calretinin, GHSR1, Pax6, Ki67, GFAP, and NeuN using the following primary antibodies (dilutions in PBS are indicated): mouse anti-GHSR1 (cat.# sc-374515, RRID:AB_10987651, Santa Cruz Biotechnology, Dallas, TX, USA, dilution 1:1000), rabbit anti-calretinin (cat.# CR 7697, RRID:AB_2619710, Swant, Burgdorf, Switzerland, dilution 1:1000), rabbit anti-Pax6 (cat.# PRB-278P, RRID:AB_291612, Covance, Princeton, NJ, USA, dilution 1:500) or mouse monoclonal anti-Pax6 (cat.# MA1-109, RRID:AB_2536820, Thermo Fisher Scientific), rabbit anti-Ki67 (cat.# VP-RM04, RRID:AB_2336545, Vector Laboratories, Burlingame, CA, USA, dilution 1:200), goat anti-GFAP (cat.# ab53554, RRID:AB_880202, Abcam, dilution 1:750), and guinea-pig anti-NeuN (cat.# ABN90, RRID:AB_11205592, Millipore, Burlington, MA, USA, dilution 1:2000). In the case of double immunostaining, the specimens were incubated sequentially in each of the primary antibodies at 4 °C for 10 h each. Following several washes in PBS, the specimens were incubated in the appropriate secondary antibodies (IgG conjugated to fluorochromes): goat anti-rabbit Alexa 488 nm (cat.# A-11008, RRID:AB_143165, Thermo Fisher Scientific, diluted 1:1000), goat anti-mouse 594 nm (cat.# A-11031, RRID:AB_144696, Thermo Fisher Scientific, diluted 1:1000), goat anti-guinea pig 594 nm (cat.# 106-585-003, RRID:AB_2337442 Jackson ImmunoResearch Labs, West Grove, PA, USA, diluted 1:500), donkey anti-goat Alexa 488 nm (cat.# A32814, RRID:AB_2762838, Thermo Fisher Scientific, diluted 1:500), and donkey anti-rabbit Alexa 594 nm (cat.# A32754, RRID:AB_2762827, Thermo Fisher Scientific, dilution 1:500) in a dark chamber at room temperature for 2 h. The samples were counterstained with the fluorescent dye 4′,6-diamidino-2-phenylindole DAPI (cat.# 10236276001, Merck, dilution 1:1000) at room temperature for 15 min. After several washes in PBS, specimens were rinsed in distilled water, embedded in fluorescent mounting medium (Carl Roth, Karlsruhe, Germany) on coverslips. For transmitted light microscopy, the incubation in primary mouse anti-GHSR1 antibody was followed by treatment with biotinylated secondary antibodies (donkey anti-mouse IgG, dilution 1:500, Jackson ImmunoResearch Labs) at room temperature for 2 h. After rinsing, the brain slices were incubated for 1 h in the Vectastain ABC-HRP Kit (6.25 μL/mL of each compound in PBS, cat.# PK-4002, Vector Laboratories) following the manufacturer's instructions. The peroxidase activity was visualised with the SG substrate kit (cat.# SK-4700, Vector Laboratories) in PBS for 5 min, at room temperature. Finally, the specimens were rinsed in PBS and mounted on coverslips.

To test the specificity of the GHSR1 antibody, we conducted two control staining procedures as follows. Prior to staining, we incubated the murine primary GHSR1 antibody with murine pituitary/hypothalamus tissue homogenate to bind the antibody with detergent-solubilised GHSR1. In particular, tissue of the pituitary gland and hypothalamus from an 11–month–old male mouse was freshly isolated and homogenised in RIPA buffer (150 mM sodium chloride, 50 mM tris-HCl, 1% Nonidet P–40, 0.5% sodium deoxycholate, pH 8.0). The homogenate was then centrifuged at 13,000 rpm (room temperature) for 10 min and the supernatant was used for further incubation steps. Approximately 120 μg of the supernatant protein were mixed with 4 ng of the GHRS1 antibody at room temperature for 30 min. This mixture was applied to fixed cerebral cortex cells at 4 °C for 10 h, followed by the standard steps of immunocytochemistry as described above. For the second control staining procedure, we applied only the Alexa 594—conjugated secondary antibody (without the primary GHSR1 antibody) to the cells (using a dilution of 1:1000) for 2 h. In both cases, the GHSR1 fluorescence signal was barely detectable with the set-up of the spinning disk microscope as described in the imaging acquisition section.

2.7. Homogenate Preparation for Dot Blot Analysis

Cerebral cortex slices (described above) were homogenised in lysis buffer containing 100 mM Tris-HCl, pH 7.4 (Carl Roth), 12 mM magnesium acetate tetrahydrate (Merck), and 6 M urea (Sigma-Aldrich, St. Louis, MO, USA) under several freezing-refreezing rounds in liquid nitrogen. The total protein amount was measured on the Genova Nano microvolume Life Science Spectrophotometer (Jenway, Staffordshire, UK) using a direct UV light detection set-up following the manufacturer's instructions. For each condition, 20 μg total protein per dot were dropped onto a nitrocellulose membrane (cat.# 10600015, 0.2 μm, GE Healthcare, Chicago, IL, USA). The membrane was rinsed in Vilber washing buffer (cat.# PU4000500, Vilber, Collégien, France) at room temperature for 30 min and then incubated with a murine antibody recognising GHSR1 (cat.# sc-374515, RRID:AB_10987651, Santa Cruz Biotechnology, dilution 1:1000, incubation overnight) or mouse β-actin antibody (cat.# A5441, RRID:AB_476744, Sigma-Aldrich, dilution 1:10,000, incubation for 1 hour) which were diluted in the Vilber Purity™ anti-mouse HRP reagent (cat.# PU4200100, Vilber) following the manufacturer's instructions. The membranes were then incubated in the Vilber Purecl™ Dura substrate (cat.# PU4400125, Vilber) according to the manufacturer's instructions and subjected to chemiluminescence detection using the Vilber Fusion FX Imager (Vilber). The duration of the exposure for the GHSR1 and β-actin signal were 10 s and 20 s, respectively. For quantification of chemiluminescent signals, the ImageJ2 software (version 2.3.0/1.53f, Fiji) was used. The membranes were then stained with Ponceau S solution (cat.# P7170-1L, Sigma-Aldrich) for 4 min, washed in distilled water for 15 min and air-dried.

2.8. Image Acquisition

After immunostaining, the specimens were subjected to fluorescence microscopy using a confocal fluorescence spinning disc microscope (Nikon, Minato, Tokyo, Japan) equipped with the PCO Edge 5.5 sCMOS camera (noise: 1.4 electrons, resolution: 5.5 megapixel, dynamic range: 22,000:1, speed: 100 fps, stabilised by Peltier cooling, Visitron Systems, Puchheim, Germany) and the VS-LMS Laser-Merge-System for CSU-X1 and 2D FRAP Option (Visitron Systems). For acquisition of images, 405 nm (3.54 mW), 488 nm (3.85 mW) and 561 nm (4.11 mW) laser wave lengths were used. Images were taken using a 40× air-magnification objective (ELWD 40×/0.6 air s plan fluor, OFN22, DIC, N1, MRH08430) or a 60× water-magnification immersion objective (60×a/1.20 WI plan Apo vc, Nikon, OFN25, DIC, N2, MRD07602). The software used for the acquisition of images was VisiView (Version 4.4.018, 16 December 2019, Visitron Systems, license # 1434). All exposure times were set to be 1000 ms with binning 2, offset 0/0, gain 0, and a non-implemented digitiser. The images were further processed and analysed with the ImageJ2 software (version 2.3.0/1.53f, Fiji). Only uncropped original images are provided in the figures.

Transmitted light microscopic images were captured using a fully motorised widefield microscope Zeiss AxioImager Z.2 (Carl Zeiss, Jena, Germany) with an AxioCam Mrm rev.3 monochrome CCD camera (Carl Zeiss), and AxioVision v.4.9 software through an EC Plan-Neofluar objective 5×. Shading from the irregular illumination field was corrected during the acquisition via the camera's built-in shading correction. The images were savedin the zvi-format (Carl Zeiss) of the AxioVision software, stitched together, and exported to an 8 bit TIFF format for further processing and analysis with the ImageJ2 software (version 2.3.0/1.53f, Fiji).

2.9. Immunogold Transmission Electron Microscopy

Formaldehyde-fixed cultures (described above) were incubated with 1% bovine serum albumin in PBS at room temperature for 30 min, followed by an incubation with the mouse monoclonal antibody against GHSR1 (diluted 1:1000 in PBS) at 4 °C for 72 h. Specimens were washed in PBS three times (20 min for each round) and incubated with a donkey polyclonal antibody coupled to 10 nm gold (cat.# ab39593, Abcam, RRID:AB_954429, dilution 1:100 in PBS) at room temperature for 1 hour. After intensive rinsing in PBS (five rounds, 5 min for each round), sections were fixed with 1% glutaraldehyde in PBS at room temperature for 10 mins. Specimens were then incubated in 0.8% NaCl and 8% glucose at room temperature for 30 min and chilled on ice for a further incubation in an aqueous mixture of 2% OsO_4 and 1% potassium hexacyanoferrate (III) (cat.# 31251, Sigma-Aldrich) for 2 h. The osmicated sections were washed in an ice-chilled aqueous mixture of 0.8% NaCl and 8% glucose for 30 min and dehydrated in 70% ethanol (2 × 15 min), 90% ethanol (2 × 15 min), 95% ethanol (2 × 15 min), 100% ethanol (2 × 30 min), and pure propylene oxide (2 × 20 min) on ice. Sections were then incubated in a propylene oxide-Araldite mixture (1:1) containing 3% accelerator on ice for 2 h and then at room temperature overnight. Sections were then transferred into Araldite containing 2% accelerator for 1 hour (at room temperature) and then the Araldite mixture was refreshed for the final embedding. The embedded sections were cured at 65°C overnight. Semithin sections (500 nm) were cut on an ultramicrotome (Ultracut R, Leica, Wetzlar, Germany), stained with 1% toluidine blue in borax (4 min) and inspected under the Leica DME light microscope (10× and 40× objectives). Ultrathin sections (55 nm) were then cut, mounted on 100-meshed Nickel grids (Ted Pella, Redding, CA, USA) and stained with an aqueous solution of 4% uranyl acetate (cat. # E22400, Science Services, München, Germany) at room temperature for 20 min. The sections were rinsed in distilled water and then stained in aqueous 2% $Pb(NO_3)_3$ (cat. # 228621, Sigma-Aldrich). After rinsing in distilled water and air drying, the sections were subjected to low voltage electron microscopy using the LVEM25 (Delong Instruments, Brno, Czech Republic).

2.10. Data Collection and Statistical Analysis

Twenty light microscopic images of cerebral cortex sections of at least 4 animal brains with transient cerebral ischemia were analysed for GHSR1 expression. The intensity of the GHSR1 signals was measured within the cells in arbitrary units (AU) and normalised on the measured cell area. The data sets were subjected to Welch's *t*-test and are presented as medians ± standard deviation (SD). The sample size of each group is indicated within the corresponding figure legend. For in vitro cultures, images of 200 to 800 cells from each marker per group were taken and subjected to analysis of the intensity signals for each condition. At least four independent cultures obtained from six animals were used per condition. The percentage of cells positive for each marker was calculated and presented as means ± standard error of the mean (SEM). The intensity of the GHSR1 fluorescence signals was measured in AU and normalised on the measured cell area—the data are presented as medians ± SD. All images used for comparison were taken at the same setting on the confocal spinning disc microscope. For dot blot analysis, the chemiluminescent signals of homogenates from organotypic slices of four animals (four independent experimental samples per condition) were processed with the ImageJ software as described above and each GHSR1 signal was referenced to the signal of the corresponding ß-actin band. The data are presented as ratios (medians ± SD). The data sets were subjected to One-Way ANOVA statistical analysis using the post hoc Tukey's Multiple Comparison Test or Kruskal-Wallis Multiple Comparison Test to assess the statistical significance of differences between the various conditions. The values of $p < 0.05$ were regarded as statistically significant. All *p*-values and sample sizes are indicated in the corresponding figure/figure legends. The statistical analyses were performed with the GraphPad Prism software (version 9.3.0, GraphPad Software Inc., San Diego, CA, USA).

3. Results

3.1. Experimental Ischemia In Vivo Leads to Increased GHSR1 Expression Levels in Cerebral Cortex Cells

There is a growing body of evidence indicating that ghrelin ameliorates neuroregeneration upon injury. Therefore, we became interested in testing whether transient cerebral ischemia, which was induced experimentally in mice, may alter the GHSR1 expression in cerebral cells within the ischemic region. To this aim, we occluded the middle cerebral artery of adult mice for 45 min and allowed afterwards reperfusion to take place, thus inducing transient cerebral ischemia. Animals were then sacrificed on day 28 after induction of stroke and subjected to immunostaining for the ghrelin receptor GHSR1. Analysis of immunostaining intensity signals revealed enhanced GHSR1 immunoreactivity in cortical cells of the stroke area when compared to the expression observed at the contralateral side (Figure 1A,B). In the ipsilateral hemisphere (occlusion site), the GSHR1 signal was detected across the cerebral cortex cell bodies, whereas contralaterally, the signal was mainly restricted to the cytoplasm (Figure 1A). The quantification analysis indicated that the condition of stroke (hypoxia) stimulated the expression of GHSR1 in cerebral cortex cells.

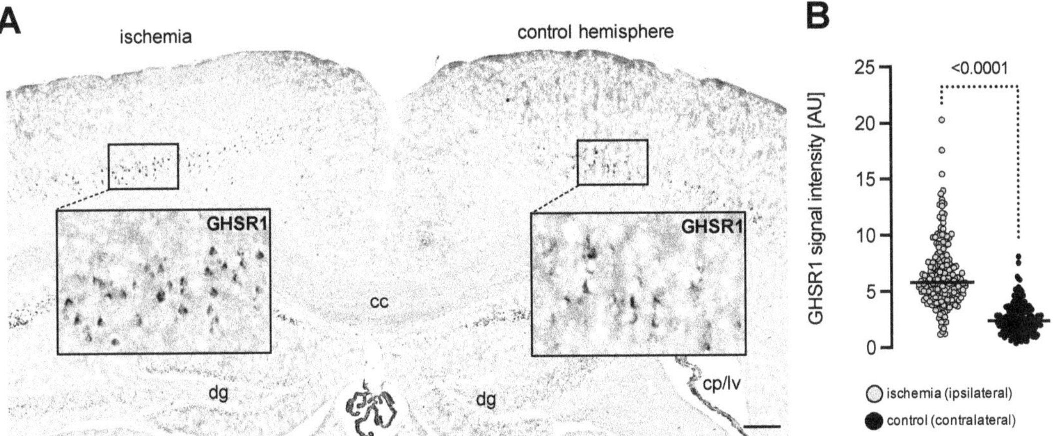

Figure 1. GHSR1 expression in cerebral cortex sections from brains of mice with experimentally induced transient cerebral ischemia. (**A,B**) Immunostaining with an antibody recognising GHSR1 (black staining) on a representative coronal brain section showing strong GHSR1-immunoreactivity at the occlusion site (ischemia/ipsilateral) in comparison to the signal of the contralateral side (control hemisphere). Note that the contralateral signal was restricted mainly to the cytoplasm (crescent-shaped appearance). Abbreviations: cc—corpus callosum, dg—dentate gyrus, cp/lv—choroid plexus/lateral ventricle. Scale bar, 300 µm. The GHSR1 signal intensity within the cells was measured in arbitrary units [AU]. Data are presented as medians ± SD analysed by Welch's *t*-test. The *p*-value is indicated; *n* = 203 measured ipsilateral cells; *n* = 197 measured contralateral cells.

3.2. Experimental Hypoxia In Vitro Leads to Increased GHSR1-Immunoreactivity in Cultured Cerebral Cortex Cells, Cerebellar Granule Neurons and Organotypic Cerebral Cortex Slices

Following the results from the experimental in vivo stroke model, we became interested in testing whether the stimulatory effect of hypoxia on the GHRS1 expression could be seen in neuronal cultures deprived of oxygen. As a proof-of-principle, we used an in vitro model of hypoxia consisting of neonatal dissociated rat cortical neurons that had been cultured for one week and then exposed to severe hypoxia for 6 h, followed by a recovery period of 24 h under normal oxygen supply. Half of the cultures were supplemented with ghrelin during these 24 h, while the other half remained untreated as a control. All samples were fixed and processed for immunostaining with antibodies recognising calretinin and GHSR1. We tested the specificity of the GHSR1 antibody in two control staining set-ups: prior to staining of the cell cultures, we incubated the murine primary GHSR1 antibody with murine pituitary/hypothalamus tissue homogenate to opsonise the antibody with detergent-solubilised GHSR1 (Supplementary Figure S1A and Methods). For the second control staining procedure, we applied only the secondary antibody, without the primary GHSR1 antibody (Supplementary Figure S1B and Methods). The signal of calretinin immunofluorescence revealed arborised neurons, which were surrounded by nonneural (fibroblasts) and glial cells (Figure 2A, arrows). The GHSR1 fluorescence signal in calretinin-positive neurons was measured (Figure 2B). Under normoxic conditions, the neuronal immunoreactivity for GHSR1 was weak at the plasma membrane, cytoplasm and nucleus (Figure 2A). Hypoxia imposed on the cell cultures led to increased GHSR1-immunoreactivity in calretinin-positive neurons (Figure 2A,B). The GHSR1 signal was increased in the neurites and nuclei of the neurons. Of note, in the nuclei as well as in the cytoplasm of hypoxic glia and fibroblasts the GHSR1 signal was also increased (Figure 2A, arrows, middle panel). When ghrelin was applied to hypoxic neurons, the fluorescence signals of GHSR1 in neurons were reduced to those measured under normoxia (Figure 2A,B).

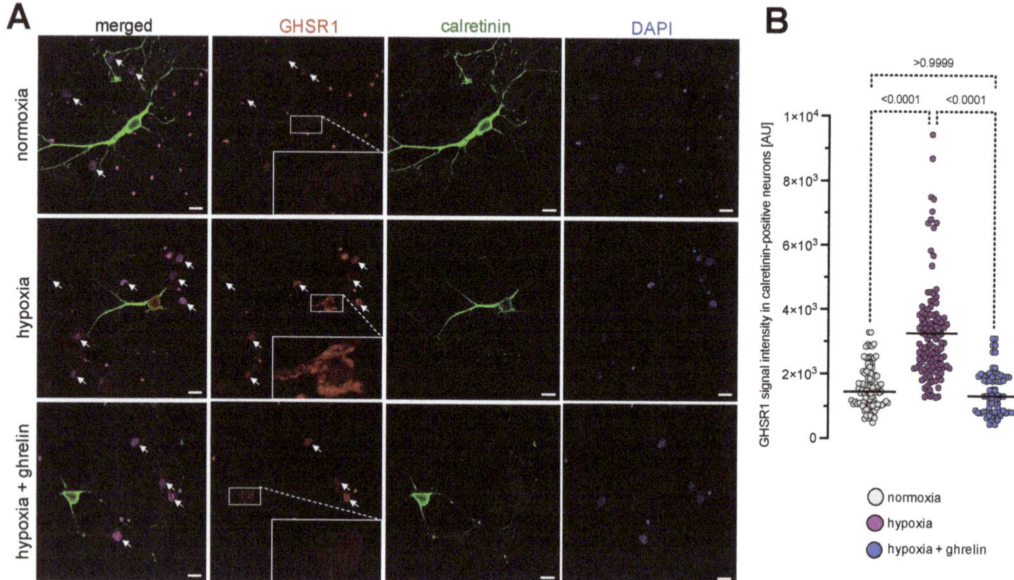

Figure 2. Hypoxia imposed on dissociated cerebral cortex cells stimulated expression of GHSR1. (**A**) Immunostaining with antibodies recognising calretinin (green) and GHSR1 (red) on neonatal dissociated rat cortical neurons that had been cultured for one week and then exposed to severe hypoxia for 6 h, followed by a recovery period of 24 h under normal oxygen supply. DAPI (blue) was used to stain the nuclei. Calretinin immunofluorescence revealed the arborisation pattern of the cerebral cortex neurons, which were surrounded by nonneural and glial cells (arrows). Under normoxic conditions, neurons expressed GHSR1 at the plasma membrane, cytoplasm and nucleus (inlets). Hypoxia imposed on the neurons led to increased expression levels of GHSR1 (**A,B**). Note that in the nuclei as well as in the cytoplasm of hypoxic glia and fibroblasts the GHSR1 signal was also increased. Application of ghrelin to hypoxic cultures reduced the expression levels of GHSR1 in neurons to normoxic levels (**A,B**). Scale bar, 30 μm. Data are presented as medians ± SD analysed by Kruskal-Wallis' Multiple Comparison Test. *p*-values are indicated; *n* = 97 for normoxia; *n* = 119 for hypoxia and *n* = 70 for hypoxia + ghrelin. See also Supplementary Figure S1.

We next asked whether these effects hold true for other types of neurons and studied, therefore, dissociated rat cerebellar granule neurons that had been cultured for 24 h and then exposed to severe hypoxia for 6 h, followed by a recovery period of 24 h under normal oxygen supply. Similar to the experimental design with cortex neurons mentioned above, half of the cultures of cerebellar granule neurons were supplemented with ghrelin during the recovery phase, while the other half remained untreated as a control. All samples were fixed and immunostained for calretinin and GHSR1 (Figure 3A). The calretinin immunostained cerebellar granule neurons appeared unipolar, often growing in clusters (Figure 3A). In normoxic cerebellar granule neurons, the expression of GHSR1 was weak and predominantly restricted to the nucleus (Figure 3A, inlets). Hypoxia led to increased GHSR1-immunoreactivity in the nucleus (Figure 3A,B). When ghrelin was applied to recovering hypoxic cerebellar granule neurons, the GHSR1 signals were reduced to those measured under normoxia (Figure 3A,B).

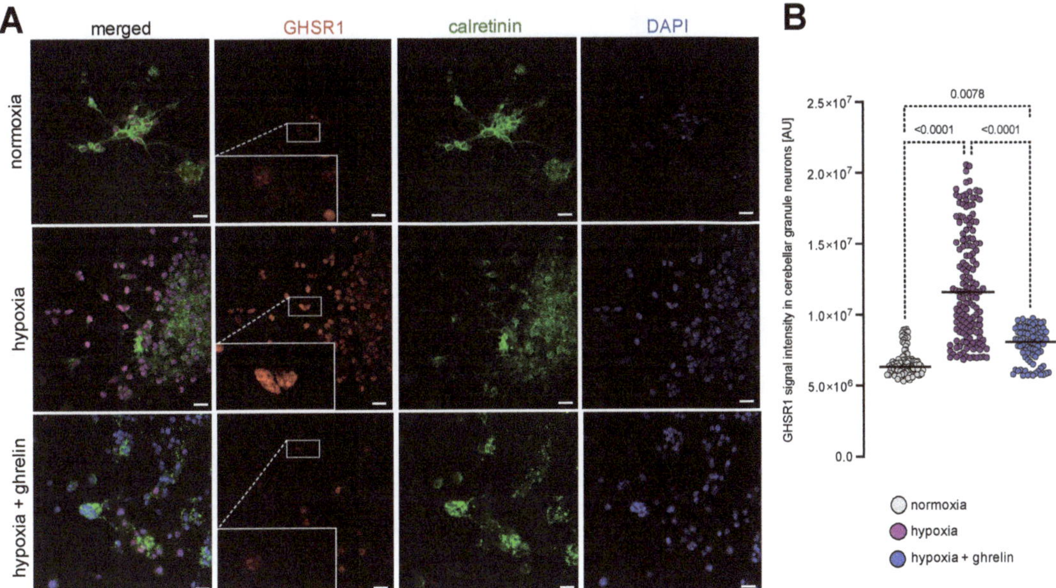

Figure 3. Hypoxia imposed on dissociated cerebellar granule neurons stimulated nuclear expression of GHSR1. (**A**) Immunostaining for calretinin (green) and GHSR1 (red) on cerebellar granule neurons that had been cultured for 24 h and then exposed to severe hypoxia for 6 h, followed by a recovery period of 24 h under normal oxygen supply. DAPI (blue) was used to stain the nuclei. Cerebellar granule neurons were unipolar, partially growing in clusters. In normoxic cerebellar granule neurons, the expression of GHSR1 was predominantly restricted to the nucleus (inlets). Hypoxia led to increased GHSR1-immunoreactivity in the nucleus (**A,B**). When ghrelin was applied to recovering hypoxic cerebellar granule neurons, the GHSR1 signals were similar to the normoxic levels (**A,B**). Scale bar, 30 μm. Data are presented as medians ± SD analysed by Kruskal-Wallis' Multiple Comparison Test. *p*-values are indicated; n = 99 for normoxia; n = 151 for hypoxia and n = 83 for hypoxia + ghrelin.

In parallel, we prepared acute organotypic slice cultures from brain hemispheres of 4-day-old Wistar rats and exposed the slices to severe hypoxia for 6 h, followed by a recovery period of 24 h under normal oxygen supply. One half of the cerebral cortex slices were supplemented with ghrelin during the recovery phase, while the other half remained untreated as a control. After fixation, the samples were immunostained for calretinin and GHSR1 (Figure 4A). In slices, the calretinin immunostained cerebral cortex neurons appeared multipolar, surrounded by a huge population of calretinin-negative cells expressing GHSR1 (Figure 4A). Under normoxia, the GHSR1 signal was homogeneously distributed across the sliced tissue (Figure 4A). Hypoxia imposed on the slices led to increased GHSR1-immunoreactivity (Figure 4A,B). When ghrelin was applied to recovering hypoxic slices, the GHSR1 signals were slightly reduced (in neurons almost to those seen under normoxia) (Figure 4A,B). However, many calretinin-negative cells were still displaying strong GHSR1 signals.

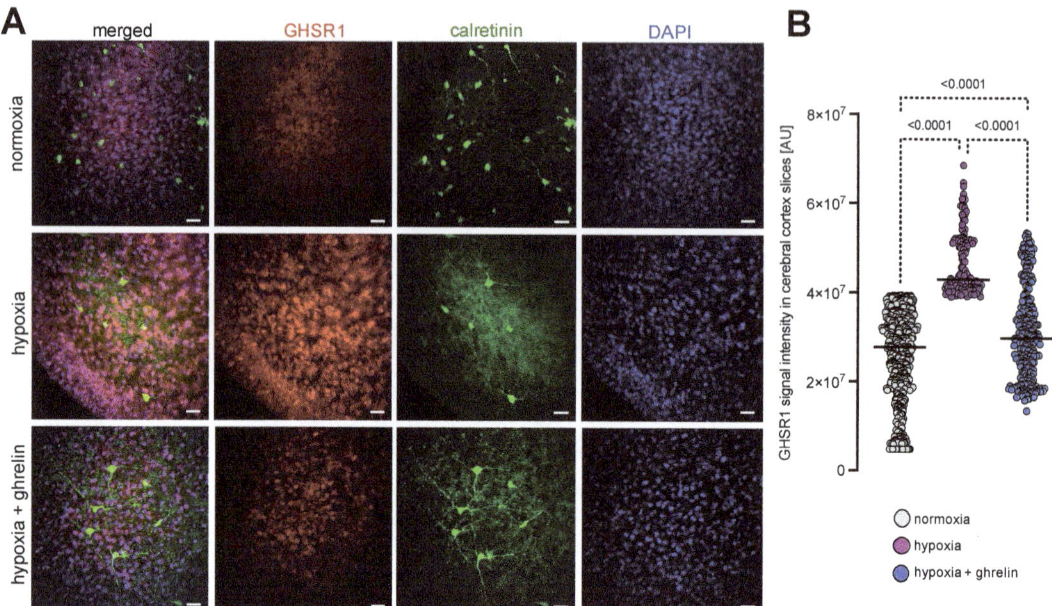

Figure 4. Hypoxia on acute cerebral cortex slices stimulated expression of GHSR1. (**A**) Calretinin (green) and GHSR1 (red) immunostaining of cerebral cortex slices. Slices were obtained from 4-day-old rats and cultured for 24 h to be then exposed to severe hypoxia for 6 h, followed by a recovery period of 24 h under normal oxygen supply. DAPI (blue) was used to stain the nuclei. In slices, calretinin-positive cerebral cortex neurons appeared multipolar, surrounded by many calretinin-negative cells expressing GHSR1. Normoxic expression of GHSR1 was homogeneously distributed across the sliced tissue. Hypoxia imposed on the slices led to increased GHSR1-immunoreactivity (**A,B**). When ghrelin was applied to recovering hypoxic slices, the GHSR1 signals were slightly reduced (in neurons almost to normoxic levels). Many calretinin-negative cells were still strongly positive for GHSR1. Scale bar, 30 μm. Data are presented as medians ± SD analysed by Kruskal-Wallis' Multiple Comparison Test; *p*-values are indicated; *n* = 562 for normoxia; *n* = 142 for hypoxia and *n* = 274 for hypoxia + ghrelin.

Furthermore, we prepared homogenates from the slices and subjected the homogenates to dot blot analysis using the GHSR1 antibody. Protein loading was controlled by immunodetection for β-actin and by staining with Ponceau S (Figures 5A and S2). The GHSR1 signal was normalised on the β-actin signal. The quantitative analysis of the dots confirmed that the expression of GHSR1 was increased in the homogenates of the slices which were subjected to hypoxia when compared to normoxic slices (Figures 5A,B and S2). Ghrelin application to slices recovering from hypoxia decreased the levels of GHSR1 (Figures 5A,B and S2).

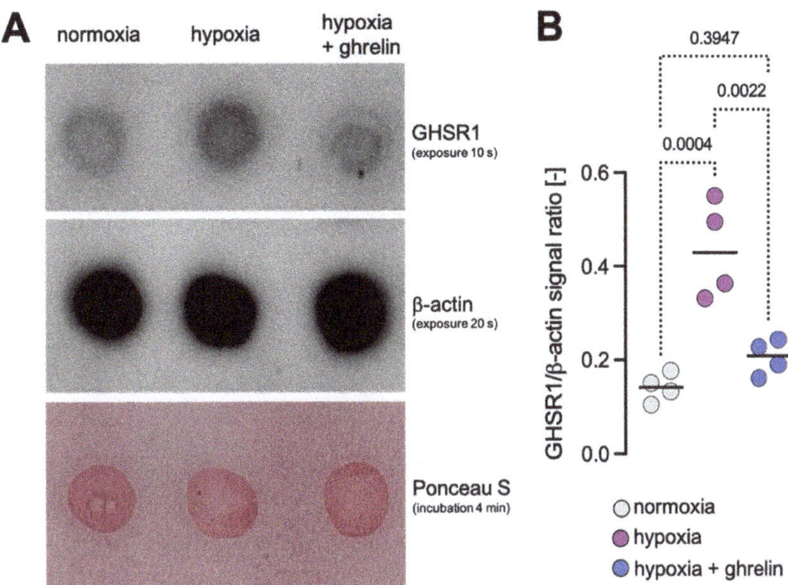

Figure 5. Dot blot analysis of homogenates from cerebral cortex slices under normoxic and hypoxic conditions. (**A**) Dot blots with homogenates from slices under normoxia, hypoxia and hypoxia followed by ghrelin treatment. Antibodies recognising GHSR1 and β-actin were used. Lower panel: Ponceau S staining. Exposure/incubation times are indicated. (**A,B**) Quantitative analysis of the dots confirmed that the expression of GHSR1 was increased in the homogenates of the slices which were subjected to hypoxia in comparison to normoxic slices. Ghrelin application to recovering hypoxic slices decreased the levels of GHSR1. Data are presented as medians ± SD analysed by One-Way ANOVA with Tukey's Multiple Comparison Test; *p*-values are indicated; *n* = 4 for normoxia; *n* = 4 for hypoxia and *n* = 4 for hypoxia + ghrelin. See also Supplementary Figure S2.

Considering the GHSR1 immunoreactivity found on the cell surface and within the cell interior, we asked whether the receptor is present in the cell nuclei and associates with the chromatin. We first mapped the murine and human GHSR1 protein sequence for nuclear localisation signals in the cNLS Mapper [28]. The predicted NLS in the mouse GHSR1 sequence (Figure 6A) and that of the human (Figure 6B) received the scores 3.2 and 4.2–4.7, respectively. Such scores suggest a potential cytoplasmatic and nuclear presence of GHSR1 [28]. Following these predictions, we performed immunogold transmission electron microscopy for GHSR1 on cerebral cortex cells under normoxia, hypoxia and hypoxia with ghrelin treatment (Figure 6C). Under normoxic conditions, we found a few immunogold grains/precipitates within the cytoplasm and the nucleus of the cells (Figure 6C, first panel), whereas, under hypoxia, the amount of the immunogold precipitates was per se higher in both the cytoplasm and nucleus (Figure 6C, second panel). In the hypoxic cells that had received ghrelin, the amount of cytoplasmic and nuclear immunogold precipitates appeared reduced (Figure 6C, third panel). Of note, in all three conditions, nuclear immunogold precipitates were found in the regions of the heterochromatin (Figure 6C, red arrows). Interestingly, under hypoxia, a few gold grains were found also within the euchromatin. These combined findings confirmed the nuclear presence and the chromatin association of GHSR1.

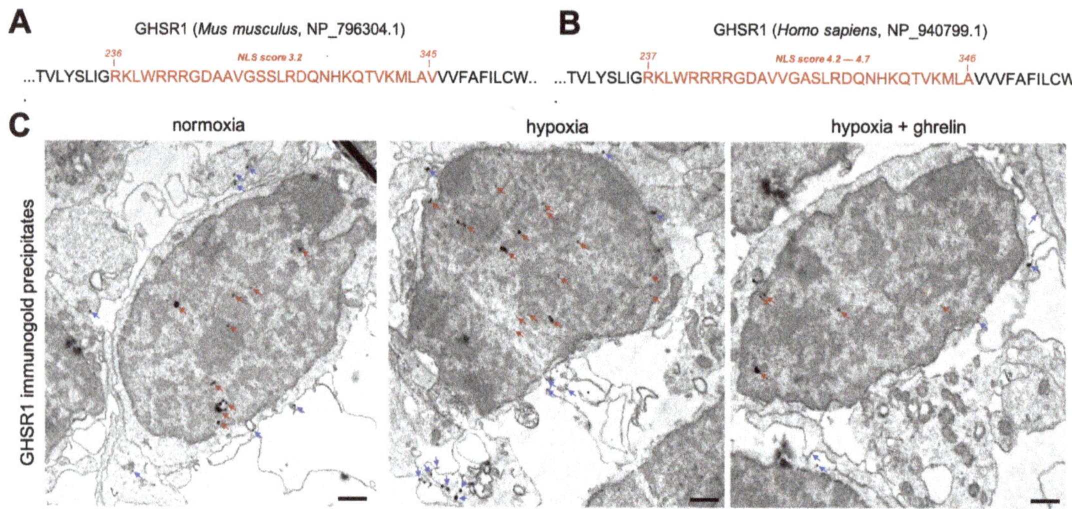

Figure 6. Nuclear presence and chromatin association of GHSR1. (**A,B**) Amino acid sequences of the murine and human GHSR1 with predicted nuclear localisation signal regions (NLS, in red). Initial and end positions of the NLS are indicated, the NLS prediction scores are given. (**C**) Transmission electron microphotographs showing the intracellular distribution of GHSR1-immunogold precipitates/grains in normoxic, hypoxic and ghrelin-treated hypoxic cerebral cortex cells. Blue arrows indicate immunogold precipitates found within the cytoplasm, on vesicles and on the plasma membrane. Red arrows indicate nuclear GHSR1-immunogold precipitates (predominantly associated with the heterochromatin). Note the increased number of cytoplasmic and nuclear gold grains under the condition of hypoxia. Scale bar, 100 nm.

3.3. Experimental Hypoxia In Vitro Leads to Decreased Numbers of Pax6 and NeuN Expressing Cerebral Cortex Cells, but Increases the Number of Ki67-Positive Cells

Based on the findings obtained from the transmission electron microscopy, we hypothesised that nuclear GHSR1 affected the expression of neurogenic and mitotic factors such as Pax6 and Ki67. To test this hypothesis, we prepared neonatal dissociated rat cerebral cortex neurons that had been cultured for 3 weeks to a mature stage to then expose them to severe hypoxia for 6 h, followed by a recovery period of 24 h under normal oxygen supply. Half of the cultures were supplemented with ghrelin during 24 h of recovery, while the other half remained untreated as a control. We immunostained the cultures for Pax6 (Figure 7A). Because of the relative difference in the cell density, we calculated the Pax6-positive cells as a percentage of all cells (counterstained with DAPI). In mature 3-week-old cultures under normoxic conditions, $55.5 \pm 3.0\%$ (mean $\pm$ SEM) of the cells expressed Pax6 (Figure 7A,B). Their proportion was decreased under hypoxia to $38.9 \pm 2.1\%$ (mean $\pm$ SEM), whereas ghrelin treatment upregulated the expression levels of Pax6 ($53.2 \pm 1.4\%$, mean $\pm$ SEM) and brought them closely to levels seen under normoxia (Figure 6B).

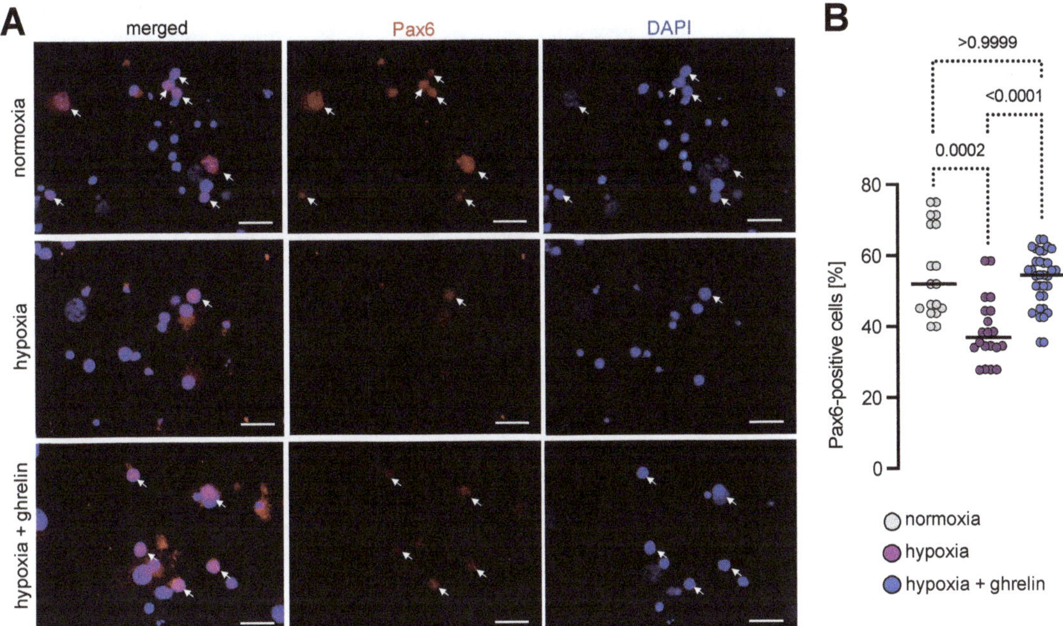

Figure 7. Hypoxia applied to dissociated cerebral cortex cells reduced the number of Pax6 expressing cells. (**A,B**) Pax6 immunostaining of dissociated cerebral cortex cells (in a mature state) under normoxic and hypoxic conditions. Approximately half of the matured cortical cultures were Pax6-positive (arrows), whereas hypoxia significantly decreased their proportion. Ghrelin treatment after exposure to hypoxia upregulated the number of Pax6 expressing cells to normoxic levels. Scale bar, 20 μm. Data are presented as medians ± SD analysed by Kruskal-Wallis' Multiple Comparison Test; *p*-values are indicated; *n* = 18 for normoxia; *n* = 20 for hypoxia and *n* = 34 for hypoxia + ghrelin.

Additionally, staining of 3-week-old cerebral cortex cell cultures for Pax6 (produced in rabbit) and GFAP was performed (Figure 8A). Under normoxic conditions, 25.6 ± 3.9% (mean ± SEM) of the cultured cells expressed both Pax6 and GFAP (Figure 8A, upper panel). In the hypoxic group 14.7 ± 2.1% (mean ± SEM) of the cells were positive for Pax6 and GFAP (Figure 8A, middle panel). In hypoxic cultures supplemented with ghrelin during the recovery period, 30.9 ± 6.0% (mean ± SEM) of all cells were double-stained for Pax6 and GFAP (Figure 8A, lower panel). In a parallel set-up of the same conditions, we used another Pax6 antibody (produced in the mouse) in combination with calretinin (Figure 8B). Normoxic neurons stained for calbindin were strongly immunoreactive for Pax6 (mouse), whereas hypoxic neurons displayed weak immunoreactivity signals (Figure 8B, upper and middle panels). Ghrelin-treated hypoxic neurons showed strong Pax6-immunoreactivity signals (Figure 8B, lower panels).

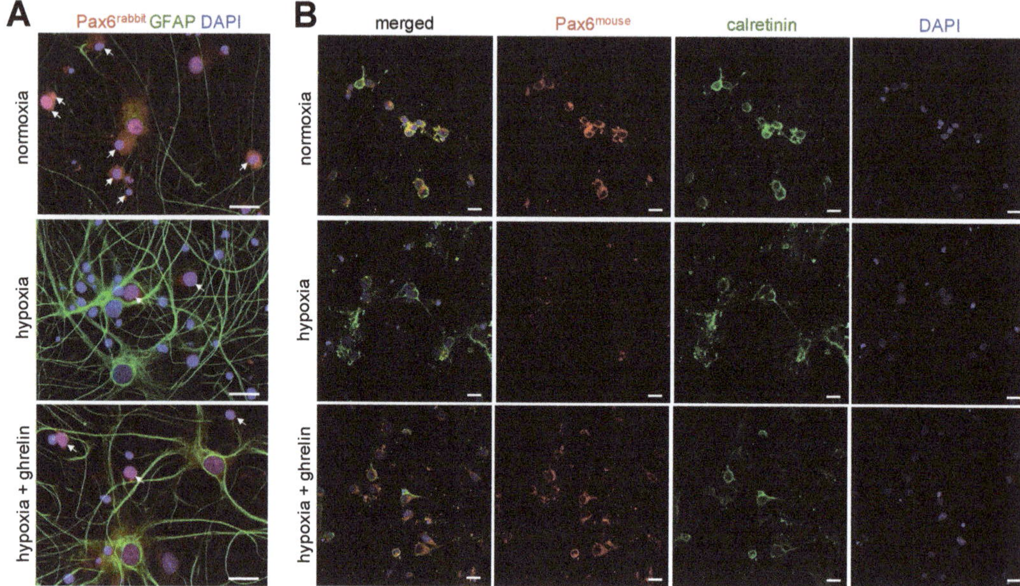

Figure 8. Hypoxia imposed on 3-week-old cerebral cortex cells led to decreased immunoreactivity for Pax6. (**A**) Pax6 (rabbit) and GFAP immunostaining of dissociated cerebral cortex cells (in a mature state) under normoxic and hypoxic conditions. Normoxic GFAP-positive cells (glial cells) were also positive for Pax6 and surrounded by many GFAP-negative cells with Pax6-immunoreactivity (arrows). Hypoxia lowered the number of GFAP/Pax6-double-positive cells, whereas, ghrelin application after hypoxia increased this number to normoxic levels (means ± SEM for normoxia: 25.6 ± 3.9%, n = 11; hypoxia: 14.7 ± 2.1%, n = 10; hypoxia + ghrelin: 30.9 ± 6.0%, n = 11; p < 0.0001, One-Way ANOVA with Tukey's Multiple Comparison Test). Note the overall weak Pax6-immunoreactivity signals in hypoxic neurons vs. the condition of normoxia and hypoxia + ghrelin. Scale bar, 20 μm. (**B**) Pax6 (mouse) and calretinin immunostaining of matured dissociated cerebral cortex cells revealed weak Pax6-immunoreactivity in hypoxic calretinin-positive neurons when compared to normoxic and ghrelin-treated hypoxic neurons. (**A,B**) DAPI was used to stain the nuclei. Analysis of the nuclear Pax6 signal intensity (medians ± SD in arbitrary units) for normoxia: 1.62 ± 0.69, n = 50; hypoxia: 0.77 ± 0.59, n = 52; hypoxia + ghrelin: 1.15 ± 0.58, n = 52; p < 0.0001, Kruskal-Wallis' Multiple Comparison Test. Scale bar, 30 μm.

We further immunostained the cultures for Ki67 (Figure 9A,B) to detect active cell proliferation. The pattern of staining varied significantly between the cells; in some of them, the staining signal was restricted to the nucleoli, whereas in the large nuclei the signal appeared dispersed throughout the entire nucleus (Figure 9A,B). Moreover, some of the nuclei showed speckled patterns of staining (e.g., glial cells, Figure 9A). Proliferating neuronal precursors displayed Ki67-positive nuclei smaller than those of the proliferating glial cells, localised in the close proximity of the GFAP-positive glial cell nets. Under normal oxygen supply, the Ki67 index (percentage of neuronal precursors positive for the Ki67 antigen) was 9.7 ± 0.8% (mean ± SEM). Interestingly, in control cultures exposed to hypoxia for 6 h and followed by 24 h of re-oxygenation, the proportion of proliferating neuronal precursors was 59.8 ± 1.4% (mean ± SEM). In the experimental group supplemented with ghrelin during the recovery period, Ki67 was detected in 15.9 ± 1.6% (mean ± SEM) of the neuronal precursor nuclei (Figure 9A,B).

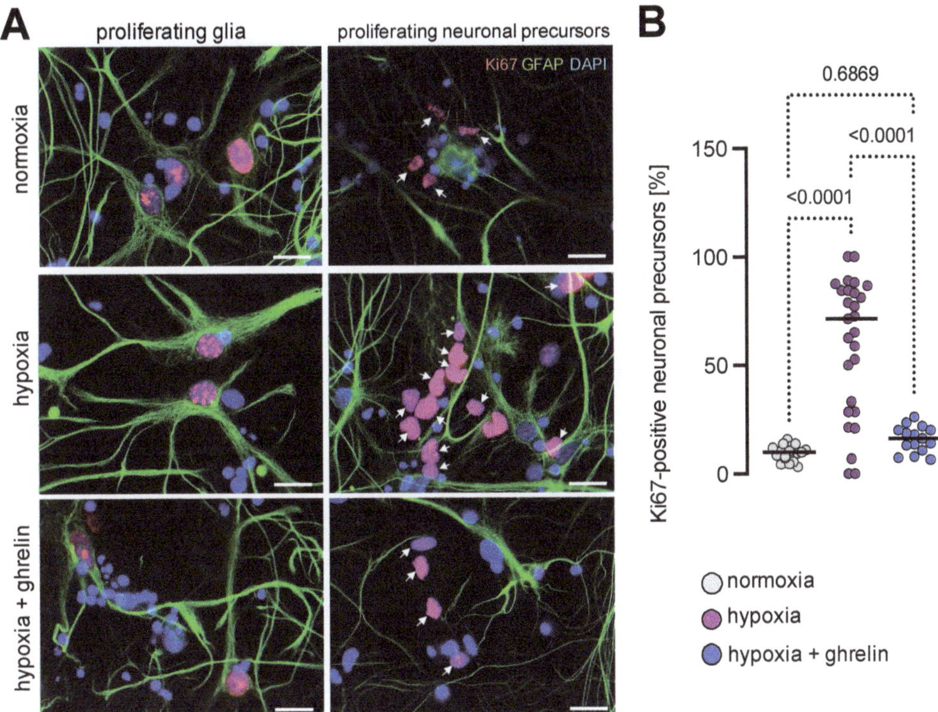

Figure 9. Hypoxia imposed on dissociated cerebral cortex neurons stimulated the proliferation of neural precursors. (**A**) Immunolabelling of dissociated cortical neurons for Ki67 (pink after merging with DAPI) and GFAP (green) revealed the complex network of glial cells and neurons. DAPI was used to stain the nuclei. Proliferating glial cells were positive for Ki67 and showed a speckled pattern of nuclear staining for the antigen, surrounded by GFAP-stained fibrils (left panels). The double staining showed also that the majority of Ki67-positive cells under hypoxic conditions were neuronal precursors (right panels, arrows). Scale bar, 20 μm. (**B**) Quantitative analysis revealed that hypoxia significantly elevated the number of Ki67-positive neuronal precursors, whereas ghrelin treatment reduced these numbers to normoxic values. Data are presented as medians ± SD analysed by Kruskal-Wallis' Multiple Comparison Test. *p*-values are indicated; *n* = 18 for normoxia; *n* = 27 for hypoxia and *n* = 15 for hypoxia + ghrelin.

Further immunostaining for visualisation of the marker for post-mitotic neurons, NeuN, revealed immunoreactivity not only in the nuclei but also in the cytoplasm of many neurons, including the initial part of the neurites (Figure 10A). Some cells exhibited a strong immunoreactivity for the antigen (Figure 10A). The comparison between the quantitative data pointed to a significant decrease in the ratio of the NeuN-positive neurons under hypoxia (Figure 10B). Post-hypoxic supplementation of ghrelin increased the number of NeuN-positive neurons to normoxic values (Figure 10A,B). In particular, we observed that hypoxia reduced significantly the ratio of NeuN-positive post-mitotic neurons (61.5 ± 3.7%, mean ± SEM) compared with the ratio which was assessed under normoxia (77.1 ± 2.8%, mean ± SEM). Ghrelin treatment during the post-hypoxic period elevated the fraction of neurons positive for NeuN up to numbers observed under normoxia (85.1 ± 3.2%, mean ± SEM).

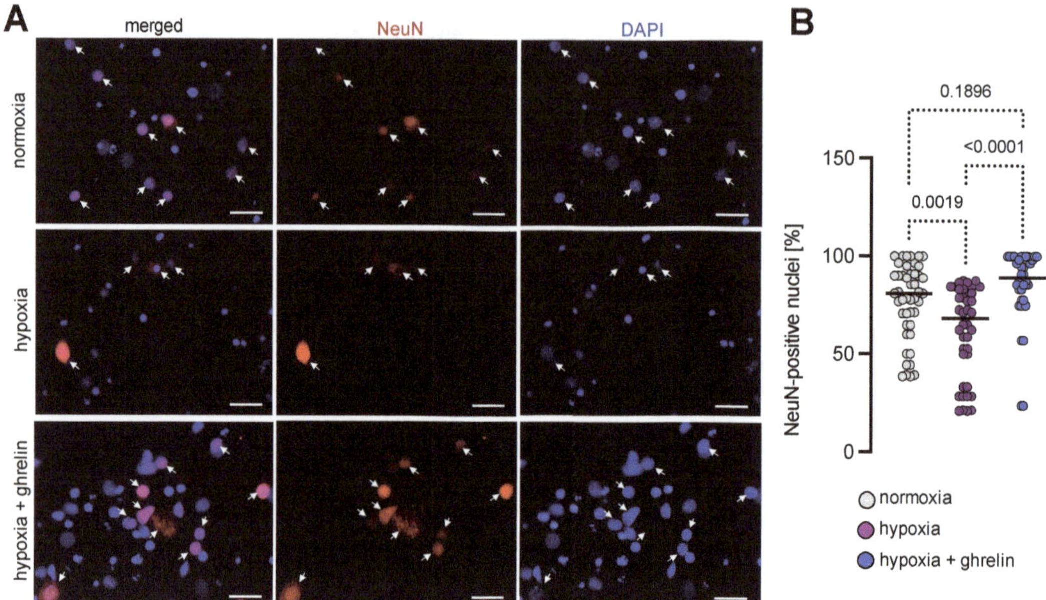

Figure 10. Hypoxia decreased the number of NeuN-expressing post-mitotic neurons. (**A**) NeuN immunolabelling of post-mitotic dissociated cortical neurons (arrows). DAPI was used to stain the nuclei. NeuN was expressed not only in the nuclei of the neurons but also in the cytoplasm of many of them. Additionally, some neurons showed a higher level of antigen expression than others. Scale bar, 20 μm. (**B**) Hypoxia decreased the ratio of NeuN-positive neurons, whereas post-hypoxic application of ghrelin increased the number of NeuN-positive neurons to normoxic values. Data are presented as medians ± SD analysed by Kruskal-Wallis' Multiple Comparison Test. *p*-values are indicated; $n = 46$ for normoxia; $n = 40$ for hypoxia and $n = 36$ for hypoxia + ghrelin.

4. Discussion

The main findings of the present study are: (i) The expression of ghrelin's receptor GHSR1 is enhanced after hypoxia-induced in vivo and In vitro; (ii) Ghrelin treatment of hypoxic cultures counteracts the increase of GHSR1; (iii) hypoxia reduces the Pax6 levels In vitro, simultaneously increasing neural progenitor proliferation.

In some areas of the nervous system, such as the hypothalamus and the spinal cord, ghrelin has been shown to play an important role in (adult) neurogenesis [29,30]. Interestingly, ghrelin is the only known endogenous ligand activating the GHSR1 [31,32]. In contrast to ghrelin, GHSR1 is highly expressed in the nervous system [33–38]. Does the expression level of GHSR1 change upon injury or stress imposed on the nervous system? Cabral et al. [39] have suggested that neuronal remodelling is GHSR1-mediated and depends on the energy balance. Moreover, experimental evidence from the past two decades has revealed that ghrelin mediates neuroregeneration upon injury (for a detailed review see [40]). Therefore, we became interested in testing whether neurons subjected to metabolic stress may alter their GHSR1 expression. We demonstrated that hypoxia imposed on the cell cultures led to increased GHSR1-immunoreactivity in the neurites and nuclei of the neurons, as well as in the cytoplasm and nuclei of hypoxic glia and fibroblasts. On the contrary, previous studies have reported downregulation of GHSR1 expression after ischemia [18,41], and controversial effects of ghrelin on GHSR1's expression—while the research group of Miao [18] reported that postischemic treatment of rats with ghrelin led to an elevation of the receptor levels, Huang et al. [41] did not observe any change. These discrepancies could be due to the methods of sampling or/and the way of exposure of the cultures or model systems to ischemia/hypoxia, or/and the concentration of ghrelin utilised for administration.

The methods of hypoxia inducement and ghrelin application in vivo reported in the literature vary significantly and lead to controversial experimental outcomes. The concentration of 0.5 μM used in our experiments complies with that reported by us [27,42] and other groups [24,26,43–45], yet differs from the concentrations utilised by Miao et al. [18] and Huang et al. [41]. Nevertheless, we abode by our previously reported concentrations because of the probability that the acute slices might require more ghrelin to recover after hypoxia: i.e., being placed on membranes, they are nourished through diffusion from the medium underneath, where ghrelin was applied. For the sake of consistency, we used 0.5 μM of ghrelin in all experiments.

Ghrelin has been shown to exert neuroprotection via attenuation of oxidative stress [41], blockade of apoptosis [18], and stimulation of synaptic plasticity [46]. Suda et al. [47] have suggested that defective GHSR1 activity of dopaminergic neurons causes marked motor impairment. Remarkably, mice deficient in GHSR1 did not display any signs of neurodegeneration [48]. We interpret the upregulation of GHSR1 upon hypoxia to be a response to hypoxia/injury/stress.

The plasma levels of ghrelin are relatively low [49], therefore, the reaction of neurons upon hypoxia with increased levels of GHSR1 could have an amplifier-like effect for low concentrations of ghrelin (see also [40]). To the best of our knowledge, the finding that hypoxia upregulates the expression levels of GHSR1 in neurons is novel. So is also the finding that GHSR1 is present in/translocates to the nucleus and associates with the chromatin. The chromatin association of GHSR1 might affect the expression of transcription factors.

Pax6 is a crucial factor determining the fate of neuronal progenitors in vitro [14] as well as in vivo, in adult brains [50]. However, the capacity of Pax6 to efficiently reprogram nonneuronal cells into neurons, observed in vitro, is limited in situ in the adult brain due to the natural postnatal glycogenic environment [51–53]. Under pathological conditions, such as hypoxia, Pax6 expression is upregulated in the neurogenic niches but not in the cortex, as shown in rats enduring prenatal hypoxia [2]. Thus, the reduced expression of Pax6 in dissociated mature cortical cells in vitro under hypoxic conditions ($38.9 \pm 2.1\%$) observed in our study is in line with the published phenomenon in vivo. On the other hand, post-hypoxic exposure to ghrelin for 24 h significantly elevated the percentage of cells expressing Pax6 ($53.2 \pm 1.4\%$), bringing it to the level at the pre-hypoxic stage ($55.5 \pm 3.0\%$). To our knowledge, this is the first study demonstrating such a modulatory effect of ghrelin on Pax6 expression in mature cortical networks. Of note, Fong et al. [54] have shown that proliferation and maturation of neuroblast cell lines can be stimulated via neurotrophins and neuroprotective factors derived from cocultured C17.2 neural precursor cells. In our study, the marker of proliferating precursor cells Pax6 was found in GFAP and calretinin-immunoreactive cells. The number of Pax6-positive cells significantly exceeded the number of Pax6/GFAP-double-labelled cells, which indirectly indicates that not only astrocytes express the transcription factor but other cell types are positive as well. Indeed, double labelling for Pax6 and calretinin confirmed that neurons in mature cortical cultures express Pax6. These results are in consent with the findings of Nacher et al. [55] that Pax6-expressing cells are also present in the adult hippocampal dentate gyrus and in the subventricular zone/rostral migratory stream, and they are proliferating precursors as well as nonproliferating resting progenitor cells or granule neurons at a very early developmental stage. Strikingly, hypoxia reduced the number of Pax6-positive cells (glial cells and neurons/neuronal precursors), whereas ghrelin treatment restored their levels. These observations raised two questions, which need further attention: does ghrelin act as a trophic factor; is the expression of Pax6 temporally and spatially regulated by ghrelin?

During development, for instance, Pax6 expression is graded in time and space: the highest level of expression in the cortical neuroepithelial cells is at the onset of neurogenesis and lack of it in the basal progenitors of the subventricular zone and in the post-mitotic neurons [56]. The same grading also refers to the Pax6-mediated control of cell cycle's duration in cortical progenitors. Primarily, Pax6 has a repressive effect on the cell cycle progression [5]. Lack of Pax6 at early stages of mice development (embryonic day 12.5) results in shortening of the progenitor cell cycle and a temporary increased production of post-mitotic neurons [57,58], whereas at later stages (embryonic day 15.5), Pax6-deficient progenitors proliferate more slowly [59]. Thus, even in the same cell type, Pax6 can obviously do both: the factor promotes and inhibits the proliferation of cerebral progenitors [17,58,60]. In the adult pancreas, Pax6-deficient beta- and alpha-cells have been shown to lose their maturation characteristics and convert into ghrelin positive cells [61]. These observations suggest that the effects of Pax6 on the cell cycle (e.g., of the cortical progenitors) are context-dependent as supported by the concept that the transcription factor affects proliferation in opposite manners via its DNA-binding subdomains [62].

The Pax6 protein comprises two distinct DNA-binding domains, one of which is regulated by alternate splicing and is able to interact with a number of co-factors [5]. Recently, Brg1/Brm associated factors complex (BAF) and Meis2 have been identified as new co-factors modulating Pax6's direct regulation of target gene expression during embryonic and adult neurogenesis [63–65]. Importantly, Pax6 is also expressed in glial progenitors in the adult brain [66,67]. We speculate that ghrelin functions as a co-factor modulating Pax6 expression, possibly affecting neural progenitor cell fate or cell cycle. However, to explore whether this modulatory effect of ghrelin is mediated directly via the transcription factor Pax6, or whether there is another yet unknown mechanism, additional experiments are needed.

The Ki67 antigen is expressed in the nuclei exclusively in proliferating cells [68,69]. Quiescent or resting cells in the G_0 phase of the cell division cycle do not express it, therefore Ki67 is considered as a marker for the so-called growth fraction of given cells, i.e., progenitor cells [70,71]. During the interphase, cells express Ki67 exclusively in the nucleus, where it is required for the normal distribution and nucleolar association of the heterochromatin [72]. During mitosis, Ki67 covers the surface of chromosomes and represents approximately one-third of their protein mass [73]. Ki67 prevents chromosomal aggregation [74] and asymmetric distribution in daughter cells [75]. The transcription factor increases during the S phase, with further escalation in the G_2 phase, and maximal intensity in the metaphase [76,77]. The half-life of Ki67 is estimated to be approximately 1 h [68]. Intriguingly, cells entering the G_0 phase when treated with growth factors can re-increase their Ki67, to re-enter the S phase [76]. During the early G_1 phase, Ki67 is multi-focally expressed throughout the nucleoplasm, while during the S and the G_2 phases, larger foci overlapping the nucleoli and the heterochromatin regions are formed [77]. After the disintegration of the nuclear envelope, part of the Ki67 protein could be observed dispersed in the cytoplasm [77]. These patterns, including the chromosomes covered by the Ki67 protein, were also observed in our experiments. Before the onset of hypoxia, we found that $9.7 \pm 0.8\%$ of the cultured neuronal progenitor cells expressed Ki67, and their proportion was surprisingly increased up to $59.8 \pm 1.4\%$ after hypoxia. Ghrelin administration downregulated the percentage of Ki67-positive neural progenitors to $15.9 \pm 1.6\%$. Our findings contradict the results from other studies showing that ghrelin administration in animal models of chronic neurodegenerative diseases, as well as endogenous ghrelin in other in vivo and in vitro experiments, triggers adult neurogenesis: ghrelin has been shown to stimulate proliferation of progenitor cells and increase the number of immature neurons in the hippocampus and the subventricular zone [78–80]. However, to the best of our knowledge, the current proof-of-principle study is the first one demonstrating that ghrelin attenuates the proliferation of progenitor cells in the cerebral cortex. Nevertheless, it is important to mention in this respect that Belaev and collaborators [81] have observed continuous proliferation of newly born neural precursors after occlusion of the middle cerebral artery of rats for 2 h. Treatment of the animals with docosahexaenoic acid (DHA) (member of the omega-3 polyunsaturated fatty acids family) 1 hour after the occlusion, followed by immunostaining two weeks post-injury, showed a neuroprotective and neurogenic effect, resulting in 88% increase of the BrdU/Ki67-positive and BrdU/NeuN-positive cortical cells in the penumbra. On the other hand, it seems that the Ki67 protein expression also depends on the degree of hypoxia, as established in experiments with glioma-derived neurospheres. Hypoxia limited to oxygen levels lower than 2% induces high proliferation rates [82], yet if the values are lowered to less than 1% oxygen, a decrease of the Ki67 antigen expression occurs [83].

Additionally, we have to carefully extrapolate the results from rodent models to non-human- and human primates due to the potential interspecies differences in adult neurogenesis. Studies on brains of adult macaque monkeys subjected to global cerebral ischemia have indicated that in both the striatum and neocortex putative newly generated neurons with long-term survival are slightly above 1% [84,85]. In adult human brains, hypoxia triggers some proliferation underneath the ependymal layer, but more than 30% of the Ki67 immunoreactivity is expressed in astrocytes, as double labelling for GFAP and Ki67 has indicated [86]. These GFAP-expressing neural stem cells can differentiate into astrocytes, oligodendrocytes and neurons in the presence [87,88] or absence of exogenous mitogens [89], thus contributing locally to the generation of new neurons as a reaction to ischemic injury. The rapid changes in the Ki67 protein expression, when comparing the normoxic neurons with those of the post-hypoxic experimental condition, can explain the significant fluctuations in the percentage of Ki67-positive cells in a relatively short time frame (24 h). Our findings are consistent with those obtained from human brains studies, where a strong upregulation of Ki67 expression has been observed in the subventricular zone of patients who died just a few days after ischemic stroke, but not in those who died six months after ischemic injury [86]. We interpret the increase in cell proliferation after ischemic injury as a transient event. Of note, Pax6 can be expressed in both neuronal and glial progenitors, and thus, the postischemic reduction of Pax6-positive cells observed in our study might include both cell types.

As Ortega et al. have reported [90], BrdU incorporation did not change after 6 h of hypoxia upon human cortical radial glial cells, but 24 h reduction of oxygen supply limited it, because hypoxia beyond 20 h of duration inflicts irreversible changes within the cells [91]. It appears that the observed effects of hypoxia vary in dependence on the experimental parameters such as location (i.e., cerebral cortex vs. subventricular zone), the severity of hypoxia, and duration of recovery.

Notably, the results from studies on NeuN in stroke models are quite controversial. Some of them report a substantial decrease of NeuN-positive cells in the infarction core and relate it to neuronal death [92], while others have related this decrease to depletion of NeuN's expression or loss of antigenicity [93]. In the present study, there was a significant change in the NeuN expression when comparing the three experimental groups, i.e., normoxia (77.1 ± 2.8%), hypoxia (61.5 ± 3.7%) and hypoxia + ghrelin (85.1 ± 3.2%). Of note, the nuclear NeuN staining was substantially reduced or even undetectable in some neurons. Mild cerebral ischemia has the same effect in in vivo experiments; however, it seems that the NeuN protein levels do not change within 24 h post-reperfusion, merely the NeuN antigenicity is reduced. Of note, the proportion of NeuN-positive neurons in the penumbra have been found to partially restore after antigen retrieval [93]. It is also possible that some special cell types or neurons with a distinct physiological/pathological status can be distinguished through the differences in NeuN/Rbfox3 expression [94]. In our experiments, ghrelin supplementation to recovering hypoxic cultures increased the proportion of neurons with nuclear expression of NeuN to pre-hypoxic values. Only more severe injuries, such as axotomy, have been shown to significantly abolish the NeuN protein levels, which then begin to restore within 7 days post-injury to reach the uninjured levels after 28 days [95]. Injured, but still viable neurons, may lose NeuN-protein expression due to downregulated protein synthesis and/or protein overconsumption [96]. Lower production of NeuN/Rbfox3 and the translocation of NeuN/Rbfox3 from the nucleus to the cytoplasm might lead to downregulation of the alternative splicing of the RNA of its target genes, and thus, change the complement of neuronal specific gene expression [97]. Differential splicing is an important mechanism for the evolutionary dynamics of the central nervous system's complexity [98] and its regular development. NeuN/Rbfox3-regulated splicing is crucial for the final neuronal differentiation during development [99], but mutations in NeuN/Rbfox3 are also associated with neuropsychiatric disorders [100]. Thus, the up-regulating effect of ghrelin on the nuclear expression of NeuN observed in this study suggests that ghrelin can influence neuronal differentiation operating on its master switch [101].

5. Conclusions

In conclusion, our in vitro findings suggest that ghrelin is involved in postischemic neuroplasticity in non-neurogenic brain areas such as the cerebral cortex. Expression of the ghrelin receptor GHSR1 is increased in hypoxic neurons and normalised whenever ghrelin is supplemented to these neurons. Moreover, ghrelin regulates Pax6 and Ki67 expression, and stimulates cytoplasmic/nuclear shuttling of NeuN in hypoxic neurons. Since these factors are key players in proliferation, neurogenesis and differentiation, their stimulation could ameliorate recovery after hypoxia. This work emphasises the relevance of additional studies on the regulatory role of ghrelin in the cell cycle of cortical progenitors and neurons exposed to hypoxia, and corroborates the therapeutic potential of ghrelin for the treatment of stroke patients.

Supplementary Materials: The following supporting information can be downloaded at: https://www.mdpi.com/article/10.3390/cells11050782/s1, Figure S1: Control immunostaining experiments for the GHSR1 antibody referring to Figure 2 of the main text; Figure S2: Uncropped dot blot images referring to Figure 5 of the main text. Areas shown in the main figures and used for quantification are indicated.

Author Contributions: I.I.S.: conceptualisation, methodology, formal analysis, investigation, resources, data curation, writing—original draft preparation, visualisation, funding acquisition, approved of the final version. A.K.: investigation, resources, data acquisition and curation, visualisation, approved of the final version. J.W.: resources, approved of the final version. T.R.D.: resources, methodology, approved of the final version. A.B.T.: resources, writing—review & editing, approved of the final version. D.L.: conceptualisation, methodology, formal analysis, investigation, resources, data curation, writing—reviewing and editing, visualisation, funding acquisition, approved of the final version. All authors have read and agreed to the published version of the manuscript.

Funding: This research was funded by the Grant DN 13/10, 2017 of the Bulgarian Ministry of Education and Science to IIS, by the European Commission Horizon 2020 Framework Programme (Project 856871—TRANSTEM) and the National Program "European Research Networks" of the Bulgarian Ministry of Education to A.B.T., and by the FoRUM grant F957N-2019 of the Ruhr University Bochum, Germany to DL.

Institutional Review Board Statement: The study was conducted according to the guidelines of the Declaration of Helsinki, and approved by the Lower Saxony State Office for Consumer Protection and Food Safety of Oldenburg (Niedersächsisches Landesamt für Verbraucherschutz und Lebensmittelsicherheit (LAVES)/Oldenburg, contract No 33.9-42502-04-11/0622).

Informed Consent Statement: Not applicable.

Data Availability Statement: The datasets in support of the findings of this study are available from the corresponding authors, upon reasonable request.

Acknowledgments: The authors would like to thank Andrew Vermeulen for proofreading the manuscript.

Conflicts of Interest: The authors declare that the research was conducted in the absence of any commercial or financial relationships that could be construed as a potential conflict of interest.

References

1. Lipton, P. Ischemic cell death in brain neurons. *Physiol. Rev.* **1999**, *79*, 1431–1568. [CrossRef] [PubMed]
2. So, K.; Chung, Y.; Yu, S.K.; Jun, Y. Regional Immunoreactivity of Pax6 in the Neurogenic Zone After Chronic Prenatal Hypoxia. *Vivo* **2017**, *31*, 1125–1129. [CrossRef]
3. Daval, J.L.; Vert, P. Apoptosis and neurogenesis after transient hypoxia in the developing rat brain. *Semin. Perinatol.* **2004**, *28*, 257–263. [CrossRef] [PubMed]
4. Arvidsson, A.; Collin, T.; Kirik, D.; Kokaia, Z.; Lindvall, O. Neuronal replacement from endogenous precursors in the adult brain after stroke. *Nat. Med.* **2002**, *8*, 963–970. [CrossRef] [PubMed]
5. Manuel, M.N.; Mi, D.; Mason, J.O.; Price, D.J. Regulation of cerebral cortical neurogenesis by the Pax6 transcription factor. *Front. Cell. Neurosci.* **2015**, *9*, 70. [CrossRef]
6. Seri, B.; Herrera, D.G.; Gritti, A.; Ferron, S.; Collado, L.; Vescovi, A.; Garcia-Verdugo, J.M.; Alvarez-Buylla, A. Composition and organization of the SCZ: A large germinal layer containing neural stem cells in the adult mammalian brain. *Cereb. Cortex* **2006**, *16* (Suppl. 1), i103–i111. [CrossRef]
7. Van Kampen, J.M.; Hagg, T.; Robertson, H.A. Induction of neurogenesis in the adult rat subventricular zone and neostriatum following dopamine D3 receptor stimulation. *Eur. J. Neurosci.* **2004**, *19*, 2377–2387. [CrossRef]
8. Bernier, P.J.; Bedard, A.; Vinet, J.; Levesque, M.; Parent, A. Newly generated neurons in the amygdala and adjoining cortex of adult primates. *Proc. Natl. Acad. Sci. USA* **2002**, *99*, 11464–11469. [CrossRef]
9. Gould, E.; Reeves, A.J.; Graziano, M.S.; Gross, C.G. Neurogenesis in the neocortex of adult primates. *Science* **1999**, *286*, 548–552. [CrossRef]
10. Carlén, M.; Cassidy, R.M.; Brismar, H.; Smith, G.A.; Enquist, L.W.; Frisén, J. Functional integration of adult-born neurons. *Curr. Biol. CB* **2002**, *12*, 606–608. [CrossRef]
11. Bi, B.; Salmaso, N.; Komitova, M.; Simonini, M.V.; Silbereis, J.; Cheng, E.; Kim, J.; Luft, S.; Ment, L.R.; Horvath, T.L.; et al. Cortical glial fibrillary acidic protein-positive cells generate neurons after perinatal hypoxic injury. *J. Neurosci. Off. J. Soc. Neurosci.* **2011**, *31*, 9205–9221. [CrossRef] [PubMed]
12. Gotz, M.; Sirko, S.; Beckers, J.; Irmler, M. Reactive astrocytes as neural stem or progenitor cells: In vivo lineage, In vitro potential, and Genome-wide expression analysis. *Glia* **2015**, *63*, 1452–1468. [CrossRef] [PubMed]
13. Walther, C.; Gruss, P. Pax-6, a murine paired box gene, is expressed in the developing CNS. *Development* **1991**, *113*, 1435–1449. [CrossRef] [PubMed]
14. Hack, M.A.; Sugimori, M.; Lundberg, C.; Nakafuku, M.; Götz, M. Regionalization and fate specification in neurospheres: The role of Olig2 and Pax6. *Mol. Cell. Neurosci.* **2004**, *25*, 664–678. [CrossRef]
15. Simpson, T.I.; Price, D.J. Pax6; a pleiotropic player in development. *BioEssays News Rev. Mol. Cell. Dev. Biol.* **2002**, *24*, 1041–1051. [CrossRef]
16. Ton, C.C.; Miwa, H.; Saunders, G.F. Small eye (Sey): Cloning and characterization of the murine homolog of the human aniridia gene. *Genomics* **1992**, *13*, 251–256. [CrossRef]
17. Sansom, S.N.; Griffiths, D.S.; Faedo, A.; Kleinjan, D.J.; Ruan, Y.; Smith, J.; van Heyningen, V.; Rubenstein, J.L.; Livesey, F.J. The level of the transcription factor Pax6 is essential for controlling the balance between neural stem cell self-renewal and neurogenesis. *PLoS Genet.* **2009**, *5*, e1000511. [CrossRef]
18. Miao, Y.; Xia, Q.; Hou, Z.; Zheng, Y.; Pan, H.; Zhu, S. Ghrelin protects cortical neuron against focal ischemia/reperfusion in rats. *Biochem. Biophys. Res. Commun.* **2007**, *359*, 795–800. [CrossRef]
19. Liu, J.; Chen, M.; Dong, R.; Sun, C.; Li, S.; Zhu, S. Ghrelin Promotes Cortical Neurites Growth in Late Stage After Oxygen-Glucose Deprivation/Reperfusion Injury. *J. Mol. Neurosci. MN* **2019**, *68*, 29–37. [CrossRef]
20. Kilkenny, C.; Browne, W.J.; Cuthill, I.C.; Emerson, M.; Altman, D.G. Improving bioscience research reporting: The ARRIVE guidelines for reporting animal research. *PLoS Biol.* **2010**, *8*, e1000412. [CrossRef]
21. Doeppner, T.R.; Herz, J.; Bähr, M.; Tonchev, A.B.; Stoykova, A. Zbtb20 Regulates Developmental Neurogenesis in the Olfactory Bulb and Gliogenesis After Adult Brain Injury. *Mol. Neurobiol.* **2019**, *56*, 567–582. [CrossRef] [PubMed]
22. Romijn, H.J.; van Huizen, F.; Wolters, P.S. Towards an improved serum-free, chemically defined medium for long-term culturing of cerebral cortex tissue. *Neurosci. Biobehav. Rev.* **1984**, *8*, 301–334. [CrossRef]
23. Matthiesen, S.; Jahnke, R.; Knittler, M.R. A Straightforward Hypoxic Cell Culture Method Suitable for Standard Incubators. *Methods Protoc.* **2021**, *4*, 25. [CrossRef] [PubMed]
24. Cowley, M.A.; Smith, R.G.; Diano, S.; Tschöp, M.; Pronchuk, N.; Grove, K.L.; Strasburger, C.J.; Bidlingmaier, M.; Esterman, M.; Heiman, M.L.; et al. The distribution and mechanism of action of ghrelin in the CNS demonstrates a novel hypothalamic circuit regulating energy homeostasis. *Neuron* **2003**, *37*, 649–661. [CrossRef]
25. Johansson, I.; Destefanis, S.; Aberg, N.D.; Aberg, M.A.; Blomgren, K.; Zhu, C.; Ghe, C.; Granata, R.; Ghigo, E.; Muccioli, G.; et al. Proliferative and protective effects of growth hormone secretagogues on adult rat hippocampal progenitor cells. *Endocrinology* **2008**, *149*, 2191–2199. [CrossRef]
26. Diano, S.; Farr, S.A.; Benoit, S.C.; McNay, E.C.; da Silva, I.; Horvath, B.; Gaskin, F.S.; Nonaka, N.; Jaeger, L.B.; Banks, W.A.; et al. Ghrelin controls hippocampal spine synapse density and memory performance. *Nat. Neurosci.* **2006**, *9*, 381–388. [CrossRef]
27. Stoyanova, I.I.; le Feber, J. Ghrelin accelerates synapse formation and activity development in cultured cortical networks. *BMC Neurosci.* **2014**, *15*, 49. [CrossRef]

28. Kosugi, S.; Hasebe, M.; Tomita, M.; Yanagawa, H. Systematic identification of cell cycle-dependent yeast nucleocytoplasmic shuttling proteins by prediction of composite motifs. *Proc. Natl. Acad. Sci. USA* **2009**, *106*, 10171–10176. [CrossRef]

29. Moon, M.; Kim, S.; Hwang, L.; Park, S. Ghrelin regulates hippocampal neurogenesis in adult mice. *Endocr. J.* **2009**, *56*, 525–531. [CrossRef]

30. Inoue, Y.; Nakahara, K.; Kangawa, K.; Murakami, N. Transitional change in rat fetal cell proliferation in response to ghrelin and des-acyl ghrelin during the last stage of pregnancy. *Biochem. Biophys. Res. Commun.* **2010**, *393*, 455–460. [CrossRef]

31. Kojima, M.; Hosoda, H.; Date, Y.; Nakazato, M.; Matsuo, H.; Kangawa, K. Ghrelin is a growth-hormone-releasing acylated peptide from stomach. *Nature* **1999**, *402*, 656–660. [CrossRef] [PubMed]

32. Bednarek, M.A.; Feighner, S.D.; Pong, S.S.; McKee, K.K.; Hreniuk, D.L.; Silva, M.V.; Warren, V.A.; Howard, A.D.; Van Der Ploeg, L.H.; Heck, J.V. Structure-function studies on the new growth hormone-releasing peptide, ghrelin: Minimal sequence of ghrelin necessary for activation of growth hormone secretagogue receptor 1a. *J. Med. Chem.* **2000**, *43*, 4370–4376. [CrossRef] [PubMed]

33. Guan, X.M.; Yu, H.; Palyha, O.C.; McKee, K.K.; Feighner, S.D.; Sirinathsinghji, D.J.; Smith, R.G.; Van der Ploeg, L.H.; Howard, A.D. Distribution of mRNA encoding the growth hormone secretagogue receptor in brain and peripheral tissues. *Brain Res. Mol. Brain Res.* **1997**, *48*, 23–29. [CrossRef]

34. Muccioli, G.; Ghè, C.; Ghigo, M.C.; Papotti, M.; Arvat, E.; Boghen, M.F.; Nilsson, M.H.; Deghenghi, R.; Ong, H.; Ghigo, E. Specific receptors for synthetic GH secretagogues in the human brain and pituitary gland. *J. Endocrinol.* **1998**, *157*, 99–106. [CrossRef]

35. Mitchell, V.; Bouret, S.; Beauvillain, J.C.; Schilling, A.; Perret, M.; Kordon, C.; Epelbaum, J. Comparative distribution of mRNA encoding the growth hormone secretagogue-receptor (GHS-R) in Microcebus murinus (Primate, lemurian) and rat forebrain and pituitary. *J. Comp. Neurol.* **2001**, *429*, 469–489. [CrossRef]

36. Zigman, J.M.; Jones, J.E.; Lee, C.E.; Saper, C.B.; Elmquist, J.K. Expression of ghrelin receptor mRNA in the rat and the mouse brain. *J. Comp. Neurol.* **2006**, *494*, 528–548. [CrossRef]

37. Andrews, Z.B. The extra-hypothalamic actions of ghrelin on neuronal function. *Trends Neurosci.* **2011**, *34*, 31–40. [CrossRef]

38. Bron, R.; Yin, L.; Russo, D.; Furness, J.B. Expression of the ghrelin receptor gene in neurons of the medulla oblongata of the rat. *J. Comp. Neurol.* **2013**, *521*, 2680–2702. [CrossRef]

39. Cabral, A.; Fernandez, G.; Tolosa, M.J.; Rey Moggia, Á.; Calfa, G.; De Francesco, P.N.; Perello, M. Fasting induces remodeling of the orexigenic projections from the arcuate nucleus to the hypothalamic paraventricular nucleus, in a growth hormone secretagogue receptor-dependent manner. *Mol. Metab.* **2020**, *32*, 69–84. [CrossRef]

40. Stoyanova, I.; Lutz, D. Ghrelin-Mediated Regeneration and Plasticity After Nervous System Injury. *Front. Cell Dev. Biol.* **2021**, *9*, 595914. [CrossRef]

41. Huang, J.; Liu, W.; Doycheva, D.M.; Gamdzyk, M.; Lu, W.; Tang, J.; Zhang, J.H. Ghrelin attenuates oxidative stress and neuronal apoptosis via GHSR-1α/AMPK/Sirt1/PGC-1α/UCP2 pathway in a rat model of neonatal HIE. *Free Radic. Biol. Med.* **2019**, *141*, 322–337. [CrossRef] [PubMed]

42. Stoyanova, I.I.; le Feber, J.; Rutten, W.L. Ghrelin stimulates synaptic formation in cultured cortical networks in a dose-dependent manner. *Regul. Pept.* **2013**, *186*, 43–48. [CrossRef] [PubMed]

43. Cecarini, V.; Bonfili, L.; Cuccioloni, M.; Keller, J.N.; Bruce-Keller, A.J.; Eleuteri, A.M. Effects of Ghrelin on the Proteolytic Pathways of Alzheimer's Disease Neuronal Cells. *Mol. Neurobiol.* **2016**, *53*, 3168–3178. [CrossRef]

44. Fry, M.; Ferguson, A.V. Ghrelin modulates electrical activity of area postrema neurons. *Am. J. Physiol. Regul. Integr. Comp. Physiol.* **2009**, *296*, R485–R492. [CrossRef] [PubMed]

45. Yanagida, H.; Morita, T.; Kim, J.; Yoshida, K.; Nakajima, K.; Oomura, Y.; Wayner, M.J.; Sasaki, K. Effects of ghrelin on neuronal activity in the ventromedial nucleus of the hypothalamus in infantile rats: An in vitro study. *Peptides* **2008**, *29*, 912–918. [CrossRef] [PubMed]

46. Stoyanova, I.I.; Hofmeijer, J.; van Putten, M.; le Feber, J. Acyl Ghrelin Improves Synapse Recovery in an In Vitro Model of Postanoxic Encephalopathy. *Mol. Neurobiol.* **2016**, *53*, 6136–6143. [CrossRef]

47. Suda, Y.; Kuzumaki, N.; Sone, T.; Narita, M.; Tanaka, K.; Hamada, Y.; Iwasawa, C.; Shibasaki, M.; Maekawa, A.; Matsuo, M.; et al. Down-regulation of ghrelin receptors on dopaminergic neurons in the substantia nigra contributes to Parkinson's disease-like motor dysfunction. *Mol. Brain* **2018**, *11*, 6. [CrossRef]

48. Albarran-Zeckler, R.G.; Brantley, A.F.; Smith, R.G. Growth hormone secretagogue receptor (GHS-R1a) knockout mice exhibit improved spatial memory and deficits in contextual memory. *Behav. Brain Res.* **2012**, *232*, 13–19. [CrossRef]

49. Erdmann, J.; Töpsch, R.; Lippl, F.; Gussmann, P.; Schusdziarra, V. Postprandial response of plasma ghrelin levels to various test meals in relation to food intake, plasma insulin, and glucose. *J. Clin. Endocrinol. Metab.* **2004**, *89*, 3048–3054. [CrossRef]

50. Kronenberg, G.; Gertz, K.; Cheung, G.; Buffo, A.; Kettenmann, H.; Götz, M.; Endres, M. Modulation of fate determinants Olig2 and Pax6 in resident glia evokes spiking neuroblasts in a model of mild brain ischemia. *Stroke* **2010**, *41*, 2944–2949. [CrossRef]

51. Grande, A.; Sumiyoshi, K.; López-Juárez, A.; Howard, J.; Sakthivel, B.; Aronow, B.; Campbell, K.; Nakafuku, M. Environmental impact on direct neuronal reprogramming in vivo in the adult brain. *Nat. Commun.* **2013**, *4*, 2373. [CrossRef] [PubMed]

52. Torper, O.; Pfisterer, U.; Wolf, D.A.; Pereira, M.; Lau, S.; Jakobsson, J.; Björklund, A.; Grealish, S.; Parmar, M. Generation of induced neurons via direct conversion in vivo. *Proc. Natl. Acad. Sci. USA* **2013**, *110*, 7038–7043. [CrossRef] [PubMed]

53. Shihabuddin, L.S.; Horner, P.J.; Ray, J.; Gage, F.H. Adult spinal cord stem cells generate neurons after transplantation in the adult dentate gyrus. *J. Neurosci.* **2000**, *20*, 8727–8735. [CrossRef] [PubMed]

54. Fong, S.P.; Tsang, K.S.; Chan, A.B.; Lu, G.; Poon, W.S.; Li, K.; Baum, L.W.; Ng, H.K. Trophism of neural progenitor cells to embryonic stem cells: Neural induction and transplantation in a mouse ischemic stroke model. *J. Neurosci. Res.* **2007**, *85*, 1851–1862. [CrossRef]

55. Nacher, J.; Varea, E.; Blasco-Ibañez, J.M.; Castillo-Gomez, E.; Crespo, C.; Martinez-Guijarro, F.J.; McEwen, B.S. Expression of the transcription factor Pax 6 in the adult rat dentate gyrus. *J. Neurosci. Res.* **2005**, *81*, 753–761. [CrossRef]

56. Englund, C.; Fink, A.; Lau, C.; Pham, D.; Daza, R.A.; Bulfone, A.; Kowalczyk, T.; Hevner, R.F. Pax6, Tbr2, and Tbr1 are expressed sequentially by radial glia, intermediate progenitor cells, and postmitotic neurons in developing neocortex. *J. Neurosci. Off. J. Soc. Neurosci.* **2005**, *25*, 247–251. [CrossRef]

57. Warren, N.; Caric, D.; Pratt, T.; Clausen, J.A.; Asavaritikrai, P.; Mason, J.O.; Hill, R.E.; Price, D.J. The transcription factor, Pax6, is required for cell proliferation and differentiation in the developing cerebral cortex. *Cereb. Cortex* **1999**, *9*, 627–635. [CrossRef]

58. Walcher, T.; Xie, Q.; Sun, J.; Irmler, M.; Beckers, J.; Öztürk, T.; Niessing, D.; Stoykova, A.; Cvekl, A.; Ninkovic, J.; et al. Functional dissection of the paired domain of Pax6 reveals molecular mechanisms of coordinating neurogenesis and proliferation. *Development* **2013**, *140*, 1123–1136. [CrossRef]

59. Estivill-Torrus, G.; Pearson, H.; van Heyningen, V.; Price, D.J.; Rashbass, P. Pax6 is required to regulate the cell cycle and the rate of progression from symmetrical to asymmetrical division in mammalian cortical progenitors. *Development* **2002**, *129*, 455–466. [CrossRef]

60. Asami, M.; Pilz, G.A.; Ninkovic, J.; Godinho, L.; Schroeder, T.; Huttner, W.B.; Götz, M. The role of Pax6 in regulating the orientation and mode of cell division of progenitors in the mouse cerebral cortex. *Development* **2011**, *138*, 5067–5078. [CrossRef]

61. Ahmad, Z.; Rafeeq, M.; Collombat, P.; Mansouri, A. Pax6 Inactivation in the Adult Pancreas Reveals Ghrelin as Endocrine Cell Maturation Marker. *PLoS ONE* **2015**, *10*, e0144597. [CrossRef] [PubMed]

62. Haubst, N.; Berger, J.; Radjendirane, V.; Graw, J.; Favor, J.; Saunders, G.F.; Stoykova, A.; Götz, M. Molecular dissection of Pax6 function: The specific roles of the paired domain and homeodomain in brain development. *Development* **2004**, *131*, 6131–6140. [CrossRef]

63. Ninkovic, J.; Steiner-Mezzadri, A.; Jawerka, M.; Akinci, U.; Masserdotti, G.; Petricca, S.; Fischer, J.; von Holst, A.; Beckers, J.; Lie, C.D.; et al. The BAF complex interacts with Pax6 in adult neural progenitors to establish a neurogenic cross-regulatory transcriptional network. *Cell Stem Cell* **2013**, *13*, 403–418. [CrossRef]

64. Agoston, Z.; Heine, P.; Brill, M.S.; Grebbin, B.M.; Hau, A.C.; Kallenborn-Gerhardt, W.; Schramm, J.; Götz, M.; Schulte, D. Meis2 is a Pax6 co-factor in neurogenesis and dopaminergic periglomerular fate specification in the adult olfactory bulb. *Development* **2014**, *141*, 28–38. [CrossRef]

65. Tuoc, T.C.; Boretius, S.; Sansom, S.N.; Pitulescu, M.E.; Frahm, J.; Livesey, F.J.; Stoykova, A. Chromatin regulation by BAF170 controls cerebral cortical size and thickness. *Dev. Cell* **2013**, *25*, 256–269. [CrossRef] [PubMed]

66. Mussa, Z.; Tome-Garcia, J.; Jiang, Y.; Akbarian, S.; Tsankova, N.M. Isolation of Adult Human Astrocyte Populations from Fresh-frozen Cortex using Fluorescence-Activated Nuclei Sorting. *J. Vis. Exp. JoVE* **2021**. [CrossRef]

67. Falcone, C.; Penna, E.; Hong, T.; Tarantal, A.F.; Hof, P.R.; Hopkins, W.D.; Sherwood, C.C.; Noctor, S.C.; Martínez-Cerdeño, V. Cortical Interlaminar Astrocytes Are Generated Prenatally, Mature Postnatally, and Express Unique Markers in Human and Nonhuman Primates. *Cereb. Cortex* **2021**, *31*, 379–395. [CrossRef] [PubMed]

68. Bruno, S.; Darzynkiewicz, Z. Cell cycle dependent expression and stability of the nuclear protein detected by Ki-67 antibody in HL-60 cells. *Cell Prolif.* **1992**, *25*, 31–40. [CrossRef]

69. Scholzen, T.; Gerdes, J. The Ki-67 protein: From the known and the unknown. *J. Cell. Physiol.* **2000**, *182*, 311–322. [CrossRef]

70. Gerdes, J.; Lemke, H.; Baisch, H.; Wacker, H.H.; Schwab, U.; Stein, H. Cell cycle analysis of a cell proliferation-associated human nuclear antigen defined by the monoclonal antibody Ki-67. *J. Immunol.* **1984**, *133*, 1710–1715.

71. Gerdes, J.; Schwab, U.; Lemke, H.; Stein, H. Production of a mouse monoclonal antibody reactive with a human nuclear antigen associated with cell proliferation. *Int. J. Cancer* **1983**, *31*, 13–20. [CrossRef] [PubMed]

72. Sun, X.; Kaufman, P.D. Ki-67: More than a proliferation marker. *Chromosoma* **2018**, *127*, 175–186. [CrossRef] [PubMed]

73. Booth, D.G.; Beckett, A.J.; Molina, O.; Samejima, I.; Masumoto, H.; Kouprina, N.; Larionov, V.; Prior, I.A.; Earnshaw, W.C. 3D-CLEM Reveals that a Major Portion of Mitotic Chromosomes Is Not Chromatin. *Mol. Cell* **2016**, *64*, 790–802. [CrossRef] [PubMed]

74. Cuylen, S.; Blaukopf, C.; Politi, A.Z.; Müller-Reichert, T.; Neumann, B.; Poser, I.; Ellenberg, J.; Hyman, A.A.; Gerlich, D.W. Ki-67 acts as a biological surfactant to disperse mitotic chromosomes. *Nature* **2016**, *535*, 308–312. [CrossRef] [PubMed]

75. Booth, D.G.; Takagi, M.; Sanchez-Pulido, L.; Petfalski, E.; Vargiu, G.; Samejima, K.; Imamoto, N.; Ponting, C.P.; Tollervey, D.; Earnshaw, W.C.; et al. Ki-67 is a PP1-interacting protein that organises the mitotic chromosome periphery. *eLife* **2014**, *3*, e01641. [CrossRef]

76. du Manoir, S.; Guillaud, P.; Camus, E.; Seigneurin, D.; Brugal, G. Ki-67 labeling in postmitotic cells defines different Ki-67 pathways within the 2c compartment. *Cytometry* **1991**, *12*, 455–463. [CrossRef] [PubMed]

77. Starborg, M.; Gell, K.; Brundell, E.; Höög, C. The murine Ki-67 cell proliferation antigen accumulates in the nucleolar and heterochromatic regions of interphase cells and at the periphery of the mitotic chromosomes in a process essential for cell cycle progression. *J. Cell Sci.* **1996**, *109 Pt 1*, 143–153. [CrossRef]

78. Moon, M.; Cha, M.Y.; Mook-Jung, I. Impaired hippocampal neurogenesis and its enhancement with ghrelin in 5XFAD mice. *J. Alzheimers Dis. JAD* **2014**, *41*, 233–241. [CrossRef]

79. Li, E.; Kim, Y.; Kim, S.; Sato, T.; Kojima, M.; Park, S. Ghrelin stimulates proliferation, migration and differentiation of neural progenitors from the subventricular zone in the adult mice. *Exp. Neurol.* **2014**, *252*, 75–84. [CrossRef]

80. Chung, H.; Li, E.; Kim, Y.; Kim, S.; Park, S. Multiple signaling pathways mediate ghrelin-induced proliferation of hippocampal neural stem cells. *J. Endocrinol.* **2013**, *218*, 49–59. [CrossRef]

81. Belayev, L.; Hong, S.H.; Menghani, H.; Marcell, S.J.; Obenaus, A.; Freitas, R.S.; Khoutorova, L.; Balaszczuk, V.; Jun, B.; Oriá, R.B.; et al. Docosanoids Promote Neurogenesis and Angiogenesis, Blood-Brain Barrier Integrity, Penumbra Protection, and Neurobehavioral Recovery After Experimental Ischemic Stroke. *Mol. Neurobiol.* **2018**, *55*, 7090–7106. [CrossRef] [PubMed]

82. Heddleston, J.M.; Li, Z.; McLendon, R.E.; Hjelmeland, A.B.; Rich, J.N. The hypoxic microenvironment maintains glioblastoma stem cells and promotes reprogramming towards a cancer stem cell phenotype. *Cell Cycle* **2009**, *8*, 3274–3284. [CrossRef] [PubMed]

83. Kolenda, J.; Jensen, S.S.; Aaberg-Jessen, C.; Christensen, K.; Andersen, C.; Brünner, N.; Kristensen, B.W. Effects of hypoxia on expression of a panel of stem cell and chemoresistance markers in glioblastoma-derived spheroids. *J. Neuro-Oncol.* **2011**, *103*, 43–58. [CrossRef] [PubMed]

84. Tonchev, A.B.; Yamashima, T.; Sawamoto, K.; Okano, H. Enhanced proliferation of progenitor cells in the subventricular zone and limited neuronal production in the striatum and neocortex of adult macaque monkeys after global cerebral ischemia. *J. Neurosci. Res.* **2005**, *81*, 776–788. [CrossRef]

85. Tonchev, A.B.; Yamashima, T.; Zhao, L.; Okano, H.J.; Okano, H. Proliferation of neural and neuronal progenitors after global brain ischemia in young adult macaque monkeys. *Mol. Cell. Neurosci.* **2003**, *23*, 292–301. [CrossRef]

86. Macas, J.; Nern, C.; Plate, K.H.; Momma, S. Increased generation of neuronal progenitors after ischemic injury in the aged adult human forebrain. *J. Neurosci. Off. J. Soc. Neurosci.* **2006**, *26*, 13114–13119. [CrossRef]

87. Chiasson, B.J.; Tropepe, V.; Morshead, C.M.; van der Kooy, D. Adult mammalian forebrain ependymal and subependymal cells demonstrate proliferative potential, but only subependymal cells have neural stem cell characteristics. *J. Neurosci. Off. J. Soc. Neurosci.* **1999**, *19*, 4462–4471. [CrossRef]

88. Nunes, M.C.; Roy, N.S.; Keyoung, H.M.; Goodman, R.R.; McKhann, G., II; Jiang, L.; Kang, J.; Nedergaard, M.; Goldman, S.A. Identification and isolation of multipotential neural progenitor cells from the subcortical white matter of the adult human brain. *Nat. Med.* **2003**, *9*, 439–447. [CrossRef]

89. Sanai, N.; Tramontin, A.D.; Quiñones-Hinojosa, A.; Barbaro, N.M.; Gupta, N.; Kunwar, S.; Lawton, M.T.; McDermott, M.W.; Parsa, A.T.; Manuel-García Verdugo, J.; et al. Unique astrocyte ribbon in adult human brain contains neural stem cells but lacks chain migration. *Nature* **2004**, *427*, 740–744. [CrossRef]

90. Ortega, J.A.; Sirois, C.L.; Memi, F.; Glidden, N.; Zecevic, N. Oxygen Levels Regulate the Development of Human Cortical Radial Glia Cells. *Cereb. Cortex* **2017**, *27*, 3736–3751. [CrossRef]

91. le Feber, J.; Erkamp, N.; van Putten, M. Loss and recovery of functional connectivity in cultured cortical networks exposed to hypoxia. *J. Neurophysiol.* **2017**, *118*, 394–403. [CrossRef] [PubMed]

92. Sugawara, T.; Lewén, A.; Noshita, N.; Gasche, Y.; Chan, P.H. Effects of global ischemia duration on neuronal, astroglial, oligodendroglial, and microglial reactions in the vulnerable hippocampal CA1 subregion in rats. *J. Neurotrauma* **2002**, *19*, 85–98. [CrossRef] [PubMed]

93. Unal-Cevik, I.; Kilinç, M.; Gürsoy-Ozdemir, Y.; Gurer, G.; Dalkara, T. Loss of NeuN immunoreactivity after cerebral ischemia does not indicate neuronal cell loss: A cautionary note. *Brain Res.* **2004**, *1015*, 169–174. [CrossRef] [PubMed]

94. Mullen, R.J.; Buck, C.R.; Smith, A.M. NeuN, a neuronal specific nuclear protein in vertebrates. *Development* **1992**, *116*, 201–211. [CrossRef]

95. McPhail, L.T.; McBride, C.B.; McGraw, J.; Steeves, J.D.; Tetzlaff, W. Axotomy abolishes NeuN expression in facial but not rubrospinal neurons. *Exp. Neurol.* **2004**, *185*, 182–190. [CrossRef]

96. Hossmann, K.A. Disturbances of cerebral protein synthesis and ischemic cell death. *Prog. Brain Res.* **1993**, *96*, 161–177. [CrossRef]

97. Duan, W.; Zhang, Y.P.; Hou, Z.; Huang, C.; Zhu, H.; Zhang, C.Q.; Yin, Q. Novel Insights into NeuN: From Neuronal Marker to Splicing Regulator. *Mol. Neurobiol.* **2016**, *53*, 1637–1647. [CrossRef]

98. Barbosa-Morais, N.L.; Irimia, M.; Pan, Q.; Xiong, H.Y.; Gueroussov, S.; Lee, L.J.; Slobodeniuc, V.; Kutter, C.; Watt, S.; Colak, R.; et al. The evolutionary landscape of alternative splicing in vertebrate species. *Science* **2012**, *338*, 1587–1593. [CrossRef]

99. Kim, K.K.; Nam, J.; Mukouyama, Y.S.; Kawamoto, S. Rbfox3-regulated alternative splicing of Numb promotes neuronal differentiation during development. *J. Cell Biol.* **2013**, *200*, 443–458. [CrossRef]

100. Lal, D.; Reinthaler, E.M.; Altmüller, J.; Toliat, M.R.; Thiele, H.; Nürnberg, P.; Lerche, H.; Hahn, A.; Møller, R.S.; Muhle, H.; et al. RBFOX1 and RBFOX3 mutations in rolandic epilepsy. *PLoS ONE* **2013**, *8*, e73323. [CrossRef]

101. Maxeiner, S.; Glassmann, A.; Kao, H.T.; Schilling, K. The molecular basis of the specificity and cross-reactivity of the NeuN epitope of the neuron-specific splicing regulator, Rbfox3. *Histochem. Cell Biol.* **2014**, *141*, 43–55. [CrossRef] [PubMed]

Review

TGF-β/Smad Signalling in Neurogenesis: Implications for Neuropsychiatric Diseases

Lih-Fhung Hiew, Chi-Him Poon (ORCID)**, Heng-Ze You and Lee-Wei Lim *** (ORCID)

Neuromodulation Laboratory, School of Biomedical Sciences, Li Ka Shing Faculty of Medicine, The University of Hong Kong, Hong Kong, China; u3005074@connect.hku.hk (L.-F.H.); chpoonac@connect.hku.hk (C.-H.P.); 17097461d@connect.polyu.hk (H.-Z.Y.)
* Correspondence: drlimleewei@gmail.com

Abstract: TGF-β/Smad signalling has been the subject of extensive research due to its role in the cell cycle and carcinogenesis. Modifications to the TGF-β/Smad signalling pathway have been found to produce disparate effects on neurogenesis. We review the current research on canonical and non-canonical TGF-β/Smad signalling pathways and their functions in neurogenesis. We also examine the observed role of neurogenesis in neuropsychiatric disorders and the relationship between TGF-β/Smad signalling and neurogenesis in response to stressors. Overlapping mechanisms of cell proliferation, neurogenesis, and the development of mood disorders in response to stressors suggest that TGF-β/Smad signalling is an important regulator of stress response and is implicated in the behavioural outcomes of mood disorders.

Keywords: neurogenesis; TGF-β; smad; depression; stress

Citation: Hiew, L.-F.; Poon, C.-H.; You, H.-Z.; Lim, L.-W. TGF-β/Smad Signalling in Neurogenesis: Implications for Neuropsychiatric Diseases. *Cells* **2021**, *10*, 1382. https://doi.org/10.3390/cells 10061382

Academic Editor: FengRu Tang

Received: 21 April 2021
Accepted: 1 June 2021
Published: 3 June 2021

Publisher's Note: MDPI stays neutral with regard to jurisdictional claims in published maps and institutional affiliations.

1. Introduction

The transforming growth factor-β (TGF-β) pathway consists of many genes involved in cell growth, differentiation, migration, and apoptosis [1,2]. TGF-β ligands bind to the TGF-β receptor kinase to form a complex that phosphorylates receptor-activated Smads (R-Smads), allowing recruitment of Smad4 that translocates to the nucleus to regulate transcriptional activity [3]. Smad-mediated TGF-β signalling is referred to as the canonical pathway. Notably, Smad3 is the primary molecule involved in canonical TGF-β signalling [4]. TGF-β also activates a myriad of other pathways, including the Erk, JNK, and MAPK pathways [5], which together are known as the non-canonical pathway. TGF-β is involved in a vast number of interactions and can have many roles depending on the cellular context. TGF-β has been found in neural progenitor cells, differentiating neurones, and mature neural cells. TGF-β exhibits both anti-tumour properties and tumourigenic properties [6] depending on the manner of its activation. Given that TGF-β is known to have proliferative effects in somatic cells [7], it was proposed that they would have a similar role in neural cells. Indeed, TGF-β induces cell cycle exit in murine hippocampal neurones [8] and has been associated with the loss of adult neurogenesis through arresting the proliferation of progenitor cells [9]. TGF-β also plays a function in various neurogenic processes, including the formation and elongation of axons [10], neurite growth [11], and initiation of neuronal migration [12]. Given the diverse roles of TGF-β and its function in the nervous system, it is clear that TGF-β signalling is also involved in neuroplasticity and neuroprotection.

Neuroplasticity broadly refers to the ability of the nervous system to responds to internal and external stimuli by reorganising its structure and function at the molecular, cellular, and organisational levels. These adaptations can be beneficial or harmful to the organism; consequently, neuropsychiatric disorders have been examined as a manifestation of deleterious neuroplasticity. Indeed, neuropsychiatric disorders have been characterised according to alterations in the limbic, fronto-striatal, and prefrontal circuits, which in turn generate disturbances in behaviour, cognition, and motivation [13]. These behavioural changes can manifest over long periods, meaning the course of treatments can be similarly drawn out. The improvements due to such treatments can also be lost over time, and each episode of illness increases the probability of relapse as the altered neural networks become more dysfunctional [14]. Neuroinflammation, neuronal survival, and proliferation are some of the factors that can negatively influence neuroplasticity. Furthermore, these processes can also be negatively affected by inflammation, as demonstrated by treatment with lipopolysaccharides from *E. coli* used to induce an immune response [15]. Neuroinflammation has also been implicated in several nervous system diseases, including multiple sclerosis, Alzheimer's Disease, Parkinson's Disease, and major depression [16]. Although inflammation is commonly associated with physical injury or infection, stress can also trigger the release of inflammatory cytokines [17], such as Interleukin-1β, Interleukin-6, and tumour necrosis factor-α (TNF-α), that interfere with the production of neural growth factors. Meanwhile, preclinical studies inhibiting interleukin-1 in mouse model showed effective reversal of this effect and alleviated stress-induced behavioural changes [18,19].

Evidence from previous work suggests an association between TGF-β signalling and the phenomena of neuroplasticity and neurogenesis, as well as their combined effects on mood disorders [20–22]. Based on these findings, TGF-β signalling mediated by Smad3 has been hypothesised to play an important role in neurogenesis in the hippocampus, and has been implicated in the development of mood disorders and the manifestation of depressive and anxiety disorders. Previous studies on mood disorders were largely based on neuroplasticity mechanisms, however, the underlying pathways are not well understood and relatively underexplored [23,24]. Thus, we hypothesise the neurogenic mechanisms of depression and anxiety involve TGF-β and its signalling molecules.

2. Non-Canonical Signalling

Non-canonical TGF-β signalling refers to molecular events that occur independent of the SMAD pathway. TGF-β activated kinase 1 (TAK1) is an important effector in non-canonical TGF-β signalling. TAK1 initiates a signalling cascade that activates c-Jun N-terminal Kinases (JNK) and p38/mitogen activated protein kinase (MAPK) in response to TNF/TGF-β signalling, as well as activates the IkB Kinase (IKK) and Nuclear Factor-κB (NF-κB) pathways. TAK1 is an important regulator of innate and adaptive immune responses [25], and is essential for the survival of haematopoietic cells and hepatocytes [26]. TAK1 is highly expressed in the brain [27] and is assumed to play a role in neural functions. Indeed, TAK1 was shown to be important in axon growth, as neurones with knockdown of TAK1 had significantly shorter axons than normal neurones. TAK1 also rescued axonal growth in the presence of JNK inhibitor SP600125. Furthermore, TAK1-negative mice exhibited embryonic lethality due to defects in the brain, supporting the essential role of TAK1 in brain development [28]. TAK1 is also a major signalling partner of TGF-β, as TGF-β receptors could only induce low levels of JNK phosphorylation in TAK1-negative cells [28].

The MAPK signalling pathway has been shown to be involved in neuronal differentiation and axon growth [29]. The MAPK pathway has also been implicated in the development of depression [30], as the inhibition of MAPK signalling inducing depressive-like behaviours [31]. Indeed, some pharmacological treatments for depression activate the MAPK signalling pathway [32]. It was shown that MAPK differentially affected mice depending on age, with juvenile mice spending significantly more time in the open arms of the elevated plus maze (EPM), indicating ablated anxiety compared with wildtype mice [33]. The involvement of MAPK signalling in mood disorders was further studied by knockout of Braf, an upstream effector of MAPK. Adult mice lacking Braf showed greater passivity in the forced swim test, indicating greater depressive behaviours [33]. Taken together, these findings indicate MAPK pathway has differential effects across the lifespan, highlighting the potential of targeting MAPK signalling in regulating the pathogenesis of mood disorders.

The c-AMP response element binding protein (CREB) is involved in the TGF-β pathway, as TGF-β was shown to activate extracellular elements that activate CREB [34]. TGF-β was also able to directly induce phosphorylation of CREB, as shown by increased phosphorylation of CREB in hippocampal cells treated with TGF-β2 compared with the control, implying that CREB can mediate the long-term effects of TGF-β2 [35]. CREB has also been implicated in hippocampal functions such as memory [36] and has been targeted by antidepressants [37]. Loss of CREB function in the brains of mice was shown to increase neurogenesis and abolish depressive-like behaviour in assays, such as the forced swim test [38]. Various other studies showed CREB is important in cell survival and maturation in the hippocampus [39,40], and can improve the antidepressant response [41]. Further studies are needed to understand the role of CREB in neurogenesis.

3. Canonical Signalling in Neurogenesis

TGF-β has been shown to play a role in inflammation and promote cell survival, induce apoptosis, and initiate proliferation and differentiation [2,42]. In mammals, TGF-β exists as three molecular isoforms (TGF-β1, β2, and β3) found in the cerebral cortex [43], olfactory epithelium [44], adult astrocytes, neurones, and microglia [45]. TGF-β is involved in the initiation of cell cycle exit and neuronal differentiation, among other functions. TGF-β is also required for neuronal survival. This is consistently supported by in vitro and in vivo data, which demonstrated lower survivability of TGF-β knockout neurones [46], and embryonic lethality, increased neuronal apoptosis and other abnormalities such as reduced synaptic integrity and microgliosis, observed in TGF-β-null mice [46,47]. Taken together, these results indicate that TGF-β plays a vital role in neuronal survival and microglial activation. Given that Smad3 is an important effector of TGF-β function, the implication of these findings on Smad3 warrants further research to delineate its functions in the brain.

Despite extensive research on the effect of TGF-β on neurogenesis (Table 1), the relationship between Smad3 signalling and neurogenesis is not well understood. Previous research on Smad3 deficiency indicated that Smad3 signalling was important in neurogenesis, given that Smad3 was found to be abundantly expressed in neurogenic zones [48–50]. The role of Smad3 in neurogenesis is further corroborated by the finding that showed colocalization of Smad3 transcript with mature neuron marker neuronal nuclei (NeuN) in the dentate gyrus of hippocampus [51]. Moreover, Smad3-null mice showed disrupted neuronal proliferation and migration [49] and expressed significantly less neurones in the dentate gyrus compared with wildtype mice [49]. Conversely, Smad3-null mice exhibited a series of alterations, for instance impaired hippocampal neurogenesis, reduced newborn neurone survival and elimination of long-term potentiation (LTP) in the medial perforant pathway by facilitating gamma-amino butyric acid (GABAergic) signalling [51,52]. These results provided evidence supporting the broad potential of Smad3 in modulating neurogenesis. Interestingly, these effects appeared to be region-specific, as increased progenitor cell population in the rostral hippocampus and intact LTP in the Schaffer collateral pathway were also found in Smad3-null mice [51]. Similarly, increased levels of TGF-β reduced neurogenesis in vitro and in vivo at rates largely similar to that in Smad3-deficient animals [9,30,53,54]. Contrary to these findings, it was shown that knockout of TGF-β receptor activin receptor-like kinase 5 (ALK5) resulted in decreased neurogenesis, whereas upregulation of ALK5 resulted in greater neurogenesis and improved memory functions [55]. This indicates that the components in the canonical pathway, especially Smad3, potentially plays a role in memory and cognitive functions and may regulate neurogenesis differentially in various parts of the brain. Furthermore, Smad3 signalling in neurones and astroglia were found to differentially regulate dendritic spine growth [56], whereas its inhibition leads to increased susceptibility to neuronal apoptosis [57], potentially increasing the risk of Parkinson's disease [25] and neurodegeneration [26]. However, the specific effects of Smad3 on behaviour have yet to be studied.

Table 1. TGF-β canonical and non-canonical signalling mechanisms in behavioural and physiological changes.

Signalling Molecule	Models	Behavioural and Physiological Changes	Mechanisms	References
TGF-β	Receptor inhibition in vivo (C57BL6 Mouse)	Increased neurogenesis	Reduction of inflammatory response mediated by B2M through attenuation of pSmad3 activity	[50]
	Knockout in vivo (C57BL6 Mouse)	Increased neuronal degeneration and microgliosis	TGF-β-related decrease in laminin-reduced survivability and increased susceptibility to apoptosis	[46]
	Chronic upregulation of TGF-β1 in vivo (C57BL6 Mouse)	Decreased immature hippocampal neurones and neurogenesis	Induced early cell cycle exit of neural progenitor cells	[54]
	Exogenous upregulation of TGF-β2 in vitro (Sprague-Dawley Rat Hippocampus)	Induction of evoked post-synaptic currents and inhibition of miniature post-synaptic currents	TGF-β-related upregulation of CREB in hippocampal neurones	[35]
	Receptor knockout in vivo (C57BL6 Mouse)	Reduction of immature neurones and neurogenesis	Increased expression of pro-apoptotic effectors; decreased expression of anti-apoptotic effectors	[55]
Smad3	Knockout in vivo (C57BL6 Mouse)	Reduction of Neurogenesis	Disruption of neuronal proliferation and migration	[49]
		Inhibition of long-term potentiation	Impairment of NMDA activity by Smad3-related increase in GABAergic signalling	[52])
		Rostral increase of proliferative cells; caudal decrease in neurogenesis	Potential compensatory mechanism in rostral DG to maintain cell numbers; increased apoptosis at intermediate cell stage reduces neurogenesis	[58]
		Decreased neuronal viability following injury	Disruption of Smad3 signalling in astrocytes	[59]
		Accelerated wound closure and decreased activation of microglia	Reduced expression of MCP-1 and reduced leukocyte activity	[60]
	Transient knockdown in vivo (Chick Embryo)	Decreased neurogenesis	Preferential activation of Smad2 targets due to loss of Smad3 activity	[48]
Smad2	Transient knockdown in vivo (Chick Embryo)	Increased neurogenesis	Preferential activation of Smad3 targets due to loss of Smad2 activity	[48]
Braf	Knockout in vivo (129S1/Sv + C57BL6 Mouse)	Increase of depressive-like behaviour in adults, decrease of anxiety in juveniles; reduction of dendritic spine growth	Disturbance of Erk/MAP signalling and alteration of serotonergic transmission	[33]
CREB	Inhibition by dominant negative mutant in vivo (129SvEv + C57BL6 Mouse)	Anti-depressant effects mediated by increase in neurogenesis	Potential mCREB interaction with non-CREB targets	[38]
		Decreased granule cell proliferation	Disruption of cAMP-CREB signalling	[39]
	Conditional knockout in vivo (129SvEv + C57BL6 Mouse)	Impairment of performance in spatial retention	Upregulation of CREB	[36]
	Transient overexpression of CREB in vivo (Sprague-Dawley Rat)	Reduction of depressive-like behaviours	Improved adaptation due to CREB-related regulation of granule cells	[41]
Gadd45b	Knockout in vivo (Mouse)	Reduction of the effectiveness of ECT in inducing neurogenesis and dendritic spine growth	Attenuation of Gadd45b-mediated demethylation in regulatory regions of BDNF and FGF1	[61]
JMJD3	Knockout in vivo (Mouse)	Disruption of neuronal migration and reduction of neurogenesis	Reduction of Dlx2 activation and H3K27me3 demethylation	[62]
			Disruption of TGF-β/Smad signalling cascade	[63]
MAPK	Transient inhibition in vivo (Mouse)	Increased behavioural despair and reduced effectiveness of anti-depressant treatment	Disruption of MEK-ERK signalling	[31]

Table 1. *Cont.*

Signalling Molecule	Models	Behavioural and Physiological Changes	Mechanisms	References
TAK1	Knockdown in vivo (Mouse)	Reduction of axonal length	Impairment of JNK activity mediated by TAK1	[64]
TrkB	Conditional knockout in vivo (Mouse)	Impairment of neurogenesis; resistance to anti-depressant treatment	Blockade of BDNF signalling	[65]

Abbreviation: B2M, β2 microglobulin; NMDA, N-methyl-D-aspartic acid; GABA, gamma-aminobutyric acid; DG, dentate gyrus; MCP-1, Monocyte chemoattractant protein-1; FGF1, Fibroblast growth factor 1; Dlx2, distal-less homeobox 2.

Besides interacting with TGF-β, Smad3 also interacts with a host of other signalling molecules such as its close analogue Smad2 (Figure 1). Although both proteins share 91% amino acid sequence, they recruit different cofactors and thus target different transcription pathways [43]. Nevertheless, they have been shown to share redundant roles in certain contexts, particularly [44,45,48,66]. Both Smad2 and Smad3 were found to cooperate and antagonize targets simultaneously, for example, Smad3 activates its targets while antagonizing Smad2 targets. This is because Smad2 targets only respond to Smad2 homodimers, whereas Smad3 targets respond to both Smad3 homodimers and heterodimers, thus increased co-expression of Smad2 and Smad3 leads to greater activation of Smad3 targets but inactivation of Smad2 targets [48]. In terms of neurogenic mechanisms, knockdown of Smad3 reduced neurogenesis, whereas knockdown of Smad2 increased neurogenesis [48]. Silencing of Smad3 in aged mice by shRNA resulted in greater neurogenesis, indicating that Smad3 plays a significant role in hippocampal degeneration in old age [49]. This suggests that Smad2 would inhibit the pro-neurogenic effects of Smad3, whereas knockout of Smad3 would result in elevated Smad2 processes, resulting in an overall reduction of neurogenesis.

TGF-β binds to the type II receptor that in turn forms a complex that phosphorylates Smad3. TGF-β receptors were found to be expressed on new cells in the neurogenic region of the dentate gyrus and the subventricular zone [49]. Similarly, Smad3 was also found to be distributed in neurogenic zones, whereas Smad3-null mice exhibited markedly decreased Bromo-deoxyuridine (BrdU)-positive cells in the dentate gyrus and subventricular zone. Both Smad3 and Smad2 pathways are activated by activin in a manner similar to activation by TGF-β. It was found that activin A was upregulated in the dentate gyrus and the CA1 region of the hippocampus following chronic paroxetine treatment [56]. Additionally, direct injection of activin A into mouse dentate gyrus significantly reduced immobility in the forced swim test, indicating it can exert antidepressant-like effects [56]. Moreover, there was increased neuronal survival and development in rat hippocampal cell cultures treated with activin. Activin also affected dendritic spine growth by modulating actin dynamics [57]. Taken together, these results imply a possible relationship between TGF-β and activin signalling. The above findings also suggest that Smads could be worth investigating, as it remains unclear how activin interacts with downstream signalling molecules to exert its antidepressant effects.

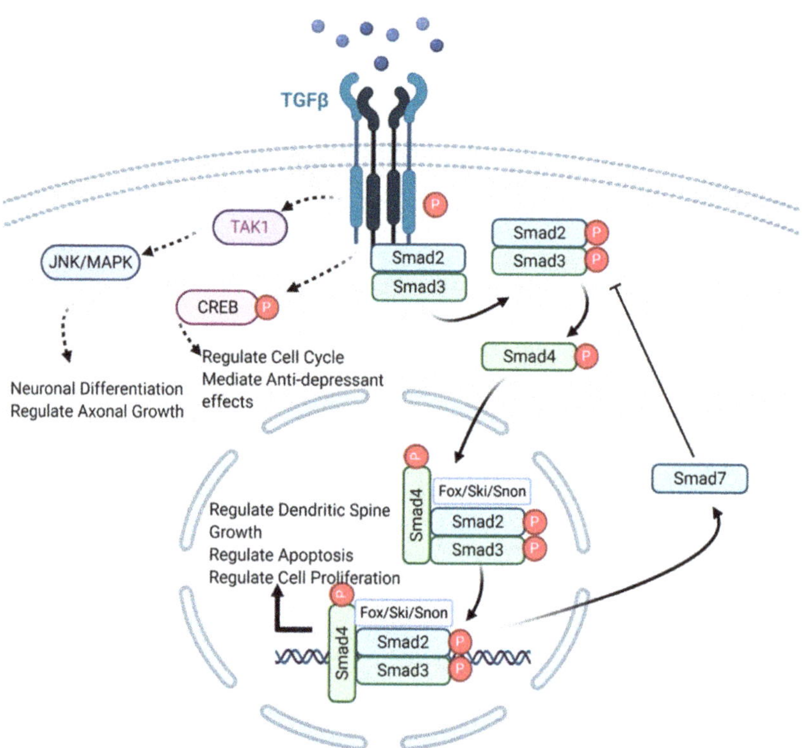

Figure 1. Illustration of canonical and non-canonical TGF-β signalling. TGF-β ligand binds with TGF-β receptor to form a complex (Created with BioRender.com, accessed on 19 April 2021). The receptor-induced phosphorylation of R-Smads leads to binding with cytoplasmic Smad2/3. Phosphorylated Smads form a complex with Smad4, which translocates to the nucleus where it binds with various transcription factors for gene transcription. Smad complexes initiate a negative feedback loop, leading to Smad7 inhibition of further phosphorylation of R-Smads. TGF-β receptors also phosphorylate TAK1 and CREB, which regulate neuronal differentiation, axonal growth, cell cycle progression, and anti-depressant effects, respectively.

4. TGF-β Signalling in Epigenetics

Smad signalling has been implicated in epigenetic functions such as Jumonji domain-containing protein D3 (JMJD3). Knockdown studies showed that spinal cord [67] and retinal [68] development are regulated by JMJD3. JMJD3 is also expressed in neural stem cells as well as doublecortin expressing neuroblasts. Postnatal JMJD3-null mice experienced significantly less neural growth and exhibited neuroblast migration disturbances [62], indicating JMJDs are required in neurogenesis. Adult neurogenesis was found to require JMJD3, as neural stem cells (NSCs) lacking JMJD3 had stunted neurogenesis and impaired differentiation with reduced oligodendrocyte production [62]. Moreover, JMJD3 knockdown impaired TGF-β signalling in NSCs. Further examination of the relationship between TGF-β and JMJD3 revealed that the functions of TGF-β in development, differentiation, and apoptosis were dependent on JMJD3 [63]. Additionally, Smad3 function was also found to involve JMJD3. Co-immunoprecipitation assays revealed that phosphorylated Smad3, but not Smad2, interacted with JMJD3. Genome-wide analysis also showed that Smad3 and JMJD3 expression overlapped at transcriptional start sites, indicating that JMJD3 is involved in Smad3 activation of gene transcription.

Growth arrest and DNA-damage inducible protein 45 (Gadd45) is another epigenetic regulator that is affected by the TGF-β signalling pathway, as shown by blocking of TGF-β and TrkB leading to the overexpression of Gadd45 in vitro [69]. Gadd45 functions as a modulator of hippocampal stem cell proliferation, and its deletion was shown to prevent demethylation of brain-derived neurotrophic factor (BDNF) leading to disrupted neurogenesis [61]. Inactivation of Smad4, which is a core component of Smad signalling, increased Gadd45 expression, implicating TGF-β/Smad interactions in the neurogenic function of Gadd45. The expression of Gadd45 was shown to increase when TGF-β receptor function was impaired, and this effect was amplified by simultaneously blocking TrkB signalling. These data indicate that TGF-β signalling is involved in DNA demethylase expression by suppressing Gadd45 expression. Further DNA methylation and gene expression analyses in the mouse genome following potassium chloride treatment revealed increased expression of certain genes, with at least six genes associated with psychiatric diseases such as autism and depression. Induction of depression in an animal model resulted in decreased Gadd45 expression in the hippocampus and prefrontal cortex, accompanied by a decrease in TGF-β2 and TGF-β3 expressions [69]. These findings show that TGF-β signalling plays a role in the regulation of epigenetic mechanisms. Given that Smad4, which operates in tandem with Smads2/3, is able to affect Gadd45 expression, the effects of impaired Smad3 phosphorylation may shed more light on the relationship between TGF-β and Gadd45.

5. Neurogenesis in Neuropsychiatric Disorders

Neurogenesis is hypothesised to play a role in depression via changes to the rate of neurogenesis, specifically the deleterious effects from stress-related neurogenesis in the dentate gyrus. Neurogenesis in the dentate gyrus can impact behavioural output and the efficacy of antidepressant treatments. New neurones heavily affects hippocampal circuitry and behaviour. Clinical evidence shows hippocampal shrinkage is linked to depression. In the pre-clinical stage, depressive behaviour results in a host of adverse effects on hippocampal neurogenesis, including reduced proliferation of neural stem cells and reduced neuronal survival. Furthermore, antidepressant treatment in rats induced hippocampal neurogenesis over time, in a pattern that mirrors the delay in antidepressant efficacy in human subjects [70]. One could associate neurogenesis with cellular reparation and plasticity, as these events have been shown in animal models given antidepressant treatments, electroconvulsive therapy, and stress reducing exercise. However, the capacity to draw a causal relationship between hippocampal neurogenesis and depression remains elusive as other studies showed a loss of neurogenesis does not necessarily lead to the development of depression [71] and stress does not always decrease neurogenesis [72].

Although a direct causal link between hippocampal neurogenesis and mood disorders seems unlikely, neurogenesis is nonetheless an important factor in the development of mood disorders. The neurogenic interactome model attempts to reconcile these opposing findings by taking into account various factors affecting neurogenesis. The neurogenic interactome considers the interactions between brain structures, the various functions of the hippocampus, and the heterogeneity of elements in the neurogenic niche [73]. For example, hippocampal connections to regions involved in stress responses and emotional memory indicates the involvement of adult neurogenesis in depression. The complex series of connections proposed by the interactome model can possibly explain the discrepancies between behaviour tests, in that each test induces different levels of stress and thus engages different components of the neurogenic interactome. The interactome also suggests that neurogenic mechanisms can respond to stressful stimuli in different ways, such as coping or adapting to stress, thus predictable stress might increase neurogenesis [74,75]. The concept of a neurogenic interactome provides a holistic framework to study the various factors in neurogenesis. It would be illuminating to determine the involvement of Smad3 in the interactome as well as the behavioural effects of disturbances in Smad3 signalling.

The pathogenesis of mood disorders has been hypothesised to be related to neuroplastic changes, particularly neuroimmune processes such as neuroinflammation, which might affect the central and periphery nervous systems to impact the neurobiology of depression. Neuroinflammatory processes are suspected to exert deleterious effects on various pathways, such as the hypothalamic-pituitary-adrenal (HPA) axis, neurogenesis, neuroimmune response, and neurocircuitry by upregulating apoptotic functions. Various mechanisms have been put forward on how the immune system regulates neurogenesis and neuroplasticity, particularly the role of T-reg cells on enhancing hippocampal neurogenesis via upregulating glial cell-derived neurotrophic factors and TGF-β. It was shown that T-reg cells are attenuated during depression, whereas T-reg cells are increased with antidepressant treatments. However, behavioural studies in animals with suppressed neurogenesis but not exposed to stressors showed that neurogenesis by itself does not produce a depressive phenotype, as there were no differences in the behaviour between animals with ablated neurogenesis and controls [76].

6. Effects of Stress on TGF-β Signalling and Neurogenesis

Stress has been shown to induce anxiety and depressive disorders by disturbing the HPA axis-regulated release of glucocorticoids [77,78]. In particular, hippocampal inhibitory regulation of the HPA axis can be disturbed by chronic stress [79,80]. Observations in various species indicated that the decreased cell proliferation and neurogenesis in the dentate gyrus following exposure to stressors was related to elevated stress hormones [81–83]. Thus, it follows that chronic stress could exacerbate depression by altering neuroplasticity and neurogenesis. Furthermore, it was shown that anti-depressant treatments could reverse the reduced neurogenesis and cell proliferation. These studies also demonstrated a causal link between hippocampal plasticity and depressive disorders. However, the finding that decreased hippocampal neurogenesis failed to increase sensitivity to stress casts doubt on the causal relationship between neurogenic effects in the hippocampus and depression. Even so, the effects of stress still constitute a major factor in the development of depressive disorders. The expression of TGF-β was shown to be reduced in the hippocampus of stressed rats, which coincided with increased expression of inflammatory cytokines such as IL-1β and lowered expression of anti-inflammatory cytokines such as IL-10 [84]. Neurogenesis was also found to be impaired, as demonstrated by the level of BrdU-stained cells in the dentate gyrus of these rats, which indicates a potential relationship between decreased TGF-β signalling, loss of neurogenic function, and development of depressive-like symptoms. In vitro studies showed neurones exposed to increased cortisol levels had decreased TGF-β expression and reduced neurogenesis, supporting this relationship [85].

A connection between stress and neurogenesis has also been observed, in that a reduction of neurogenesis precipitated the development of stress-related disorders such as post-traumatic stress disorder [86]. This connection is reflected in several animal models such as social defeat paradigm and unpredictable chronic mild stress to mimic symptoms of depression and stress disorders [76]. Neurogenic functions in the dentate gyrus are known to be sensitive to stress, showing reduced cell proliferation in response to stress or to elevated serum glucocorticoid levels, but induced cell division with reduced glucocorticoid levels [87]. A loss of neurogenic function in the dentate gyrus inhibits regular functioning of the neural circuitry, and thus prevents the formation of new cognition and associations, perpetuating the development of depression [88]. Indeed, adult neurogenesis is capable of neuroplasticity, which facilitates the acquisition and separation of closely spaced memories [89]. It has been shown that decreased neurogenesis results in impaired pattern separation and learning, whereas increased neurogenesis improves them. Apart from the obvious functions in learning and memory, pattern separation may also be important in recognizing danger and stress [90]. Mice exposed to social stressors undergo increased neurogenesis, possibly indicating superior recognition of negative stimuli, whereas preventing neurogenesis diminished social avoidance [72]. Stress also affects the amygdala by disrupting endocannabinoid signalling, resulting in pathological anxiety [91]. Stress has also been shown to increase dendritic spine length in the basolateral amygdala, which may be indicative of spine instability during plasticity responses, and such morphological changes correspond with increased anxiety [92].

The hippocampus has been implicated in the development of depressive disorders. Neuroimaging studies and post-mortem examinations showed an association between reduced hippocampal volume and the duration of depression [93–95]. Hippocampal neurogenesis was also found to be necessary for the function of certain antidepressants. The effects of imipramine and fluoxetine were abolished when hippocampal neurogenesis was ablated by irradiation. The importance of hippocampal neurogenesis in drug treatments was further reinforced by the observation that antidepressants also increased hippocampal neurogenesis. Hippocampal activity can also affect neurotrophin release, as demonstrated by the release of TGF-β and BDNF following induced depolarization [69]. In animals exposed to prenatal stress and in patients suffering from childhood trauma, the hippocampus was shown to express greater amounts of FoxO, a proinflammatory transcription factor that responds to insulin signalling as well as oxidative stress [96]. The upregulation of FoxO1 in the presence of cortisol mediated cortisol-induced reduction of neuronal proliferation [97].

The amygdala has also been implicated in the pathogenesis of anxiety disorders, likely caused by hyperexcitability of neurones in the amygdala [98,99]. Further study on the amygdala revealed the endocannabinoid system is a key pathway that influences the regulation of stress and anxiety. Disruption of endocannabinoid signalling induced neurobehavioural effects similar to those produced by exposure to stress, such as increased activation of the HPA axis and heightened anxiety behaviours [100,101]. Interestingly, these effects could be replicated by direct disruption of endocannabinoid signalling in the amygdala [102]. Chronic stress via impaired LMO4 function resulted in halting endocannabinoid signalling in the amygdala, which caused anxiety behaviours, supporting the involvement of the amygdala in regulating anxiety states [92].

7. Discussion and Perspectives

The TGF-β pathway has been implicated in a wide range of processes. This is evident by reports showing increased TGF-β in circulation predisposes an individual to hypertension [103] and polymorphisms in TGF-β result in variable risk for the development of oesophageal cancer [104]. Intriguingly, TGF-β demonstrated paradoxical effects in oncology studies, in which TGF-β exhibited essential roles in tumorigenesis as well as tumour suppression. Histology on clinical samples of gastric carcinoma revealed that TGF-β itself gradually increases in expression as carcinoma progresses and is correlated with advanced stages of carcinoma while Smad molecules are associated with earlier stages [105]. Further substantiating the role of TGF-β in tumorigenesis is the finding that exogenous upregulation of TGF-β induces tumours with metastatic properties in zebrafish [106]. Interestingly, loss of TGF-β function in carcinomas appears to contribute to tumour suppressant properties [107,108]. Specifically, TGF-β exercises anti-proliferative effects through Smad signalling by increasing expression of Cyclin-dependent kinase (Cdk), which in turn results in cell cycle arrest [109]. Indeed, aberration in Smad function of gastric cancer cells renders TGF-β mediated growth inhibition ineffective [110]. The role that TGF-β plays in the cell cycle is no less apparent in the brain, directing axon growth during the developmental stage with cells lacking TGF-β failing to form axons [12]. Furthermore, the endogenous levels of TGF-β in circulation varies among the population, and this produces differences in phenotype. For instance, overexpression of TGF-β seems to heighten its antiproliferative effects, as the development of several cancers are ablated by increase of TGF-β [111–113]. These findings indicate that cell proliferation mechanisms can be affected through measured manipulation of TGF-β and its associated signalling molecules, as such there is potential in the TGF-β pathway modulation to alleviate disorders involving premature apoptosis and impaired cell proliferation.

TGF-β plays an important role in the central nervous system as a response to injury and trauma and functions as a signal to initiate cellular repair and protection [114,115]. TGF-β has been shown to be upregulated significantly following spinal lesioning and was observed in astrocytes, motor neurons, and surrounding epithelial cells. This is in tandem with astrocyte mobilisation around the site of injury [116], which remains for up to a year post injury [117]. TGF-β is also implicated in immune dysregulation disorders such as Multiple Sclerosis (MS), in which a drastic increase in TGF-β expression was observed during an MS attack and increased TGF-β circulation observed in patients afflicted with the disease [118]. Furthermore, increased TGF-β activity was also observed in neurodegenerative diseases. It has been posited that the increase in TGF-β is a protective response against neuronal loss that occur in these diseases. Evidence compiled in a review of the role of TGF-β in neurodegeneration seems to support this view as aging and chronic inflammation impair TGF-β/Smad signalling and promotes overactivation of microglia, which implicates loss of TGF-β signalling in plaque and tangle formation [119]. Indeed, post-mortem analysis of Alzheimer's and Parkinson's disease patients showed that TGF-β ligands were elevated in cerebrospinal fluid [120,121].

The potential for TGF-β to be applied in a therapeutic context has been considered in the context of oncology where anti-TGF-β treatment has been proposed to target tumour cells and microenvironment [122]. This concept is expanded upon in studies where various TGF-β inhibitors have been tested against fibrotic diseases and cancer, showing encouraging results [123]. The possible negative effects of TGF-β inhibition therapy have been examined as well, and it was found that adverse changes to major organs and lymphocyte function were limited in mice treated with anti-TGF-β antibodies chronically [124]. Even a lifetime inhibition of TGF-β receptors has not produced significant adversities in mice [125]. However, a high degree of inhibition has been observed to result in epithelial hyperplasia and carcinoma in certain conditions [126]. Similarly, TGF-β upregulation has also been discussed as a potential treatment to Alzheimer's Disease [127], but relatively little has been discussed regarding the therapeutic potential of TGF-β outside of the context of cancer.

Although there is a link between neurogenesis and depression, several studies [128–130] indicated this was an indirect relationship. In support of neurogenesis playing a role in the aetiology of depression, clinical studies showed a correlation between depression and neuronal degeneration in the hippocampus [131,132] and reduced hippocampal volume [93–95,133]. Furthermore, some established depression treatments, such as 5-HT, tianeptine, and electroconvulsive therapy, have been observed to stimulate neurogenesis [70,133,134]. Preclinical studies using animal models showed antidepressants could induce neurogenesis [135,136], which is necessary to mediate the antidepressant effect as the abolition of neurogenesis resulted loss of antidepressant function [137]. Consistent with these observations, exposure to stressors was found to inhibit neurogenesis, which recovered after antidepressant treatment [82]. All these studies provide a convincing case for the involvement of neurogenesis in the aetiology of depression. However, this narrative is complicated by some studies that showed neurogenesis is not correlated with depressive-like symptoms, or is not a causal factor in the development of depression and associated behaviours [138]. Furthermore, antidepressant treatments might act through neurogenesis-independent mechanisms to achieve disease attenuation [139]. Yet other studies showed that neuronal proliferation, which is a standard measure of neurogenesis, may be preserved despite decreased cell survivability [140]. Taken together, our current understanding points to the hypothesis that suppressed neurogenesis predisposes to depression but does not necessarily produce depressive symptoms [141–143].

Analysis of human hippocampal neurogenesis showed that neurogenesis takes place at a roughly equivalent rate to that in middle-age rodents [144] and persists even into old age [145], indicating that neurogenesis plays a role in normal brain function in humans. The relationship between hippocampal neurogenesis and mood dysregulation is still unclear, although stress appears to impair hippocampal neurogenesis and is associated with depressive behaviours [146]. This observation is in line with the findings that hippocampal neurogenesis regulated the endocrine stress response, and that loss of neurogenesis in the dentate gyrus resulted in hypersecretion of corticosteroids in response to stress leading to reduced clearance of glucocorticoids [147]. Nonetheless, discrepancies across studies make it difficult to define the precise role of hippocampal neurogenesis in the aetiology of depression. The proposed concept of the 'neurogenic interactome' attempts to reconcile these discrepancies. The theory posits three key elements: (1) Interconnectivity between the hippocampus and regions such as the HPA and limbic system, (2) interrelations between mood and memory, and (3) complex interplay among heterogenous components of the neurogenic niche influencing the observed response in behaviour and drug effectiveness [73].

The observation that the inhibition of neurogenesis does not replicate the stress-induced hippocampal volume loss or behavioural changes indicates that neurogenesis is not the sole mechanism in the manifestation of depressive behaviours, and it is likely that stress affects other contributory factors. Regarding the loss of neurogenesis, previous work found it could be induced by autophagic cell death of neural stem cells. It was shown that deletion of Atg7, an important molecule in the formation of autophagosomes, resulted in neuroprotective effects in chronic restraint stress (CRS) and cortisol injection [148]. This may explain why the inhibition of neurogenesis was insufficient to reproduce the morphological changes seen in stressed animals. Clinical research showed that depression-related changes to hippocampal volume largely occurred in the posterior hippocampus [149,150]. In rodents, the dorsal hippocampus also showed stress-related volume losses [151], while the ventral hippocampus was largely unaffected [71]. These changes were suggested to involve variation in glucocorticoid receptor distribution in the dorsal and ventral hippocampus [152], and variation in amygdala input to the dorsal and ventral hippocampus [153,154]. However, the volume changes were not shown to be the result of reduced neurogenesis, indeed, other factors such as loss of interneurones and microvasculature [155,156], and reduction and shrinkage of astrocytes in the hippocampus [157–159] were shown to play a part.

In conclusion, the TGF-β pathway is involved in many functions and is required for embryonic viability; however, its role in neurogenic mechanisms and their implication on mood regulation are largely unknown. Based on the available evidence, the TGF-β pathway potentially exerts effects on neurogenesis via the canonical pathway. These effects may in turn play a role in the development of mood disorders such as depression and pathological anxiety. Inactivation of Smad3 in animal models of depression induced by chronic restraint stress reveals TGF-β/Smad3 signalling together with non-canonical pathway components such as TAK1 and Erk play potential roles in hippocampal neurogenesis.

Author Contributions: Conceptualization and writing—original draft preparation, L.-F.H. and L.-W.L.; writing—review and editing, L.-F.H., C.-H.P., H.-Z.Y., and L.-W.L.; funding acquisition and supervision, L.-W.L. All authors have read and agreed to the published version of the manuscript.

Funding: This scientific work was supported by the Hong Kong Research Grant Council, and research funding from The University of Hong Kong (Seed Fund for Basic Research and Seed Fund for Translational & Applied Research) awarded to L.-W.L.

Conflicts of Interest: The authors declare no conflict of interest.

References

1. Derynck, R.; Akhurst, R.J. Differentiation plasticity regulated by TGF-beta family proteins in development and disease. *Nat. Cell Biol.* **2007**, *9*, 1000–1004. [CrossRef]
2. Massague, J.; Blain, S.W.; Lo, R.S. TGFbeta signaling in growth control, cancer, and heritable disorders. *Cell* **2000**, *103*, 295–309. [CrossRef]
3. Ten Dijke, P.; Hill, C.S. New insights into TGF-beta-Smad signalling. *Trends Biochem. Sci.* **2004**, *29*, 265–273. [CrossRef]
4. Nakao, A.; Afrakhte, M.; Moren, A.; Nakayama, T.; Christian, J.L.; Heuchel, R.; Itoh, S.; Kawabata, M.; Heldin, N.E.; Heldin, C.H.; et al. Identification of Smad7, a TGFbeta-inducible antagonist of TGF-beta signalling. *Nature* **1997**, *389*, 631–635. [CrossRef] [PubMed]
5. Derynck, R.; Zhang, Y.E. Smad-dependent and Smad-independent pathways in TGF-beta family signalling. *Nature* **2003**, *425*, 577–584. [CrossRef] [PubMed]
6. Massague, J. TGFbeta in Cancer. *Cell* **2008**, *134*, 215–230. [CrossRef]
7. Lyons, R.M.; Moses, H.L. Transforming growth factors and the regulation of cell proliferation. *Eur. J. Biochem.* **1990**, *187*, 467–473. [CrossRef]

8.   Vogel, T.; Ahrens, S.; Buttner, N.; Krieglstein, K. Transforming growth factor beta promotes neuronal cell fate of mouse cortical and hippocampal progenitors in vitro and in vivo: Identification of Nedd9 as an essential signaling component. *Cereb. Cortex* **2010**, *20*, 661–671. [CrossRef]

9.   Wachs, F.P.; Winner, B.; Couillard-Despres, S.; Schiller, T.; Aigner, R.; Winkler, J.; Bogdahn, U.; Aigner, L. Transforming growth factor-beta1 is a negative modulator of adult neurogenesis. *J. Neuropathol. Exp. Neurol.* **2006**, *65*, 358–370. [CrossRef] [PubMed]

10.   Abe, K.; Chu, P.J.; Ishihara, A.; Saito, H. Transforming growth factor-beta 1 promotes re-elongation of injured axons of cultured rat hippocampal neurons. *Brain Res.* **1996**, *723*, 206–209. [CrossRef]

11.   Knoferle, J.; Ramljak, S.; Koch, J.C.; Tonges, L.; Asif, A.R.; Michel, U.; Wouters, F.S.; Heermann, S.; Krieglstein, K.; Zerr, I.; et al. TGF-beta 1 enhances neurite outgrowth via regulation of proteasome function and EFABP. *Neurobiol. Dis.* **2010**, *38*, 395–404. [CrossRef] [PubMed]

12.   Yi, J.J.; Barnes, A.P.; Hand, R.; Polleux, F.; Ehlers, M.D. TGF-beta signaling specifies axons during brain development. *Cell* **2010**, *142*, 144–157. [CrossRef]

13.   Beauregard, M.; Levesque, J.; Bourgouin, P. Neural correlates of conscious self-regulation of emotion. *J. Neurosci.* **2001**, *21*, RC165. [CrossRef]

14.   Patten, S.B. Sensitization: The sine qua non of the depressive disorders? *Med. Hypotheses* **2008**, *71*, 872–875. [CrossRef] [PubMed]

15.   Ekdahl, C.T.; Claasen, J.H.; Bonde, S.; Kokaia, Z.; Lindvall, O. Inflammation is detrimental for neurogenesis in adult brain. *Proc. Natl. Acad. Sci. USA* **2003**, *100*, 13632–13637. [CrossRef] [PubMed]

16.   Wang, B.; Jin, K. Current perspectives on the link between neuroinflammation and neurogenesis. *Metab. Brain Dis.* **2015**, *30*, 355–365. [CrossRef] [PubMed]

17.   Garcia-Bueno, B.; Caso, J.R.; Leza, J.C. Stress as a neuroinflammatory condition in brain: Damaging and protective mechanisms. *Neurosci. Biobehav. Rev.* **2008**, *32*, 1136–1151. [CrossRef]

18.   Menachem-Zidon, O.B.; Goshen, I.; Kreisel, T.; Menahem, Y.B.; Reinhartz, E.; Hur, T.B.; Yirmiya, R. Intrahippocampal transplantation of transgenic neural precursor cells overexpressing interleukin-1 receptor antagonist blocks chronic isolation-induced impairment in memory and neurogenesis. *Neuropsychopharmacology* **2008**, *33*, 2251–2262. [CrossRef]

19.   Goshen, I.; Kreisel, T.; Ben-Menachem-Zidon, O.; Licht, T.; Weidenfeld, J.; Ben-Hur, T.; Yirmiya, R. Brain interleukin-1 mediates chronic stress-induced depression in mice via adrenocortical activation and hippocampal neurogenesis suppression. *Mol. Psychiatry* **2008**, *13*, 717–728. [CrossRef] [PubMed]

20.   Dias, J.M.; Alekseenko, Z.; Applequist, J.M.; Ericson, J. Tgfbeta signaling regulates temporal neurogenesis and potency of neural stem cells in the CNS. *Neuron* **2014**, *84*, 927–939. [CrossRef] [PubMed]

21.   Kandasamy, M.; Lehner, B.; Kraus, S.; Sander, P.R.; Marschallinger, J.; Rivera, F.J.; Trumbach, D.; Ueberham, U.; Reitsamer, H.A.; Strauss, O.; et al. TGF-beta signalling in the adult neurogenic niche promotes stem cell quiescence as well as generation of new neurons. *J. Cell. Mol. Med.* **2014**, *18*, 1444–1459. [CrossRef]

22.   Meyers, E.A.; Kessler, J.A. TGF-beta Family Signaling in Neural and Neuronal Differentiation, Development, and Function. *Cold Spring Harb. Perspect. Biol.* **2017**, *9*, a022244. [CrossRef] [PubMed]

23.   Drevets, W.C. Neuroplasticity in mood disorders. *Dialogues Clin. Neurosci.* **2004**, *6*, 199–216. [PubMed]

24.   Manji, H.K.; Moore, G.J.; Rajkowska, G.; Chen, G. Neuroplasticity and cellular resilience in mood disorders. *Mol. Psychiatry* **2000**, *5*, 578–593. [CrossRef] [PubMed]

25.   Sato, S.; Sanjo, H.; Takeda, K.; Ninomiya-Tsuji, J.; Yamamoto, M.; Kawai, T.; Matsumoto, K.; Takeuchi, O.; Akira, S. Essential function for the kinase TAK1 in innate and adaptive immune responses. *Nat. Immunol.* **2005**, *6*, 1087–1095. [CrossRef]

26.   Tang, M.; Wei, X.; Guo, Y.; Breslin, P.; Zhang, S.; Zhang, S.; Wei, W.; Xia, Z.; Diaz, M.; Akira, S.; et al. TAK1 is required for the survival of hematopoietic cells and hepatocytes in mice. *J. Exp. Med.* **2008**, *205*, 1611–1619. [CrossRef] [PubMed]

27.   Yamaguchi, K.; Shirakabe, K.; Shibuya, H.; Irie, K.; Oishi, I.; Ueno, N.; Taniguchi, T.; Nishida, E.; Matsumoto, K. Identification of a member of the MAPKKK family as a potential mediator of TGF-beta signal transduction. *Science* **1995**, *270*, 2008–2011. [CrossRef]

28.   Yu, J.; Zhang, F.; Wang, S.; Zhang, Y.; Fan, M.; Xu, Z. TAK1 is activated by TGF-beta signaling and controls axonal growth during brain development. *J. Mol. Cell Biol.* **2014**, *6*, 349–351. [CrossRef] [PubMed]

29.   Zhong, J.; Li, X.; McNamee, C.; Chen, A.P.; Baccarini, M.; Snider, W.D. Raf kinase signaling functions in sensory neuron differentiation and axon growth in vivo. *Nat. Neurosci.* **2007**, *10*, 598–607. [CrossRef] [PubMed]

30.   Coyle, J.T.; Duman, R.S. Finding the intracellular signaling pathways affected by mood disorder treatments. *Neuron* **2003**, *38*, 157–160. [CrossRef]

31.   Duman, C.H.; Schlesinger, L.; Kodama, M.; Russell, D.S.; Duman, R.S. A role for MAP kinase signaling in behavioral models of depression and antidepressant treatment. *Biol. Psychiatry* **2007**, *61*, 661–670. [CrossRef] [PubMed]

32.   Einat, H.; Yuan, P.; Gould, T.D.; Li, J.; Du, J.; Zhang, L.; Manji, H.K.; Chen, G. The role of the extracellular signal-regulated kinase signaling pathway in mood modulation. *J. Neurosci.* **2003**, *23*, 7311–7316. [CrossRef] [PubMed]

33.   Wefers, B.; Hitz, C.; Holter, S.M.; Trumbach, D.; Hansen, J.; Weber, P.; Putz, B.; Deussing, J.M.; de Angelis, M.H.; Roenneberg, T.; et al. MAPK signaling determines anxiety in the juvenile mouse brain but depression-like behavior in adults. *PLoS ONE* **2012**, *7*, e35035. [CrossRef] [PubMed]

34.   Lhuillier, L.; Dryer, S.E. Developmental regulation of neuronal KCa channels by TGFbeta 1: Transcriptional and posttranscriptional effects mediated by Erk MAP kinase. *J Neurosci.* **2000**, *20*, 5616–5622. [CrossRef] [PubMed]

35. Fukushima, T.; Liu, R.Y.; Byrne, J.H. Transforming growth factor-beta2 modulates synaptic efficacy and plasticity and induces phosphorylation of CREB in hippocampal neurons. *Hippocampus* **2007**, *17*, 5–9. [CrossRef]
36. Balschun, D.; Wolfer, D.P.; Gass, P.; Mantamadiotis, T.; Welzl, H.; Schutz, G.; Frey, J.U.; Lipp, H.P. Does cAMP response element-binding protein have a pivotal role in hippocampal synaptic plasticity and hippocampus-dependent memory? *J. Neurosci.* **2003**, *23*, 6304–6314. [CrossRef] [PubMed]
37. Nibuya, M.; Nestler, E.J.; Duman, R.S. Chronic antidepressant administration increases the expression of cAMP response element binding protein (CREB) in rat hippocampus. *J. Neurosci.* **1996**, *16*, 2365–2372. [CrossRef] [PubMed]
38. Gur, T.L.; Conti, A.C.; Holden, J.; Bechtholt, A.J.; Hill, T.E.; Lucki, I.; Malberg, J.E.; Blendy, J.A. cAMP response element-binding protein deficiency allows for increased neurogenesis and a rapid onset of antidepressant response. *J. Neurosci.* **2007**, *27*, 7860–7868. [CrossRef] [PubMed]
39. Nakagawa, S.; Kim, J.E.; Lee, R.; Chen, J.; Fujioka, T.; Malberg, J.; Tsuji, S.; Duman, R.S. Localization of phosphorylated cAMP response element-binding protein in immature neurons of adult hippocampus. *J. Neurosci.* **2002**, *22*, 9868–9876. [CrossRef]
40. Nakagawa, S.; Kim, J.E.; Lee, R.; Malberg, J.E.; Chen, J.; Steffen, C.; Zhang, Y.J.; Nestler, E.J.; Duman, R.S. Regulation of neurogenesis in adult mouse hippocampus by cAMP and the cAMP response element-binding protein. *J. Neurosci.* **2002**, *22*, 3673–3682. [CrossRef]
41. Chen, A.C.; Shirayama, Y.; Shin, K.H.; Neve, R.L.; Duman, R.S. Expression of the cAMP response element binding protein (CREB) in hippocampus produces an antidepressant effect. *Biol. Psychiatry* **2001**, *49*, 753–762. [CrossRef]
42. Dennler, S.; Goumans, M.J.; ten Dijke, P. Transforming growth factor beta signal transduction. *J. Leukoc. Biol.* **2002**, *71*, 731–740. [PubMed]
43. Brown, K.A.; Pietenpol, J.A.; Moses, H.L. A tale of two proteins: Differential roles and regulation of Smad2 and Smad3 in TGF-beta signaling. *J. Cell. Biochem.* **2007**, *101*, 9–33. [CrossRef] [PubMed]
44. Bernard, D.J. Both SMAD2 and SMAD3 mediate activin-stimulated expression of the follicle-stimulating hormone beta subunit in mouse gonadotrope cells. *Mol. Endocrinol.* **2004**, *18*, 606–623. [CrossRef] [PubMed]
45. Li, Q.; Pangas, S.A.; Jorgez, C.J.; Graff, J.M.; Weinstein, M.; Matzuk, M.M. Redundant roles of SMAD2 and SMAD3 in ovarian granulosa cells in vivo. *Mol. Cell. Biol.* **2008**, *28*, 7001–7011. [CrossRef]
46. Brionne, T.C.; Tesseur, I.; Masliah, E.; Wyss-Coray, T. Loss of TGF-beta 1 leads to increased neuronal cell death and microgliosis in mouse brain. *Neuron* **2003**, *40*, 1133–1145. [CrossRef]
47. Kulkarni, A.B.; Huh, C.G.; Becker, D.; Geiser, A.; Lyght, M.; Flanders, K.C.; Roberts, A.B.; Sporn, M.B.; Ward, J.M.; Karlsson, S. Transforming growth factor beta 1 null mutation in mice causes excessive inflammatory response and early death. *Proc. Natl. Acad. Sci. USA* **1993**, *90*, 770–774. [CrossRef]
48. Miguez, D.G.; Gil-Guinon, E.; Pons, S.; Marti, E. Smad2 and Smad3 cooperate and antagonize simultaneously in vertebrate neurogenesis. *J. Cell Sci.* **2013**, *126*, 5335–5343. [CrossRef]
49. Wang, Y.; Symes, A.J. Smad3 deficiency reduces neurogenesis in adult mice. *J. Mol. Neurosci.* **2010**, *41*, 383–396. [CrossRef]
50. Yousef, H.; Conboy, M.J.; Morgenthaler, A.; Schlesinger, C.; Bugaj, L.; Paliwal, P.; Greer, C.; Conboy, I.M.; Schaffer, D. Systemic attenuation of the TGF-beta pathway by a single drug simultaneously rejuvenates hippocampal neurogenesis and myogenesis in the same old mammal. *Oncotarget* **2015**, *6*, 11959–11978. [CrossRef] [PubMed]
51. Tapia-Gonzalez, S.; Munoz, M.D.; Cuartero, M.I.; Sanchez-Capelo, A. Smad3 is required for the survival of proliferative intermediate progenitor cells in the dentate gyrus of adult mice. *Cell Commun. Signal.* **2013**, *11*, 1–15. [CrossRef] [PubMed]
52. Munoz, M.D.; Antolin-Vallespin, M.; Tapia-Gonzalez, S.; Sanchez-Capelo, A. Smad3 deficiency inhibits dentate gyrus LTP by enhancing GABAA neurotransmission. *J. Neurochem.* **2016**, *137*, 190–199. [CrossRef]
53. Schmierer, B.; Tournier, A.L.; Bates, P.A.; Hill, C.S. Mathematical modeling identifies Smad nucleocytoplasmic shuttling as a dynamic signal-interpreting system. *Proc. Natl. Acad. Sci. USA* **2008**, *105*, 6608–6613. [CrossRef]
54. Buckwalter, M.S.; Yamane, M.; Coleman, B.S.; Ormerod, B.K.; Chin, J.T.; Palmer, T.; Wyss-Coray, T. Chronically increased transforming growth factor-beta1 strongly inhibits hippocampal neurogenesis in aged mice. *Am. J. Pathol.* **2006**, *169*, 154–164. [CrossRef]
55. He, Y.; Zhang, H.; Yung, A.; Villeda, S.A.; Jaeger, P.A.; Olayiwola, O.; Fainberg, N.; Wyss-Coray, T. ALK5-dependent TGF-beta signaling is a major determinant of late-stage adult neurogenesis. *Nat. Neurosci.* **2014**, *17*, 943–952. [CrossRef]
56. Ganea, K.; Menke, A.; Schmidt, M.V.; Lucae, S.; Rammes, G.; Liebl, C.; Harbich, D.; Sterlemann, V.; Storch, C.; Uhr, M.; et al. Convergent animal and human evidence suggests the activin/inhibin pathway to be involved in antidepressant response. *Transl. Psychiatry* **2012**, *2*, e177. [CrossRef] [PubMed]
57. Shoji-Kasai, Y.; Ageta, H.; Hasegawa, Y.; Tsuchida, K.; Sugino, H.; Inokuchi, K. Activin increases the number of synaptic contacts and the length of dendritic spine necks by modulating spinal actin dynamics. *J. Cell Sci.* **2007**, *120*, 3830–3837. [CrossRef]
58. Tapia-Gonzalez, S.; Giraldez-Perez, R.M.; Cuartero, M.I.; Casarejos, M.J.; Mena, M.A.; Wang, X.F.; Sanchez-Capelo, A. Dopamine and alpha-synuclein dysfunction in Smad3 null mice. *Mol. Neurodegener.* **2011**, *6*, 1–23. [CrossRef]
59. Villapol, S.; Wang, Y.; Adams, M.; Symes, A.J. Smad3 deficiency increases cortical and hippocampal neuronal loss following traumatic brain injury. *Exp. Neurol.* **2013**, *250*, 353–365. [CrossRef] [PubMed]
60. Wang, Y.; Moges, H.; Bharucha, Y.; Symes, A. Smad3 null mice display more rapid wound closure and reduced scar formation after a stab wound to the cerebral cortex. *Exp. Neurol.* **2007**, *203*, 168–184. [CrossRef]

61. Ma, D.K.; Jang, M.H.; Guo, J.U.; Kitabatake, Y.; Chang, M.L.; Pow-Anpongkul, N.; Flavell, R.A.; Lu, B.; Ming, G.L.; Song, H. Neuronal activity-induced Gadd45b promotes epigenetic DNA demethylation and adult neurogenesis. *Science* **2009**, *323*, 1074–1077. [CrossRef] [PubMed]
62. Park, D.H.; Hong, S.J.; Salinas, R.D.; Liu, S.J.; Sun, S.W.; Sgualdino, J.; Testa, G.; Matzuk, M.M.; Iwamori, N.; Lim, D.A. Activation of neuronal gene expression by the JMJD3 demethylase is required for postnatal and adult brain neurogenesis. *Cell Rep.* **2014**, *8*, 1290–1299. [CrossRef]
63. Estaras, C.; Akizu, N.; Garcia, A.; Beltran, S.; de la Cruz, X.; Martinez-Balbas, M.A. Genome-wide analysis reveals that Smad3 and JMJD3 HDM co-activate the neural developmental program. *Development* **2012**, *139*, 2681–2691. [CrossRef] [PubMed]
64. Yu, C.Y.; Gui, W.; He, H.Y.; Wang, X.S.; Zuo, J.; Huang, L.; Zhou, N.; Wang, K.; Wang, Y. Neuronal and astroglial TGFbeta-Smad3 signaling pathways differentially regulate dendrite growth and synaptogenesis. *Neuromolecular Med.* **2014**, *16*, 457–472. [CrossRef]
65. Li, Y.; Luikart, B.W.; Birnbaum, S.; Chen, J.; Kwon, C.H.; Kernie, S.G.; Bassel-Duby, R.; Parada, L.F. TrkB Regulates Hippocampal Neurogenesis and Governs Sensitivity to Antidepressive Treatment. *Neuron* **2008**, *59*, 399–412. [CrossRef] [PubMed]
66. Jia, S.; Ren, Z.; Li, X.; Zheng, Y.; Meng, A. Smad2 and Smad3 are required for mesendoderm induction by transforming growth factor-beta/nodal signals in zebrafish. *J. Biol. Chem.* **2008**, *283*, 2418–2426. [CrossRef]
67. Akizu, N.; Estaras, C.; Guerrero, L.; Marti, E.; Martinez-Balbas, M.A. H3K27me3 regulates BMP activity in developing spinal cord. *Development* **2010**, *137*, 2915–2925. [CrossRef]
68. Iida, A.; Iwagawa, T.; Kuribayashi, H.; Satoh, S.; Mochizuki, Y.; Baba, Y.; Nakauchi, H.; Furukawa, T.; Koseki, H.; Murakami, A.; et al. Histone demethylase Jmjd3 is required for the development of subsets of retinal bipolar cells. *Proc. Natl. Acad. Sci. USA* **2014**, *111*, 3751–3756. [CrossRef]
69. Grassi, D.; Franz, H.; Vezzali, R.; Bovio, P.; Heidrich, S.; Dehghanian, F.; Lagunas, N.; Belzung, C.; Krieglstein, K.; Vogel, T. Neuronal Activity, TGFbeta-Signaling and Unpredictable Chronic Stress Modulate Transcription of Gadd45 Family Members and DNA Methylation in the Hippocampus. *Cereb. Cortex* **2017**, *27*, 4166–4181. [CrossRef]
70. Malberg, J.E.; Eisch, A.J.; Nestler, E.J.; Duman, R.S. Chronic antidepressant treatment increases neurogenesis in adult rat hippocampus. *J. Neurosci.* **2000**, *20*, 9104–9110. [CrossRef]
71. Jayatissa, M.N.; Henningsen, K.; Nikolajsen, G.; West, M.J.; Wiborg, O. A reduced number of hippocampal granule cells does not associate with an anhedonia-like phenotype in a rat chronic mild stress model of depression. *Stress* **2010**, *13*, 95–105. [CrossRef] [PubMed]
72. Lagace, D.C.; Donovan, M.H.; DeCarolis, N.A.; Farnbauch, L.A.; Malhotra, S.; Berton, O.; Nestler, E.J.; Krishnan, V.; Eisch, A.J. Adult hippocampal neurogenesis is functionally important for stress-induced social avoidance. *Proc. Natl. Acad. Sci. USA* **2010**, *107*, 4436–4441. [CrossRef]
73. Eisch, A.J.; Petrik, D. Depression and hippocampal neurogenesis: A road to remission? *Science* **2012**, *338*, 72–75. [CrossRef]
74. Lyons, D.M.; Buckmaster, P.S.; Lee, A.G.; Wu, C.; Mitra, R.; Duffey, L.M.; Buckmaster, C.L.; Her, S.; Patel, P.D.; Schatzberg, A.F. Stress coping stimulates hippocampal neurogenesis in adult monkeys. *Proc. Natl. Acad. Sci. USA* **2010**, *107*, 14823–14827. [CrossRef]
75. Parihar, V.K.; Hattiangady, B.; Kuruba, R.; Shuai, B.; Shetty, A.K. Predictable chronic mild stress improves mood, hippocampal neurogenesis and memory. *Mol. Psychiatry* **2011**, *16*, 171–183. [CrossRef]
76. Petrik, D.; Lagace, D.C.; Eisch, A.J. The neurogenesis hypothesis of affective and anxiety disorders: Are we mistaking the scaffolding for the building? *Neuropharmacology* **2012**, *62*, 21–34. [CrossRef] [PubMed]
77. Caspi, A.; Sugden, K.; Moffitt, T.E.; Taylor, A.; Craig, I.W.; Harrington, H.; McClay, J.; Mill, J.; Martin, J.; Braithwaite, A.; et al. Influence of life stress on depression: Moderation by a polymorphism in the 5-HTT gene. *Science* **2003**, *301*, 386–389. [CrossRef]
78. Heuser, I. Anna-Monika-Prize paper. The hypothalamic-pituitary-adrenal system in depression. *Pharmacopsychiatry* **1998**, *31*, 10–13. [CrossRef]
79. Mizoguchi, K.; Ishige, A.; Aburada, M.; Tabira, T. Chronic stress attenuates glucocorticoid negative feedback: Involvement of the prefrontal cortex and hippocampus. *Neuroscience* **2003**, *119*, 887–897. [CrossRef]
80. Herman, J.P.; Spencer, R. Regulation of hippocampal glucocorticoid receptor gene transcription and protein expression in vivo. *J. Neurosci.* **1998**, *18*, 7462–7473. [CrossRef]
81. Gould, E.; McEwen, B.S.; Tanapat, P.; Galea, L.A.; Fuchs, E. Neurogenesis in the dentate gyrus of the adult tree shrew is regulated by psychosocial stress and NMDA receptor activation. *J. Neurosci.* **1997**, *17*, 2492–2498. [CrossRef]
82. Malberg, J.E.; Duman, R.S. Cell proliferation in adult hippocampus is decreased by inescapable stress: Reversal by fluoxetine treatment. *Neuropsychopharmacology* **2003**, *28*, 1562–1571. [CrossRef] [PubMed]
83. Pham, K.; Nacher, J.; Hof, P.R.; McEwen, B.S. Repeated restraint stress suppresses neurogenesis and induces biphasic PSA-NCAM expression in the adult rat dentate gyrus. *Eur. J. Neurosci.* **2003**, *17*, 879–886. [CrossRef]
84. You, Z.; Luo, C.; Zhang, W.; Chen, Y.; He, J.; Zhao, Q.; Zuo, R.; Wu, Y. Pro- and anti-inflammatory cytokines expression in rat's brain and spleen exposed to chronic mild stress: Involvement in depression. *Behav. Brain Res.* **2011**, *225*, 135–141. [CrossRef]
85. Anacker, C.; Cattaneo, A.; Luoni, A.; Musaelyan, K.; Zunszain, P.A.; Milanesi, E.; Rybka, J.; Berry, A.; Cirulli, F.; Thuret, S.; et al. Glucocorticoid-related molecular signaling pathways regulating hippocampal neurogenesis. *Neuropsychopharmacology* **2013**, *38*, 872–883. [CrossRef]
86. Bremner, J.D.; Elzinga, B.; Schmahl, C.; Vermetten, E. Structural and functional plasticity of the human brain in posttraumatic stress disorder. *Prog. Brain Res.* **2008**, *167*, 171–186. [CrossRef]

87. Gould, E.; Cameron, H.A.; Daniels, D.C.; Woolley, C.S.; McEwen, B.S. Adrenal hormones suppress cell division in the adult rat dentate gyrus. *J. Neurosci.* **1992**, *12*, 3642–3650. [CrossRef] [PubMed]
88. Jacobs, B.L.; van Praag, H.; Gage, F.H. Adult brain neurogenesis and psychiatry: A novel theory of depression. *Mol. Psychiatry* **2000**, *5*, 262–269. [CrossRef] [PubMed]
89. Nakashiba, T.; Cushman, J.D.; Pelkey, K.A.; Renaudineau, S.; Buhl, D.L.; McHugh, T.J.; Rodriguez Barrera, V.; Chittajallu, R.; Iwamoto, K.S.; McBain, C.J.; et al. Young dentate granule cells mediate pattern separation, whereas old granule cells facilitate pattern completion. *Cell* **2012**, *149*, 188–201. [CrossRef]
90. Anacker, C.; Pariante, C.M. Can adult neurogenesis buffer stress responses and depressive behaviour? *Mol. Psychiatry* **2012**, *17*, 9–10. [CrossRef]
91. Hill, M.N.; Patel, S. Translational evidence for the involvement of the endocannabinoid system in stress-related psychiatric illnesses. *Biol. Mood Anxiety Disord.* **2013**, *3*, 1–14. [CrossRef] [PubMed]
92. Qin, Z.; Zhou, X.; Pandey, N.R.; Vecchiarelli, H.A.; Stewart, C.A.; Zhang, X.; Lagace, D.C.; Brunel, J.M.; Beique, J.C.; Stewart, A.F.; et al. Chronic stress induces anxiety via an amygdalar intracellular cascade that impairs endocannabinoid signaling. *Neuron* **2015**, *85*, 1319–1331. [CrossRef] [PubMed]
93. Bremner, J.D.; Narayan, M.; Anderson, E.R.; Staib, L.H.; Miller, H.L.; Charney, D.S. Hippocampal volume reduction in major depression. *Am. J. Psychiatry* **2000**, *157*, 115–118. [CrossRef] [PubMed]
94. MacQueen, G.M.; Campbell, S.; McEwen, B.S.; Macdonald, K.; Amano, S.; Joffe, R.T.; Nahmias, C.; Young, L.T. Course of illness, hippocampal function, and hippocampal volume in major depression. *Proc. Natl. Acad. Sci. USA* **2003**, *100*, 1387–1392. [CrossRef]
95. Sheline, Y.I.; Wang, P.W.; Gado, M.H.; Csernansky, J.G.; Vannier, M.W. Hippocampal atrophy in recurrent major depression. *Proc. Natl. Acad. Sci. USA* **1996**, *93*, 3908–3913. [CrossRef]
96. Wang, X.W.; Yu, Y.; Gu, L. Dehydroabietic acid reverses TNF-alpha-induced the activation of FOXO1 and suppression of TGF-beta1/Smad signaling in human adult dermal fibroblasts. *Int. J. Clin. Exp. Pathol.* **2014**, *7*, 8616–8626.
97. Cattaneo, A.; Cattane, N.; Malpighi, C.; Czamara, D.; Suarez, A.; Mariani, N.; Kajantie, E.; Luoni, A.; Eriksson, J.G.; Lahti, J.; et al. FoxO1, A2M, and TGF-beta1: Three novel genes predicting depression in gene X environment interactions are identified using cross-species and cross-tissues transcriptomic and miRNomic analyses. *Mol. Psychiatry* **2018**, *23*, 2192–2208. [CrossRef] [PubMed]
98. Etkin, A.; Prater, K.E.; Schatzberg, A.F.; Menon, V.; Greicius, M.D. Disrupted amygdalar subregion functional connectivity and evidence of a compensatory network in generalized anxiety disorder. *Arch. Gen. Psychiatry* **2009**, *66*, 1361–1372. [CrossRef]
99. Kim, M.J.; Loucks, R.A.; Palmer, A.L.; Brown, A.C.; Solomon, K.M.; Marchante, A.N.; Whalen, P.J. The structural and functional connectivity of the amygdala: From normal emotion to pathological anxiety. *Behav. Brain Res.* **2011**, *223*, 403–410. [CrossRef]
100. Patel, S.; Roelke, C.T.; Rademacher, D.J.; Cullinan, W.E.; Hillard, C.J. Endocannabinoid signaling negatively modulates stress-induced activation of the hypothalamic-pituitary-adrenal axis. *Endocrinology* **2004**, *145*, 5431–5438. [CrossRef] [PubMed]
101. Shonesy, B.C.; Bluett, R.J.; Ramikie, T.S.; Baldi, R.; Hermanson, D.J.; Kingsley, P.J.; Marnett, L.J.; Winder, D.G.; Colbran, R.J.; Patel, S. Genetic disruption of 2-arachidonoylglycerol synthesis reveals a key role for endocannabinoid signaling in anxiety modulation. *Cell Rep.* **2014**, *9*, 1644–1653. [CrossRef] [PubMed]
102. Gunduz-Cinar, O.; Hill, M.N.; McEwen, B.S.; Holmes, A. Amygdala FAAH and anandamide: Mediating protection and recovery from stress. *Trends Pharmacol. Sci.* **2013**, *34*, 637–644. [CrossRef] [PubMed]
103. Cambien, F.; Ricard, S.; Troesch, A.; Mallet, C.; Generenaz, L.; Evans, A.; Arveiler, D.; Luc, G.; Ruidavets, J.B.; Poirier, O. Polymorphisms of the transforming growth factor-beta 1 gene in relation to myocardial infarction and blood pressure. The Etude Cas-Temoin de l'Infarctus du Myocarde (ECTIM) Study. *Hypertension* **1996**, *28*, 881–887. [CrossRef] [PubMed]
104. Jin, G.; Deng, Y.; Miao, R.; Hu, Z.; Zhou, Y.; Tan, Y.; Wang, J.; Hua, Z.; Ding, W.; Wang, L.; et al. TGFB1 and TGFBR2 functional polymorphisms and risk of esophageal squamous cell carcinoma: A case-control analysis in a Chinese population. *J. Cancer Res. Clin. Oncol.* **2008**, *134*, 345–351. [CrossRef] [PubMed]
105. Kim, S.H.; Lee, S.H.; Choi, Y.L.; Wang, L.H.; Park, C.K.; Shin, Y.K. Extensive alteration in the expression profiles of TGFB pathway signaling components and TP53 is observed along the gastric dysplasia-carcinoma sequence. *Histol. Histopathol.* **2008**, *23*, 1439–1452. [CrossRef]
106. Yan, C.; Yang, Q.; Gong, Z. Transgenic expression of tgfb1a induces hepatic inflammation, fibrosis and metastasis in zebrafish. *Biochem. Biophys. Res. Commun.* **2019**, *509*, 175–181. [CrossRef] [PubMed]
107. Elliott, R.L.; Blobe, G.C. Role of transforming growth factor Beta in human cancer. *J. Clin. Oncol.* **2005**, *23*, 2078–2093. [CrossRef] [PubMed]
108. Hata, A.; Shi, Y.; Massague, J. TGF-beta signaling and cancer: Structural and functional consequences of mutations in Smads. *Mol. Med. Today* **1998**, *4*, 257–262. [CrossRef]
109. Jahn, S.C.; Law, M.E.; Corsino, P.E.; Law, B.K. TGF-beta antiproliferative effects in tumor suppression. *Front. Biosci. (Schol. Ed.)* **2012**, *4*, 749–766. [CrossRef] [PubMed]
110. Han, S.U.; Kim, H.T.; Seong, D.H.; Kim, Y.S.; Park, Y.S.; Bang, Y.J.; Yang, H.K.; Kim, S.J. Loss of the Smad3 expression increases susceptibility to tumorigenicity in human gastric cancer. *Oncogene* **2004**, *23*, 1333–1341. [CrossRef]
111. Boulanger, C.A.; Smith, G.H. Reducing mammary cancer risk through premature stem cell senescence. *Oncogene* **2001**, *20*, 2264–2272. [CrossRef]
112. Pierce, D.F., Jr.; Gorska, A.E.; Chytil, A.; Meise, K.S.; Page, D.L.; Coffey, R.J., Jr.; Moses, H.L. Mammary tumor suppression by transforming growth factor beta 1 transgene expression. *Proc. Natl. Acad. Sci. USA* **1995**, *92*, 4254–4258. [CrossRef] [PubMed]

113. Siegel, P.M.; Shu, W.; Cardiff, R.D.; Muller, W.J.; Massague, J. Transforming growth factor beta signaling impairs Neu-induced mammary tumorigenesis while promoting pulmonary metastasis. *Proc. Natl. Acad. Sci. USA* **2003**, *100*, 8430–8435. [CrossRef] [PubMed]

114. Buckwalter, M.S.; Wyss-Coray, T. Modelling neuroinflammatory phenotypes in vivo. *J. Neuroinflamm.* **2004**, *1*, 1–12. [CrossRef] [PubMed]

115. Finch, C.E.; Laping, N.J.; Morgan, T.E.; Nichols, N.R.; Pasinetti, G.M. TGF-beta 1 is an organizer of responses to neurodegeneration. *J. Cell. Biochem.* **1993**, *53*, 314–322. [CrossRef]

116. O'Brien, M.F.; Lenke, L.G.; Lou, J.; Bridwell, K.H.; Joyce, M.E. Astrocyte response and transforming growth factor-beta localization in acute spinal cord injury. *Spine* **1994**, *19*, 2321–2329. [CrossRef]

117. Buss, A.; Pech, K.; Kakulas, B.A.; Martin, D.; Schoenen, J.; Noth, J.; Brook, G.A. TGF-beta1 and TGF-beta2 expression after traumatic human spinal cord injury. *Spinal Cord* **2008**, *46*, 364–371. [CrossRef] [PubMed]

118. Nicoletti, F.; Di Marco, R.; Patti, F.; Reggio, E.; Nicoletti, A.; Zaccone, P.; Stivala, F.; Meroni, P.L.; Reggio, A. Blood levels of transforming growth factor-beta 1 (TGF-beta1) are elevated in both relapsing remitting and chronic progressive multiple sclerosis (MS) patients and are further augmented by treatment with interferon-beta 1b (IFN-beta1b). *Clin. Exp. Immunol.* **1998**, *113*, 96–99. [CrossRef]

119. Estrada, L.D.; Oliveira-Cruz, L.; Cabrera, D. Transforming Growth Factor Beta Type I Role in Neurodegeneration: Implications for Alzheimer s Disease. *Curr. Protein Pept. Sci.* **2018**, *19*, 1180–1188. [CrossRef]

120. Chao, C.C.; Hu, S.; Frey, W.H., 2nd; Ala, T.A.; Tourtellotte, W.W.; Peterson, P.K. Transforming growth factor beta in Alzheimer's disease. *Clin. Diagn. Lab. Immunol.* **1994**, *1*, 109–110. [CrossRef]

121. Vawter, M.P.; Dillon-Carter, O.; Tourtellotte, W.W.; Carvey, P.; Freed, W.J. TGFbeta1 and TGFbeta2 concentrations are elevated in Parkinson's disease in ventricular cerebrospinal fluid. *Exp. Neurol.* **1996**, *142*, 313–322. [CrossRef]

122. Saunier, E.F.; Akhurst, R.J. TGF beta inhibition for cancer therapy. *Curr. Cancer Drug Targets* **2006**, *6*, 565–578. [CrossRef]

123. Prud'homme, G.J. Pathobiology of transforming growth factor beta in cancer, fibrosis and immunologic disease, and therapeutic considerations. *Lab. Investig.* **2007**, *87*, 1077–1091. [CrossRef]

124. Ruzek, M.C.; Hawes, M.; Pratt, B.; McPherson, J.; Ledbetter, S.; Richards, S.M.; Garman, R.D. Minimal effects on immune parameters following chronic anti-TGF-beta monoclonal antibody administration to normal mice. *Immunopharmacol. Immunotoxicol.* **2003**, *25*, 235–257. [CrossRef] [PubMed]

125. Yang, Y.A.; Dukhanina, O.; Tang, B.; Mamura, M.; Letterio, J.J.; MacGregor, J.; Patel, S.C.; Khozin, S.; Liu, Z.Y.; Green, J.; et al. Lifetime exposure to a soluble TGF-beta antagonist protects mice against metastasis without adverse side effects. *J. Clin. Investig.* **2002**, *109*, 1607–1615. [CrossRef] [PubMed]

126. Forrester, E.; Chytil, A.; Bierie, B.; Aakre, M.; Gorska, A.E.; Sharif-Afshar, A.R.; Muller, W.J.; Moses, H.L. Effect of conditional knockout of the type II TGF-beta receptor gene in mammary epithelia on mammary gland development and polyomavirus middle T antigen induced tumor formation and metastasis. *Cancer Res.* **2005**, *65*, 2296–2302. [CrossRef] [PubMed]

127. Wyss-Coray, T. Tgf-Beta pathway as a potential target in neurodegeneration and Alzheimer's. *Curr. Alzheimer Res.* **2006**, *3*, 191–195. [CrossRef]

128. Sahay, A.; Drew, M.R.; Hen, R. Dentate gyrus neurogenesis and depression. *Prog. Brain Res.* **2007**, *163*, 697–722. [CrossRef]

129. Sahay, A.; Hen, R. Adult hippocampal neurogenesis in depression. *Nat. Neurosci.* **2007**, *10*, 1110–1115. [CrossRef]

130. Vollmayr, B.; Mahlstedt, M.M.; Henn, F.A. Neurogenesis and depression: What animal models tell us about the link. *Eur. Arch. Psychiatry Clin. Neurosci.* **2007**, *257*, 300–303. [CrossRef]

131. Gold, P.W.; Goodwin, F.K.; Chrousos, G.P. Clinical and biochemical manifestations of depression. Relation to the neurobiology of stress (2). *N. Engl. J. Med.* **1988**, *319*, 413–420. [CrossRef] [PubMed]

132. Sapolsky, R.M. The possibility of neurotoxicity in the hippocampus in major depression: A primer on neuron death. *Biol. Psychiatry* **2000**, *48*, 755–765. [CrossRef]

133. Davidson, R.J.; Lewis, D.A.; Alloy, L.B.; Amaral, D.G.; Bush, G.; Cohen, J.D.; Drevets, W.C.; Farah, M.J.; Kagan, J.; McClelland, J.L.; et al. Neural and behavioral substrates of mood and mood regulation. *Biol. Psychiatry* **2002**, *52*, 478–502. [CrossRef]

134. Czeh, B.; Michaelis, T.; Watanabe, T.; Frahm, J.; de Biurrun, G.; van Kampen, M.; Bartolomucci, A.; Fuchs, E. Stress-induced changes in cerebral metabolites, hippocampal volume, and cell proliferation are prevented by antidepressant treatment with tianeptine. *Proc. Natl. Acad. Sci. USA* **2001**, *98*, 12796–12801. [CrossRef]

135. Anacker, C.; Zunszain, P.A.; Cattaneo, A.; Carvalho, L.A.; Garabedian, M.J.; Thuret, S.; Price, J.; Pariante, C.M. Antidepressants increase human hippocampal neurogenesis by activating the glucocorticoid receptor. *Mol. Psychiatry* **2011**, *16*, 738–750. [CrossRef]

136. Perera, T.D.; Coplan, J.D.; Lisanby, S.H.; Lipira, C.M.; Arif, M.; Carpio, C.; Spitzer, G.; Santarelli, L.; Scharf, B.; Hen, R.; et al. Antidepressant-induced neurogenesis in the hippocampus of adult nonhuman primates. *J. Neurosci.* **2007**, *27*, 4894–4901. [CrossRef] [PubMed]

137. Santarelli, L.; Saxe, M.; Gross, C.; Surget, A.; Battaglia, F.; Dulawa, S.; Weisstaub, N.; Lee, J.; Duman, R.; Arancio, O.; et al. Requirement of hippocampal neurogenesis for the behavioral effects of antidepressants. *Science* **2003**, *301*, 805–809. [CrossRef]

138. Vollmayr, B.; Simonis, C.; Weber, S.; Gass, P.; Henn, F. Reduced cell proliferation in the dentate gyrus is not correlated with the development of learned helplessness. *Biol. Psychiatry* **2003**, *54*, 1035–1040. [CrossRef]

139. Czeh, B.; Welt, T.; Fischer, A.K.; Erhardt, A.; Schmitt, W.; Muller, M.B.; Toschi, N.; Fuchs, E.; Keck, M.E. Chronic psychosocial stress and concomitant repetitive transcranial magnetic stimulation: Effects on stress hormone levels and adult hippocampal neurogenesis. *Biol. Psychiatry* **2002**, *52*, 1057–1065. [CrossRef]
140. Lee, K.J.; Kim, S.J.; Kim, S.W.; Choi, S.H.; Shin, Y.C.; Park, S.H.; Moon, B.H.; Cho, E.; Lee, M.S.; Choi, S.H.; et al. Chronic mild stress decreases survival, but not proliferation, of new-born cells in adult rat hippocampus. *Exp. Mol. Med.* **2006**, *38*, 44–54. [CrossRef]
141. Bennett, A.J.; Lesch, K.P.; Heils, A.; Long, J.C.; Lorenz, J.G.; Shoaf, S.E.; Champoux, M.; Suomi, S.J.; Linnoila, M.V.; Higley, J.D. Early experience and serotonin transporter gene variation interact to influence primate CNS function. *Mol. Psychiatry* **2002**, *7*, 118–122. [CrossRef]
142. Coe, C.L.; Kramer, M.; Czeh, B.; Gould, E.; Reeves, A.J.; Kirschbaum, C.; Fuchs, E. Prenatal stress diminishes neurogenesis in the dentate gyrus of juvenile rhesus monkeys. *Biol. Psychiatry* **2003**, *54*, 1025–1034. [CrossRef]
143. Kaufman, J.; Yang, B.Z.; Douglas-Palumberi, H.; Grasso, D.; Lipschitz, D.; Houshyar, S.; Krystal, J.H.; Gelernter, J. Brain-derived neurotrophic factor-5-HTTLPR gene interactions and environmental modifiers of depression in children. *Biol. Psychiatry* **2006**, *59*, 673–680. [CrossRef]
144. Spalding, K.L.; Bergmann, O.; Alkass, K.; Bernard, S.; Salehpour, M.; Huttner, H.B.; Bostrom, E.; Westerlund, I.; Vial, C.; Buchholz, B.A.; et al. Dynamics of hippocampal neurogenesis in adult humans. *Cell* **2013**, *153*, 1219–1227. [CrossRef] [PubMed]
145. Boldrini, M.; Fulmore, C.A.; Tartt, A.N.; Simeon, L.R.; Pavlova, I.; Poposka, V.; Rosoklija, G.B.; Stankov, A.; Arango, V.; Dwork, A.J.; et al. Human Hippocampal Neurogenesis Persists throughout Aging. *Cell Stem Cell* **2018**, *22*, 589–599.e585. [CrossRef] [PubMed]
146. Toda, T.; Parylak, S.L.; Linker, S.B.; Gage, F.H. The role of adult hippocampal neurogenesis in brain health and disease. *Mol. Psychiatry* **2019**, *24*, 67–87. [CrossRef]
147. Snyder, J.S.; Soumier, A.; Brewer, M.; Pickel, J.; Cameron, H.A. Adult hippocampal neurogenesis buffers stress responses and depressive behaviour. *Nature* **2011**, *476*, 458–461. [CrossRef]
148. Jung, S.; Choe, S.; Woo, H.; Jeong, H.; An, H.K.; Moon, H.; Ryu, H.Y.; Yeo, B.K.; Lee, Y.W.; Choi, H.; et al. Autophagic death of neural stem cells mediates chronic stress-induced decline of adult hippocampal neurogenesis and cognitive deficits. *Autophagy* **2020**, *16*, 512–530. [CrossRef]
149. Neumeister, A.; Wood, S.; Bonne, O.; Nugent, A.C.; Luckenbaugh, D.A.; Young, T.; Bain, E.E.; Charney, D.S.; Drevets, W.C. Reduced hippocampal volume in unmedicated, remitted patients with major depression versus control subjects. *Biol. Psychiatry* **2005**, *57*, 935–937. [CrossRef]
150. Sivakumar, P.T.; Kalmady, S.V.; Venkatasubramanian, G.; Bharath, S.; Reddy, N.N.; Rao, N.P.; Kovoor, J.M.; Jain, S.; Varghese, M. Volumetric analysis of hippocampal sub-regions in late onset depression: A 3 tesla magnetic resonance imaging study. *Asian J. Psychiatry* **2015**, *13*, 38–43. [CrossRef]
151. Pinto, V.; Costa, J.C.; Morgado, P.; Mota, C.; Miranda, A.; Bravo, F.V.; Oliveira, T.G.; Cerqueira, J.J.; Sousa, N. Differential impact of chronic stress along the hippocampal dorsal-ventral axis. *Brain Struct. Funct.* **2015**, *220*, 1205–1212. [CrossRef] [PubMed]
152. Ahima, R.S.; Harlan, R.E. Charting of type II glucocorticoid receptor-like immunoreactivity in the rat central nervous system. *Neuroscience* **1990**, *39*, 579–604. [CrossRef]
153. Aggleton, J.P. A description of the amygdalo-hippocampal interconnections in the macaque monkey. *Exp. Brain Res.* **1986**, *64*, 515–526. [CrossRef] [PubMed]
154. Pikkarainen, M.; Ronkko, S.; Savander, V.; Insausti, R.; Pitkanen, A. Projections from the lateral, basal, and accessory basal nuclei of the amygdala to the hippocampal formation in rat. *J. Comp. Neurol.* **1999**, *403*, 229–260. [CrossRef]
155. Czeh, B.; Simon, M.; Schmelting, B.; Hiemke, C.; Fuchs, E. Astroglial plasticity in the hippocampus is affected by chronic psychosocial stress and concomitant fluoxetine treatment. *Neuropsychopharmacology* **2006**, *31*, 1616–1626. [CrossRef]
156. Czeh, B.; Varga, Z.K.; Henningsen, K.; Kovacs, G.L.; Miseta, A.; Wiborg, O. Chronic stress reduces the number of GABAergic interneurons in the adult rat hippocampus, dorsal-ventral and region-specific differences. *Hippocampus* **2015**, *25*, 393–405. [CrossRef]
157. Cobb, J.A.; O'Neill, K.; Milner, J.; Mahajan, G.J.; Lawrence, T.J.; May, W.L.; Miguel-Hidalgo, J.; Rajkowska, G.; Stockmeier, C.A. Density of GFAP-immunoreactive astrocytes is decreased in left hippocampi in major depressive disorder. *Neuroscience* **2016**, *316*, 209–220. [CrossRef]
158. Czeh, B.; Abumaria, N.; Rygula, R.; Fuchs, E. Quantitative changes in hippocampal microvasculature of chronically stressed rats: No effect of fluoxetine treatment. *Hippocampus* **2010**, *20*, 174–185. [CrossRef]
159. Rajkowska, G.; Stockmeier, C.A. Astrocyte pathology in major depressive disorder: Insights from human postmortem brain tissue. *Curr. Drug Targets* **2013**, *14*, 1225–1236. [CrossRef]

Review

The Influence of Gut Microbiota on Neurogenesis: Evidence and Hopes

Fiorella Sarubbo [1,2], Virve Cavallucci [3,4,*] and Giovambattista Pani [3,4,*]

1 Faculty of Science, University of the Balearic Islands UIB, 07122 Palma, Spain; fiorella.sarubbo@uib.es
2 Research Unit, Son Llàtzer University Hospital, Health Research Institute of the Balearic Islands (IdISBa), 07198 Palma, Spain
3 Fondazione Policlinico Universitario A. Gemelli IRCCS, 00168 Rome, Italy
4 Institute of General Pathology, Università Cattolica del Sacro Cuore, 00168 Rome, Italy
* Correspondence: v.cavallucci@gmail.com (V.C.); giovambattista.pani@unicatt.it (G.P.)

Abstract: Adult neurogenesis (i.e., the life-long generation of new neurons from undifferentiated neuronal precursors in the adult brain) may contribute to brain repair after damage, and participates in plasticity-related processes including memory, cognition, mood and sensory functions. Among the many intrinsic (oxidative stress, inflammation, and ageing), and extrinsic (environmental pollution, lifestyle, and diet) factors deemed to impact neurogenesis, significant attention has been recently attracted by the myriad of saprophytic microorganismal communities inhabiting the intestinal ecosystem and collectively referred to as the gut microbiota. A growing body of evidence, mainly from animal studies, reveal the influence of microbiota and its disease-associated imbalances on neural stem cell proliferative and differentiative activities in brain neurogenic niches. On the other hand, the long-claimed pro-neurogenic activity of natural dietary compounds endowed with antioxidants and anti-inflammatory properties (such as polyphenols, polyunsaturated fatty acids, or pro/prebiotics) may be mediated, at least in part, by their action on the intestinal microflora. The purpose of this review is to summarise the available information regarding the influence of the gut microbiota on neurogenesis, analyse the possible underlying mechanisms, and discuss the potential implications of this emerging knowledge for the fight against neurodegeneration and brain ageing.

Keywords: gut microbiota; gut-brain axis; adult neurogenesis; ageing; neural stem cells; neurodegeneration; nutrients; antioxidants; polyphenols

Citation: Sarubbo, F.; Cavallucci, V.; Pani, G. The Influence of Gut Microbiota on Neurogenesis: Evidence and Hopes. *Cells* **2022**, *11*, 382. https://doi.org/10.3390/cells11030382

Academic Editor: FengRu Tang

Received: 10 December 2021
Accepted: 20 January 2022
Published: 23 January 2022

Publisher's Note: MDPI stays neutral with regard to jurisdictional claims in published maps and institutional affiliations.

1. Introduction

The search for new therapeutic targets against brain ageing and associated neurodegenerative diseases represents one of the most urgent and challenging issues in current biomedicine, due to the increasing proportion of the elderly population worldwide [1]. Among the leading causes of ageing and neurodegeneration, the limited renewal capacity of brain cells [2], alterations of brain vasculature [3,4] and neuronal/glial dysfunction [5] are accompanied by an age-dependent decline in the brain damage repair systems, which include adult neurogenesis [6–10]. Neurogenesis can be defined as the generation of new neurons, glial cells and other neural lineages from neural stem cells (NSCs) and neural progenitor cells (NPCs) [11,12]. This process includes the maturation, migration and functional integration of NSCs or NPSs into the preexisting neuronal network [13,14]. When it occurs in adult life, it is known as adult neurogenesis (AN). Although NSCs are present in several brain regions, the subgranular zone of the hippocampus and the subventricular zone of the lateral ventricle are the main AN niches [15]. AN in other adult brain regions (e.g., the neocortex, striatum, amygdala and substantia nigra) is limited under normal physiological conditions, but could be induced after injury [16]. Maintenance of neurogenesis contributes to brain repair after damage and is believed to play a role in stress-responses and higher functions involving brain plasticity such as memory and cognition [6,17–20], mood [21], or perceptual (e.g., olfactory) learning [22,23]. Accordingly, an impairment in neurogenesis, as seen during ageing or in pathological conditions [24], has been associated with seizures [25,26], depression [27], and decline of learning abilities [28]. Impaired neurogenesis may occur because of a reduction in the number and/or function of NSCs and NPCs [29]. This may be due to the synergic action of several mechanisms operating in the brain in ageing or neurodegenerative conditions: inflammation [30,31], oxidative stress [32], or toxic substances like short-chain fatty acids (SCFAs), branched chain amino acids and peptidoglycans, originating from an altered intestinal microbiota [33]. Gut-resident microbial communities are in turn modulated by extrinsic factors, such as lifestyle and diet; importantly, imbalances affecting this complex ecosystem can impact the permeability of the body barriers, including the blood brain barrier (BBB) and the enteric barrier, so as to allow the passage of potentially noxious substances to brain tissue along the so-called gut-brain axis (GBA) [34,35].

To counteract the deterioration of neurogenesis, mechanisms that could be exogenously regulated, such as the composition of gut microbiota, are of particular interest. Gut microbiota is comprised of several species of microorganisms, including bacteria, yeast, and viruses [36], cohabiting in a delicate balance whose disruption (dysbiosis) can lead to aberrant neural and glial reactivity accompanied by loss of neurogenic ability [37]. Thus, a functional relationship links microbiota, GBA and neurogenesis [5,34], and alterations in this axis not only affect the neural regulation of the gastrointestinal tract, but, also contribute to several brain disturbances, such as mood (e.g., depression, anxiety) and neurodevelopmental (e.g., autism) [38,39] and cognitive disorders (e.g., Alzheimer's disease) [34,40–42]. Therefore, in establishing a bidirectional connection between enteric microbes and the brain, GBA exploits several anatomic structures, systems, and metabolic routes [34], such as the neuroendocrine (by the hypothalamic–pituitary–adrenal (HPA) axis) and neuroimmune systems, the sympathetic and parasympathetic arms of the autonomic nervous system, including the enteric nervous system, the vagus nerve [43], and the immune system. Not surprisingly, therefore, the GBA has been portrayed as a "second brain" [34].

As an additional layer of complexity, many factors can influence microbiota composition, including infection, mode of birth delivery, use of antibiotic medications, the nature of nutritional provision, environmental stressors, host genetics and ageing [44,45]. Among the potential therapeutic approaches aimed at the microbiota to target GBA and neurogenesis, diet composition appears particularly attractive for its feasibility. For instance, natural antioxidants and anti-inflammatory molecules, such as dietary polyphenols, have long been investigated as potential adjuvants to support AN [46]. In simple terms, maintaining a healthy brain across the lifespan [47] may simply require "good" intestinal bacteria and the right diet to keep them going.

The purpose of this review is to summarise the currently available information regarding the influence of the gut microbiota on AN and the potential of microbiota-centred interventions as a strategy against brain ageing and neurodegeneration. To address the topic we followed the classical methodological frameworks for a state-of-the-art review [48]. A list of keywords (neurogenesis, ageing, neural stem cells, gut-microbiota, gut-brain-axis, nutrients, polyphenols, and neurodegeneration) was initially identified. Then, different keyword combinations, each containing the term "neurogenesis", were used to interrogate the following sources: PubMed, Embase, Medline, Scopus, Web of Knowledge and Google Scholar. Articles published in English and indexed as original articles, meta-analysis reviews, narrative reviews, clinical cases and comments to the editor, with qualitative and quantitative data, were included in the analysis. Although the time range was not limited, the most recent publications were prioritized.

2. Evidence for the Connection between Intestinal Microbiota and Neurogenesis

Several clinical and experimental studies point to a functional connection between intestinal microbiota and neurogenesis through the GBA. This emerging evidence implies that microbiota composition may represent both a causative determinant and a therapeutic target in diseases where neurogenesis plays a key role [47,49–52]. Experimental data in support of the influence of microbiota on AN can be grouped in four general domains: (a) data from Germ-free (GF) animals; (b) data on substances derived from bacterial fermentation of food; (c) changes in bacteria homeostasis due to exogenous factors (e.g., antibiotics or stress); (d) consequences of dietary changes (Figure 1):

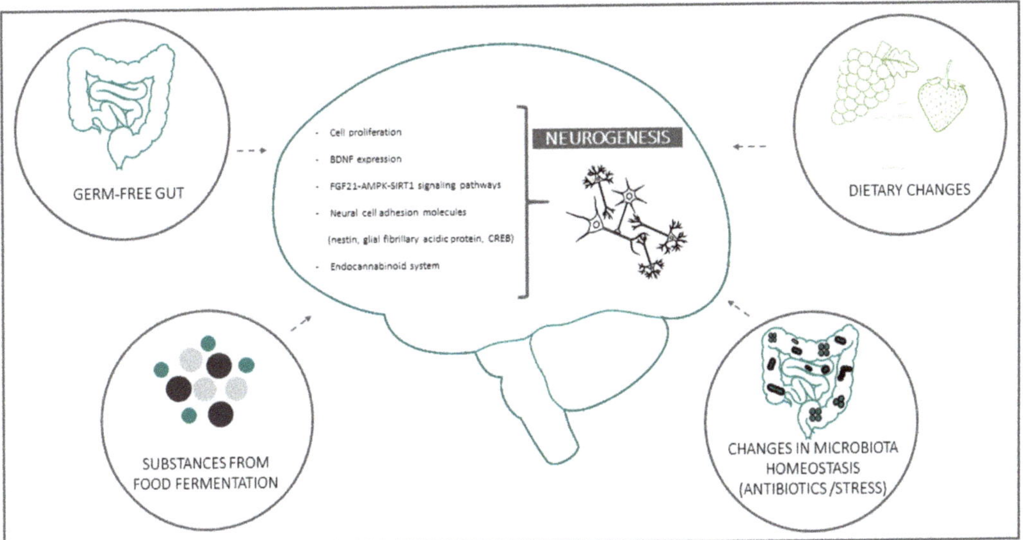

Figure 1. Schematic representation of the four main experimental models used to investigate the functional linkages between intestinal bacteria and adult (mainly hippocampal) neurogenesis. Biochemical and functional parameters employed in most studies for the evaluation of neurogenesis and its microbiota-induced modifications are listed in the central, brain-shaped field.

(a) GF gut: GF animal models, usually mice or rats, grown up without any exposure to microorganisms, constitute an essential tool in studying the influence of the gut microbiota on brain function; not surprisingly, one of the first studies that highlighted the effect of the microbiota on neurogenesis was conducted on this model.

Using bromo-2-deoxyuridine (BrdU) immunohistochemistry, it was shown that, compared to conventionally raised mice, GF and GF–colonized mice exhibited a trend to increased cell proliferation, predominantly in the dorsal hippocampus, accompanied by alterations in the hippocampal brain-derived neurotrophic factor (BDNF) [19]. In agreement, another study reported an altered expression of synaptic plasticity-related genes, with significantly lower BDNF mRNA expression in the hippocampus, amygdala, and cingulate cortex in GF mice; of note, these areas participate in neurogenesis and are key components of the neural circuitry underlying behaviour. Along similar lines, Kundu and colleagues [49], investigated the effects of transplanting the gut microbiota from young or old donor mice into young GF recipient mice [49]. They found that the transplant-induced hippocampal AN is in parallel with the activation of the pro-neurogenic FGF21-AMPK-SIRT1 signalling pathway. Moreover, it has been observed that intestinal bacteria and components of the bacterial cell wall maintain the adult enteric neuron system and nitrergic neurons by promoting intestinal neurogenesis via the Toll-like Receptor 2 (TLR2) [53].

(b) Substances produced by food fermentation: Converging lines of evidence point to the potential role of food fermentation substances produced by gut bacteria on the modulation of AN. This is the case of the SCFA butyrate, synthesized from non-absorbed carbohydrates by colonic microbiota [54]. In an animal model of ischemia it was demonstrated that the histone deacetylase inhibitor, sodium butyrate, stimulates the incorporation of BrdU in the subgranular and the subventricular zone of the hippocampus, striatum, and frontal cortex in rats subjected to permanent cerebral ischemia. This treatment also increased the number of cells expressing the polysialic acid-neural cell adhesion molecule, nestin, the glial fibrillary acidic protein, the phospho-cAMP response element-binding protein (CREB), and BDNF in various brain regions after brain ischemia [55]. Accordingly, it was also demonstrated that oral sodium butyrate impacts brain metabolism and hippocampal neurogenesis in pigs [56].

(c) Changes in bacteria homeostasis due to exogenous factors (e.g., antibiotics or stress): Prompted by the emerging notion that the intestinal ecosystem can influence the vegetative and cognitive functions of the host [57,58], several studies have focused on the impact of antibiotics on microbiota and gut-brain communication. In mice, depleting gut microbiota with antibiotics, from weaning onward, induces cognitive deficits, specifically in memory retention, and leads to a significant reduction of BDNF in the adult brain, maybe by the involvement in AN [59]. However, while consistent with the observed changes, a specific impact of microbiota depletion on neurogenesis was not directly demonstrated in this study. This aspect was instead specifically addressed by Môhle et al. [60], who reported a long-lasting impairment in neurogenesis, accompanied by behaviour deficits in antibiotic-treated mice. It is worthy of note that these alterations were partially restored by exercise (running) and probiotics administration. Mechanistically, the above treatments increased the number of Ly6C(hi) monocytes [60], a cell population involved in immune surveillance and host defense upon infections and inflammation. Moreover, elimination of Ly6Chi monocytes by antibody depletion or by using knockout mice resulted in decreased neurogenesis, whereas the adoptive transfer of Ly6Chi monocytes was able to preserve neurogenesis after antibiotic treatment [60].

Besides antibiotics, the homeostasis of intestinal microbiota can also be affected by other drugs and stress factors. Chronic stress can impact gut microbiota diversity, promoting an increase in pathogenic bacteria at the expense of beneficial ones (dysbiosis). This imbalance, in turn, affects lipid metabolism and decreases the endocannabinoid signalling system, thus reducing hippocampal AN. Of note, dysbiosis frequently accompanies ageing and may lead to chronic inflammation and a decrease in pro-neurogenic bacterial metabolites (such as SCFAs) in the senescent intestine [61].

(d) Dietary changes: High-fat or choline-deficient diets produce a specific gut microbiota signature in the small intestine and cecum, marked by increased propionate and butyrate synthesis, mitochondrial biogenesis and generation of reactive oxidative species (ROS) downstream of SCFAs. All of these variations affect NSCs fate, leading to premature differentiation and depletion of the NSC pool in the AN niches of high-fat or choline-deficient-fed mice, ultimately impairing AN [47]. On the other hand, dietary or probiotic interventions have been indicated as effective therapeutic approaches to fight stress-associated neurological disturbances operating through the GBA [17]. Importantly, a clinical study on bacterial strains known to boost neurogenesis in mice reported improved cognitive functions in adult patients with major depression; while the involvement of neurogenesis in the effects observed in human subjects can be only indirectly inferred; the consistency with results gleaned in the preclinical setting is intriguing [62,63]. Furthermore, with regard to probiotics, it was found that in a rat model of early-life stress, maternal separation caused a marked decrease in hippocampal BDNF, while the probiotic *Bifidobacterium breve 6330* increased BDNF to levels observed in control animals, suggesting that BDNF might be involved in the regulation of anxiety through microbiome-GBA [64]. Thus, diet and probiotics represent major environmental determinants of the gut flora composition [65] and, as such, constitute potential tools for the restoration and maintenance of brain homeostasis. Further information regarding extrinsic modulators of neurogenesis is found in Section 3.2.

GBA: Physiological Architecture of the Communication Way between the Intestinal Microbiota and the Brain

The communication between the gut microbiota and brain through the GBA is the result of a long-term symbiosis and co-evolution process which involves immunological, endocrine, neurological, and metabolic signalling pathways [37]. The physiological mechanisms and elements underlying this neurogenesis-impacting communication involve: (a) The parasympathetic system, mainly the vagus nerve; (b) The monoaminergic system; (c) The neuroendocrine system, mainly the HPA; (d) The immune system; (e) biochemical metabolites from microbiota metabolism (Figure 2).

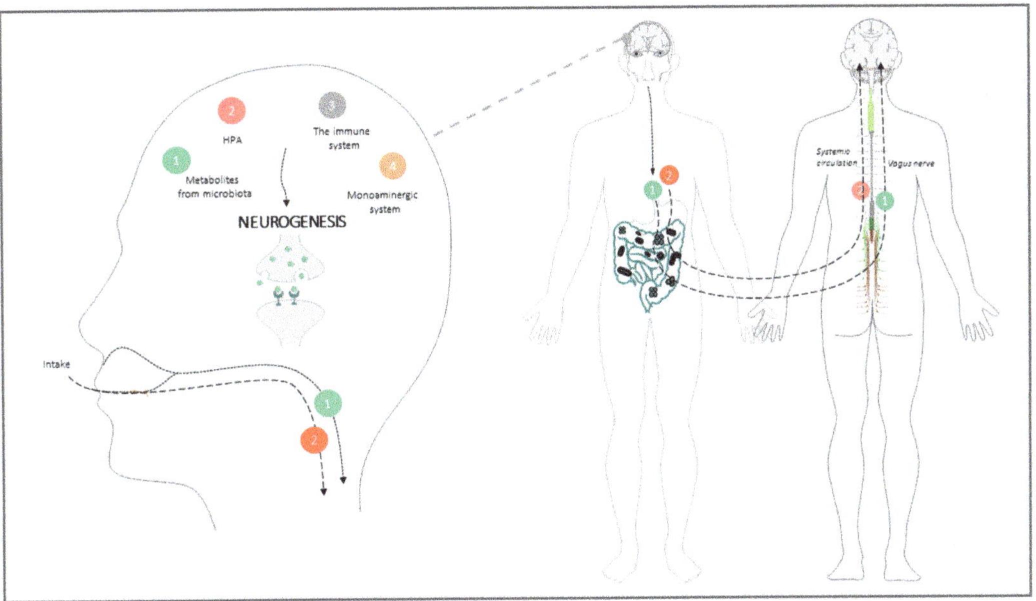

Figure 2. Anatomic-functional structure of the GBA. Dietary factors modify microbiota composition and metabolism leading to the formation of secondary compounds that signal to the brain. Microbial inputs are either relayed by the vagus nerve afferent fibers (green route) or travel through the systemic circulation and the BBB to the SNC, where they perturb the HPA (red route). Vagus nerve afferences and HPA, together with signals from the monoaminergic system and immune/inflammatory molecules modify neurogenic circuits. More details regarding the brain-gut limb of the axis are in the main text.

(a)   The parasympathetic system: Neural transmission through the vagus nerve can be activated or inactivated by microbial factors synthesized in the intestine, and represents the main communication route between gut and brain. For instance, SCFAs produced during food fermentation can evoke CNS responses by activating vagal chemoreceptors. One remarkable example of these central responses is the "inflammatory reflex", whereby efferent inputs travelling through the vagus nerve inhibit the release of pro-inflammatory cytokines in the periphery. Anti-inflammatory signals in turn preserve intestinal barrier integrity, and, by doing so, indirectly affect hippocampal plasticity and neurogenesis [66]. On the afferent side of the vagus nerve-mediated communication, microbial signals may have a role in supporting cell growth, differentiation and survival during neural development [67]. In fact, vagotomized mice displayed a decrease of BDNF mRNA in all areas of the hippocampus together with a reduced proliferation and survival of newborn cells, and a decreased number of immature neurons, especially those with complex dendritic morphology [68]. Thus, gut microbiota may modulate brain BDNF expression (and by extension, AN) through neuronal inputs relayed by the vagus nerve. Of note, altered BDNF expression is a distinctive feature of several neurodegenerative disorders, e.g., Alzheimer's disease [69].

(b)   The monoaminergic system: Neural-mediated effects of the gut microbiome on hippocampal neurogenesis may also involve the monoaminergic (especially serotoninergic) system [37]. Serotonin and its precursor tryptophan are important signalling molecules in both the CNS and the gastrointestinal tract. Serotonin exerts modulatory effects on stress, anxiety, mood, and cognition [70,71]; moreover, it participates in hippocampal homeostasis and promotes hippocampal AN [72]. Investigating the impact of intestinal microbiota on the hippocampal serotonergic system, Agus et al. 2018 reviewed this topic and explain that in several experiments, compared to conventionally raised controls, GF male mice had elevated serotonin levels and increased plasma concentrations of tryptophan [73]. The authors also found that fecal transfer from mice exposed to mild chronic stress to healthy controls impaired the pro-neurogenic effects of fluoxetine, a standard selective serotonin reuptake inhibitor previously known to promote the proliferation, differentiation, and survival of progenitor cells in the hippocampus, and to influence the plasticity of newly generated neurons [74].

(c)   The HPA axis: Evidence exists that metabolites released by intestinal microbes may enter the bloodstream, cross the BBB and directly reach the brain, where they cause hormonal interference as well as cognitive changes [75]. Perturbation of the HPA axis in turn results in intestinal dysfunction via excess release of glucocorticoids (cortisol and corticosterone), mineralocorticoids (aldosterone) and catecholamines (dopamine, epinephrine and noradrenaline) [76]. The impact on HPA and the neuroendocrine axis likely represents a major mechanism for microbiota-related brain and behavioural changes [75].

Neural stem and progenitor cells (NSPC) express the glucocorticoid and mineralcorticoid receptors, and several in vitro studies have highlighted a direct inhibitory effect of corticosteroids (especially at high doses) on the proliferative and differentiative capacity of neural precursors [77,78]. Mechanistically, dexamethasone, via the glucocorticoid receptor, induces the expression of DKK1 (Dikkopf1), an inihibitor of neurogenic signalling through the Wnt pathway, in human NPSC [79]. Additionally, high dose cortisol, possibly through the serum and glucocorticoid inducible kinase 1, inhibited Hedghog signalling in human hippocampal precursor cells, while downstream reactivation of the same pathway by the *smoothened* agonist purmorphamine cancelled the inhibitory effect of cortisol on neuronal differentiation [78]. Also of note, as in other cell types, cortisol elicits neural progenitor resistance to insulin/insulin-like growth factor (IGF) signalling and blunts the activation of the major downstream effectors ERK1 and AkT-mTOR. Accordingly, high dose IGF1 reversed the differentiation and survival defect displayed by cortisol-exposed rat embryonic NSC [80]. Given the relevance of the above pathways for (adult and embryonic) neurogenesis, and their deep interconnection (i.e., Akt and Wnt signalling converge on the inhibition of GSK3 beta), it is conceivable that GBA-controlled neurormonal stress responses impact directly on niche-derived signals and the downstream intracellular pathways that normally govern neurogenesis.

(d)   The immune system: microbes govern the induction, training and function of the host immune system. Reciprocally, the immune system maintains the symbiotic relationship of the host with the biological diversity present in the microbiome. When operating optimally, the immune system-microbiota alliance allows the induction of protective responses to pathogens and the maintenance of tolerance to innocuous antigens [81].

The microglia constitute the most abundant innate immune cell population of the CNS; microglial cells belong to the macrophage lineage, and comprise between 10% and a 15% of all glial cells. Microglia is involved in CNS homeostasis, antigen presentation, phagocytosis, and control of inflammation throughout life [82]. A recent study provided direct evidence that microbiota can influence neurogenesis by modulating the brain immune system specifically through this cell population [83]. In weaning mice administered a low dose of dextran sodium sulphate to induce acute colonic inflammation, Salvo et al. 2020 found that alterations of intestinal bacterial populations were paralleled by behavioural deficits, diminished neurogenesis in adulthood, and increased hippocampal expression of genes encoding pattern recognition receptor and T-helper 17 cell-related cytokines. Moreover, hippocampal microglia displayed an activated phenotype in these animals, as revealed by increased expression of the gene encoding the ionized calcium-binding adapter molecule 1(Iba1) [83]. Bacterial-derived SCFAs have been shown to have a key role in microglial maturation and its efficient functioning. GF mice display a reduction in both microglial maturity and number, with morphological microglia abnormalities compared to control mice. Likewise, mice treated with antibiotics show decreased microglial maturity [84]. Furthermore, the microglia of GF mice do not exhibit an activated phenotype in response to the intrusion of bacteria and viruses, which highlights the critical role of microbiota in mounting an appropriate immune response in the CNS. Additionally, the immature phenotype of microglia was also observed following antibiotic-induced microbiota depletion in specific pathogen-free mice, albeit with no significant change in the cell number [85]. Importantly, the administration of SCFAs, food fermentation products of microbial metabolism, normalized microglia functions in GF mice, indicating that the gut microbiota is essential for the normal structure and function of this unique immune cell population.

Besides microglia, astrocytes also participate in the CNS-microbiota cross-talk. Astrocytes have significant roles in ion homeostasis, neurotransmitter clearance, maintenance of the BBB, support of neuronal signalling and relevant to this article, protection against neuroinflammation [86].

Finally, as an additional line of immune/inflammatory communication between gut microbes and the brain, evidence exists that bacteria and immunoregulatory factors released from peripheral sites under the influence of the microbiome can damage the BBB and alter its physical integrity and transport selectivity, or induce the local release of neuroimmune substances from the barrier cells, ultimately leading to mental disturbances [86] in a fashion that does not directly involve resident immune cells.

NPCs express receptors for a wide range of cytokines/chemokines, and the establishment of a pro-inflammatory environment in NSC niches is clearly detrimental towards neurogenic activities [87]. Accordingly, by transcriptomic profiling of human subependymal zone (SEZ, a major adult neurogenic area) post-mortem tissue samples of a wide age range, Bitar and colleagues recently reported a marked age-dependent increase of the inflammatory signature, in parallel with a decrease of neurogenesis-related profiles [88]. Moreover, 3D mixed cultures of temporal lobe biopsies from epileptic patients revealed enrichment of inflammatory cells and elevated levels of IL-1 and HMGBI (high mobility group box 1), concomitant with impaired neurogenesis in vitro. Interestingly, the pharmacological blockade of the two cytokines significantly improved precursor proliferation and differentiation in 2D and 3D cultures, consistent with the notion that inflammatory signals impair neurogenesis [89]. While the effect of inflammation and related oxidative stress on rodent models of neurogenesis has long been established, these recent papers have the merit of convincingly extending this concept to human samples. Of note, unlike IL-1 and HMGBI, inflammatory mediators related to cell damage and microglial activation, other "neuropoietic" cytokines such as LIF, CNTF and CT-1 promote NSC self-renewal and progenitor cell division and differentiation in mouse brain, likely through the Janus kinase-signal transducer and activator of the transcription (JAK/STAT) pathway [90].

(e)   Chemical mediators from microbioma metabolism: communication between the gut microbiota and CNS also occurs through microbial-derived intermediates, the most relevant being SCFAs [33], tryptophan metabolites [73] and secondary bile acids [91]. Among SCFAs, acetate, propionate, and butyrate are the most abundantly present in the intestinal lumen. These molecules interact directly with enteroendocrine cells, mucosal immune cells, and vagal nerve terminations to propagate bottom-up signalling [92]. Moreover, since they can cross the BBB and bind to brain G protein-coupled receptors, SCFAs can act on both the peripheral and central nervous systems [29], thus exerting their immunomodulatory and anti-inflammatory influence on brain function at multiple levels [85]. More specifically, SCFAs modulate the release of neuropeptides such as serotonin and peptide YY [93], which is potentially relevant for GBA and involved in AN. In addition, SCFAs can affect the integrity of the BBB and regulate the secretion of 5-hydroxytryptamine by enterochromaffin cells, which indirectly impacts emotion and memory. Most relevant to neurogenesis, physiological levels of SCFAs were found to directly stimulate the growth and differentiation of human neural progenitor cells in vitro, a finding corroborated by the increased expression of several neurogenesis-related genes (ATR, BCL2, BID CASP8, CDK2, VEGFA, E2F1, FAS, NDN) in the same experimental setting [94]. These effects on NPCs could be mediated, at least in part, by the SCFA-stimulated endogenous synthesis of serotonin in adult NSC [95]; unfortunately, to the best of our knowledge, this intriguing possibility has yet to be tested.

3. Main Modulators of the Microbiota with Impact on Neurogenesis through the GBA

3.1. Intrinsic Modulators: Ageing, Oxidative Stress and Inflammation

A large body of literature supports the notion that neurogenesis can be influenced by several pathophysiological conditions including neuroinflammation, oxidative stress, brain injury, adverse early-life stress, and ageing. Relevant to the present article, many of these diverse and often synergistic factors may involve gut microbiota in the causal chain of events leading to impaired proliferation and differentiation of neural precursors in the brain. As an example, in mice deprived of social interactions post waning, changes in gut microbiota composition were found to be associated with learning and memory defects, reduced hippocampal levels of IL-6 and IL-10, and impaired neurogenesis [96]. Other studies have shown that gut microbial composition alterations are frequently associated with neuroinflammation, reduced hippocampal neurogenesis and behaviour disorders such as depression. For instance, Diviccaro et al., 2019 [97] observed in rats treated with finasteride, an inhibitor of the enzyme 5-α-reductase involved in steroid metabolism, a long-term inhibitory effect on neurogenesis accompanied by changes in gut microbiota and an inflammatory state, together with a depressive-like behavioural profile [97]. Furthermore, lower levels of butyrate-producing bacteria in the gut [98] and inflammation of different tissues were found in spontaneously hypertensive rats, suggesting a role for dysbiosis-related cytokine derangement in vascular dysfunction. Interestingly, primary astrocyte cultures from these animals exhibit high basal levels of specific inflammatory cytokines, and butyrate treatment reduces inflammatory markers, again pointing to microbial metabolites as regulators of neuroinflammation [99].

Thus, changes in the composition and metabolic properties of intestinal bacterial communities have consistently been associated with brain inflammation, impaired neurogenesis, and downstream behavioural defects in rodent models of disease; while microbe-triggered neuroinflammation may arguably represent a mechanistic explanation for these findings, however, the actual cause-effect relationships and their molecular underpinnings require further investigation.

In agreement with Harman's free radical/oxidative stress theory [100] of ageing, oxidative stress is alleged as the primary contributor to neurogenic and cognitive decline in the elderly [101]. The rate of oxidative damage increases in senescent tissues, in parallel with a decrease in the efficiency of the antioxidant and repair mechanisms [102–105]. In a current model, AN decline in the ageing brain is principally caused by oxidative stress and neuroinflammation, which interferes with the pro-neurogenic signalling pathways and factors implicated in self-renewal and differentiation of NSCs, like SIRT1, NF-κB, Notch, Nrf2, and Wnt/β-catenin. Along parallel lines of evidence, important changes in the gut microbiota have been described during ageing. More specifically, age-related gut dysbiosis is deemed a major contributor to the global inflammatory state of the elderly (known as inflammaging). Mechanistically, dysbiosis leads to the release of endotoxins such as lipopolysaccharides and other proinflammatory metabolic products into the systemic circulation via an increase the intestinal permeability, and hence to the CNS through a damaged BBB. In particular, dysbiosis could concur to increase the intestinal and BBB permeability in neurodegenerative disease (e.g., Parkinson's disease), enhancing the entrance of microbiota-produced substances into the CNS [106] (Figure 3). In ageing and neurodegenerative conditions, microglia exhibits a highly activated phenotype and secretes neurotoxic pro-inflammatory mediators, nitric oxide, cytokines and chemokines (e.g., interleukin IL-1β, IL-6, IL-8, tumor necrosis factor-α (TNFα), transforming growth factor-β), and ROS [107,108]. On the other hand, the "aged"-type gut microbiota is accompanied by increased levels of several cytokines (e.g., IL6, IL-10, TNF-α, TGF-β), the activation of TLR2, NF-κB and mTOR, and decreased levels of cyclin E and CDK2. Specifically, in aged humans and in centenarians, gut bacteria of the phylum *Proteobacteria* exhibit a positive correlation with IL-6 and IL-8, while *Ruminococcus lactaris* correlates with low IL-8 [109]. Thus, even in a small quantity, typical microbial alterations produced in a senescent intestine are associated with gut and brain inflammation [81,110,111].

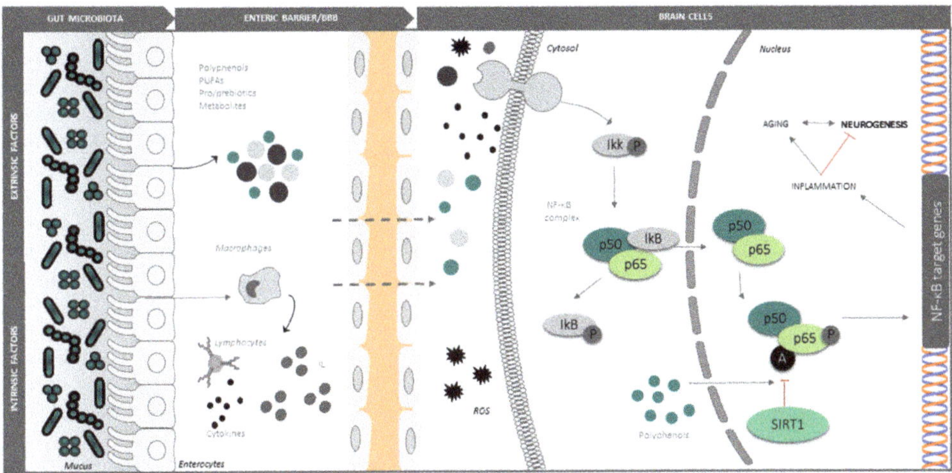

Figure 3. Central role of the SIRT1-NF-kB signalling pathway in neuroinflammation and its modulation by dietary, microbial and immune factors from the intestine. Cytokines, bacterial wall components and ROS activate NF-kB, while polyphenols inhibit proinflammatory signalling via SIRT-1 mediated inhibitory deacetylation of the factor. Both enteric and BBB act as microbe-modifiable checkpoints in the intestine-gut communication. See the text for details.

Several lines of evidence point to the NF-kB signalling pathway [112] and the Notch signalling cascade [113] as central points for the activation of microglia and by extension to neuroinflammation and its downstream pathologic consequences including impaired neurogenesis. NF-κB is expressed both in neurons and glia [114,115], and following stress signals, such as an accumulation of ROS or proinflammatory molecules, the coordinate action of protein kinases that phosphorylate the NF-kB inhibitor IkB [116], and protein deacetylases like SIRT-1 (a potent deacetylator of the lysine 310 of RelA/p65 subunit), modulate the nuclear translocation of the p50/p65 factor and its transcriptional activity at the promoter regions of proinflammatory genes [117–121]. In particular, deacetylation by SIRT1 inhibits the proinflammatory transcriptional program activated by NF-kB, making this circuit attractive for the interventional control of neuroinflammation [122–124]. As an example, in c-Rel knockout mice, the unbalanced activity of aberrantly acetylated RelA in the basal ganglia accelerates the senescence of dopaminergic neurons, triggering Parkinson's-like changes of the substantia nigra with neuroinflammation, and accumulation of alpha-synuclein and iron [125]. Thus, the manipulation of NF-κB acetylation and downstream signalling may be valuable in neurodegenerative disorders, including Alzheimer's and Parkinson's disease [126]. Considerable interest exists in developing efficient NF-κB inhibitors for neurodegenerative diseases. Strategies that block molecules upstream of the NF-κB pathway or the associated signalling adapters or those that target the IκB inhibitory proteins have been shown to exhibit significant propensity for systemic and off-target toxicities. Strategies that directly target p65/p50 dimers are likely to regain the homeostasis. Since elevated p65 is highly expressed only in pathologically activated cells, selective targeting of this NF-κB subunit may yield therapeutic drugs with better safety profiles. In recent years, chemical derivatives of natural compounds that inhibit NF-κB have been evaluated for therapeutic potential in neurodegenerative diseases. Mechanistically, the active chemical moiety of many natural compounds such as the diterpenes have been shown to form adducts with select residues of p65, compromising its DNA binding and transactivation ability [127]. Along these lines of investigation, Lim et al. isolated anti-inflammatory *Lactobacillus johnsonii CJLJ103* (LJ) from human fecal microbiota and provided it orally to mice treated with LPS. LJ administration improved LPS-induced memory impairment, inhibited NF-κB activation, enhanced CREB phosphorylation, and increased BDNF expression in the hippocampus [128]. Other studies have shown that gut microbiota alteration by extrinsic stress increases NF-kB activation and TNF-α expression, inducing memory impairment in animal models. Conversely, by restoring gut microbiota composition, an attenuation of the neuroinflammation symptoms in the hippocampus was observed [129]. NF-κB overactivation contributes to the transcriptional signature of ageing [120,130], being relevant for neurogenesis, blocking the reprogramming of aged cells into the pluripotent stem cell (iPSCs) [131,132]. On the contrary, genetic and pharmacological inhibition of the NF-κB signalling pathway prevents age-associated characteristics in different mouse models from accelerated ageing and prevents cell differentiation in favor of a pluripotent state [114,133]. Similarly, the activation of Notch signalling promotes gliogenesis and supports self-renewal of NSCs [134], thus contributing to the maintenance of the undifferentiated state and the active self-renewing growth of NSCs [135]. Notably, the Notch signalling pathway is also involved in the regulation of microglia activation in response to stress situations such as hypoxia by increasing the expression and translocation of intracellular Notch receptor domain (NICD) together with RBP-Jκ and target gene Hes-1 expression, partly through the crosstalk with NF-κB/TLR4/MyD88/TRAF6/pathways. Specifically, Notch inhibition reduced NF-κB/p65 expression and translocation [135]. Intriguingly, Notch and Wnt/β-catenin signalling pathways are essential for the maintenance of intestinal functions and homeostasis [136,137], and are modulated by both microbiota and diet [136].

Relevant to the present article, of the few gastrointestinal tract-derived microbes so far studied, all appear capable of triggering an NF-κB-miRNA-146a signalling pathway to promote neuropathological changes such as amyloidogenesis, apoptosis, inflammatory neurodegeneration, synaptic and neurotropic defects. Furthermore, microbial components secreted in the gut (e.g., neurotoxic exudates, endotoxins and exotoxins, fragilysin, select lipoglycans, lipopolysaccharide LPS, and microRNA-like small non-coding RNAs) affect the entire neural system [138,139]. Therefore the modulation of these factors and their impact on the Wnt/ β-catenin, Notch and NF-κB signalling may serve at multiple levels of the GBA as a prospective target for inhibiting microglia activation and inflammation, with the aim of improving neurogenesis in neurodegeneration and ageing [113].

3.2. Extrinsic Modulators

The microbiota-GBA-neurogenesis circuit is amenable to control by extrinsic modulators that could be exploited therapeutically. These include lifestyle, physical exercise, and above all diet modifications, for example through caloric restriction or diet component adjustments.

3.2.1. Antioxidants and Anti-Inflammatories: Polyphenols

Several studies indicate that dietary antioxidants can attenuate oxidative damage and preserve cognitive function [140] in the ageing brain by suppressing the expression of senescence-related genes [140,141]. A decrease in AN accompanies age-related cognitive decline and correlates with reduced concentrations of antioxidants in both serum [142] and brain tissue [142], thus providing a strong rationale for antioxidant supplementation [143]. Polyphenols (e.g., resveratrol) are plant-derived natural compounds endowed with antioxidant and anti-inflammatory properties and are able to cross the enteric and BBB [144–146]. The antioxidant activities of polyphenols are manifold: they (a) directly quench ROS [147], (b) inhibit the major ROS-forming enzymes, including monoamine oxidase or xanthine oxidase [148], (c) chelate metal ions (iron and copper) involved in ROS reactions [149], and (d) regulate the redox metal homeostasis and prevent the metal deposition and neurotoxicity, with important implications for neurodegenerative diseases as dementia, Alzheimer's disease and Parkinson's disease [150]. Most important, polyphenols can modulate signalling pathways and factors relevant to neurogenesis, including SIRT1, NF-κB, Nrf2 and Wnt/β-catenin [46], and are under active investigation as adjuvants in neurodegenerative disorders with impaired AN. At the opposite end of the GBA, polyphenols modify microbiota composition by favouring the expansion of some bacterial species and inhibiting others. Furthermore, gut bacteria can metabolize polyphenols into neurotransmitters and bioactive metabolites with pro-survival and anti-inflammatory effects on neurons [85]. The effect of polyphenols on intestinal microbes and their recently highlighted importance as intermediate substrates for microbial synthesis of bioactive factors adds a new dimension to the long described neuroprotective and proneurogenic action of these natural compounds, opening them to further research aimed at microbiota and GBA-centered interventions against neurodegeneration and brain ageing symptoms [93].

3.2.2. Polyunsaturated Fatty Acids (PUFAs)

PUFAs are essential unsaturated fatty acids with more than one double bond (C=C). They are important nutrients that must be obtained primarily from the diet or from supplements as mammals cannot synthesize them de novo. Long chain PUFAs can be divided into two main biologically important groups: n-6 PUFAs (omega-6) and n-3 PUFAs (omega-3), with their first double bond at C6 or C3 counting from the methyl C, respectively. Besides, Linoleic acid (n-6 PUFA) and α-linolenic acid (n-3 PUFA) are essential fatty acids, as humans can't produce them and are the precursor of other important PUFAs [151,152]. Among the nutritionally important PUFAs, α-linolenic acid, eicosapentaenoic acid and docosahexaenoic acid are highly concentrated in the brain and have antioxidant, anti-inflammatory and antiapoptotic effects [153]. Relevant to the present article, dietary PUFAs have a significant impact on the intestinal microbial ecosystem [154]. In mice engineered to overproduce n-6 PUFAs and to increase the n-6 to n-3 ratio, signs of systemic inflammation have been reported in association with evidence of intestinal dysbiosis (increased Proteobacteria while reduced Bacteroides and Actinobacteria) and abnormal gut permeability. In the same study, instead, transgenic enhancement of n-3 PUFAs tissue content inhibited LPS induced inflammation, preserved and stimulated the growth of Bifidobacterium, and preserved the intestinal barrier [155]. While the authors focused their attention on metabolic endotoxemia as the main dysbiosis-associated disorder in their transgenic animals, this work outlines a possible PUFA-microbiota-brain signalling circuitry to be further explored.

3.2.3. Probiotics/Prebiotics

Natural bioactive molecules such as probiotics and prebiotics can modify the gut microbiota composition and are therefore attracting increasing interest for the adjuvant treatment of an array of dysbiosis-associated disorders, from intestinal inflammation to autism spectrum disorders [156]. According to a criteria established by the WHO, a probiotic is defined as a live organism which provides a benefit to the host when provided in adequate quantities [157]. Numerous organisms meet the criteria, but the leading ones are *S. boulardii*, the Gram-negative *E. coli* strain Nissle 1917, various lactic-acid-producing lactobacilli strains, and a number of bifidobacteria strains (for a list see Table 1 of [158]). Instead, prebiotics are defined by the International Scientific Association of Probiotics and Prebiotics as a selectively fermented ingredient that results in specific changes in the composition and/or activity of the gastrointestinal microbiota, thus conferring benefits upon host health [159]. The majority of them are a subset of carbohydrate groups, mostly oligosaccharide carbohydrates such as fructo-oligosaccharides and galacto-oligosaccharides. The effects of prebiotics on human health are mediated through their degradation products by microorganisms; fermentation of prebiotics by gut microbiota produces SCFAs, including lactic acid, propionic acid, and butyric acid. This last, for example, influences intestinal epithelial development. Since SCFAs can diffuse to blood circulation through enterocytes, prebiotics have the ability to affect not only the gastrointestinal tract but also distant site organs such as the brain [160], for example by the vagus nerve [161]. Some prebiotics, such as fructo-oligosaccharides and galacto-oligosaccharides, have regulatory effects on BDNF, neurotransmitters (e.g., D-serine), and synaptic proteins (e.g., synaptophysin and NMDA receptor subunits) [162,163]. Other beneficial effects of probiotics involve the maintenance of immune homeostasis through the block of pathogen growth via the release of lactic acid, the inhibition of systemic immune responses, and the preservation of intestinal barrier integrity [164,165]. Relevant to the present article, the administration of probiotics has been reported to impact cognition and behaviour; for example the combination of *Lactobacillus helveticus* and *Bifidobacterium longum* decreases anxiety in rats and humans [166]. In aged rats, the administration of VSL#3, a probiotic mixture, reduces inflammation by reducing IL-10 protein expression and contributes to the increase BDNF and synapsin mRNA in the hippocampus [167]. In a study investigating the effect of *Lactobacillus* strains on cognitive decline in ageing-accelerated mice, a diet supplemented with *Lactobacillus paracasei* K71 improved cognitive performance probably through an upregulation of hippocampal BDNF expression [168]. In rats chronically fed with a chronic high-fat diet, the supplementation with *Lactobacillus paracasei* HII01was able to increase hippocampal plasticity and attenuate brain mitochondrial dysfunction and microglial activation [169]. The administration of SLAB51, a probiotic formulation of lactic acid bacteria and bifidobacteria, activates SIRT1 and promotes antioxidant and neuroprotective effects in a mouse model of Alzheimer's disease [170]. All together these results seem to indicate that probiotics may be beneficial in maintaining brain function; although a solid link with neurogenesis is still missing, the generation of SCFAs modulation of oxidative stress and inflammation and the upregulation of hippocampal BDNF all represent potential "pro-neurogenic" activities of probiotics and prebiotics to be further investigated in the appropriate experimental models [61].

3.2.4. Physical Exercise

Physical exercise and the gut microbiome have been independently described to prevent age-related global brain atrophy and both increase brain volume in the frontal lobes and left superior temporal lobe, which are important for cognition, attention and memory [171]. Data suggest that within the hippocampus, the dentate gyrus is susceptible to exercise intervention, with an increase in exercise-related neurogenesis. Specifically in humans, aerobic exercise was linked with the upregulation of serum levels of BDNF (a mediator of neurogenesis in the dentate gyrus) and with greater hippocampal volume and a subsequent decrease in psychological disorders (e.g., depression, anxiety) [172]. Moderate-intensity aerobic exercise training has also been described to improve mood and brain functional activation in older adults aged 60–79 years [173,174]. To date, specific mechanisms linking exercise with GBA and the brain have not been described; however, exercise appears to alter the gut biodiversity in both quantitative and qualitative ways [175]. Aerobic exercise has been posited to increase microbiome diversity, alter functional metabolism and modify the by-products released by intestinal bacteria [176]. More specifically, it has been demonstrated that aerobic exercise improves microbiome diversity in humans increasing genera of the *Firmicutes* phylum (specifically, *Faecalibacterium prausnitzii* and species from the genus *Oscillospira*, *Lachnospira*, and *Coprococcus*), which produce an enriched profile of SCFAs [177]. Some of the identified SFCAs (acetate, propionate, and butyrate, which are produced by gut microbes from protein, fibers, and nondigestible starches [178]), are essential to neuro-immunoendocrine regulation. For example, SCFAs influence neuroinflammation by affecting glial cell morphology and function as well as by modulating the levels of neurotrophic factors, increasing neurogenesis, contributing to the biosynthesis of serotonin, and improving neuronal homeostasis and function [179]. Additionally, exercise-induced changes in intestinal microbiota may improve the function of the gut-vascular barrier (GVD), possibly by increasing the intestinal content of the bile acid analogue and farnesoid X receptor agonist obeticholic acid. In view of the recently discovered functional linkage between GVD with the BBB [180], a positive impact of these changes on brain function and in particular on neurogenic activities in BBB-supported niches could easily be envisaged [181]. At the moment, however, the possibility that this gut-centered circuitry contributes to neurogenesis regulation by physical exercise remains speculative.

4. Future Challenges and Conclusions

Although much evidence points to a pivotal role for intestinal microbiota and bacteria-derived metabolites in the gut-brain communication axis and in particular in the modulation of AN, the consolidation of this critical knowledge and its possible translation to the clinical practice still face a number of daunting challenges. The first challenge deals with the lack of reliable approaches to reproducibly assess AN in humans, and above all in living individuals. This limitation applies to the entire AN research field, to the point that it was not until recently that the question as to whether AN is relevant to human health has begun to be settled [182,183]. Even in the most rigorous studies, adult neurogenesis is assessed by the immuno-histochemical detection of putative markers for neural precursors and immature neurons whose reliability is mainly inferred from animal models. Additionally, since no prospective labelling is possible in human samples, uncertainty exists as to whether neuroblasts/immature neurons will fully differentiate and integrate long term in the preexisting neural network [184]. Research on human adult neurogenesis and its modulation by endogenous and exogenous factors including nutrition, microbiota and lifestyle would benefit greatly from the possibility of detecting and measuring the process non-invasively. In this respect, initial enthusiasm resulting from the possibility (reported some 15 years ago) of detecting NPCs by nuclear resonance spectrometry brain imaging in vivo [185] has not been followed up by convincing evidence. The development of biochemical markers and/or imaging approaches to quantify new neurons/neural precursors and evaluate their functionality, coupled with advances in metagenomics and metabolomics techniques for the individualized characterization of intestinal microflora (for which this article is concerned) will hopefully help overcome these seemingly insurmountable obstacles.

A second challenge comes from the elucidation of the specific molecular mechanisms underlying the influence of gut microbiota on neurogenesis, even in the experimental (i.e., animal) setting. In fact, many of the currently hypothesized mechanism are chains of logical connections between phenomena (i.e., microbiota regulation by dietary factors → microbial control of neuroinflammation → modulation of neurogenesis by inflammaging) occurring at the two ends of the GBA, without any mechanistic linkage being truly demonstrated. Related to this, distinguishing direct (i.e., affecting neuronal precursors) and indirect (i.e., resulting from systemic neurohormonal or even behavioural response) effects of intestinal bacteria on neurogenesis will likely require further theoretical and methodological efforts.

Specific future lines of research could clarify how selected molecular targets involved in neurogenesis, such as NF-κB and SIRT1, are susceptible to modification by diet components or microbiome metabolites, as well as how multiple molecular pathways act in a synergic crosstalk. Overall, an integrative approach appears advisable based on the study of the topic from different perspectives. More basic research on animal and cell culture models is needed to characterize compounds potentially active on neurogenesis and gut microbiota in terms of dosing, bioassimilation and the combinatorial effect, both in physiological and pathological settings. On the other hand, more clinical research is needed (with the limitations acknowledged above), to clinically test the efficacy of promising molecules, particularly antioxidants and anti-inflammatory natural compounds of microbial origin, for their neuroprotective and possibly neurogenic effects. The ultimate goal (or hope) of this future research will be to establish the basis for the discovery of advanced therapeutics and the identification of novel biomarkers, which may help with early intervention and the prevention of brain disorders in which neurogenesis likely plays an essential role.

Author Contributions: G.P. and V.C. proposed the topic and critically reviewed/edited the manuscript. F.S. designed the article outline, retrieved and analysed the literature, and prepared the first draft. All authors have read and agreed to the published version of the manuscript.

Funding: This research received no external funding. V.C. is supported by the Italian Ministry of Health (Young Researchers Grant GR-2016-02363179).

Institutional Review Board Statement: Not applicable.

Informed Consent Statement: Not applicable.

Data Availability Statement: Not applicable.

Conflicts of Interest: The authors declare that they have no conflict of interest.

Abbreviations

AN	adult neurogenesis
BBB	blood brain barrier
BDNF	brain-derived neurotrophic factor
BrdU	bromo-2-deoxyuridine
CREB	cAMP response element-binding protein
GBA	gut-brain axis
GVD	gut-vascular barrier
CNS	central nervous system
GF	germ-free
HPA	hypothalamic–pituitary–adrenal axis
IGF	insulin-like growth factor
IL	interleukin
NPCs	neural progenitor cells
NSCs	neural stem cells
LJ	*Lactobacillus johnsonii CJLJ103*
PUFAs	Polyunsaturated fatty acids
ROS	reactive oxygen species
SEZ	subependymal zone
SCFAs	short-chain fatty acids
TNFα	tumor necrosis factor-α
TLR2	toll-like receptor

References

1. Wimo, A.; Jönsson, L.; Bond, J.; Prince, M.; Winblad, B. The worldwide economic impact of dementia 2010. *Alzheimer's Dement.* **2013**, *9*, 1–11.e3. [CrossRef] [PubMed]
2. Cotman, C.W. *The Aging Mind: Opportunities in Cognitive Research*; National Academies Press: Washington, DC, USA, 2000.
3. Bondolfi, L.; Ermini, F.; Long, J.; Ingram, D.; Jucker, M. Impact of age and caloric restriction on neurogenesis in the dentate gyrus of C57BL/6 mice. *Neurobiol. Aging* **2004**, *25*, 333–340. [CrossRef]
4. Mattson, M.P.; Arumugam, T.V. Hallmarks of Brain Aging: Adaptive and Pathological Modification by Metabolic States. *Cell Metab.* **2018**, *27*, 1176–1199. [CrossRef] [PubMed]
5. Hanslik, K.; Marino, K.; Ulland, T. Modulation of Glial Function in Health, Aging, and Neurodegenerative Disease. *Front Cell Neurosci.* **2021**, *15*, 718324. [CrossRef] [PubMed]
6. Apple, D.M.; Solano-Fonseca, R.; Kokovay, E. Neurogenesis in the aging brain. *Biochem. Pharmacol.* **2017**, *141*, 77–85. [CrossRef]
7. Klempin, F.; Kempermann, G. Adult hippocampal neurogenesis and aging. *Eur. Arch. Psychiatry Clin. Neurosci.* **2007**, *257*, 271–280. [CrossRef]
8. Riddle, D.; Lichtenwalner, R. *Brain Aging: Models, Methods, and Mechanisms*; Riddle, D.R., Ed.; CRC Press/Taylor & Francis: Boca Raton, FL, USA, 2007.
9. Isaev, N.K.; Stelmashook, E.V.; Genrikhs, E.E. Neurogenesis and brain aging. *Rev. Neurosci.* **2019**, *30*, 573–580. [CrossRef]

10. Mathews, K.J.; Allen, K.M.; Boerrigter, D.; Ball, H.; Shannon Weickert, C.; Double, K.L. Evidence for reduced neurogenesis in the aging human hippocampus despite stable stem cell markers. *Aging Cell* **2017**, *16*, 1195–1199. [CrossRef]
11. Gage, F.H.F. Mammalian neural stem cells. *Science* **2000**, *287*, 1433–1438. [CrossRef]
12. Zhao, C.; Deng, W.; Gage, F. Mechanisms and functional implications of adult neurogenesis. *Cell* **2008**, *132*, 645–660. [CrossRef]
13. Alvarez-Buylla, A.; Lim, D.A. For the long run: Maintaining germinal niches in the adult brain. *Neuron* **2004**, *41*, 683–686. [CrossRef]
14. Riquelme, P.A.; Drapeau, E.; Doetsch, F. Brain micro-ecologies: Neural stem cell niches in the adult mammalian brain. *Philos. Trans. R. Soc. Lond. B Biol. Sci.* **2008**, *363*, 123–137. [CrossRef] [PubMed]
15. Cavallucci, V.; Fidaleo, M.; Pani, G. Neural Stem Cells and Nutrients: Poised Between Quiescence and Exhaustion. *Trends. Endocrinol. Metab.* **2016**, *27*, 756–769. [CrossRef] [PubMed]
16. Mirescu, C.; Gould, E. Stress and adult neurogenesis. *Hippocampus* **2006**, *16*, 233–238. [CrossRef] [PubMed]
17. Chevalier, G.; Siopi, E.; Guenin-Macé, L.; Pascal, M.; Laval, T.; Rifflet, A.; Boneca, I.G.; Demangel, C.; Colsch, B.; Pruvost, A.; et al. Effect of gut microbiota on depressive-like behaviors in mice is mediated by the endocannabinoid system. *Nat. Commun.* **2020**, *11*, 6363. [CrossRef] [PubMed]
18. Colangelo, A.; Cirillo, G.; Alberghina, L.; Papa, M.; Westerhoff, H. Neural plasticity and adult neurogenesis: The deep biology perspective. *Neural Regen. Res.* **2019**, *14*, 201–205. [CrossRef]
19. Ogbonnaya, E.S.; Clarke, G.; Shanahan, F.; Dinan, T.G.; Cryan, J.F.; O'Leary, O.F. Adult Hippocampal Neurogenesis Is Regulated by the Microbiome. *Biol. Psychiatry* **2015**, *78*, e7–e9. [CrossRef]
20. Deng, W.; Aimone, J.B.; Gage, F.H. New neurons and new memories: How does adult hippocampal neurogenesis affect learning and memory? *Nat. Rev. Neurosci.* **2010**, *11*, 339–350. [CrossRef]
21. Anacker, C.; Hen, R. Adult hippocampal neurogenesis and cognitive flexibility—Linking memory and mood. *Nat. Rev. Neurosci.* **2017**, *18*, 335–346. [CrossRef]
22. Lazarini, F.; Lledo, P.M. Is adult neurogenesis essential for olfaction? *Trends Neurosci.* **2011**, *34*, 20–30. [CrossRef]
23. Braun, S.M.G.; Jessberger, S. Adult neurogenesis: Mechanisms and functional significance. *Development* **2014**, *141*, 1983–1986. [CrossRef] [PubMed]
24. Mattson, M.P. Neuroprotective signaling and the aging brain: Take away my food and let me run. *Brain Res.* **2000**, *886*, 47–53. [CrossRef]
25. Jessberger, S.; Nakashima, K.; Clemenson, G.D.; Mejia, E.; Mathews, E.; Ure, K.; Ogawa, S.; Sinton, C.M.; Gage, F.H.; Hsieh, J. Epigenetic Modulation of Seizure-Induced Neurogenesis and Cognitive Decline. *J. Neurosci.* **2007**, *27*, 5967–5975. [CrossRef] [PubMed]
26. Kron, M.M.; Zhang, H.; Parent, J.M. The developmental stage of dentate granule cells dictates their contribution to seizure-induced plasticity. *J. Neurosci.* **2010**, *30*, 2051–2059. [CrossRef]
27. Dokter, M.; von Bohlen und Halbach, O. Neurogenesis within the adult hippocampus under physiological conditions and in depression. *Neural Regen. Res.* **2012**, *7*, 552–559.
28. Leuner, B.; Gould, E.; Shors, T.J. Is there a link between adult neurogenesis and learning? *Hippocampus* **2006**, *16*, 216–224. [CrossRef]
29. Walgrave, H.; Balusu, S.; Snoeck, S.; Vanden Eynden, E.; Craessaerts, K.; Thrupp, N.; Wolfs, L.; Horré, K.; Fourne, Y.; Ronisz, A.; et al. Restoring miR-132 expression rescues adult hippocampal neurogenesis and memory deficits in Alzheimer's disease. *Cell Stem Cell* **2021**, *28*, 1805–1821. [CrossRef]
30. Franceschi, C.; Bonafè, M.; Valensin, S.; Olivieri, F.; De Luca, M.; Ottaviani, E.; De Benedictis, G. Inflamm-aging. An evolutionary perspective on immunosenescence. *Ann. N. Y. Acad. Sci.* **2000**, *908*, 244–254. [CrossRef]
31. Ekdahl, C.T.; Claasen, J.H.; Bonde, S.; Kokaia, Z.; Lindvall, O. Inflammation is detrimental for neurogenesis in adult brain. *Proc. Natl. Acad. Sci. USA* **2003**, *100*, 13632–13637. [CrossRef]
32. Taupin, P. A dual activity of ROS and oxidative stress on adult neurogenesis and Alzheimer's disease. *Cent. Nerv. Syst. Agents Med. Chem.* **2010**, *10*, 16–21. [CrossRef]
33. van de Wouw, M.; Boehme, M.; Lyte, J.M.; Wiley, N.; Strain, C.; O'Sullivan, O.; Clarke, G.; Stanton, C.; Dinan, T.G.; Cryan, J.F. Short-chain fatty acids: Microbial metabolites that alleviate stress-induced brain-gut axis alterations. *J. Physiol.* **2018**, *596*, 4923–4944. [CrossRef] [PubMed]
34. Carabotti, M.; Scirocco, A.; Maselli, M.A.; Severi, C. The gut-brain axis: Interactions between enteric microbiota, central and enteric nervous systems. *Ann. Gastroenterol.* **2015**, *28*, 203–209. [PubMed]
35. Saffrey, M.J. Cellular changes in the enteric nervous system during ageing. *Dev. Biol.* **2013**, *382*, 344–355. [CrossRef] [PubMed]
36. Rooks, M.G.; Garrett, W.S. Gut microbiota, metabolites and host immunity. *Nat. Rev. Immunol.* **2016**, *16*, 341–352. [CrossRef]
37. Liu, C.; Yang, S.Y.; Wang, L.; Zhou, F. The gut microbiome: Implications for neurogenesis and neurological diseases. *Neural Regen. Res.* **2022**, *17*, 53–58.
38. Finegold, S.M.; Downes, J.; Summanen, P.H. Microbiology of regressive autism. *Anaerobe* **2012**, *18*, 260–262. [CrossRef]
39. Patusco, R.; Ziegler, J. Role of probiotics in managing gastrointestinal dysfunction in children with autism spectrum disorder: AN update for practitioners. *Adv. Nutr.* **2018**, *9*, 637–650. [CrossRef]

40. Rinninella, E.; Raoul, P.; Cintoni, M.; Franceschi, F.; Miggiano, G.A.D.; Gasbarrini, A.; Mele, M.C. What is the healthy gut microbiota composition? A changing ecosystem across age, environment, diet, and diseases. *Microorganisms* **2019**, *7*, 14. [CrossRef]

41. Köhler, C.; Maes, M.; Slyepchenko, A.; Berk, M.; Solmi, M.; Lanctôt, K.; Carvalho, A. The Gut-Brain Axis, Including the Microbiome, Leaky Gut and Bacterial Translocation: Mechanisms and Pathophysiological Role in Alzheimer's Disease. *Curr. Pharm. Des.* **2016**, *22*, 6152–6166. [CrossRef]

42. Liu, P.; Wu, L.; Peng, G.; Han, Y.; Tang, R.; Ge, J.; Zhang, L.; Jia, L.; Yue, S.; Zhou, K.; et al. Altered microbiomes distinguish Alzheimer's disease from amnestic mild cognitive impairment and health in a Chinese cohort. *Brain Behav. Immun.* **2019**, *80*, 633–643. [CrossRef]

43. Rao, M.; Gershon, M.D. The bowel and beyond: The enteric nervous system in neurological disorders. *Nat. Rev. Gastroenterol. Hepatol.* **2016**, *13*, 517–528. [CrossRef] [PubMed]

44. Claesson, M.J.; Jeffery, I.B.; Conde, S.; Power, S.E.; O'connor, E.M.; Cusack, S.; Harris, H.M.B.; Coakley, M.; Lakshminarayanan, B.; O'sullivan, O.; et al. Gut microbiota composition correlates with diet and health in the elderly. *Nature* **2012**, *488*, 178–184. [CrossRef] [PubMed]

45. Santoro, A.; Zhao, J.; Wu, L.; Carru, C.; Biagi, E.; Franceschi, C. Microbiomes other than the gut: Inflammaging and age-related diseases. *Semin. Immunopathol.* **2020**, *42*, 589–605. [CrossRef] [PubMed]

46. Sarubbo, F.; Moranta, D.; Pani, G. Dietary polyphenols and neurogenesis: Molecular interactions and implication for brain ageing and cognition. *Neurosci. Biobehav. Rev.* **2018**, *90*, 456–470. [CrossRef]

47. Ribeiro, M.F.; Santos, A.A.; Afonso, M.B.; Rodrigues, P.M.; Sá Santos, S.; Castro, R.E.; Rodrigues, C.M.P.; Solá, S. Diet-dependent gut microbiota impacts on adult neurogenesis through mitochondrial stress modulation. *Brain Commun.* **2020**, *2*, fcaa165. [CrossRef]

48. Grant, M.; Booth, A. A typology of reviews: An analysis of 14 review types and associated methodologies. *Health Inf. Libr. J.* **2009**, *26*, 91–108. [CrossRef]

49. Kundu, P.; Lee, H.U.; Garcia-Perez, I.; Tay, E.X.Y.; Kim, H.; Faylon, L.E.; Martin, K.A.; Purbojati, R.; Drautz-Moses, D.I.; Ghosh, S.; et al. Neurogenesis and prolongevity signaling in young germ-free mice transplanted with the gut microbiota of old mice. *Sci. Transl. Med.* **2019**, *11*, eaau4760. [CrossRef]

50. Pearson-Leary, J.; Zhao, C.; Bittinger, K.; Eacret, D.; Luz, S.; Vigderman, A.S.; Dayanim, G.; Bhatnagar, S. The gut microbiome regulates the increases in depressive-type behaviors and in inflammatory processes in the ventral hippocampus of stress vulnerable rats. *Mol. Psychiatry* **2020**, *25*, 1068–1079. [CrossRef] [PubMed]

51. Guida, F.; Turco, F.; Iannotta, M.; De Gregorio, D.; Palumbo, I.; Sarnelli, G.; Furiano, A.; Napolitano, F.; Boccella, S.; Luongo, L.; et al. Antibiotic-induced microbiota perturbation causes gut endocannabinoidome changes, hippocampal neuroglial reorganization and depression in mice. *Brain Behav. Immun.* **2018**, *67*, 230–245. [CrossRef]

52. Cerdó, T.; Diéguez, E.; Campoy, C. Impact of gut microbiota on neurogenesis and neurological diseases during infancy. *Curr. Opin. Pharmacol.* **2020**, *50*, 33–37. [CrossRef]

53. Yarandi, S.S.; Kulkarni, S.; Saha, M.; Sylvia, K.E.; Sears, C.L.; Pasricha, P.J. Intestinal Bacteria Maintain Adult Enteric Nervous System and Nitrergic Neurons via Toll-like Receptor 2-induced Neurogenesis in Mice. *Gastroenterology* **2020**, *159*, 200–213. [CrossRef] [PubMed]

54. Canani, R.B.; Di Costanzo, M.; Leone, L.; Pedata, M.; Meli, R.; Calignano, A. Potential beneficial effects of butyrate in intestinal and extraintestinal diseases. *World J. Gastroenterol.* **2011**, *17*, 1519–1528. [CrossRef] [PubMed]

55. Kim, H.J.; Leeds, P.; Chuang, D.M. The HDAC inhibitor, sodium butyrate, stimulates neurogenesis in the ischemic brain. *J. Neurochem.* **2009**, *110*, 1226–1240. [CrossRef] [PubMed]

56. Val-Laillet, D.; Guérin, S.; Coquery, N.; Nogret, I.; Formal, M.; Romé, V.; Le Normand, L.; Meurice, P.; Randuineau, G.; Guilloteau, P.; et al. Oral sodium butyrate impacts brain metabolism and hippocampal neurogenesis, with limited effects on gut anatomy and function in pigs. *FASEB J.* **2018**, *32*, 2160–2171. [CrossRef]

57. Heijtz, R.D.; Wang, S.; Anuar, F.; Qian, Y.; Björkholm, B.; Samuelsson, A.; Hibberd, M.L.; Forssberg, H.; Pettersson, S. Normal gut microbiota modulates brain development and behavior. *Proc. Natl. Acad. Sci. USA* **2011**, *108*, 3047–3052. [CrossRef]

58. Mayer, E.A. Gut feelings: The emerging biology of gut-"brain communication. *Nat. Rev. Neurosci.* **2011**, *12*, 453–466. [CrossRef]

59. Desbonnet, L.; Clarke, G.; Traplin, A.; O'Sullivan, O.; Crispie, F.; Moloney, R.D.; Cotter, P.D.; Dinan, T.G.; Cryan, J.F. Gut microbiota depletion from early adolescence in mice: Implications for brain and behaviour. *Brain Behav. Immun.* **2015**, *48*, 165–173. [CrossRef]

60. Möhle, L.; Mattei, D.; Heimesaat, M.M.; Bereswill, S.; Fischer, A.; Alutis, M.; French, T.; Hambardzumyan, D.; Matzinger, P.; Dunay, I.R.; et al. Ly6Chi Monocytes Provide a Link between Antibiotic-Induced Changes in Gut Microbiota and Adult Hippocampal Neurogenesis. *Cell Rep.* **2016**, *15*, 1945–1956. [CrossRef]

61. Romo-Araiza, A.; Ibarra, A. Prebiotics and probiotics as potential therapy for cognitive impairment. *Med. Hypotheses* **2020**, *134*, 109410. [CrossRef]

62. Rudzki, L.; Ostrowska, L.; Pawlak, D.; Małus, A.; Pawlak, K.; Waszkiewicz, N.; Szulc, A. Probiotic *Lactobacillus Plantarum* 299v decreases kynurenine concentration and improves cognitive functions in patients with major depression: A double-blind, randomized, placebo controlled study. *Psychoneuroendocrinology* **2019**, *100*, 213–222. [CrossRef]

63. Ishikawa, R.; Fukushima, H.; Nakakita, Y.; Kado, H.; Kida, S. Dietary heat-killed *Lactobacillus brevis* SBC8803 (SBL88™) improves hippocampus-dependent memory performance and adult hippocampal neurogenesis. *Neuropsychopharmacol. Rep.* **2019**, *39*, 140–145. [CrossRef] [PubMed]

64. O'Sullivan, E.; Barrett, E.; Grenham, S.; Fitzgerald, P.; Stanton, C.; Ross, R.P.; Quigley, E.M.M.; Cryan, J.F.; Dinan, T.G. BDNF expression in the hippocampus of maternally separated rats: Does *Bifidobacterium breve* 6330 alter BDNF levels? *Benef. Microbes* **2011**, *2*, 199–207. [CrossRef] [PubMed]

65. Rothschild, D.; Weissbrod, O.; Barkan, E.; Kurilshikov, A.; Korem, T.; Zeevi, D.; Costea, P.I.; Godneva, A.; Kalka, I.N.; Bar, N.; et al. Environment dominates over host genetics in shaping human gut microbiota. *Nature* **2018**, *555*, 210–215. [CrossRef] [PubMed]

66. Chesnokova, V.; Pechnick, R.N.; Wawrowsky, K. Chronic peripheral inflammation, hippocampal neurogenesis, and behavior. *Brain Behav. Immun.* **2016**, *58*, 1–8. [CrossRef]

67. Ichim, G.; Tauszig-Delamasure, S.; Mehlen, P. Neurotrophins and cell death. *Exp. Cell Res.* **2012**, *318*, 1221–1228. [CrossRef]

68. O'Leary, O.F.; Ogbonnaya, E.S.; Felice, D.; Levone, B.R.; Conroy, L.C.; Fitzgerald, P.; Bravo, J.A.; Forsythe, P.; Bienenstock, J.; Dinan, T.G.; et al. The vagus nerve modulates BDNF expression and neurogenesis in the hippocampus. *Eur. Neuropsychopharmacol.* **2018**, *28*, 307–316. [CrossRef]

69. Nagahara, A.H.; Tuszynski, M.H. Potential therapeutic uses of BDNF in neurological and psychiatric disorders. *Nat. Rev. Drug Discov.* **2011**, *10*, 209–219. [CrossRef]

70. Szapacs, M.E.; Mathews, T.A.; Tessarollo, L.; Ernest Lyons, W.; Mamounas, L.A.; Andrews, A.M. Exploring the relationship between serotonin and brain-derived neurotrophic factor: Analysis of BDNF protein and extraneuronal 5-HT in mice with reduced serotonin transporter or BDNF expression. *J. Neurosci. Methods* **2004**, *140*, 81–92. [CrossRef]

71. Asan, E.; Steinke, M.; Lesch, K.P. Serotonergic innervation of the amygdala: Targets, receptors, and implications for stress and anxiety. *Histochem. Cell Biol.* **2013**, *139*, 785–813. [CrossRef]

72. Alenina, N.; Klempin, F. The role of serotonin in adult hippocampal neurogenesis. *Behav. Brain Res.* **2015**, *277*, 49–57. [CrossRef]

73. Agus, A.; Planchais, J.; Sokol, H. Gut Microbiota Regulation of Tryptophan Metabolism in Health and Disease. *Cell Host Microbe* **2018**, *23*, 716–724. [CrossRef] [PubMed]

74. Siopi, E.; Chevalier, G.; Katsimpardi, L.; Saha, S.; Bigot, M.; Moigneu, C.; Eberl, G.; Lledo, P.M. Changes in Gut Microbiota by Chronic Stress Impair the Efficacy of Fluoxetine. *Cell Rep.* **2020**, *30*, 3682–3690. [CrossRef] [PubMed]

75. Huo, R.; Zeng, B.; Zeng, L.; Cheng, K.; Li, B.; Luo, Y.; Wang, H.; Zhou, C.; Fang, L.; Li, W.; et al. Microbiota modulate anxiety-like behavior and endocrine abnormalities in hypothalamic-pituitary-adrenal axis. *Front. Cell. Infect. Microbiol.* **2017**, *7*, 489. [CrossRef] [PubMed]

76. Rein, T.; Ambrée, O.; Fries, G.R.; Rappeneau, V.; Schmidt, U.; Touma, C. The hypothalamic-pituitary-adrenal axis in depression: Molecular regulation, pathophysiological role, and translational implications. *Neurobiol. Depress. Road Nov. Ther.* **2019**, 89–96. [CrossRef]

77. Odaka, H.; Adachi, N.; Numakawa, T. Impact of glucocorticoid on neurogenesis. *Neural Regen. Res.* **2017**, *12*, 1028–1035.

78. Anacker, C.; Cattaneo, A.; Luoni, A.; Musaelyan, K.; Zunszain, P.A.; Milanesi, E.; Rybka, J.; Berry, A.; Cirulli, F.; Thuret, S.; et al. Glucocorticoid-related molecular signaling pathways regulating hippocampal neurogenesis. *Neuropsychopharmacology* **2013**, *38*, 872–883. [CrossRef]

79. Moors, M.; Bose, R.; Johansson-Haque, K.; Edoff, K.; Okret, S.; Ceccatelli, S. Dickkopf 1 mediates glucocorticoid-induced changes in human neural progenitor cell proliferation and differentiation. *Toxicol. Sci.* **2012**, *125*, 488–495. [CrossRef]

80. Odaka, H.; Numakawa, T.; Yoshimura, A.; Nakajima, S.; Adachi, N.; Ooshima, Y.; Inoue, T.; Kunugi, H. Chronic glucocorticoid exposure suppressed the differentiation and survival of embryonic neural stem/progenitor cells: Possible involvement of ERK and PI3K/Akt signaling in the neuronal differentiation. *Neurosci. Res.* **2016**, *113*, 28–36. [CrossRef]

81. Belkaid, Y.; Hand, T.W. Role of the microbiota in immunity and inflammation. *Cell* **2014**, *157*, 121–141. [CrossRef]

82. Lenz, K.M.; Nelson, L.H. Microglia and beyond: Innate immune cells as regulators of brain development and behavioral function. *Front. Immunol.* **2018**, *9*, 698. [CrossRef]

83. Salvo, E.; Stokes, P.; Keogh, C.E.; Brust-Mascher, I.; Hennessey, C.; Knotts, T.A.; Sladek, J.A.; Rude, K.M.; Swedek, M.; Rabasa, G.; et al. A murine model of pediatric inflammatory bowel disease causes microbiota-gut-brain axis deficits in adulthood. *Am. J. Physiol.—Gastrointest. Liver Physiol.* **2020**, *319*, G361–G374. [CrossRef]

84. Erny, D.; De Angelis, A.L.H.; Jaitin, D.; Wieghofer, P.; Staszewski, O.; David, E.; Keren-Shaul, H.; Mahlakoiv, T.; Jakobshagen, K.; Buch, T.; et al. Host microbiota constantly control maturation and function of microglia in the CNS. *Nat. Neurosci.* **2015**, *18*, 965–977. [CrossRef] [PubMed]

85. Sivaprakasam, S.; Prasad, P.D.; Singh, N. Benefits of short-chain fatty acids and their receptors in inflammation and carcinogenesis. *Pharmacol. Ther.* **2016**, *164*, 144–151. [CrossRef] [PubMed]

86. Khakh, B.S.; Sofroniew, M.V. Diversity of astrocyte functions and phenotypes in neural circuits. *Nat. Neurosci.* **2015**, *18*, 942–952. [CrossRef] [PubMed]

87. Gonzalez-Perez, O.; Gutierrez-Fernandez, F.; Lopez-Virgen, V.; Collas-Aguilar, J.; Quinones-Hinojosa, A.; Garcia-Verdugo, J.M. Immunological regulation of neurogenic niches in the adult brain. *Neuroscience* **2012**, *226*, 270–281. [CrossRef] [PubMed]

88. Bitar, M.; Weissleder, C.; North, H.F.; Clearwater, M.S.; Zalucki, O.; Halliday, G.M.; Webster, M.J.; Piper, M.; Weickert, C.S.; Barry, G. Identifying gene expression profiles associated with neurogenesis and inflammation in the human subependymal zone from development through aging. *Sci. Rep.* **2022**, *12*, 40. [CrossRef]

89. Zaben, M.; Haan, N.; Sharouf, F.; Ahmed, A.; Sundstrom, L.E.; Gray, W.P. IL-1β and HMGB1 are anti-neurogenic to endogenous neural stem cells in the sclerotic epileptic human hippocampus. *J. Neuroinflamm.* **2021**, *18*, 218. [CrossRef]

90. Bauer, S. Cytokine Control of Adult Neural Stem Cells. *Ann. N. Y. Acad. Sci.* **2009**, *1153*, 48–56. [CrossRef]

91. MahmoudianDehkordi, S.; Arnold, M.; Nho, K.; Ahmad, S.; Jia, W.; Xie, G.; Louie, G.; Kueider-Paisley, A.; Moseley, M.A.; Thompson, J.W.; et al. Altered bile acid profile associates with cognitive impairment in Alzheimer's disease—An emerging role for gut microbiome. *Alzheimer's Dement.* **2019**, *15*, 76–92. [CrossRef]

92. Baj, A.; Moro, E.; Bistoletti, M.; Orlandi, V.; Crema, F.; Giaroni, C. Glutamatergic signaling along the microbiota-gut-brain axis. *Int. J. Mol. Sci.* **2019**, *20*, 1482. [CrossRef]

93. Filosa, S.; Di Meo, F.; Crispi, S. Polyphenols-gut microbiota interplay and brain neuromodulation. *Neural Regen. Res.* **2018**, *13*, 2055–2059. [PubMed]

94. Yang, L.L.; Millischer, V.; Rodin, S.; MacFabe, D.F.; Villaescusa, J.C.; Lavebratt, C. Enteric short-chain fatty acids promote proliferation of human neural progenitor cells. *J. Neurochem.* **2020**, *154*, 635–646. [CrossRef]

95. Benninghoff, J.; Gritti, A.; Rizzi, M.; Lamorte, G.; Schloesser, R.J.; Schmitt, A.; Robel, S.; Genius, J.; Moessner, R.; Riederer, P.; et al. Serotonin depletion hampers survival and proliferation in neurospheres derived from adult neural stem cells. *Neuropsychopharmacology* **2010**, *35*, 893–903. [CrossRef]

96. Dunphy-Doherty, F.; O'Mahony, S.M.; Peterson, V.L.; O'Sullivan, O.; Crispie, F.; Cotter, P.D.; Wigmore, P.; King, M.V.; Cryan, J.F.; Fone, K.C.F. Post-weaning social isolation of rats leads to long-term disruption of the gut microbiota-immune-brain axis. *Brain Behav. Immun.* **2018**, *68*, 261–273. [CrossRef]

97. Diviccaro, S.; Giatti, S.; Borgo, F.; Barcella, M.; Borghi, E.; Trejo, J.L.; Garcia-Segura, L.M.; Melcangi, R.C. Treatment of male rats with finasteride, an inhibitor of 5alpha-reductase enzyme, induces long-lasting effects on depressive-like behavior, hippocampal neurogenesis, neuroinflammation and gut microbiota composition. *Psychoneuroendocrinology* **2019**, *99*, 206–215. [CrossRef] [PubMed]

98. Yang, T.; Santisteban, M.M.; Rodriguez, V.; Li, E.; Ahmari, N.; Carvajal, J.M.; Zadeh, M.; Gong, M.; Qi, Y.; Zubcevic, J.; et al. Gut Dysbiosis is Linked to Hypertension. *Hypertension* **2015**, *65*, 1331–1340. [CrossRef]

99. Yang, T.; Rodriguez, V.; Malphurs, W.L.; Schmidt, J.T.; Ahmari, N.; Sumners, C.; Martyniuk, C.J.; Zubcevic, J. Butyrate regulates inflammatory cytokine expression without affecting oxidative respiration in primary astrocytes from spontaneously hypertensive rats. *Physiol. Rep.* **2018**, *6*, e13732. [CrossRef]

100. Harman, D. Aging: A theory based on free radical and radiation chemistry. *J. Gerontol.* **1956**, *11*, 298–300. [CrossRef]

101. Figueira, T.R.; Barros, M.H.; Camargo, A.A.; Castilho, R.F.; Ferreira, J.C.B.; Kowaltowski, A.J.; Sluse, F.E.; Souza-Pinto, N.C.; Vercesi, A.E. Mitochondria as a Source of Reactive Oxygen and Nitrogen Species: From Molecular Mechanisms to Human Health. *Antioxid. Redox Signal.* **2013**, *18*, 2029–2074. [CrossRef]

102. Rizvi, S.I.; Maurya, P.K. Alterations in antioxidant enzymes during aging in humans. *Mol. Biotechnol.* **2007**, *37*, 58–61. [CrossRef] [PubMed]

103. Rizvi, S.I.; Maurya, P.K. Markers of oxidative stress in erythrocytes during aging in humans. *Ann. N. Y. Acad. Sci.* **2007**, *1100*, 373–382. [CrossRef]

104. Reiter, R. Melatonin, active oxygen species and neurological damage. *Drug News Perspect.* **1998**, *11*, 291–296. [CrossRef]

105. Reiter, R. Oxidative damage in the central nervous system: Protection by melatonin. *Prog. Neurobiol.* **1998**, *56*, 359–384. [CrossRef]

106. Baizabal-Carvallo, J.F.; Alonso-Juarez, M. The Link between Gut Dysbiosis and Neuroinflammation in Parkinson's Disease. *Neuroscience* **2020**, *432*, 160–173. [CrossRef]

107. Rubio-Perez, J.M.; Morillas-Ruiz, J.M. A review: Inflammatory process in Alzheimer's disease, role of cytokines. *Sci. World J.* **2012**, *2012*, 756357. [CrossRef]

108. Rodríguez-Gómez, J.A.; Kavanagh, E.; Engskog-Vlachos, P.; Engskog, M.K.R.; Herrera, A.J.; Espinosa-Oliva, A.M.; Joseph, B.; Hajji, N.; Venero, J.L.; Burguillos, M.A. Microglia: Agents of the CNS Pro-Inflammatory Response. *Cells* **2020**, *9*, 1717. [CrossRef]

109. Biagi, E.; Nylund, L.; Candela, M.; Ostan, R.; Bucci, L.; Pini, E.; Nikkïla, J.; Monti, D.; Satokari, R.; Franceschi, C.; et al. Through ageing, and beyond: Gut microbiota and inflammatory status in seniors and centenarians. *PLoS ONE* **2010**, *5*, e10667. [CrossRef]

110. Shintouo, C.M.; Mets, T.; Beckwee, D.; Bautmans, I.; Ghogomu, S.M.; Souopgui, J.; Leemans, L.; Meriki, H.D.; Njemini, R. Is inflammageing influenced by the microbiota in the aged gut? A systematic review. *Exp. Gerontol.* **2020**, *141*, 111079. [CrossRef]

111. Cerovic, M.; Forloni, G.; Balducci, C. Neuroinflammation and the Gut Microbiota: Possible Alternative Therapeutic Targets to Counteract Alzheimer's Disease? *Front. Aging Neurosci.* **2019**, *11*, 284. [CrossRef]

112. Kopitar-Jeraia, N. Innate immune response in brain, nf-kappab signaling and cystatins. *Front. Mol. Neurosci.* **2015**, *8*, 73.

113. Yao, L.; Kan, E.M.; Kaur, C.; Dheen, S.T.; Hao, A.; Lu, J.; Ling, E.A. Notch-1 signaling regulates microglia activation via NF-κB pathway after hypoxic exposure in vivo and in vitro. *PLoS ONE* **2013**, *8*, e78439. [CrossRef] [PubMed]

114. Tilstra, J.; Robinson, A.; Wang, J.; Gregg, S.; Clauson, C.; Reay, D.; Nasto, L.; St Croix, C.; Usas, A.; Vo, N.; et al. NF-κB inhibition delays DNA damage-induced senescence and aging in mice. *J. Clin. Investig.* **2012**, *122*, 2601–2612. [CrossRef] [PubMed]

115. Meberg, P.; Kinney, W.; Valcourt, E.; Routtenberg, A. Gene expression of the transcription factor NF-kB in hippocampus: Regulation by synaptic activity. *Mol. Brain Res* **1996**, *38*, 179–190. [CrossRef]

116. Baldwin, A. The NF-kB and IkB proteins: New discoveries and insights. *Annu. Rev. Immunol.* **1996**, *14*, 649–681. [CrossRef]

117. Chen, L.-F.; Greene, W.C. Regulation of distinct biological activities of the NF-kappaB transcription factor complex by acetylation. *J. Mol. Med.* **2003**, *81*, 549–557. [CrossRef]

118. Kaltschmidt, C.; Kaltschmidt, I.; Neumann, H.; Wekerle, H.; Baeuerle, P. Constitutive NF-KB Activity in Neurons. *Mol. Cell. Biol.* **1994**, *14*, 3981–3992.

119. Oeckinghaus, A.; Ghosh, S. The NF-kappaB family of transcription factors and its regulation. *Cold Spring Harb. Perspect. Biol.* **2009**, *1*, a000034. [CrossRef]

120. Adler, A.; Sinha, S.; Kawahara, T.; Zhang, J.; Segal, E.; Chang, H. Motif module map reveals enforcement of aging by continual NF-κB activity. *Genes Dev.* **2007**, *21*, 3244–3257. [CrossRef]

121. Kauppinen, A.; Suuronen, T.; Ojala, J.; Kaarniranta, K.; Salminen, A. Antagonistic crosstalk between NF-κB and SIRT1 in the regulation of inflammation and metabolic disorders. *Cell. Signal.* **2013**, *25*, 1939–1948. [CrossRef]

122. Siebenlist, U.; Franzoso, G.; Brown, K. Structure, regulation and function of NF-kB. *Annu. Rev. Cell Biol.* **1994**, *10*, 405–455. [CrossRef]

123. Kwon, H.; Brent, M.; Getachew, R.; Jayakumar, P.; Chen, D.; Schnolzer, M.; McBurney, M.; Marmorstein, R.; Greene, W.; Ott, M. Human immunodeficiency virus type 1 Tat protein inhibits the SIRT1 deacetylase and induces T cell hyperactivation. *Cell Host Microbe* **2008**, *3*, 158–167. [CrossRef] [PubMed]

124. Spencer, J.; Vafeiadou, K.; Williams, R.; Vauzour, D.; Spencer, J.; Vafeiadou, K.; Williams, R.; Vauzour, D. Neuroinflammation: Modulation by flavonoids and mechanisms of action. *Mol. Asp. Med.* **2012**, *33*, 83–97. [CrossRef] [PubMed]

125. Lanzillotta, A.; Porrini, V.; Bellucci, A.; Benarese, M.; Branca, C.; Parrella, E.; Spano, P.F.; Pizzi, M. NF-κB in innate neuroprotection and age-related neurodegenerative diseases. *Front. Neurol.* **2015**, *6*, 98. [CrossRef] [PubMed]

126. Mattson, M.P.; Camandola, S. NF-kappaB in neuronal plasticity and neurodegenerative disorders. *J. Clin. Investig.* **2001**, *107*, 247–254. [CrossRef]

127. Srinivasan, M.; Lahiri, D.K. Significance of NF-κB as a pivotal therapeutic target in the neurodegenerative pathologies of Alzheimer's disease and multiple sclerosis. *Expert. Opin. Ther. Targets* **2015**, *19*, 471–487. [CrossRef]

128. Lim, S.M.; Jang, H.M.; Jeong, J.J.; Han, M.J.; Kim, D.H. *Lactobacillus johnsonii* CJLJ103 attenuates colitis and memory impairment in mice by inhibiting gut microbiota lipopolysaccharide production and NF-κB activation. *J. Funct. Foods* **2017**, *34*, 359–368. [CrossRef]

129. Jang, H.M.; Lee, K.E.; Lee, H.J.; Kim, D.H. Immobilization stress-induced *Escherichia coli* causes anxiety by inducing NF-κB activation through gut microbiota disturbance. *Sci. Rep.* **2018**, *8*, 13897. [CrossRef]

130. Quivy, V.; Van Lint, C. Regulation at multiple levels of NF-kapa B-mediated transactivation by protein acetylation. *Biochem. Pharmacol.* **2004**, *68*, 1221–1229. [CrossRef]

131. Soria-Valles, C.; Osorio, F.; Gutierrez-Fernandez, A.; De Los Angeles, A.; Bueno, C.; Menendez, P.; Martin-Subero, J.; Daley, G.; Freije, J.; Lopez-Otin, C. NF-kappaB activation impairs somatic cell reprogramming in ageing. *Nat. Cell Biol.* **2015**, *17*, 1004–1013. [CrossRef]

132. Soria-Valles, C.; Osorio, F.; López-Otín, C. Reprogramming aging through DOT1L inhibition. *Cell Cycle* **2015**, *14*, 3345–3346. [CrossRef]

133. Osorio, F.; Bárcena, C.; Soria-Valles, C.; Ramsay, A.; de Carlos, F.; Cobo, J.; Fueyo, A.; Freije, J.; López-Otín, C. Nuclear lamina defects cause ATM-dependent NF-κB activation and link accelerated aging to a systemic inflammatory response. *Genes Dev.* **2012**, *26*, 2311–2324. [CrossRef] [PubMed]

134. Zhou, Z.D.; Kumari, U.; Xiao, Z.C.; Tan, E.K. Notch as a molecular switch in neural stem cells. *IUBMB Life* **2010**, *62*, 618–623. [CrossRef] [PubMed]

135. Artavanis-Tsakonas, S. Notch Signaling: Cell Fate Control and Signal Integration in Development. *Science* **1999**, *284*, 770–776. [CrossRef]

136. Ahmed, I.; Chandrakesan, P.; Tawfik, O.; Xia, L.; Anant, S.; Umar, S. Critical roles of notch and Wnt/β-catenin pathways in the regulation of hyperplasia and/or colitis in response to bacterial infection. *Infect. Immun.* **2012**, *80*, 3107–3127. [CrossRef] [PubMed]

137. Troll, J.V.; Hamilton, M.K.; Abel, M.L.; Ganz, J.; Bates, J.M.; Stephens, W.Z.; Melancon, E.; van der Vaart, M.; Meijer, A.H.; Distel, M.; et al. Microbiota promote secretory cell determination in the intestinal epithelium by modulating host Notch signaling. *Development* **2018**, *145*, dev155317. [CrossRef] [PubMed]

138. Peng, C.; Ouyang, Y.; Lu, N.; Li, N. The NF-κB Signaling Pathway, the Microbiota, and Gastrointestinal Tumorigenesis: Recent Advances. *Front. Immunol.* **2020**, *11*, 1387. [CrossRef] [PubMed]

139. Zhao, Y.; Lukiw, W.J. Microbiome-mediated upregulation of microRNA-146a in sporadic Alzheimer's disease. *Front. Neurol.* **2018**, *9*, 145. [CrossRef] [PubMed]

140. Lau, F.; Shukitt-Hale, B.; Joseph, J. The beneficial effects of fruit polyphenols on brain aging. *Neurobiol. Aging* **2005**, *26*, 128–132. [CrossRef]

141. Liu, J.; Killilea, D.; Ames, B. Age-associated mitochondrial oxidative decay: Improvement of carnitine acetyltransferase substrate-binding affinity and activity in brain by feeding old rats acetyl-L-carnitine and/or R.-alpha-lipoic acid. *Proc. Natl. Acad. Sci. USA* **2002**, *99*, 1876–1881. [CrossRef]

142. Rinaldi, P.; Polidori, M.; Metastasio, A.; Mariani, E.; Mattioli, P.; Cherubini, A.; Catani, M.; Cecchetti, R.; Senin, U.; Mecocci, P. Plasma antioxidants are similarly depleted in mild cognitive impairment and in Alzheimer's disease. *Neurobiol. Aging* **2003**, *24*, 915–919. [CrossRef]

143. Yuan, T.F.; Gu, S.; Shan, C.; Marchado, S.; Arias-Carrión, O. Oxidative Stress and Adult Neurogenesis. *Stem Cell Rev. Rep.* **2015**, *11*, 333–364. [CrossRef]

144. Corredor, C. *Metabolismo, Nutrición y Shock*, 4th ed.; "Antioxidantes" en Patiño, J.F., Ed.; Editorial Médica Panamericana: Bogotá, Colombia, 2006; pp. 293–306.

145. Joseph, J.; Cole, G.; Head, E.; Ingram, D. Nutrition, brain aging, and neurodegeneration. *J. Neurosci.* **2009**, *29*, 12795–12801. [CrossRef]

146. Sarubbo, F.; Ramis, M.R.; Aparicio, S.; Ruiz, L.; Esteban, S.; Miralles, A.; Moranta, D. Improving effect of chronic resveratrol treatment on central monoamine synthesis and cognition in aged rats. *Age* **2015**, *37*, 9777. [CrossRef] [PubMed]

147. Hollman, P.C.H.; Cassidy, A.; Comte, B.; Heinonen, M.; Richelle, M.; Richling, E.; Serafini, M.; Scalbert, A.; Sies, H.; Vidry, S. The biological relevance of direct antioxidant effects of polyphenols for cardiovascular health in humans is not established. *J. Nutr.* **2011**, *141*, 989S–1009S. [CrossRef]

148. Sandoval-Acuña, C.; Ferreira, J.; Speisky, H. Polyphenols and mitochondria: An update on their increasingly emerging ROS-scavenging independent actions. *Arch. Biochem. Biophys.* **2014**, *559*, 75–90. [CrossRef]

149. Pandey, K.B.; Rizvi, S.I. Plant polyphenols as dietary antioxidants in human health and disease. *Oxid. Med. Cell. Longev.* **2009**, *2*, 270–278. [CrossRef]

150. Lakey-Beitia, J.; Berrocal, R.; Rao, K.S.; Durant, A.A. Polyphenols as Therapeutic Molecules in Alzheimer's Disease Through Modulating Amyloid Pathways. *Mol. Neurobiol.* **2015**, *51*, 466–479. [CrossRef]

151. Michalak, A.; Mosińska, P.; Fichna, J. Polyunsaturated fatty acids and their derivatives: Therapeutic value for inflammatory, functional gastrointestinal disorders, and colorectal cancer. *Front. Pharmacol.* **2016**, *7*, 459. [CrossRef]

152. Iizuka, K. The Role of Carbohydrate Response Element-Binding Protein in the Development of Liver Diseases. *Int. J. Mol. Sci.* **2021**, *22*, 12058. [CrossRef]

153. Crupi, R.; Marino, A.; Cuzzocrea, S. n-3 Fatty Acids: Role in Neurogenesis and Neuroplasticity. *Curr. Med. Chem.* **2013**, *20*, 2953–2963. [CrossRef]

154. Marrone, M.C.; Coccurello, R. Dietary fatty acids and microbiota-brain communication in neuropsychiatric diseases. *Biomolecules* **2019**, *10*, 12. [CrossRef] [PubMed]

155. Kaliannan, K.; Wang, B.; Li, X.Y.; Kim, K.J.; Kang, J.X. A host-microbiome interaction mediates the opposing effects of omega-6 and omega-3 fatty acids on metabolic endotoxemia. *Sci. Rep.* **2015**, *5*, 11276. [CrossRef] [PubMed]

156. Svoboda, E. Could the gut microbiome be linked to autism? *Nature* **2020**, *577*, S14–S15. [CrossRef] [PubMed]

157. Araya, M.; Morelli, L.; Reid, G.; Sanders, M.; Stanton, C.; Pineiro, M.; Ben Embarek, P. *Joint FAO/WHO Working Group Report on Drafting Guidelines for the Evaluation of Probiotcs in Food*; World Health Organization, Food and Agriculture Organization of the United Nations: London, ON, Canada, 2002.

158. Gareau, M.G.; Sherman, P.M.; Walker, W.A. Probiotics and the gut microbiota in intestinal health and disease. *Nat. Rev. Gastroenterol. Hepatol.* **2010**, *7*, 503–514. [CrossRef]

159. Gibson, G.R.; Scott, K.P.; Rastall, R.A.; Tuohy, K.M.; Hotchkiss, A.; Dubert-Ferrandon, A.; Gareau, M.; Murphy, E.F.; Saulnier, D.; Loh, G.; et al. Dietary prebiotics: Current status and new definition. *Food Sci. Technol. Bull. Funct. Foods* **2010**, *7*, 1–19. [CrossRef]

160. Davani-Davari, D.; Negahdaripour, M.; Karimzadeh, I.; Seifan, M.; Mohkam, M.; Masoumi, S.J.; Berenjian, A.; Ghasemi, Y. Prebiotics: Definition, types, sources, mechanisms, and clinical applications. *Foods* **2019**, *8*, 92. [CrossRef]

161. Forsythe, P.; Bienenstock, J.; Kunze, W.A. Microbial Endocrinology: The Microbiota-Gut-Brain Axis in Health and Disease Chapter 17. *Adv. Exp. Med. Biol.* **2014**, *817*, 3–24.

162. Savignac, H.M.; Corona, G.; Mills, H.; Chen, L.; Spencer, J.P.E.; Tzortzis, G.; Burnet, P.W.J. Prebiotic feeding elevates central brain derived neurotrophic factor, N-methyl-d-aspartate receptor subunits and d-serine. *Neurochem. Int.* **2013**, *63*, 756–764. [CrossRef]

163. Williams, S.; Chen, L.; Savignac, H.M.; Tzortzis, G.; Anthony, D.C.; Burnet, P.W. Neonatal prebiotic (BGOS) supplementation increases the levels of synaptophysin, GluN2A-subunits and BDNF proteins in the adult rat hippocampus. *Synapse* **2016**, *70*, 121–124. [CrossRef]

164. Logan, A.C.; Katzman, M. Major depressive disorder: Probiotics may be an adjuvant therapy. *Med. Hypotheses* **2005**, *64*, 533–538. [CrossRef]

165. Franco-Robles, E.; López, M.G. Implication of fructans in health: Immunomodulatory and antioxidant mechanisms. *Sci. World J.* **2015**, *2015*, 289267. [CrossRef] [PubMed]

166. Messaoudi, M.; Lalonde, R.; Violle, N.; Javelot, H.; Desor, D.; Nejdi, A.; Bisson, J.F.; Rougeot, C.; Pichelin, M.; Cazaubiel, M.; et al. Assessment of psychotropic-like properties of a probiotic formulation (*Lactobacillus helveticus* R0052 and *Bifidobacterium longum* R0175) in rats and human subjects. *Br. J. Nutr.* **2011**, *105*, 755–764. [CrossRef] [PubMed]

167. Distrutti, E.; O'Reilly, J.A.; McDonald, C.; Cipriani, S.; Renga, B.; Lynch, M.A.; Fiorucci, S. Modulation of intestinal microbiota by the probiotic VSL#3 resets brain gene expression and ameliorates the age-related deficit in LTP. *PLoS ONE* **2014**, *9*, e106503.

168. Corpuz, H.M.; Ichikawa, S.; Arimura, M.; Mihara, T.; Kumagai, T.; Mitani, T.; Nakamura, S.; Katayama, S. Long-term diet supplementation with *Lactobacillus paracasei* K71 prevents age-related cognitive decline in senescence-accelerated mouse prone 8. *Nutrients* **2018**, *10*, 762. [CrossRef]

169. Chunchai, T.; Thunapong, W.; Yasom, S.; Wanchai, K.; Eaimworawuthikul, S.; Metzler, G.; Lungkaphin, A.; Pongchaidecha, A.; Sirilun, S.; Chaiyasut, C.; et al. Decreased Microglial Activation Through Gut-brain Axis by Prebiotics, Probiotics, or Synbiotics Effectively Restored Cognitive Function in Obese-insulin Resistant Rats. *J. Neuroinflamm.* **2018**, *15*, 11. [CrossRef]

170. Bonfili, L.; Cecarini, V.; Cuccioloni, M.; Angeletti, M.; Berardi, S.; Scarpona, S.; Rossi, G.; Eleuteri, A.M. SLAB51 Probiotic Formulation Activates SIRT1 Pathway Promoting Antioxidant and Neuroprotective Effects in an AD Mouse Model. *Mol. Neurobiol.* **2018**, *55*, 7987–8000. [CrossRef]
171. Colcombe, S.J.; Erickson, K.I.; Scalf, P.E.; Kim, J.S.; Prakash, R.; McAuley, E.; Elavsky, S.; Marquez, D.X.; Hu, L.; Kramer, A.F. Aerobic exercise training increases brain volume in aging humans. *J. Gerontol. Ser. A Biol. Sci. Med. Sci.* **2006**, *61*, 1166–1170. [CrossRef]
172. Erickson, K.I.; Voss, M.W.; Prakash, R.S.; Basak, C.; Szabo, A.; Chaddock, L.; Kim, J.S.; Heo, S.; Alves, H.; White, S.M.; et al. Exercise training increases size of hippocampus and improves memory. *Proc. Natl. Acad. Sci. USA* **2011**, *108*, 73017–73022. [CrossRef]
173. Gomez-Pinilla, F.; Hillman, C. The influence of exercise on cognitive abilities. *Compr. Physiol.* **2013**, *3*, 4032–4428.
174. Laske, C.; Banschbach, S.; Stransky, E.; Bosch, S.; Straten, G.; MacHann, J.; Fritsche, A.; Hipp, A.; Niess, A.; Eschweiler, G.W. Exercise-induced normalization of decreased BDNF serum concentration in elderly women with remitted major depression. *Int. J. Neuropsychopharmacol.* **2010**, *13*, 595–602. [CrossRef]
175. Bermon, S.; Petriz, B.; Kajeniene, A.; Prestes, J.; Castell, L.; Franco, O.L. The microbiota: An exercise immunology perspective. *Exerc. Immunol. Rev.* **2015**, *21*, 70–79. [PubMed]
176. Dalton, A.; Mermier, C.; Zuhl, M. Exercise influence on the microbiome-gut-brain axis. *Gut Microbes* **2019**, *10*, 555–568. [CrossRef] [PubMed]
177. Monda, V.; Villano, I.; Messina, A.; Valenzano, A.; Esposito, T.; Moscatelli, F.; Viggiano, A.; Cibelli, G.; Chieffi, S.; Monda, M.; et al. Exercise Modifies the Gut Microbiota with Positive Health Effects. *Oxid. Med. Cell. Longev.* **2017**, *2017*, 3831972. [CrossRef] [PubMed]
178. Macfarlane, G.T.; Macfarlane, S. Bacteria, colonic fermentation, and gastrointestinal health. *J. AOAC Int.* **2012**, *95*, 50–60. [CrossRef] [PubMed]
179. Silva, Y.P.; Bernardi, A.; Frozza, R.L. The Role of Short-Chain Fatty Acids From Gut Microbiota in Gut-Brain Communication. *Front. Endocrinol.* **2020**, *11*, 25. [CrossRef] [PubMed]
180. Mouries, J.; Brescia, P.; Silvestri, A.; Spadoni, I.; Sorribas, M.; Wiest, R.; Mileti, E.; Galbiati, M.; Invernizzi, P.; Adorini, L.; et al. Microbiota-driven gut vascular barrier disruption is a prerequisite for non-alcoholic steatohepatitis development. *J. Hepatol.* **2019**, *71*, 1216–1228. [CrossRef] [PubMed]
181. Pozhilenkova, E.A.; Lopatina, O.L.; Komleva, Y.K.; Salmin, V.V.; Salmina, A.B. Blood-brain barrier-supported neurogenesis in healthy and diseased brain. *Rev. Neurosci.* **2017**, *28*, 397–415. [CrossRef] [PubMed]
182. Tobin, M.K.; Musaraca, K.; Disouky, A.; Shetti, A.; Bheri, A.; Honer, W.G.; Kim, N.; Dawe, R.J.; Bennett, D.A.; Arfanakis, K.; et al. Human Hippocampal Neurogenesis Persists in Aged Adults and Alzheimer's Disease Patients. *Cell Stem Cell* **2019**, *24*, 974–982. [CrossRef]
183. Lima, S.M.A.; Gomes-Leal, W. Neurogenesis in the hippocampus of adult humans: Controversy "fixed" at last. *Neural Regen. Res.* **2019**, *14*, 1917–1918.
184. Bergmann, O.; Spalding, K.L.; Frisén, J. Adult neurogenesis in humans. *Cold Spring Harb. Perspect. Biol.* **2015**, *7*, a018994. [CrossRef]
185. Manganas, L.N.; Zhang, X.; Li, Y.; Hazel, R.D.; Smith, S.D.; Wagshul, M.E.; Henn, F.; Benveniste, H.; Djuric, P.M.; Enikolopov, G.; et al. Magnetic resonance spectroscopy identifies neural progenitor cells in the live human brain. *Science* **2007**, *318*, 980–985. [CrossRef] [PubMed]

Review

Therapeutic Potential of Complementary and Alternative Medicines in Peripheral Nerve Regeneration: A Systematic Review

Yoon-Yen Yow [1,*], Tiong-Keat Goh [1], Ke-Ying Nyiew [1], Lee-Wei Lim [2,*], Siew-Moi Phang [3,4], Siew-Huah Lim [5], Shyamala Ratnayeke [1] and Kah-Hui Wong [6,*]

1   Department of Biological Sciences, School of Medicine and Life Sciences, Sunway University, Petaling Jaya 47500, Malaysia; tiongkeatgoh@gmail.com (T.-K.G.); kynyiew@gmail.com (K.-Y.N.); shyamalar@sunway.edu.my (S.R.)
2   Neuromodulation Laboratory, School of Biomedical Sciences, Li Ka Shing Faculty of Medicine, The University of Hong Kong, 21 Sassoon Road, L4 Laboratory Block, Hong Kong
3   Institute of Ocean and Earth Sciences, Universiti Malaya, Kuala Lumpur 50603, Malaysia; phang@um.edu.my
4   Faculty of Applied Sciences, UCSI University, Cheras, Kuala Lumpur 56000, Malaysia
5   Department of Chemistry, Faculty of Science, Universiti Malaya, Kuala Lumpur 50603, Malaysia; shlim80@um.edu.my
6   Department of Anatomy, Faculty of Medicine, Universiti Malaya, Kuala Lumpur 50603, Malaysia
*   Correspondence: yoonyeny@sunway.edu.my (Y.-Y.Y.); limlw@hku.hk (L.-W.L.); wkahhui@um.edu.my (K.-H.W.); Tel.: +603-7491-8622 (Y.-Y.Y.); +852-3917-6830 (L.-W.L.); +603-7967-4729 (K.-H.W.)

Citation: Yow, Y.-Y.; Goh, T.-K.; Nyiew, K.-Y.; Lim, L.-W.; Phang, S.-M.; Lim, S.-H.; Ratnayeke, S.; Wong, K.-H. Therapeutic Potential of Complementary and Alternative Medicines in Peripheral Nerve Regeneration: A Systematic Review. *Cells* **2021**, *10*, 2194. https://doi.org/10.3390/cells10092194

Academic Editor: FengRu Tang

Received: 24 July 2021
Accepted: 20 August 2021
Published: 25 August 2021

Publisher's Note: MDPI stays neutral with regard to jurisdictional claims in published maps and institutional affiliations.

Abstract: Despite the progressive advances, current standards of treatments for peripheral nerve injury do not guarantee complete recovery. Thus, alternative therapeutic interventions should be considered. Complementary and alternative medicines (CAMs) are widely explored for their therapeutic value, but their potential use in peripheral nerve regeneration is underappreciated. The present systematic review, designed according to guidelines of Preferred Reporting Items for Systematic Review and Meta-Analysis Protocols, aims to present and discuss the current literature on the neuroregenerative potential of CAMs, focusing on plants or herbs, mushrooms, decoctions, and their respective natural products. The available literature on CAMs associated with peripheral nerve regeneration published up to 2020 were retrieved from PubMed, Scopus, and Web of Science. According to current literature, the neuroregenerative potential of *Achyranthes bidentata*, *Astragalus membranaceus*, *Curcuma longa*, *Panax ginseng*, and *Hericium erinaceus* are the most widely studied. Various CAMs enhanced proliferation and migration of Schwann cells *in vitro*, primarily through activation of MAPK pathway and FGF-2 signaling, respectively. Animal studies demonstrated the ability of CAMs to promote peripheral nerve regeneration and functional recovery, which are partially associated with modulations of neurotrophic factors, pro-inflammatory cytokines, and anti-apoptotic signaling. This systematic review provides evidence for the potential use of CAMs in the management of peripheral nerve injury.

Keywords: complementary and alternative medicines; natural products; peripheral nerve injury; nerve repair; nerve regeneration; functional recovery

1. Introduction

Peripheral nerve injury (PNI) can result in partial or total loss of motor, sensory and autonomic functions at denervated regions, leading to temporary or life-long disability [1]. In addition to reduced quality of life, functional deficits from PNI have a substantial economic impact on the affected individuals [2]. A recent study found that, over nine years (from 2009 to 2018), more than 550,000 individuals were afflicted by PNI in the United

States. Moreover, the incidence rate has more than doubled throughout that period of time [3]. Such injuries are primarily due to vehicular and traumatic accidents, lacerations, and iatrogenic causes [4–6].

Despite progressive advances in our understanding of the processes and mechanisms of nerve injury, effective nerve repair and regeneration approaches that ensure complete functional recovery remain scarce [7]. Nerve autograft is considered the gold standard for repairing peripheral nerve defects [8]. However, this method is restricted by limited donor nerves and donor site morbidity, while successful recovery rates remain unsatisfactory [9]. Consequently, alternative strategies for enhancing nerve repairs have been proposed, including the application of nerve conduits and the addition of growth factors [10,11]. Likewise, the exploration of novel therapeutics, even combinatorial therapies, capable of enhancing axonal regeneration and promoting functional recovery, are of great interest.

PNI often results in neuropathic pain, and when conventional treatments are inadequate in providing relief, patients may turn to complementary and alternative medicines (CAMs), such as herbal medicines and nutritional supplements [12]. Indeed, medicinal plants, including the *Acorus calamus* [13], *Curcuma longa* [14], and *Ginkgo biloba* [15], have displayed ameliorating effects in animal models of neuropathic pain. Research on the potential of medicinal plants in the treatment of PNI is prompted by the notion that plants are great sources of natural products (NPs), which are small molecules produced by living organisms. Many NPs are the focus of drug development, as it is generally believed that they are largely devoid of adverse effects compared to synthetic drugs [16,17]. NPs also have the advantage of being evolutionary-driven, thus they are more likely to possess tremendous chemical and structural diversity that facilitates efficient engagement with biologically relevant targets and receptors, making them more biologically active [18]. In fact, many small-molecule drugs that have been approved by regulatory agencies were derived from natural sources [19], including Taxol from *Taxus brevifolia* [20] and Vinblastine from *Catharanthus roseus* [21].

However, compared to the extensive research on naturally derived products for other non-communicable and infectious diseases, NPs remain largely unexplored in the field of nerve repair and regeneration. A review published nearly half a decade ago has shed light on the neuroprotective effects of NPs in PNI models [22]. This review presents current research findings and evaluates the role of CAMs, focusing on plants or herbs, mushrooms, and decoctions, as well as their NPs, in peripheral nerve regeneration, to highlight their therapeutic potential for the management of PNI.

2. Materials and Methods

This systematic review was designed according to guidelines of Preferred Reporting Items for Systematic Review and Meta-Analysis Protocols (PRISMA-P) [23].

2.1. Search Strategy and Data Extraction

A literature search was performed to find all relevant publications up to 25 October 2020 across three electronic databases, PubMed, Scopus, and Web of Science. The following keywords were used to search each respective database: (("peripheral* nerve* regenera*" OR "peripheral* nerve* repair*" OR "neuroregenera*") AND ("alga*" OR "seaweed*" OR "plant" OR "natural product*" OR "mushroom" OR "Basidiomycete*" OR "herb*" OR "Traditional Chinese Medicine*" OR "alternative medicine" OR "complementary medicine*")).

2.2. Eligibility Criteria

Research articles describing the use of plants or herbs, mushrooms, algae, decoction, and their natural products in peripheral nerve repair and regeneration, written in English, and having full-text availability were considered. Articles not representing original research studies and NPs derived from sources other than plants, herbs, algae, and mushrooms were excluded (e.g., *Lumbricus rubellus*—earthworm). Retrieved articles were screened based on their title, abstract, and full-text to determine their eligibility for inclusion in this review.

3. Results

A preliminary search across the three databases yielded 560 records, of which 215 were duplicates (Figure 1). Together with 18 other records identified by other means, the remaining articles were screened based on the eligibility criteria, resulting in 289 additional records being excluded, leaving 56 records remaining and their findings being included in the qualitative synthesis (Figure 1). The studies investigated the neuroregenerative potential of 25 species of plants, three different mushrooms, and four traditional Chinese medicine decoctions, of which 18 known NPs were characterized. None of the studies investigated the potential of algae in peripheral nerve regeneration.

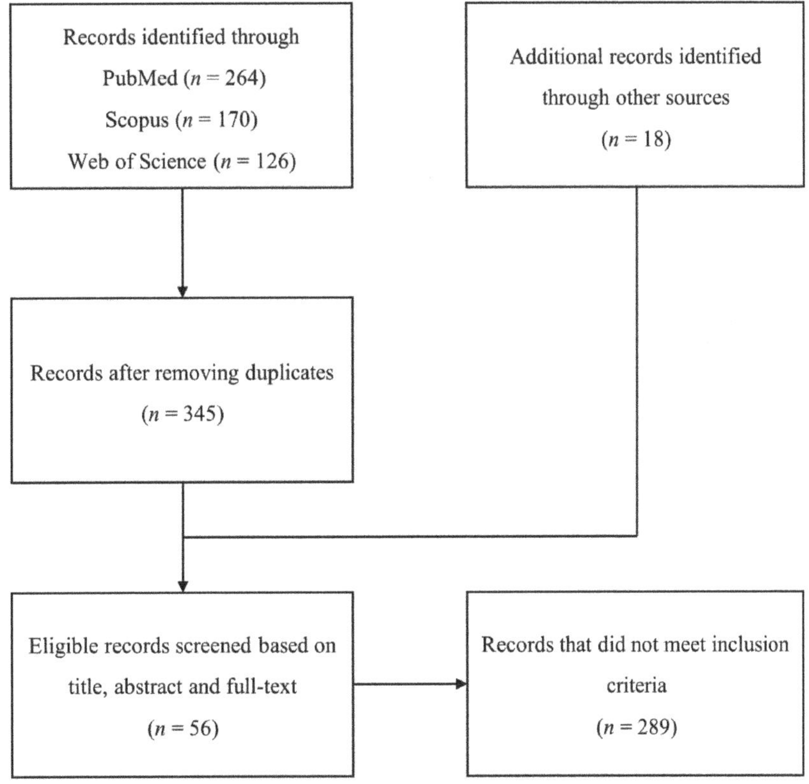

Figure 1. Flow diagram of the literature search procedure for the selection of studies up to 25 October 2020 on the use of plants, mushrooms, algae, decoctions, and their natural products (NPs) in peripheral nerve repair and regeneration. Only articles written in English, and having full-text availability were included. Articles not representing original research studies and NPs derived from sources other than plants, herbs, algae, and mushrooms were excluded.

Among the 58 records, the majority of the reported findings were from *in vivo* studies (38 records) that used mainly histological and electrophysiological evaluation to examine peripheral nerve regeneration in rat models of sciatic nerve injury (SNI). In contrast, 11 records were *in vitro* studies, which included reports of the promoting effects of plants, mushrooms, decoctions, and their natural products on the proliferation and migration of Schwann cells (SCs), and on neurite outgrowth in dorsal root ganglion (DRG) explants and neurons. Additionally, nine records included both *in vitro* and *in vivo* studies. In terms of the mechanisms of the biological effects, regulation of the mitogen-activated protein kinase (MAPK) pathway was reported to be highly involved across these studies.

4. Discussion

4.1. Current Therapeutic Approaches against Peripheral Nerve Injuries

Peripheral nerves are prone to injury because of their delicate structures and superficial location throughout the human body. The prevalence of PNI together with its societal impact poses a health concern that needs to be addressed properly. Current treatment strategies for PNI are divided into surgical and non-surgical approaches that can be effective when applied appropriately [24]. Surgical techniques, including suturing of severed nerves and nerve grafting, do yield successful outcomes but are sometimes not feasible due to limitations such as the timing of surgery, size of nerve gaps, and donor site morbidity [25,26]. Consequently, other promising alternatives have emerged in recent years and have been receiving increasing attention, such as the utilization of different nerve conduits capable of housing and delivering biological cues whilst enhancing and guiding nerve regeneration 11, growth factor treatments [27], and cell-based therapies [28]. In contrast, non-surgical options for the management of PNI are far more limited, including approved medications on the market, electrical nerve stimulation [29], and the application of phytochemicals and secondary metabolites. The latter is widespread in other areas of research including cancer [30] and neurological disorders [31], but are far less prevalent in the field of peripheral nerve regeneration.

4.2. Mechanisms of Peripheral Nerve Injury and Regeneration

Nerve bundles are primarily composed of axons covered with myelin sheaths produced by Schwann cells with fibroblasts scattered in between the nerve fibers. During peripheral nerve injury, instantaneous tissue damage occurs at the site of the lesion together with the accumulation of galectin-3 macrophages, whereas nerve stumps that are distally located undergo cellular variation despite not being directly affected [32]. After an axonal injury, Wallerian degeneration occurs, followed by axonal regeneration, and eventually end-organ reinnervation (see Figure 2) [33]. Wallerian degeneration takes place 24 to 48 h following nerve injury. Axons begin to disintegrate and growth factors such as nerve growth factor (NGF) and brain-derived neurotrophic factor (BDNF) are released by SCs in the segment distal to the injured site. Galectin-3 macrophages are then recruited to the distal end, which contributes to myelin degradation and removal of remaining debris [34]. Growth factors are also retrogradely transported proximally toward the cell body. Subsequent removal of deteriorated myelin and axonal matter leads to the proliferation and alignment of SCs, forming the bands of Büngner that further guide the regenerating axons from the proximal to the distal site [35]. Axonal regeneration in humans is known to occur at a rate of approximately 1 mm per day [36], which would require months or even years for severe nerve injuries to fully recover. Moreover, poor functional recovery can occur due to a number of reasons, including progressive failure of axonal regeneration, disruption of SC function in providing a growth-supportive environment, and misdirection of regenerating axons [36].

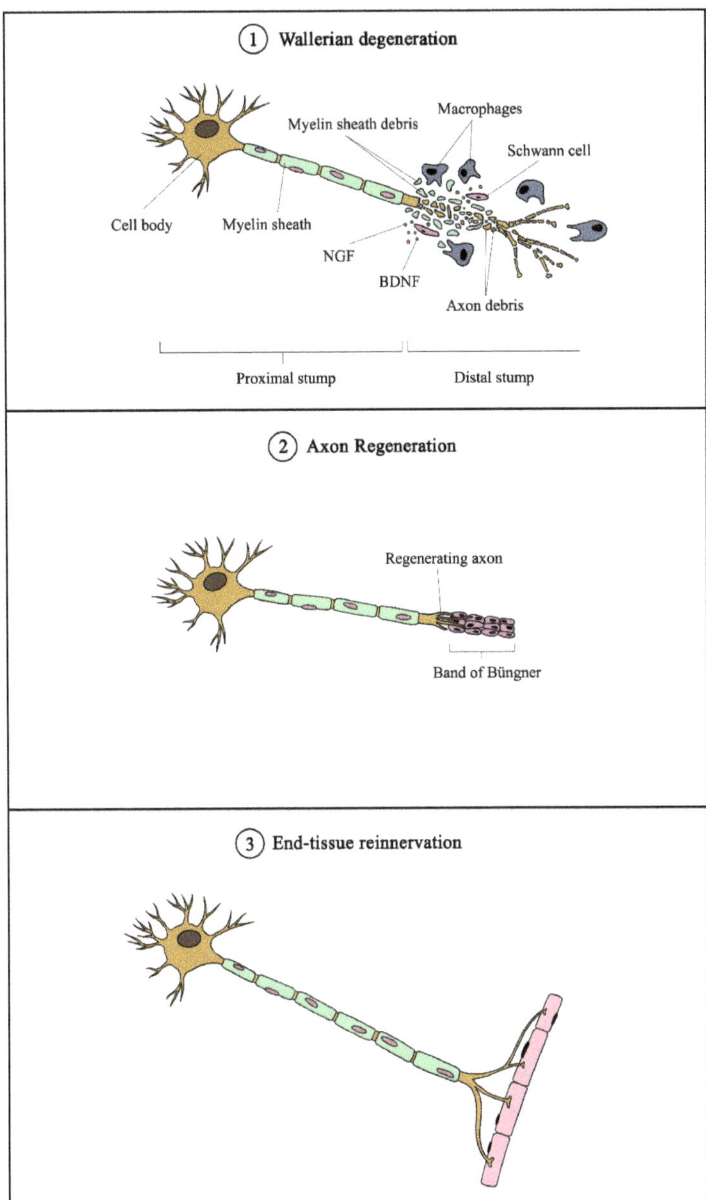

Figure 2. Overview of mechanism of peripheral nerve injury and regeneration. Following nerve injury, Wallerian degeneration occurs, in which axons begin to disintegrate at the distal end, and growth factors (such as NGF and BDNF) are released by Schwann cells. Galectin-3 macrophages are recruited to remove axonal debris and degrade myelin sheaths. Subsequently, SCs align to form the Band of Büngner, which guides the regenerating axons from the proximal to distal sites. Eventually, the regenerated axons innervate the end tissue to complete the recovery process. NGF—nerve growth factor; BDNF—brain-derived neurotrophic factor.

4.3. Role of Schwann Cells in Nerve Regeneration

Schwann cells are supportive glial cells that are known to play a pivotal role in the proper functioning and maintenance of peripheral nerves. They are responsible for producing the basal lamina that determines the polarity of SCs and myelinating axons [37]. The myelin sheaths on axons allow the conduction of action potentials at high velocity via the formation of specialized nodes of Ranvier [38]. The high plasticity of SCs allows them to further develop into repair phenotypes in response to nerve injury (Figure 3). Following nerve injury, SCs can re-differentiate into repair SCs that align themselves to form bands of Büngner. This in turn allows axons to emerge from growth cones proximal to the injured site, which then elongate along the bands until the target organ is reinnervated. The repair SCs also participate in the removal of axon and myelin debris, and they can recruit macrophages to assist in the process [39]. In addition, repair SCs can also secrete neurotrophic factors that help promote cellular survival, proliferation, and differentiation, which are all essential for peripheral nerve repair [40]. Due to the importance of SCs in promoting peripheral nerve regeneration, it is expected that any disruption in SC proliferation, such as that caused by impairment in cyclin D1, will affect nerve regeneration following injury [41]. However, findings from past studies suggest that axonal regeneration is independent of SC proliferation [42,43]. Nevertheless, considering the association of SCs with axonal elongation and myelination, it is reasonable to hypothesize that enhanced SC proliferation may lead to greater regenerative potential. Hence, numerous studies have attempted to investigate the effects of NPs in promoting the proliferation and migration ability of SCs (Table 1).

Table 1. Summary of plants, mushrooms, and decoctions their natural products relating to peripheral nerve regeneration.

Source	Molecule(s)/ Ingredients	Experimental Model	Effective Concentration	Application Method	Biological Effect	Mechanism	Reference
				PLANT			
Achyranthes bidentata	Polypeptides	*In vitro* (SCs isolated from the sciatic nerves of 1-day old SD rats)	0.1 µg/mL	Incubation	Promoted migration of SCs	Upregulation of NOX4/DUOX2-derived ROS production	[44]
	Polypeptides	*In vitro* (DRG explants harvested from spinal and peripheral roots of postnatal day 1 SD rats)	0.01, 0.1, 1 µg/mL (dose-dependent manner)	Incubation	Promoted neurite outgrowth from cultured DRG explants/neurons	Activation of ERK1/2	[45]
		In vivo (Adult New Zealand rabbits)	6.0 mg/kg	Intravenous injection	Enhanced nerve regeneration and functional restoration after crush injury to rabbit common peroneal nerve (increased CMAP, density, diameter and thickness of myelinated fibers, and number of motor neurons in anterior horn)	N/A	
	Polypeptides (Fraction K)	*In vitro* (DRG explants harvested from spinal and peripheral roots of postnatal day 1 SD rats)	50, 250 ng/mL (dose-dependent manner)	Incubation	Promoted neurite outgrowth in DRG explant and neurons	Activation of ERK1/2	[46]
		In vivo (ICR mice)	10 mg/kg	Intravenous injection	Promoted peripheral nerve regeneration in mice after SNI (increased diameter and thickness of myelinated fibers, CSA of gastrocnemius muscle fibers, SFI, and CMAP)	N/A	
	Polypeptides	*In vivo* (SD rats)	2 mg in 0.2 mL saline	Intraperitoneal injection	Promoted functional and histological recovery after rat sciatic nerve crush (increased SFI, CMAP, MNCV, myelin thickness, lamellae number, CSA of gastrocnemius muscle fibers)	Modulation of mRNA expression of GAP-43, neurotrophic factors (NGF, BDNF, CNTF), and neurotrophic factor receptors (TrkA, TrkB)	[47]
	Polypeptides	*In vivo* (ICR mice)	1, 4, 16 mg/kg (dose-independent manner)	Tail vein injection	Promoted functional and histological recovery after rat sciatic nerve crush (increased SFI, CMAP, MNCV, number, and diameter of myelinated fibers, axon diameter, myelin thickness, lamellae number, CSA of gastrocnemius muscle fibers)	N/A	[48]
	Aqueous extract	*In vivo* (Adult New Zealand rabbits)	10, 20 mg/kg (dose-dependent manner)	Intravenous injection	Promoted peripheral nerve regeneration in the crushed common peroneal nerve in rabbits (increased CMAP, CSA of tibialis posterior muscle, number of regenerated myelinated nerve fibers, and motoneurons in anterior horn of the spinal cord)	N/A	[49]

Table 1. *Cont.*

Source	Molecule(s)/ Ingredients	Experimental Model	Effective Concentration	Application Method	Biological Effect	Mechanism	Reference
Alpinate Oxyphyllae Fructus (*Alpinia oxyphylla* Miq)	Protocatechuic acid	*In vitro* (RSC96 SCs)	1 mM	Incubation	Promoted proliferation and survival of RSC96 SCs	Upregulation of IGF-1 and activation of PI3K/Akt signaling	[50]
	Aqueous extract	*In vitro* (RSC96 SCs)	Proliferation: 20, 60, 200 µg/mL (dose-independent manner Migration: 20–200 µg/mL (dose-dependent manner	Incubation	Promoted proliferation and migration of RSC96 SCs	Upregulation of PAs (uPA, tPA) and MMP2/9 mediated through the activation of MAPK pathway (ERK1/2, JNK, p38)	[51]
		In vivo (SD rats)	30, 60, 100, 150, 200 µg/mL (dose-independent manner)	Injection into a silicone rubber tube bridging a 15mm sciatic nerve defect	Promoted peripheral nerve regeneration in rats with SNI		
	Astragaloside IV	*In vivo* (BALB/c mice)	2.5, 5, 10 mg/kg (dose-dependent manner)	Intraperitoneal injection	Promoted sciatic nerve regeneration and functional recovery in mice (increased number and diameter of myelinated nerve fibers, MNCV, CMAP)	Upregulation of GAP-43 expression	[52]
	Astragaloside IV	*In vivo* (SD rats)	50 µM	Injection into a silicone rubber tube bridging a 15mm sciatic nerve defect	Promoted peripheral nerve regeneration in rats with SNI (increased number of myelinated axons and CMAP)	N/A	[53]
Astragalus membranaceus	Extract	*In vivo* (SD rats)	3 g/kg in 0.01 M of PBS	Intragastric gavage	Promoted peripheral nerve regeneration in rats with SNI (increased MNCV and latency, fluorogold labeling in the DRG, mean axonal density, percentage of CGRP area ratio, and macrophage density)	Modulation of local growth factors (FGF, NGF, PDGF, TGF-β) and immunoregulatory factors (IL-1, IFN-γ)	[54]
	Aqueous extract	*In vitro* (RSC96 SCs)	Proliferation: 12.5, 125, 250, 500 µg/mL (optimal at 12.5 µg/mL) Migration: 1.25, 12.5, 125, 250, 500 µg/mL (optimal at 1.25 µg/mL)	Incubation	Promoted proliferation and migration of RSC96 SCs	Proliferation: Increased cyclin protein A, D1, and E via ERK and p38 signaling pathways Migration: Activation of FGF-2 signaling, leading to upregulation of uPA and downregulation of PAI-1	[55]
Centella asiatica	Hydro-ethanolic extract	*In vivo* (SD rats)	400 µg/mL	Nerve conduit developed using decellularized artery seeded with *C. asiatca*-neurodifferentiated mesenchymal stem cells bridging a 15mm sciatic nerve defect	Promoted nerve regeneration and functional restoration in rats with SNI (increased CMAP, latency, MNCV, confirmation of angiogenesis, increased MBP expression, and number of myelinated axons)	N/A	[56]

Table 1. *Cont.*

Source	Molecule(s)/ Ingredients	Experimental Model	Effective Concentration	Application Method	Biological Effect	Mechanism	Reference
Citrus medica var. *sarcodactylis*	Aqueous extract	*In vitro* (RSC96 SCs)	0.85, 1.7, 2.55, 3.4, 4.25 μg/mL (dose-dependent manner)	Incubation	Promoted proliferation and migration of RSC96 SCs	Proliferation: Upregulation of cyclin A and B1 Migration: Activation of FGF-2 signaling, leading to the upregulation of uPA and MMP-9	[57]
Codonopsis pilosula	Aqueous extract	*In vitro* (RSC96 SCs)	20, 40, 60, 80, 100 μg/mL (dose-independent manner)	Incubation	Promoted proliferation and migration of RSC96 SCs	Proliferation: Enhanced IGF-I signaling pathway, cell cycle controlling protein expressions (cyclin A, D1, E) and MAPK pathway (ERK, p38) Migration: Stimulated FGF-2-uPA-MMP9 migration pathway	[58]
Crocus sativus	Crocin	*In vivo* (Wistar rats)	20, 80 mg/kg	Intraperitoneal injection	Promoted functional recovery in rats with SNI (Increased SFI, reduced plasma MDA levels, alleviated histological changes due to a crushing injury)	N/A	[59]
Curcuma longa	Alcoholic extract	*In vivo* (Wistar rats)	100 mg/kg (3, 6, or 9 times across 28 days)	Intraperitoneal injection	Protected against peripheral nerve degeneration in rats with SNI (Increased number of intact neurons in the right ventral horn of spinal cord region)	N/A	[60]
	Curcumin	*In vivo* (SD rats)	100 mg/kg (dissolved in olive oil)	Oral gavage	Promoted peripheral nerve regeneration in rats with SNI (increased mean cell volume, total volume and surface of DRG cells, total number, diameter, and area of myelinated nerve fibers)	N/A	[61]
	Curcumin	*In vivo* (SD rats)	100 mg/kg (dissolved in olive oil)	Oral gavage	Promoted functional recovery (improved SFI) and protective effect on DRG (increased volume and number of A- and B- cells, number of satellite cells) in rats with SNI	N/A	[62]
	Curcumin	*In vivo* (SD rats)	50, 100, 300 mg/kg	Intraperitoneal injection	Promoted peripheral nerve regeneration in rats with SNI (increased number of motoneurons, number and diameter of myelinated axons, SFI, MNCV, amplitude of CMAP, muscle fiber area and reduced latency of CMAP, mechanical withdrawal threshold, thermal withdrawal latency)	N/A	[63]
	Curcumin	*In vitro* (SCs isolated from S100β-DsRed transgenic mice)	0.04-1 μM (0.1 μM having the highest proliferative effect)	Incubation	Promoted proliferation and migration of SCs	Proliferation: Modulated by ERK and p38 kinase pathways	[64]

Table 1. *Cont.*

Source	Molecule(s)/ Ingredients	Experimental Model	Effective Concentration	Application Method	Biological Effect	Mechanism	Reference
Curcuma longa (curcumin); from honeybees (propolis)	Curcumin, propolis	*In vivo* (Wistar rats)	Curcumin (100 mg/kg) Propolis (200 mg/kg)	Administration through a nasogastric tube	Promoted functional recovery in rats with SNI (Increased SFI and amplitude of CMAP, reduced latency time)	N/A	[65]
Dioscoreae rhizoma	Aqueous extract	*In vivo* (SD rats)	10 mg/mL	Applied directly into the crush site	Promoted peripheral nerve regeneration in rats with SNI (increased number of DRG sensory neurons and motor neurons in the spinal cord)	Increasing protein levels of GAP-43 and Cdc2	[66]
Epimedium	Icariin	*In vivo* (SD rats)	20 mg in 5 mL	Injection into a poly(lactic-co-glycolic acid) biological conduit sleeve bridging a 5mm sciatic nerve defect	Promoted peripheral nerve regeneration in rats with SNI (increased sciatic nerve conduction velocity and number of myelinated fibers)	N/A	[67]
Epimedium	Epimedium extract, icariin	*In vivo* (SD rats)	4.873 mg/mL	Intragastric administration	Promoted peripheral nerve regeneration in rats with SNI (Increased SFI, nerve regeneration based on nerve pinch test, MNCV, muscle wet weight)	N/A	[68]
Gardenia jasminoides Ellis	Genipin	*In vivo* (SD rats)	3% aqueous gelatin solution fixed with 3% genipin	Injection into a silicone rubber tube bridging a 10mm sciatic nerve defect	Promoted peripheral nerve regeneration in rats with SNI	N/A	[69]
Gastrodia elata Blume	Gastrodin	*In vitro* (RSC96 SCs)	50, 100, 200 μM (dose-dependent manner)	Incubation	Promoted proliferation of RSC96 SCs in a dose- and time-dependent manner	Inhibition of ERK1/2 phosphorylation and activation of Akt phosphorylation	[70]
	Ginkgo biloba extract (EGb 761)	*In vivo* (SD rats)	50 mg/kg	Intraperitoneal injection paired with an 18mm acellular nerve allograft bridging a 15mm sciatic nerve defect	Promoted peripheral nerve regeneration in rats with SNI (increased density of regenerated axons, muscle mass, axon number and diameter, expression of CD34 and NF200)	Increasing expression of angiogenesis-related genes (*Vegf, Sox18, Prom1, IL-6*)	[71]
Ginkgo biloba	Ginkgo biloba extract (EGb 761)	*In vitro* (SCs isolated from spinal nerves of 1-day old SD rats)	1, 10, 20, 50, 100 μg/mL (dose-dependent manner)	Incubation	Promoted cell attachment and survival of SCs		
Ginkgo biloba	Ginkgo biloba extract (EGb 761)	*In vivo* (SD rats)	10, 50 μg/mL	Injection into poly(DL-lactic acid-co-glycolic acid) conduit seeded with Schwann cells bridging a 12mm sciatic nerve defect	Promoted histological and functional recovery in rats with SNI (increased number and area of myelinated axons, increased CMAP)	N/A	[72]

Table 1. *Cont.*

Source	Molecule(s)/ Ingredients	Experimental Model	Effective Concentration	Application Method	Biological Effect	Mechanism	Reference
Ginseng	Ginsenoside Rg1	*In vitro* (RSC96 SCs)	Ginseng: 100, 200, 300, 400, 500 µg/mL Ginsenoside: 5, 10, 15, 20, 25 µg/mL) (Dose-dependent manner for both)	Incubation	Promoted proliferation and migration of RSC96 SCs	Proliferation: Enhancing protein expression of IGF-I pathway regulators (IGF-IR, PI3K, p-Akt, p-Bad, Bcl-2), cell cycle controlling proteins (cyclin D1, E, A), and MAPK signaling pathway (ERK, JNK, p38) Migration: Stimulating the FGF-2-uPA-MMP9 migrating pathway	[73]
	Ginsenoside Rg1	*In vivo*(SD rats)	1.5 mg/kg	Intraperitoneal injection	Promoted peripheral nerve regeneration in rats with SNI (increased number of motoneurons, number, and diameter of myelinated axons, SFI, MNCV, improved CMAP latency and amplitude, the increased average percentage of muscle fiber)	N/A	[74]
	Ginsenoside Re	*In vitro* (SCs isolated from sciatic nerves of 3-day old SD rats)	0.5 mg/mL	Incubation	Promoted proliferation and migration of SCs	Phosphorylation of ERK1/2 and JNK 1/2	[75]
		In vivo (SD rats)	2.0 mg/kg	Intraperitoneal injection	Promoted peripheral nerve regeneration in rats with SNI (increased SFI, TSI, PCNA expression level, improved pathological changes due to crushing injury, GAP43, and S-100 expression)		
Green tea	(-)-Epigallocatechin-3-gallate (EGCG)	*In vivo* (Wistar rats)	50 mg/kg	Intraperitoneal injection	Promoted functional recovery (improved outcomes of foot position, toe spreading, extensor postural thrust, hopping reflex, von Frey hair, Randall–Sellito, hotplate, and tail-flick tests), improved morphological recovery in skeletal muscle tissues muscles, and protection towards muscle fibers in rats with SNI	Protection of muscle fibers from cellular death through activation of an anti-apoptotic signaling pathway (modulation of Bax, Bcl-2, and p53 expression)	[76]
	(-)-Epigallocatechin-3-gallate (EGCG)	*In vivo* (Wistar rats)	50 mg/kg	Intraperitoneal injection	Promoted peripheral nerve regeneration in rats with SNI (improved nerve morphology and functional recovery assessed by foot position, extensor postural thrust test, and withdrawal reflex threshold)	Reversal of Bax, Bcl-2, and survivin mRNA expression induced by sciatic nerve injury	[77]
Can be found in a wide variety of plants	Syringic acid	*In vitro* (RSC96 SCs)	600 µM	Incubation	Promoted proliferation and migration of RSC96 SCs	Downregulation of miR-451-5p	[78]

Table 1. *Cont.*

Source	Molecule(s)/ Ingredients	Experimental Model	Effective Concentration	Application Method	Biological Effect	Mechanism	Reference
Can be found in a wide variety of plants	Ursolic acid	*In vivo* (BALB/c mice)	2.5, 5, 10 mg/kg (dose-dependent manner)	Intraperitoneal injection	Promoted peripheral nerve regeneration in rats with SNI (increased number and diameter of myelinated nerve fibers and soleus muscle mass)	Increasing S100 protein expression levels	[79]
Lycium barbarum	Polysaccharide	*In vitro* (1) PC12 cells (2) Rat SCs (3) DRG neurons isolated from the embryo of 14-day pregnant rat	10, 30, 50 mg/mL (optimal at 30 mg/mL)	Incorporated into core-shell structured nanofibrous scaffolds by coaxial electrospinning	(1) Promoted proliferation and neuronal differentiation of PC12 cells (2) Promoted proliferation and myelination of SCs (3) Promoted neurite outgrowth of DRG neurons	N/A	[80]
Can be found in a wide variety of plants	Quercetin	*In vitro* (RSC96 SCs)	0.1, 1, 10 μg/mL	Incubation	Promoted proliferation of RSC96 SCs	N/A	[81]
		In vivo (SD rats)	0.1, 1, 10 μg/mL	Injection into a silicone rubber tube bridging a 15mm sciatic nerve defect	Promoted peripheral nerve regeneration in rats with SNI (increased count and density of myelinated axons, and resulted in larger area and amplitude of CMAP)		
Morus sp.	Cortex Mori Radicis (aqueous extract)	*In vivo* (SD rats)	100 mg/kg	Gastrointestinal administration	Reduced blood glucose levels, improved nerve functions (thermal latency and mechanical threshold), reversed the loss of Nissl bodies and induced neurite outgrowth in DRG neurons, and restored the response of growth cones to NGF in diabetic rats	Neurite outgrowth: Increased expression of TRPC1, reduced Ca^{2+} influx, and activation of PI3K/Akt signaling	[82]
Pueraria lobata	Puerarin	*In vitro* (RSC96 SCs)	1, 10, 100 μM (dose-independent manner)	Incubation	Promoted growth of SCs	N/A	[83]
		In vivo (SD rats)	1, 10, 100 μM (dose-independent manner)	Injection into a silicone rubber tube bridging a 15mm sciatic nerve defect	Promoted peripheral nerve regeneration in rats with SNI (increased density of myelinated axons, CMAP, and MNCV)		
	Serum metabolites (obtained from rats fed with *Pueraria lobata* extract)	*In vitro* (PC12 cells)	0.01, 0.1, 1 unit	Incubation	Enhanced NGF-mediated neurite outgrowth and expression of synapsin I in PC12 cells	N/A	[84]
		In vivo (SD rats)	0.01, 0.1, 1 unit	Injection into silicone rubber chamber bridging a 10mm sciatic nerve defect	Promoted peripheral nerve regeneration in rats with SNI (increased mean values of myelinated axon number, endoneurial area, and total nerve area)		
Radix Hedysari	Aqueous extract	*In vivo* (SD rats)	1 g/mL	Oral gavage paired with biodegradable chitin conduit bridging a 2mm sciatic nerve defect	Promoted peripheral nerve regeneration in rats with SNI (increased MNCV, fiber and axon diameter, g-ratio)	N/A	[85]
	Polysaccharides	*In vivo* (SD rats)	0.25 g/mL	Oral gavage	Promoted peripheral nerve regeneration in rats with sciatic nerve defect (increased SFI, TFI, PFI values, MNCV, and number of regenerated myelinated nerve fibers)	N/A	[86]

Table 1. *Cont.*

Source	Molecule(s)/ Ingredients	Experimental Model	Effective Concentration	Application Method	Biological Effect	Mechanism	Reference
Rhodiola rosea L.	Salidroside	*In vivo* (SD rats)	5, 10 mg/kg	Intraperitoneal injection	Promoted peripheral nerve regeneration in rats with SNI (increased number and diameter of myelinated axons, number of motoneurons, SFI, amplitude of CMAP, MNCV)	N/A	[87]
Scutellaria baicalensis Georgi	Baicalin	*In vitro* (RSC96 SCs)	5, 10, 20 μM (dose-dependent manner)	Incubation	Promoted proliferation of RSC96 SCs	Modulation of neurotrophic factors (GDNF, BDNF, CNTF) and S100β	[88]
Trigonella foenum-graecum (fenugreek)	IND01 (Fenugreek seed extract)	*In vivo* (Wistar rats)	50, 100, 200 mg/kg	Oral administration	Promoted peripheral nerve regeneration in rats with: (1) partial sciatic nerve ligation (ameliorated thermal hyperalgesia, improved motor function test scores) (2) SNI (ameliorated thermal hyperalgesia, improved motor function test scores, increased MNCV)	N/A	[89]
Tripterygium wilfordii Hook. F.	Triptolide	*In vivo* (SD rats)	100 μg/kg	Intraperitoneal injection	Promoted peripheral nerve regeneration in rats with SNI (increased number of motoneurons, number of myelinated axons, diameter of nerve fibers, SFI, CMAP amplitude, MNCV, muscle fiber area)	Reduction of TNF-α, IL-β, and IL-6 expression	[90]
MUSHROOM							
Amanita muscaria	Muscimol	*In vivo* (SD rats)	400 μg/mL	Applied directly to the right L5 DRG	Promoted peripheral nerve regeneration in rats with SNI (prevented the development of thermal and mechanical hypersensitivity and mechanical allodynia, improved basal membrane integrity, and increased nerve fibers)	Normalization of PMP22 protein expression level by GABAergic modulation in the ipsilateral DRG	[91]
Hericium erinaceus	Aqueous extract	*In vivo* (SD rats)	10 mL/kg	Oral administration	Promoted peripheral nerve regeneration in rats following peroneal nerve crush	Activation of signaling pathways (Akt, MAPK, c-Jun, c-Fos) and protein synthesis	[92]
	Polysaccharide	*In vivo* (SD rats)	30 mg/mL/kg	Oral administration	Promoted sensory functional recovery following peroneal nerve crush in rats (reduced withdrawal reflex latency)	Activation of Akt and p38 MAPK signaling and increased expression of RECA-1	[93]
	Aqueous extract	*In vivo* (SD rats)	10, 20 mL/kg	Oral administration	Promoted peripheral nerve regeneration in rats following peroneal nerve crush (increased PFI, improved axon morphology, and development of neuromuscular junction)	N/A	[94]

Table 1. *Cont.*

Source	Molecule(s)/ Ingredients	Experimental Model	Effective Concentration	Application Method	Biological Effect	Mechanism	Reference
Lignosus rhinocerotis	Aqueous extract	*In vivo* (SD rats)	500, 1000 mg/kg	Oral administration	Promoted motor and sensory functional recovery in rats with SNI (improved WRL and toe-spreading reflex)	N/A	[95]
DECOCTION							
Bogijetong	(1) Bogijietong decoction (18 ingredients) (2) A reconstituted formulation of BGJTD (BeD) with 4 ingredients (3) *Angelica gigas* (an ingredient in BeD)	*In vitro* (Primary neurons isolated from DRG at lumbar levels 4 and 5 in adult rats)	400 mg/kg	Incubation	Promoted neurite outgrowth of DRG neurons	(1) BGJTD and BeD: Downregulation of TNF-α and p38, upregulation of p-ERK1/2; (2) *Angelica gigas*: Regulation of ERK1/2 activity and TNF-α production	[96]
		In vivo (SD rats and BALB/c mice)	400 mg/kg	Oral administration	Reduced latency time in rats		
Buyang Huanwu	Buyang Huanwu decoction (16 ingredients: Modified formulation)	*In vivo* (SD rats)	1800 mg/kg	Oral administration paired with silicone rubber tube bridging a 10mm sciatic nerve defect	Promoted peripheral nerve regeneration in rats with sciatic nerve defect (increased nerve formation, myelinated axons, and endoneurial area)	N/A	[97]
Jiaweibugan	Jiaweibugan decoction (9 ingredients)	*In vivo* (Wistar rats)	28.6 g/kg	Intragastric administration	Protective effect on peripheral nerve injury by playing an anti-oxidative role in a diabetic rat model (increased MNCV and serum levels of glutathione, decreased serum levels of MDA)	Downregulation of NF-kB p65 and p38 MAPK mRNA expression	[98]
Qian-Zheng-San	Qian-Zheng-San (3 ingredients: *Typhonii rhizoma, Bombyx batryticatus, Scorpio*)	*In vivo* (SD rats)	1.75 g/mL	Oral gavage	Promoted peripheral nerve regeneration in rats with sciatic nerve defect (Increased SFI, MNCV, muscle wet weight, number of regenerated axons, axon diameter, nerve fiber diameter, myelin thickness, number of motor neurons in the lumbar spinal cord anterior horn)	N/A	[99]

Akt—protein kinase B; Bad—Bcl-2 associated agonist of cell death; Bax—Bcl-2-associated X protein; Bcl-2—B-cell lymphoma 2; BDNF—brain-derived neurotrophic factor; BGJTD—Bogijetong decoction; Cdc2—cell division cycle protein 2 homolog; CGRP—calcitonin gene-related peptide; CMAP—compound muscle action potential; CNTF—ciliary neurotrophic factor; CSA—cross-sectional area; DRG—dorsal root ganglion; DUOX2—dual oxidase 2; ERK—extracellular signal-regulated kinase; FGF—fibroblast growth factor; GABA—γ-aminobutyric acid; GAP-43—growth associated protein 43; GDNF—glial cell-derived neurotrophic factor; ICR mice—Institute of Cancer Research mice; IFN—interferon; IGF-1—insulin-like growth factor 1; IGF-IR—insulin-like growth factor 1 receptor; IL—interleukin; JNK—c-Jun N-terminal kinase; MAPK—mitogen-activated protein kinase; MBP—myelin basic protein; MDA—malondialdehyde; MMP—matrix metallopeptidase; MNCV—motor nerve conduction velocity; NF-κB—nuclear factor kappa B; NGF—nerve growth factor; NOX4—nicotinamide adenine dinucleotide phosphate oxidase 4; PAI-1—plasminogen activator inhibitor-1; PBS—phosphate buffered saline; PC12—pheochromocytoma cells; PCNA—proliferating cell nuclear antigen; PDGF—platelet-derived growth factor; PFI—peroneal nerve function index; PI3K—phosphoinositide 3-kinase; PMP22—peripheral myelin protein 22; Prom1—prominin 1; RECA-1—mouse monoclonal endothelial cell antibody; ROS—reactive oxygen species; RSC96 SC—RSC96 Schwann cell; SCs—Schwann cells; SD rats—Sprague-Dawley rats; SFI—sciatic function index; SNI—sciatic nerve injury; Sox18—sex-determining region Y-box transcription factor 18; TGF-β—transforming growth factor beta; TNF-α—tumor necrosis factor alpha; tPA—tissue plasminogen activator; Trk—tropomyosin receptor kinase; TRPC1—classical transient receptor potential 1; TSI—toe spread index; uPA—urokinase plasminogen activator; Vegf—vascular endothelial growth factor; WRL—withdrawal reflex latency.

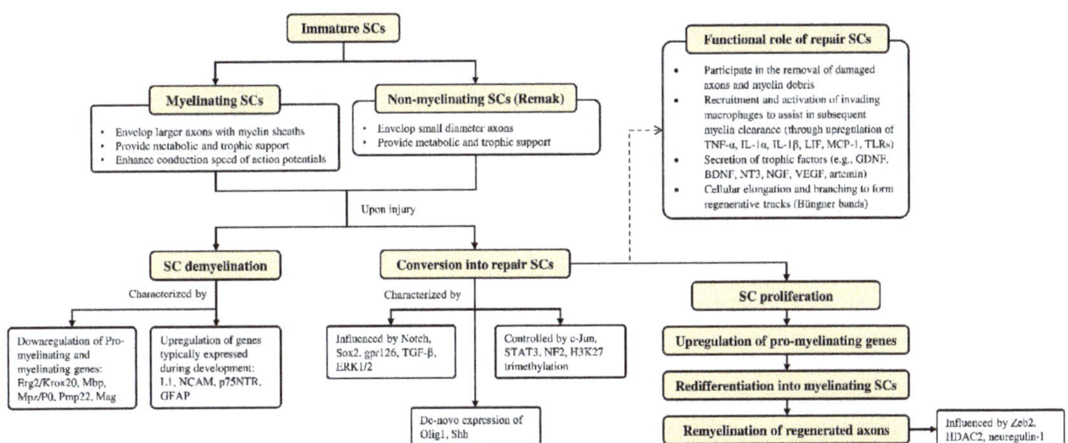

Figure 3. Overview of Schwann cell plasticity and their roles following peripheral nerve injury. Immature SCs develop into either myelinated or non-myelinated forms depending on the type of axon association. Upon nerve injury, SCs are capable of converting into a repair phenotype alongside the demyelination process that is mediated by different genes and transcriptional mechanisms. These events promote neuronal survival and enhance axonal regeneration following injury. Subsequently, repair SCs can be reprogrammed back to remyelinate regenerated axons. Further details on SC plasticity are presented in the reviews by Jessen & Mirsky [39] and Nocera & Jacob [100]. BDNF—brain-derived neurotrophic factor; Erg2/Krox20—early growth response 2; ERK—extracellular signal-regulated protein kinase; GDNF—glial cell-derived neurotrophic factor; GFAP—glial fibrillary acidic protein; gpr126—adhesion G protein-coupled receptor G6; H3K27—methylation of histone H3 on lysine 27; HDAC2—histone deacetylase 2; IL—interleukin; L1—L1 cell adhesion molecule; LIF—leukemia inhibitory factor; Mag—myelin associated glycoprotein; Mbp—myelin basic protein; MCP-1—monocyte chemotactic protein 1; Mpz/P0—myelin protein zero; NCAM—neural cell adhesion molecule; NF2—neurofibromatosis 2; NGF—nerve growth factor; NT3—neurotrophin-3; Olig1—oligodendrocyte transcription factor 1; p75NTR—p75 neurotrophin receptor; Pmp22—peripheral myelin protein 22; SCs—Schwann cells; Shh—Sonic Hedgehog; Sox2—(sex determining region Y)-box 2; STAT3—signal transducer and activator of transcription 3; TGF-β—transforming growth factor-β; TLRs—Toll-like receptors; TNF-α—tumor necrosis factor-α; VEGF—vascular endothelial growth factor; Zeb2—zinc finger E-box-binding homeobox 2.

4.4. Experimental Strategies and Neuroprotective Effects of Complementary and Alternative Medicines (CAMs) against Peripheral Nerve Injury

4.4.1. CAMs with Neuroregenerative Potential

Due to the limitations of conventional therapies for PNIs, much attention has been dedicated to finding alternative approaches in treating PNIs. To date, studies have explored the potential of 20 species of plants, three species of mushrooms, and four types of decoctions in promoting peripheral nerve regeneration (Table 1). Notably, the neuroregenerative potential of *Achyranthes bidentata* [44–49], *Astragalus membranaceus* [52–55], *Curcuma longa* [60–65], *Panax ginseng* [73–75], and *Hericium erinaceus* [92–94] have been most studied. A total of 18 natural products have been identified across the studies, and their chemical structures are shown in Table 2. Among those, ursolic acid, syringic acid, and quercetin are the NPs that can be found across a variety of plant species [78,79,81,101–103]. Decoctions are usually made according to traditional formulae. However, among the decoctions discussed in this study, the Bogijetong decoction is a relatively modern formulation that was specifically developed to treat neuropathic pain [96].

Table 2. Chemical structures of natural products and their respective sources.

Sources	Natural Product	Chemical Structure
Alpinate Oxyphyllae Fructus (*Alpinia oxyphylla* Miq)	Protocatechuic acid	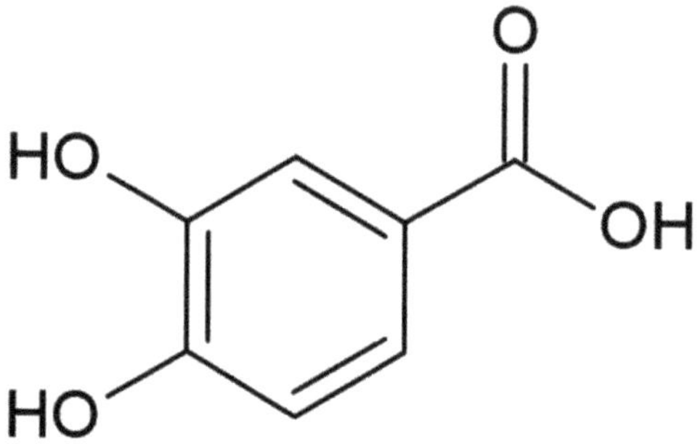
Astragalus membranaceus	Astragaloside IV	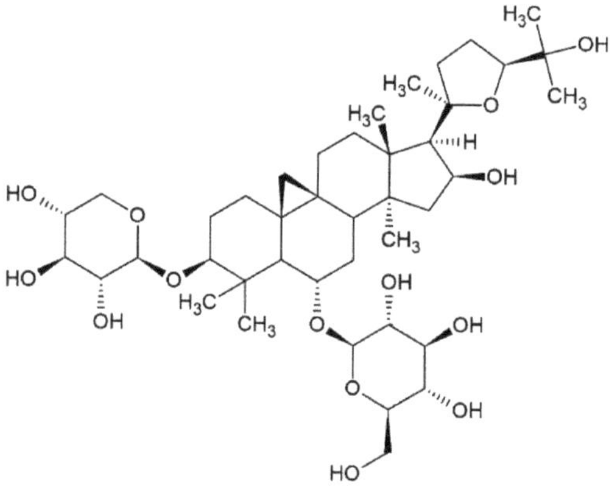
Crocus sativus	Crocin	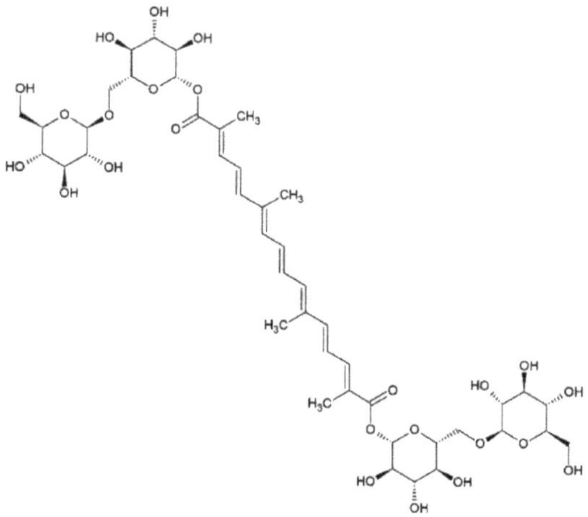

Table 2. *Cont.*

Sources	Natural Product	Chemical Structure
Curcuma longa	Curcumin	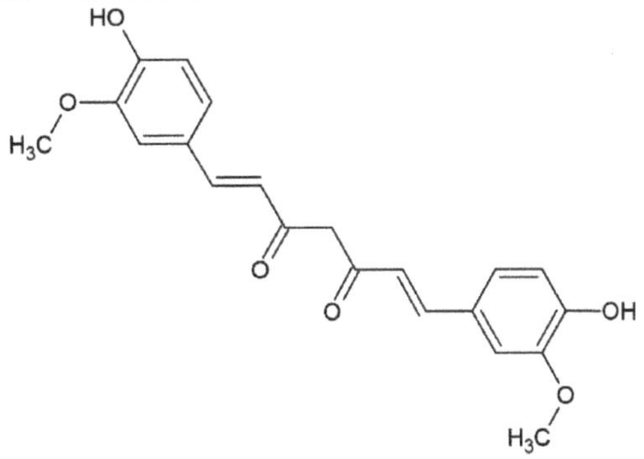
Epimedium	Icariin	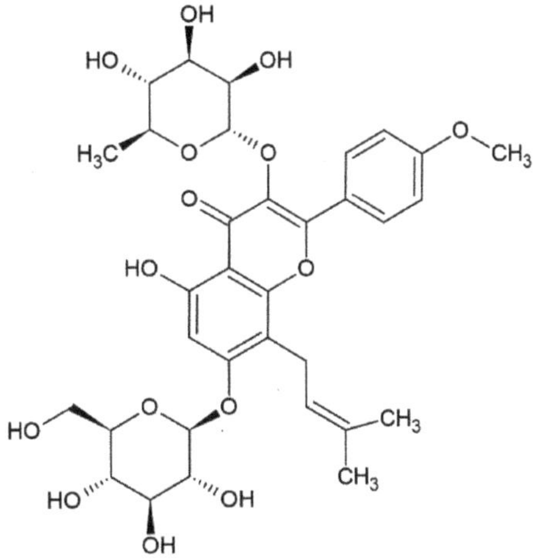
Gardenia jasminoides Ellis	Genipin	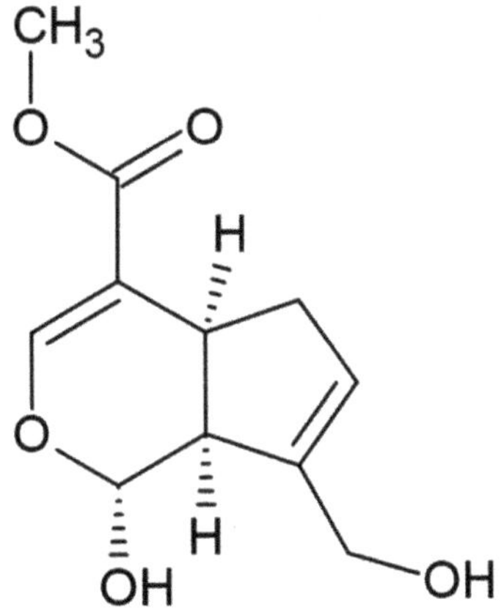

Table 2. Cont.

Sources	Natural Product	Chemical Structure
Gastrodia elata Blume	Gastrodin	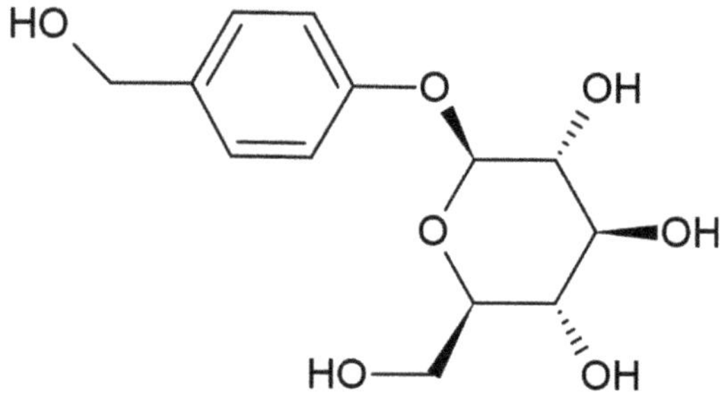
Ginseng	Ginsenoside Rg1	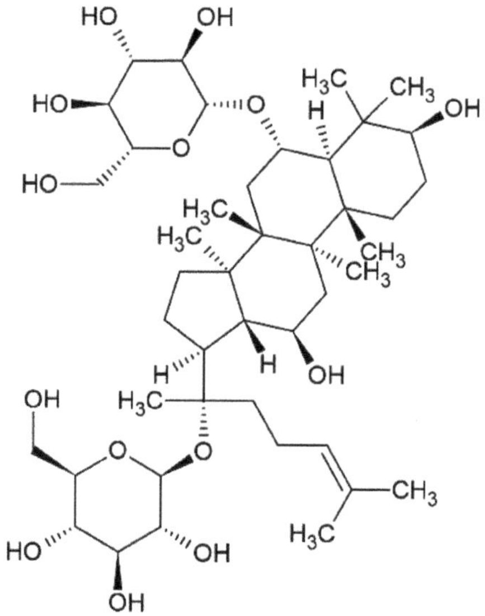
	Ginsenoside Re	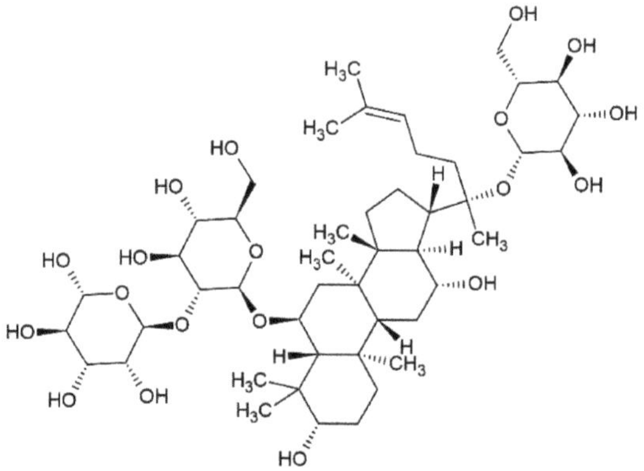

Table 2. *Cont.*

Sources	Natural Product	Chemical Structure
Green tea	(-)-Epigallocatechin-3-gallate (EGCG)	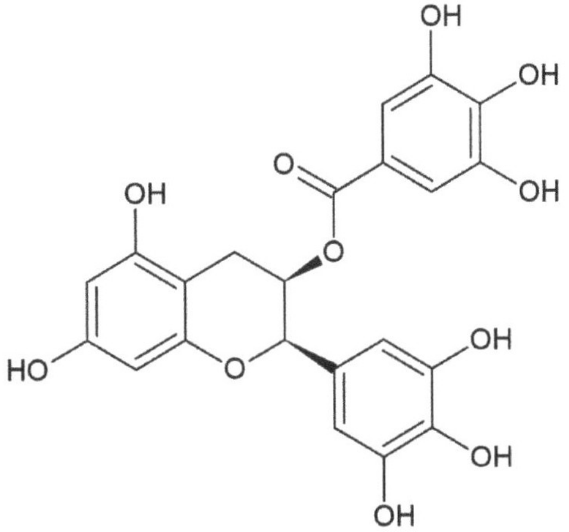
Isolated from a variety of plants (e.g., grapes, *Vitis vinifera*; olive, *Olea europaea*; radish, *Raphanus sativus*; pumpkin, *Cucurbita pepo* [101])	Syringic acid	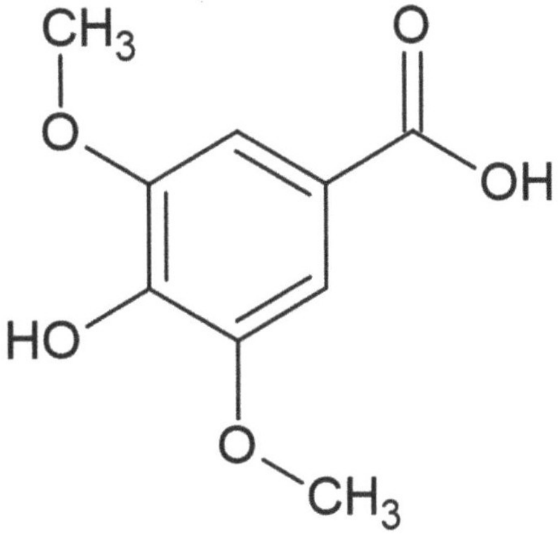
Isolated from a variety of plants (e.g., apple, *Malus domestica*; cranberry, *Vaccinium oxycoccus*; peppermint, *Mentha piperita*; and thyme, *Thymus vulgaris* [102])	Ursolic acid	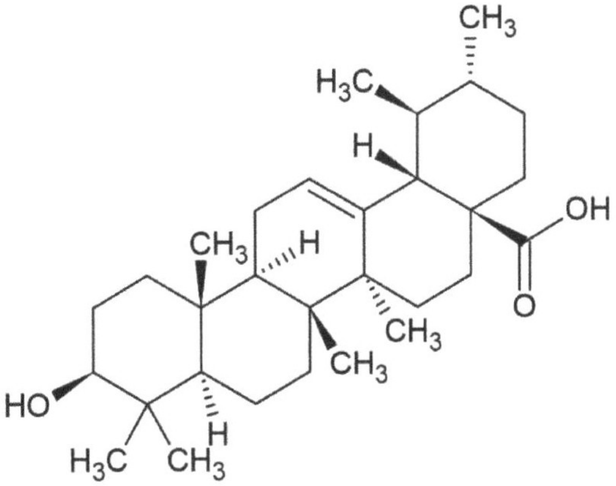

Table 2. *Cont.*

Sources	Natural Product	Chemical Structure
Isolated from a variety of plants (e.g., apple, *Malus domestica*; caper, *Capparis spinosa*; onion, *Allium cepa*; tomato, *Solanum lycopersicum*; and grapes, *Vitis vinifera* [103])	Quercetin	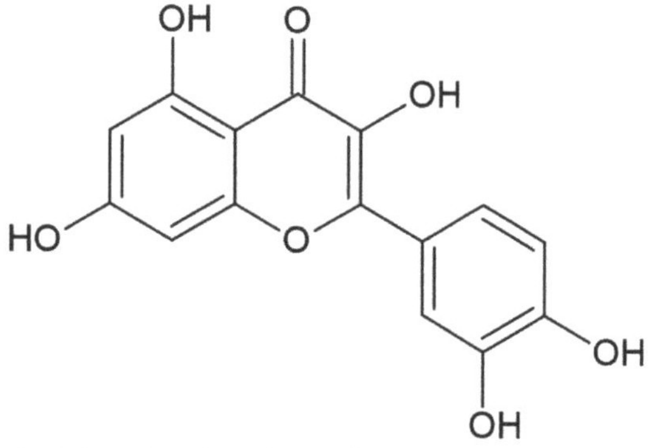
Pueraria lobata	Puerarin	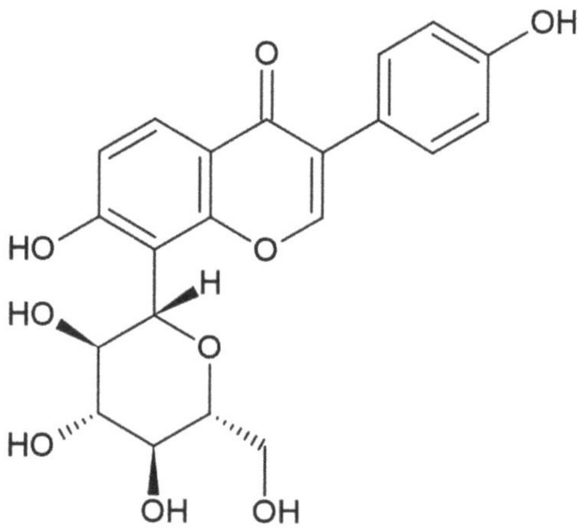
Rhodiola rosea L.	Salidroside	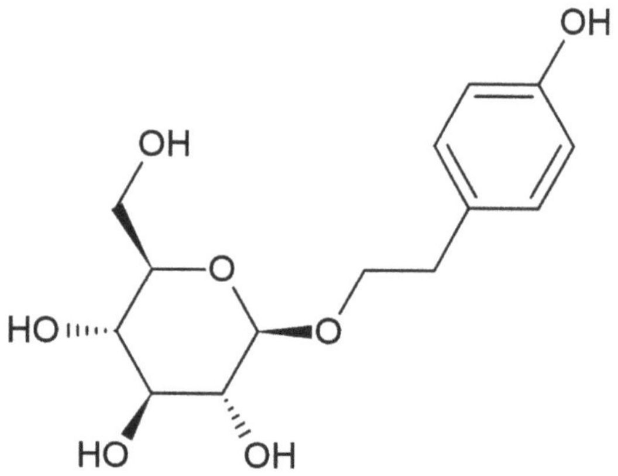

Table 2. *Cont.*

Sources	Natural Product	Chemical Structure
Scutellaria baicalensis Georgi	Baicalin	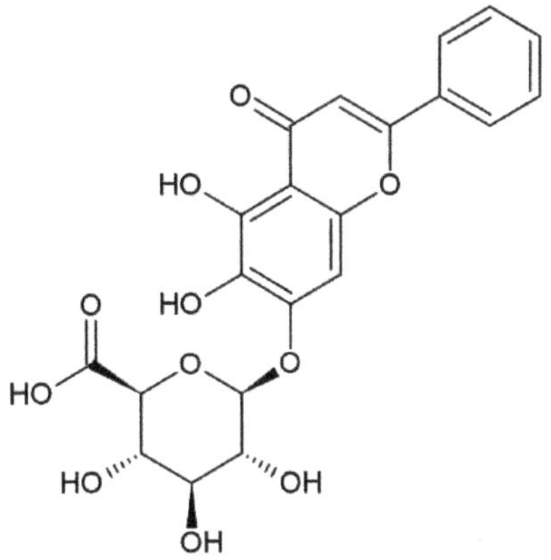
Tripterygium wilfordii Hook.F.	Triptolide	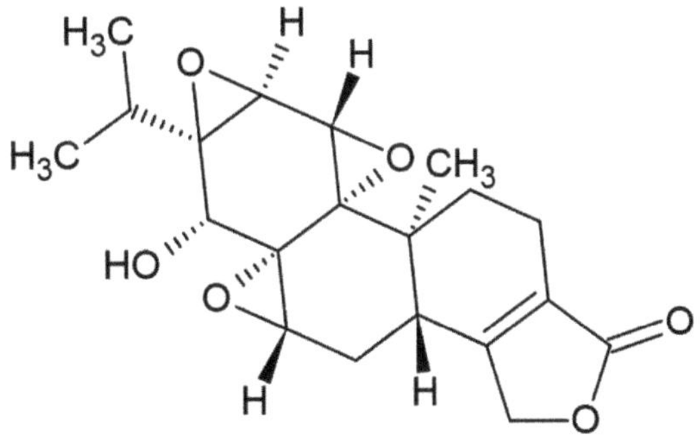
Amanita muscaria	Muscimol	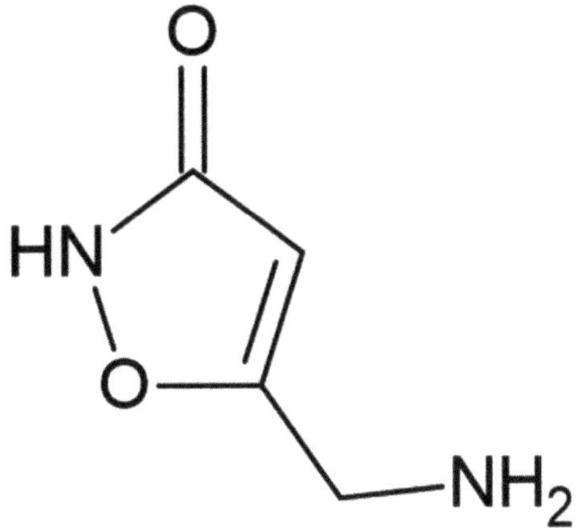

4.4.2. *In Vitro* Studies on Neuroregenerative Potential of CAMs

Figure 4 summarizes the *in vitro* studies on neuroregenerative properties of complementary and alternative medicines. Most of the studies were in Schwann cells, with a few using DRG explants, neurons, and PC12 cells (rat pheochromocytoma). Some CAMs were reported to induce proliferation, differentiation, and neurite outgrowth in PC12 cells. Similarly, neurite outgrowth was also promoted in DRG neurons through modulation of the extracellular signal-regulated kinase (ERK), p38, and tumor necrosis factor-α (TNF-α). Polypeptides isolated from *Achyranthes bidentata* have demonstrated the ability to promote neurite outgrowth in DRG neurons through the activation of ERK1/2 [45,46]. These findings resemble an earlier study that also reported neurite growth in DRG neurons induced by CD95 through ERK activation [104]. The Bogijetong decoction and its reconstituted formulation BeD elicited similar neuroprotective effects through downregulation of p38 and TNF-α [96] It was previously shown that TNF-α could inhibit neurite outgrowth in cultured DRG neurons [105,106], whereas the application of a TNF-α antagonist supported axonal regeneration following nerve injury [107].

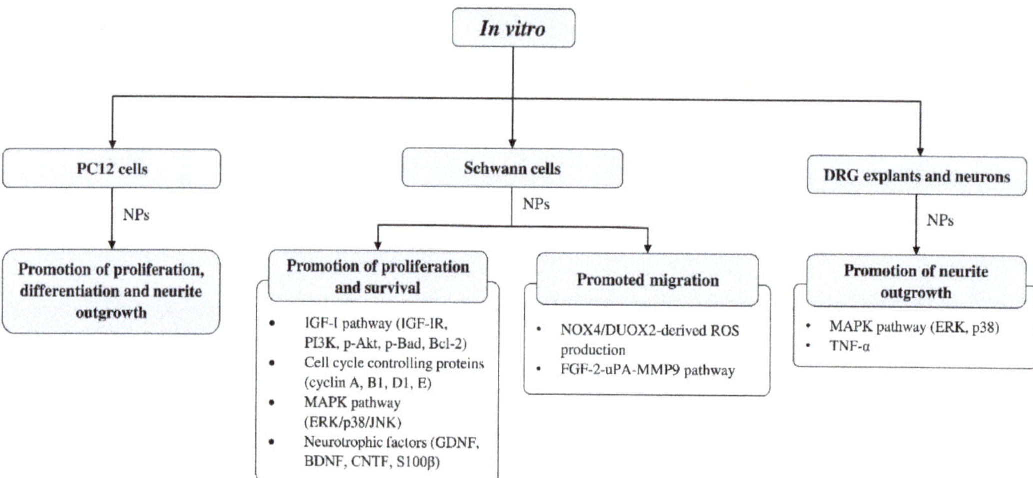

Figure 4. Overview of *in vitro* studies that demonstrated the effects of natural products relating to peripheral nerve regeneration across different cell types with associated mechanisms. Akt—protein kinase B; Bad—Bcl-2 associated agonist of cell death; Bcl-2—B-cell lymphoma 2; BDNF—brain-derived neurotrophic factor; CNTF—ciliary neurotrophic factor; DRG—dorsal root ganglion; DUOX2—dual oxidase 2; ERK—extracellular signal-regulated kinase; FGF—fibroblast growth factor; GDNF—glial cell-derived neurotrophic factor; IGF-I—insulin-like growth factor 1; IGF-IR—insulin-like growth factor 1 receptor; JNK—c-Jun N-terminal kinase; MAPK—mitogen-activated protein kinase; MMP9—matrix metallopeptidase 9; NOX4—nicotinamide adenine dinucleotide phosphate (NADPH) oxidase 4; NPs—natural products; PC12—pheochromocytoma cells; PI3K—phosphoinositide 3-kinase; ROS—reactive oxygen species; S100β—S100 calcium-binding protein β; TNF-α—tumor necrosis factor-α; uPA—urokinase plasminogen activator.

Effects of CAMs on Schwann Cell Activity *In Vitro*

The studies examining the effects of complementary and alternative medicines and their related natural products on Schwann cells are primarily focused on promoting their proliferation and survival. The molecular mechanisms that were investigated in these studies include signaling pathways such as IGF-I and MAPK, as well as cell cycle controlling proteins and various neurotrophic factors (Figure 4). Past studies have demonstrated that ERK is required for proper myelination of SCs during development [108,109], and ERK signaling was rapidly activated following nerve injury, contributing to SC differentiation [110]. Moreover, evidence suggests that nerve regeneration following injury is closely associated with ERK [111,112], and ERK inhibition leads to impaired regenerative capability [111,113]. On the other hand, inhibition of p38 MAPK prevented SC demyelination and dedifferentiation, indicating its role in promoting the breakdown of myelin following nerve injury [114]. It is not unexpected that cyclins are associated with SC proliferation, as these proteins control cell cycle progression through the interaction of cyclin-dependent kinases. For instance, cyclin D is associated with Cdk4 or Cdk6 in the G1 phase, cyclin A participates with Cdk1 or Cdk2 in the S phase, cyclin E is involved with Cdk2 in G1 and S phases, cyclin B and Cdk1 regulates M phase [115,116].

Protocatechuic acid isolated from *Alpinia oxyphylla* Miq [50] and the aqueous extract of *Codonopsis pilosula* [58] were found to promote SCs proliferation by further enhancing IGF-I (insulin-like growth factor 1) signaling. The IGF-I growth factor is known to play a crucial role in neuromuscular recovery following injury. It is reported to be involved in promoting G1/S cell cycle progression [117] and survival of SCs [118] *in vitro*, and to facilitate peripheral nerve regeneration *in vivo* [119–122]. One study reported baicalin, a flavonoid that possesses various neuroprotective effects [123], induced proliferation of SCs through the modulation of neurotrophic factors including glial cell-derived neurotrophic factor (GDNF), BDNF, and ciliary neurotrophic factor (CTNF) [88]. These neurotrophic factors are integral to many aspects of nerve regeneration, as evident in past studies that showed their roles in myelin formation [124,125] and axonal regeneration [126,127].

In addition to promoting the proliferation of SCs, some NPs may promote the migratory ability of SCs, which is essential for regeneration and remyelination following nerve injury. Aqueous extracts of *Alpinia oxyphylla* Miq [51], *Astragalus membranaceus* [55], *Citrus medica* var. *sarcodactylis* [57], *Codonopsis pilosula* [58], and ginsenoside Rg1 isolated from ginseng [73] enhanced SC migration through the activation of FGF-2 signaling. The role of FGF-2 in the repair and regeneration of tissues [128] and its involvement in cell migration [129,130] is widely documented. A recently published study reported that another subfamily member, FGF5, is also involved in regulating SC migration and adhesion [131]. Besides FGF-2 signaling, another study investigating polypeptides of *A. bidentata* revealed that the upregulation of NOX4/DUOX2-derived reactive oxygen species (ROS) production was responsible for promoting SC migration [44]. Excessive accumulation of ROS production has been linked to neurodegenerative disorders [132] and peripheral neuropathy [133], but moderate levels of ROS may prove beneficial by acting as signal messengers in regulating biological processes, including cell adhesion and migration [134,135]. Syringic acid was shown to promote the proliferation and migration of SCs *in vitro*. Although the expression of several microRNAs was affected by syringic acid, further analysis suggested that the plant polyphenol promoted SC proliferation and migration mainly by suppressing the microRNA miR-451-5p [78].

4.4.3. *In Vivo* Studies on Neuroregenerative Potential of CAMs

Current *in vivo* studies on the potential of complementary and alternative medicines in peripheral nerve regeneration are limited to rodent models (Figure 5 and Table 1). Most of the studies involved different strains of rats and mice, with only two studies using rabbits as their animal models. Models of peripheral nerve injury used in the studies include diabetic peripheral neuropathy, peroneal nerve injury, and sciatic nerve injury. The effects of CAMs on peripheral nerve regeneration were evaluated by functional recovery indexes (e.g., PFI; sciatic function index, SFI; tibial function index, TFI; CMAP; MNCV; and WRL) and histological examinations (e.g., number, diameter, the thickness of myelinated fibers and regenerated axons; the number of motoneurons; and muscle mass).

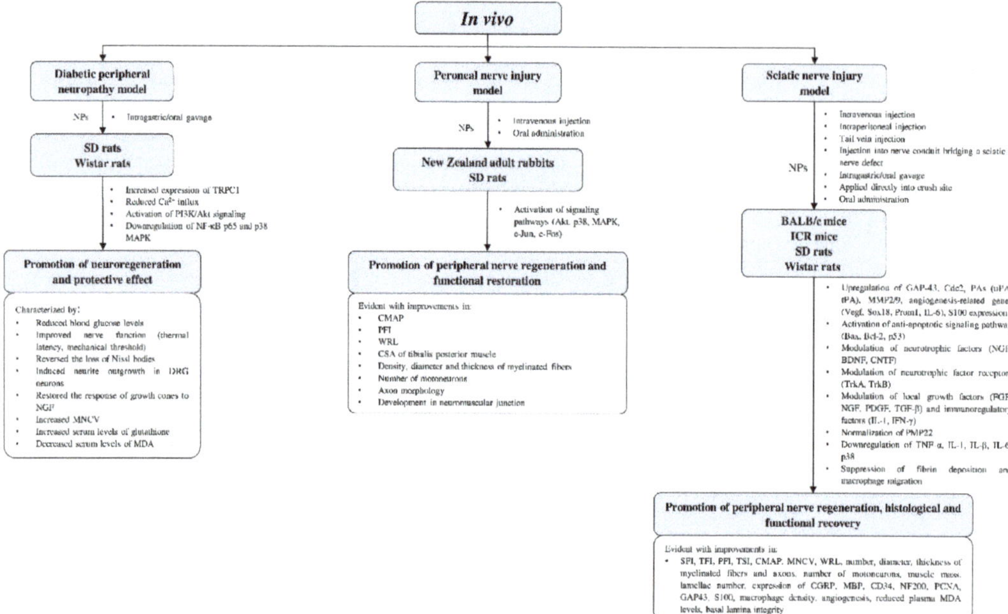

Figure 5. Overview of *in vivo* studies that demonstrated the effects of natural products relating to peripheral nerve regeneration across different experimental models with associated mechanisms. Akt—protein kinase B; Bax—Bcl-2-associated X protein; Bcl-2—B-cell lymphoma 2; BDNF—brain-derived neurotrophic factor; Cdc2—cell division control protein; CGRP—calcitonin gene-related peptide; CMAP—compound muscle action potential; CNTF—ciliary neurotrophic factor; CSA—cross-sectional area; DRG—dorsal root ganglion; FGF—fibroblast growth factor; GAP-43—growth associated protein 43; ICR—Institute of Cancer Research; IFN-γ—interferon-γ; IL—interleukin; MAPK—mitogen-activated protein kinase; MBP—myelin basic protein; MDA—malondialdehyde; MMP2/9—matrix-metalloproteinase-2/9; MNCV—motor nerve conduction velocity; NF-κB—nuclear factor kappa B; NGF—nerve growth factor; NPs—natural products; PAs—plasminogen activators; PCNA—proliferating cell nuclear antigen; PDGF—platelet-derived growth factor; PFI—peroneal function index; PI3K—phosphoinositide 3-kinase; PMP22—peripheral myelin protein 22; Prom1—prominin 1; SD—Sprague-Dawley; SFI—sciatic function index; Sox18—sex-determining region Y-box transcription factor 18; TFI—tibial function index; TGF-β—transforming growth factor-β; TNF-α—tumor necrosis factor-α; tPA—tissue plasminogen activator; Trk—tropomyosin receptor kinase; TRPC1—transient receptor potential cation channel subfamily C member 1; TSI—toe spread index; uPA—urokinase plasminogen activator; Vegf—vascular endothelial growth factor; WRL—withdrawal reflex latency.

Diabetic Peripheral Neuropathy Model

In the diabetic neuropathy (DPN) model, aqueous extract of Cortex Mori Radicis had anti-diabetic and neuroregenerative effects, as evidenced by reduced blood glucose levels, induced neurite outgrowth, restoration of the loss of Nissl bodies, and a response in the growth cones of DRG neurons [82]. The authors identified that the observed effects were mediated by the activation of the PI3K/Akt pathway and increased expression of TRPC1, which in turn reduced Ca2+ influx. The PI3K/Akt pathway is a crucial intracellular signaling pathway that governs cell proliferation, survival, and metabolism [136], its protective role against DPN has been previously hinted at [137,138]. The transient receptor potential (TRPC) is a family of Ca2+-permeable channels that participates in axonal regeneration [139]. In particular, TRPC1 and TRPC4 were shown to induce neurite outgrowth in PC12 cells and DRG neurons [140,141]. In another study, administration of Jiaweibugan decoction in a DPN model ameliorated changes in motor nerve conduction velocity (MNCV), and malondialdehyde (MDA), and glutathione levels through an anti-oxidative pathway via downregulating NF-κB p65 and p38 MAPK [98]. The activation of p38 MAPK, which belongs to a family of kinases that are responsive to stress stimuli, further activates NF-κB leading to inflammation, a driving factor of DPN [142,143].

Peroneal Nerve Injury Model

In the peroneal nerve injury model, aqueous extract and polypeptides of *A. bidentata* were shown to enhance nerve regeneration [45,49], as indicated by increased density and diameter of myelinated fibers, and numbers of motor neurons. Although behavioral analyses were not performed in the studies, improvements in compound muscle action potential (CAMP) demonstrated the ability of *A. bidentata* aqueous extract and polypeptides to promote functional recovery. Aqueous extract and polysaccharides from *Hericium erinaceus* also exhibited nerve regeneration and functional recovery following peroneal nerve crush [92–94], as evidenced by the improvements in the peroneal function index (PFI), withdrawal reflex latency (WRL) and axon morphology, and the development of neuromuscular junction. These findings were supported by the activation of Akt, p38, c-Jun, and c-Fos, which is in line with other studies that showed the importance of these signaling events for axonal regeneration [144–146].

Sciatic Nerve Injury Model

The sciatic nerve injury (SNI) model is the most commonly used model in the study of the effects of complementary and alternative medicines on peripheral nerve regeneration, and many studies have investigated the underlying mechanisms or molecular pathways through which CAMs elicit their neuroregenerative properties. For instance, polypeptides of *A. bidentata* [47], astragaloside IV isolated from *A. membranaceus* [52], and aqueous extract of *Dioscoreae rhizoma* [66] promoted nerve regeneration via upregulation of GAP-43 expression. The GAP-43 protein is highly associated with the development and plasticity of the nervous system [147]. Its expression is known to be elevated following nerve injury [148] and is involved in the neurite outgrowth of hippocampal neurons [149]. Similarly, modulation of other neurotrophic factors such as NGF, BDNF, CNTF [47,54], and pro-inflammatory cytokines including IL-1, IL-6, IL-β, and TNF-α [54,90] were also involved in promoting nerve regeneration as well. Although an inflammatory response following injury is necessary for the regenerative process [150], prolonged inflammation can impede recovery and may even lead to the development of neuropathic pain [151], which reflects the double-edged property of inflammation and the importance of proper regulation. Additionally, a study on *Ginkgo biloba* extract showed that it promoted axonal angiogenesis through the modulation of related genes, including Vegf, Sox18, Prom1, and IL-6 [71]. Studies have also demonstrated the participation of Vegf [152,153], Prom1 [154], and another subfamily gene, Sox11 in sciatic nerve regeneration [155], and the restorative role of IL-6 has also been implied in DPN and central nervous system models [156,157]. Muscimol prevented hyperalgesia through the modulation of PMP22 [91], a key component of the basal lamina. The expression of PMP22 is correlated with myelin formation and nerve regeneration [158,159]. Studies investigating EGCG in an SNI model showed that it had neuroprotective and regenerative effects, partly due to the modulation of the apoptotic machinery, including Bax, Bcl-2, p53, and survivin [76,77]. The subsequent loss of neurons after PNI is known to be closely related to apoptosis [160] which is partly influenced by p53 and Bax [161], while the association of survivin in nerve injury has also been documented [162].

4.4.4. Involvement of CAMs in Combinatorial Approaches for the Treatment of PNI

There is increasing evidence that the successful repair and regeneration of nerves will require not just a single treatment strategy, but a multifaceted strategy involving different disciplines. Studies adopting combinatorial approaches have yielded interesting findings. For example, *Lycium barbarum* polysaccharide incorporated into core-shell structured nanofibrous scaffolds by coaxial electrospinning showed proliferative effects in PC12, SCs, and DRG neurons [80]. In two separate studies, puerarin, the active component extracted from *Pueraria lobata* roots, as well as rat serum metabolites of *P. lobata* enhanced the neuroregenerative effects of silicone rubber nerve chambers. Increases in myelinated axons and structurally mature regenerated axons were observed, while muscle reinnervation led to functional recovery, as indicated by an increase in action potential and nerve conduction [83,84]. Similar results were obtained with Buyang Huanwu decoction being administered as a co-treatment alongside silicone rubber nerve chambers, which led to more prominent axonal regeneration [97]. In an SNI model, a magnetic nanocomposite scaffold produced from using magnetic nanoparticles and biodegradable chitosan-glycerophosphate polymer enhanced SC viability, nerve regeneration, and functional recovery when paired with an applied magnetic field [163]. The use of nerve guiding conduits gained popularity over the years. They have been used to isolate regenerating axons from fibrotic tissues, to protect them from mechanical forces, and to guide new-forming tissue as well as condensing growth factors secreted by SCs [164]. The concept was initiated with a simple hollow design but has since advanced to innovative ways of redesigning nerve conduits to further extend their original capabilities 11. The attractive characteristics of modern nerve conduits offer tremendous potentials. These nerve conduits are occasionally paired with other strategies for improving nerve outcomes. For instance, Chang et al. [165] developed a natural biodegradable multi-channeled scaffold with aligned electrospun nanofibers and a neurotrophic gradient, which resulted in superior nerve recovery and less muscle atrophy compared with nerve autografts. Hussin et al. [56] used *Centella asiatica* (L.) to neurodifferentiate mesenchymal stem cells. This was subsequently developed with decellularized artery as a nerve conduit, which demonstrated functional restoration in an SNI model similar to that of reversed autograft.

4.5. Limitations and Future Prospects

As mentioned earlier, PNI represents a significant health issue while the effectiveness of current treatment approaches is highly subjective. Hence, substantial effort is required to discover and establish proper methods for the management of PNI. Present studies have shown promising findings in utilizing various applications including nerve conduits [166], stem cell therapy [167], phytochemicals [22], and electrical stimulation [168] for treating PNI, and their potential may subsequently be improved when paired together. Evidence from *in vitro* and *in vivo* studies have delineated the neuroregenerative properties of various CAMs, and the underlying mechanisms have been investigated (as summarized in Figure 6), although they still remain incompletely understood and require further elucidation. Subsequently, pre-clinical and clinical studies on existing potential candidates and approaches should be supported to drive the development of future therapeutics.

Existing studies on the effect of complementary medicines in treating PNI are preliminary findings with limited information (Table 1). The majority of studies investigated crude extracts or specific fractions of extracts, with only 24 out of the 56 studies managing to identify the exact NPs responsible for the observed effects. Additionally, 25 studies did not report the underlying mechanisms for the resultant effects of NPs, especially at the *in vivo* stage. This situation highlights the need for greater efforts among the scientific community to fully investigate the purported effects of NPs. Another issue is the route and method of administration *in vivo*. It is known that oral administration is generally economical and relatively safe, but the resultant efficacy may not be reliable due to uncontrollable animal habits and behaviors [169]. In contrast, gavage or injection routes typically require some form of restraint, which may result in animal stress that may influence study outcomes. The administration routine also varied across studies, with the treatments lasting from a few days to months. Moreover, treatment frequency also influences experimental outcomes. Although it is difficult to standardize animal handling procedures, these factors should be taken into account with carefully designed studies.

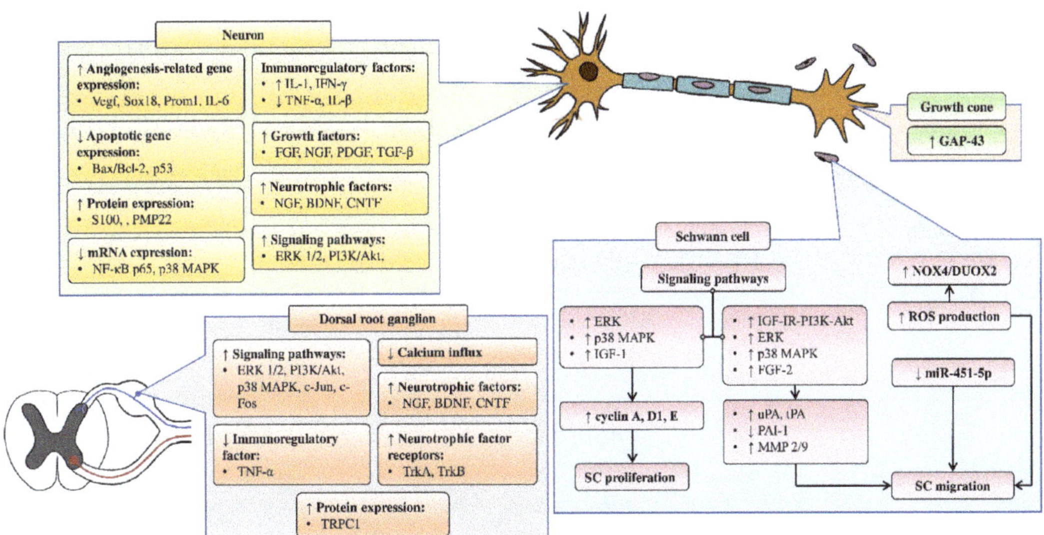

Figure 6. Summary of the molecular mechanisms associated with the neuroregenerative effects of CAMs. Vegf—vascular endothelial growth factor; Sox18—sex-determining region Y-box transcription factor 18; Prom1—prominin 1; IL—interleukin; IFN-γ—interferon-γ; Bax—Bcl-2-associated X protein; Bcl-2—B-cell lymphoma 2; Trk—tropomyosin receptor kinase; PMP22—peripheral myelin protein 22; FGF—fibroblast growth factor; NGF—nerve growth factor; PDGF—platelet-derived growth factor; TGF-β—transforming growth factor-β; NF-κB—nuclear factor kappa B; MAPK—mitogen-activated protein kinase; PI3K—phosphoinositide 3-kinase; Akt—protein kinase B; BDNF—brain-derived neurotrophic factor; CNTF—ciliary neurotrophic factor; TNF-α—tumor necrosis factor-α; TRPC1—transient receptor potential cation channel subfamily C member 1; GAP-43—growth-associated protein 43; PAI-1—plasminogen activator inhibitor-1; MMP2/9—matrix-metalloproteinase-2/9; tPA—tissue plasminogen activator; uPA—urokinase plasminogen activator.

In this review, the majority of studies on NPs as a treatment for PNI were based on plants and herbs, with a few studies on mushrooms such as *Amanita muscaria*, *Hericium erinaceus*, and *Lignosus rhinocerotis*, as well as some decoctions. This is unsurprising, considering that phytochemicals are highly studied for drug development, which should shed more light on this area of research [170–172]. However, the use of NPs for peripheral nerve repair and regeneration is still largely overlooked and could be an untapped potential source for promising drug candidates. For instance, a previous study demonstrated that various mushrooms including *Agaricus blazei* Murrill, *Antrodia cinnamomea*, *Ganoderma lucidum*, and *Hirsutella sinensis* could activate intracellular signaling kinases ERK, JNK, and p38, which are associated with peripheral nerve regeneration [173]. Another study showed that *G. lucidum*, *Ganoderma neo-japonicum*, and *Grifola frondosa* promoted neuritogenesis via the involvement of the MAPK signaling pathway [174]. Aside from exploring untapped sources of NPs, future research may also simultaneously examine the efficiency of CAMs or NPs with known neuroregenerative properties to compare their ability to promote regeneration of peripheral nerves.

The use of algae in peripheral nerve regeneration merits attention. Algae are well-known for their diverse applications in food nutrition [175], biofuels [176], cosmetics [177], and pharmaceuticals [178,179]. Recent studies have also demonstrated that algae could have potential in the treatment of neurological disorders, including Parkinson's and Alzheimer's disease [180,181]. However, the potential uses of algae in peripheral nerve regeneration have yet to be explored, despite evidence showing the ability of macroalgae to promote neurite outgrowth in hippocampal neurons [182–184]. More recently, a study showed that *Gracilaria manilaensis* induced the proliferation of neurite-bearing cells in the rat pheochromocytoma cell line, which is believed to mimic the neuroactivity of NGF [185]. Thus, investigation on the nerve regenerative potential of other NPs holds much promise.

5. Conclusions

Peripheral nerve injury remains a challenge, while future prospects are leaning towards multi-combinatorial approaches. Natural products are highly appreciated for their therapeutic value, and there is a growing body of evidence in their potential for peripheral nerve regeneration. The present findings showed that various NPs promote the proliferation and migration of SCs, most commonly through the activation of MAPK and FGF-2 signaling pathways, respectively. Promotion of peripheral nerve regeneration was also observed in rodent models, partly through the modulation of neurotrophic factors, pro-inflammatory cytokines, and anti-apoptotic signaling. Hence, NPs could play key roles in nerve repair and regeneration in the near future, especially when paired with other innovative approaches such as modern nerve conduits.

Author Contributions: Conceptualization, Y.-Y.Y., K.-H.W. and L.-W.L.; data curation, Y.-Y.Y., T.-K.G. and writing—original draft preparation, Y.-Y.Y., T.-K.G. and K.-Y.N.; review and editing, Y.-Y.Y., K.-Y.N., K.-H.W., L.-W.L., S.-M.P., S.-H.L. and S.R.; supervision, Y.-Y.Y., K.-H.W. and L.-W.L.; project administration, Y.-Y.Y. and K.-H.W.; funding acquisition, Y.-Y.Y. All authors have read and agreed to the published version of the manuscript.

Funding: This work was funded by Fundamental Research Grant Scheme FRGS/1/2019/STG05/ SYUC/02/1 from the Ministry of Higher Education Malaysia.

Acknowledgments: The authors are thankful to Sunway University for providing the necessary internet and library facilities for literature searching.

Conflicts of Interest: The authors declare no conflict of interest.

References

1. Navarro, X. Functional evaluation of peripheral nerve regeneration and target reinnervation in animal models: A critical overview. *Eur. J. Neurosci.* **2016**, *43*, 271–286. [CrossRef]
2. Wojtkiewicz, D.M.; Saunders, J.; Domeshek, L.; Novak, C.B.; Kaskutas, V.; Mackinnon, S.E. Social impact of peripheral nerve injuries. *Hand* **2015**, *10*, 161–167. [CrossRef]
3. Li, N.Y.; Onor, G.I.; Lemme, N.J.; Gil, J.A. Epidemiology of peripheral nerve injuries in sports, exercise, and recreation in the United States, 2009–2018. *Phys. Sportsmed.* **2020**, *49*, 1–8. [CrossRef]
4. Scholz, T.; Krichevsky, A.; Sumarto, A.; Jaffurs, D.; Wirth, G.A.; Paydar, K.; Evans, G.R.D. Peripheral nerve injuries: An international survey of current treatments and future perspectives. *J. Reconstr. Microsurg.* **2009**, *25*, 339–344. [CrossRef]
5. Antoniadis, G.; Kretschmer, T.; Pedro, M.T.; König, R.W.; Heinen, C.P.G.; Richter, H.P. Iatrogenic nerve injuries - prevalence, diagnosis and treatment. *Dtsch. Arztebl. Int.* **2014**, *111*, 273–279. [CrossRef]
6. Ciaramitaro, P.; Mondelli, M.; Logullo, F.; Grimaldi, S.; Battiston, B.; Sard, A.; Scarinzi, C.; Migliaretti, G.; Faccani, G.; Cocito, D. Traumatic peripheral nerve injuries: Epidemiological findings, neuropathic pain and quality of life in 158 patients. *J. Peripher. Nerv. Syst.* **2010**, *15*, 120–127. [CrossRef] [PubMed]
7. Grinsell, D.; Keating, C.P. Peripheral nerve reconstruction after injury: A review of clinical and experimental therapies. *BioMed Res. Int.* **2014**, *2014*, 698256. [CrossRef]
8. Ray, W.Z.; Mackinnon, S.E. Management of nerve gaps: Autografts, allografts, nerve transfers, and end-to-side neurorrhaphy. *Exp. Neurol.* **2010**, *223*, 77–85. [CrossRef] [PubMed]
9. Wang, E.W.; Zhang, J.; Huang, J.H. Repairing peripheral nerve injury using tissue engineering techniques. *Neural Regen. Res.* **2015**, *10*, 1393–1394. [CrossRef] [PubMed]
10. Houschyar, K.S.; Momeni, A.; Pyles, M.N.; Cha, J.Y.; Maan, Z.N.; Duscher, D.; Jew, O.S.; Siemers, F.; van Schoonhoven, J. The role of current techniques and concepts in peripheral nerve repair. *Plast. Surg. Int.* **2016**, *2016*, 4175293. [CrossRef]
11. Carvalho, C.R.; Oliveira, J.M.; Reis, R.L. Modern trends for peripheral nerve repair and regeneration: Beyond the hollow nerve guidance conduit. *Front. Bioeng. Biotechnol.* **2019**, *7*, 337. [CrossRef]
12. Brunelli, B.; Gorson, K.C. The use of complementary and alternative medicines by patients with peripheral neuropathy. *J. Neurol. Sci.* **2004**, *218*, 59–66. [CrossRef]
13. Muthuraman, A.; Singh, N.; Jaggi, A.S. Effect of hydroalcoholic extract of *Acorus calamus* on tibial and sural nerve transection-induced painful neuropathy in rats. *J. Nat. Med.* **2011**, *65*, 282–292. [CrossRef]

14. Zhao, X.; Xu, Y.; Zhao, Q.; Chen, C.R.; Liu, A.M.; Huang, Z.L. Curcumin exerts antinociceptive effects in a mouse model of neuropathic pain: Descending monoamine system and opioid receptors are differentially involved. *Neuropharmacology* **2012**, *62*, 843–854. [CrossRef] [PubMed]
15. Kim, Y.S.; Park, H.J.; Kim, T.K.; Moon, D.E.; Lee, H.J. The effects of *Ginkgo biloba* extract EGB 761 on mechanical and cold allodynia in a rat model of neuropathic pain. *Anesth. Analg.* **2009**, *108*, 1958–1963. [CrossRef] [PubMed]
16. Calixto, J.B. Efficacy, safety, quality control, marketing and regulatory guidelines for herbal medicines (phytotherapeutic agents). *Braz. J. Med. Biol. Res.* **2000**, *33*, 179–189. [CrossRef]
17. Karimi, A.; Majlesi, M.; Rafieian-Kopaei, M. Herbal versus synthetic drugs; beliefs and facts. *J. Nephropharmacology* **2015**, *4*, 27–30.
18. Lahlou, M. The success of natural products in drug discovery. *Pharmacol. Pharm.* **2013**, *04*, 17–31. [CrossRef]
19. Newman, D.J.; Cragg, G.M. Natural products as sources of new drugs over the nearly four decades from 01/1981 to 09/2019. *J. Nat. Prod.* **2020**, *83*, 770–803. [CrossRef]
20. Weaver, B.A. How taxol/paclitaxel kills cancer cells. *Mol. Biol. Cell* **2014**, *25*, 2677–2681. [CrossRef]
21. Noble, R.L. The discovery of the vinca alkaloids—Chemotherapeutic agents against cancer. *Biochem. Cell Biol.* **1990**, *68*, 1344–1351. [CrossRef]
22. Araújo-Filho, H.G.; Quintans-Júnior, L.J.; Barreto, A.S.; Almeida, J.R.G.S.; Barreto, R.S.S.; Quintans, J.S.S. Neuroprotective effect of natural products on peripheral nerve degeneration: A systematic review. *Neurochem. Res.* **2016**, *41*, 647–658. [CrossRef]
23. Moher, D.; Shamseer, L.; Clarke, M.; Ghersi, D.; Liberati, A.; Petticrew, M.; Shekelle, P.; Stewart, L.A.; Group, P.-P. Preferred reporting items for systematic review and meta-analysis protocols (PRISMA-P) 2015 statement. *Syst. Rev.* **2015**, *4*, 1–19. [CrossRef] [PubMed]
24. Hussain, G.; Wang, J.; Rasul, A.; Anwar, H.; Qasim, M.; Zafar, S.; Aziz, N.; Razzaq, A.; Hussain, R.; de Aguilar, J.L.G.; et al. Current status of therapeutic approaches against peripheral nerve injuries: A detailed story from injury to recovery. *Int. J. Biol. Sci.* **2020**, *16*, 116–134. [CrossRef] [PubMed]
25. Millesi, H. Bridging defects: Autologous nerve grafts. *Acta Neurochir. Suppl.* **2007**, *100*, 37–38. [CrossRef] [PubMed]
26. Griffin, J.W.; Hogan, M.C.V.; Chhabra, A.B.; Deal, D.N. Peripheral nerve repair and reconstruction. *J. Bone Joint Surg. Am.* **2013**, *95*, 2144–2151. [CrossRef]
27. Li, R.; Li, D.H.; Zhang, H.Y.; Wang, J.; Li, X.K.; Xiao, J. Growth factors-based therapeutic strategies and their underlying signaling mechanisms for peripheral nerve regeneration. *Acta Pharmacol. Sin.* **2020**, *41*, 1289–1300. [CrossRef]
28. Kubiak, C.A.; Grochmal, J.; Kung, T.A.; Cederna, P.S.; Midha, R.; Kemp, S.W.P. Stem-cell–based therapies to enhance peripheral nerve regeneration. *Muscle Nerve* **2020**, *61*, 449–459. [CrossRef]
29. Gordon, T.; English, A.W. Strategies to promote peripheral nerve regeneration: Electrical stimulation and/or exercise. *Eur. J. Neurosci.* **2016**, *43*, 336–350. [CrossRef]
30. Choudhari, A.S.; Mandave, P.C.; Deshpande, M.; Ranjekar, P.; Prakash, O. Phytochemicals in cancer treatment: From preclinical studies to clinical practice. *Front. Pharmacol.* **2020**, *10*, 1614. [CrossRef]
31. Kumar, G.P.; Khanum, F. Neuroprotective potential of phytochemicals. *Pharmacogn. Rev.* **2012**, *6*, 81–90. [CrossRef]
32. Rotshenker, S. Wallerian degeneration: The innate-immune response to traumatic nerve injury. *J. Neuroinflammation* **2011**, *8*, 1–14. [CrossRef] [PubMed]
33. Menorca, R.M.G.; Fussell, T.S.; Elfar, J.C. Peripheral nerve trauma: Mechanisms of injury and recovery. *Hand Clin.* **2013**, *29*, 317–330. [CrossRef]
34. Perry, V.H.; Brown, M.C.; Gordon, S. The macrophage response to central and peripheral nerve injury: A possible role for macrophages in regeneration. *J. Exp. Med.* **1987**, *165*, 1218–1223. [CrossRef]
35. Tetzlaff, W. Tight junction contact events and temporary gap junctions in the sciatic nerve fibres of the chicken during Wallerian degeneration and subsequent regeneration. *J. Neurocytol.* **1982**, *11*, 839–858. [CrossRef]
36. Sulaiman, W.; Gordon, T. Neurobiology of peripheral nerve injury, regeneration, and functional recovery: From bench top research to bedside application. *Ochsner J.* **2013**, *13*, 100–108. [PubMed]
37. Simons, M.; Trotter, J. Wrapping it up: The cell biology of myelination. *Curr. Opin. Neurobiol.* **2007**, *17*, 533–540. [CrossRef] [PubMed]
38. Arancibia-Carcamo, I.L.; Attwell, D. The node of Ranvier in CNS pathology. *Acta Neuropathol.* **2014**, *128*, 161–175. [CrossRef]
39. Jessen, K.R.; Mirsky, R. The success and failure of the Schwann cell response to nerve injury. *Front. Cell. Neurosci.* **2019**, *13*, 1–33. [CrossRef]
40. Arthur-Farraj, P.J.; Latouche, M.; Wilton, D.K.; Quintes, S.; Chabrol, E.; Banerjee, A.; Woodhoo, A.; Jenkins, B.; Rahman, M.; Turmaine, M.; et al. c-Jun reprograms Schwann cells of injured nerves to generate a repair cell essential for regeneration. *Neuron* **2012**, *75*, 633–647. [CrossRef]
41. Atanasoski, S.; Shumas, S.; Dickson, C.; Scherer, S.S.; Suter, U. Differential cyclin D1 requirements of proliferating Schwann cells during development and after injury. *Mol. Cell. Neurosci.* **2001**, *18*, 581–592. [CrossRef]
42. Kim, H.A.; Pomeroy, S.L.; Whoriskey, W.; Pawlitzky, I.; Benowitz, L.I.; Sicinski, P.; Stiles, C.D.; Roberts, T.M. A developmentally regulated switch regenerative growth of Schwann cells through cyclin D1. *Neuron* **2000**, *26*, 405–416. [CrossRef]
43. Yang, D.P.; Zhang, D.P.; Mak, K.S.; Bonder, D.E.; Scott, L.; Kim, H.A. Schwann cell proliferation during Wallerian degeneration is not necessary for regeneration and remyelination of the peripheral nerves: Axon-dependent removal of newly generated Schwann cells by apoptosis. *Mol. Cell. Neurosci.* **2008**, *38*, 80–88. [CrossRef] [PubMed]

44. Song, H.; Zhao, H.; Yang, L.; Li, L.; Zhang, T.; Pan, J.; Meng, Y.; Shen, W.; Yuan, Y. *Achyranthes bidentata* polypeptides promotes migration of Schwann cells via NOX4/DUOX2-dependent ROS production in rats. *Neurosci. Lett.* **2019**, *696*, 99–107. [CrossRef]

45. Cheng, Q.; Yuan, Y.; Sun, C.; Gu, X.; Cao, Z.; Ding, F. Neurotrophic and neuroprotective actions of *Achyranthes bidentata* polypeptides on cultured dorsal root ganglia of rats and on crushed common peroneal nerve of rabbits. *Neurosci. Lett.* **2014**, *562*, 7–12. [CrossRef] [PubMed]

46. Cheng, Q.; Jiang, C.; Wang, C.; Yu, S.; Zhang, Q.; Gu, X.; Ding, F. The *Achyranthes bidentata* polypeptide k fraction enhances neuronal growth in vitro and promotes peripheral nerve regeneration after crush injury *in vivo*. *Neural Regen. Res.* **2014**, *9*, 2142–2150. [CrossRef]

47. Wang, Y.; Shen, W.; Yang, L.; Zhao, H.; Gu, W.; Yuan, Y. The protective effects of *Achyranthes bidentata* polypeptides on rat sciatic nerve crush injury causes modulation of neurotrophic factors. *Neurochem. Res.* **2012**, *38*, 538–546. [CrossRef] [PubMed]

48. Yuan, Y.; Shen, H.; Yao, J.; Hu, N.; Ding, F.; Gu, X. The protective effects of *Achyranthes bidentata* polypeptides in an experimental model of mouse sciatic nerve crush injury. *Brain Res. Bull.* **2010**, *81*, 25–32. [CrossRef] [PubMed]

49. Ding, F.; Cheng, Q.; Gu, X. The repair effects of *Achyranthes bidentata* extract on the crushed common peroneal nerve of rabbits. *Fitoterapia* **2008**, *79*, 161–167. [CrossRef]

50. Ju, D.T.; Liao, H.E.; Shibu, M.A.; Ho, T.J.; Padma, V.V.; Tsai, F.J.; Chung, L.C.; Day, C.H.; Lin, C.C.; Huang, C.Y. Nerve Regeneration potential of protocatechuic acid in RSC96 Schwann cells by induction of cellular proliferation and migration through IGF-IR-PI3K-Akt signaling. *Chin. J. Physiol.* **2015**, *58*, 412–419. [CrossRef]

51. Chang, Y.M.; Ye, C.X.; Ho, T.J.; Tsai, T.N.; Chiu, P.L.; Tsai, C.C.; Lin, Y.M.; Kuo, C.H.; Tsai, F.J.; Tsai, C.H.; et al. *Alpinia oxyphylla* Miquel fruit extract activates MAPK-mediated signaling of PAs and MMP2/9 to induce Schwann cell migration and nerve regeneration. *Int. J. Artif. Organs* **2014**, *37*, 402–413. [CrossRef] [PubMed]

52. Zhang, X.H.; Chen, J.J. The mechanism of astragaloside IV promoting sciatic nerve regeneration. *Neural Regen. Res.* **2013**, *8*, 2256–2265. [CrossRef] [PubMed]

53. Cheng, C.Y.; Yao, C.H.; Liu, B.S.; Liu, C.J.; Chen, G.W.; Chen, Y.S. The role of astragaloside in regeneration of the peripheral nerve system. *J. Biomed. Mater. Res. A* **2006**, *76*, 463–469. [CrossRef] [PubMed]

54. Chen, Y.S.; Chen, C.C.; Chang, L.C.; Yao, C.H.; Hsu, Y.M.; Lin, J.H.; Yang, T.Y.; Chen, Y.H. Increased calcitonin gene-related peptide and macrophages are involved in *Astragalus membranaceus*-mediated peripheral nerve regeneration in rats. *Am. J. Chin. Med.* **2018**, *46*, 69–86. [CrossRef]

55. Fang, W.K.; Ko, F.Y.; Wang, H.L.; Kuo, C.H.; Chen, L.M.; Tsai, F.J.; Tsai, C.H.; Chen, Y.S.; Kuo, W.W.; Huang, C.Y. The proliferation and migration effects of huangqi on RSC96 Schwann cells. *Am. J. Chin. Med.* **2009**, *37*, 945–959. [CrossRef] [PubMed]

56. Hussin, H.M.; Lawi, M.M.; Haflah, N.H.M.; Kassim, A.Y.M.; Idrus, R.B.H.; Lokanathan, Y. *Centella asiatica* (L.)-neurodifferentiated mesenchymal stem cells promote the regeneration of peripheral nerve. *Tissue Eng. Regen. Med.* **2020**, *17*, 237–251. [CrossRef]

57. Huang, C.Y.; Kuo, W.W.; Shibu, M.A.; Hsueh, M.F.; Chen, Y.S.; Tsai, F.J.; Yao, C.H.; Lin, C.C.; Pan, L.F.; Ju, D.T. *Citrus medica* var. sarcodactylis (foshou) activates fibroblast growth factor-2 signaling to induce migration of RSC96 Schwann cells. *Am. J. Chin. Med.* **2014**, *42*, 443–452. [CrossRef] [PubMed]

58. Chen, H.T.; Tsai, Y.L.; Chen, Y.S.; Jong, G.P.; Chen, W.K.; Wang, H.L.; Tsai, F.J.; Tsai, C.H.; Lai, T.Y.; Tzang, B.S.; et al. Dangshen (*Codonopsis pilosula*) activates IGF-I and FGF-2 pathways to induce proliferation and migration effects in RSC96 Schwann cells. *Am. J. Chin. Med.* **2010**, *38*, 359–372. [CrossRef]

59. Tamaddonfard, E.; Farshid, A.A.; Ahmadian, E.; Hamidhoseyni, A. Crocin enhanced functional recovery after sciatic nerve crush injury in rats. *Iran. J. Basic Med. Sci.* **2013**, *16*, 83–90. [CrossRef] [PubMed]

60. Tehranipour, M.; Javaheri, R. Neuroprotective effect of *Curcuma longa* alcoholic extract on peripheral nerves degeneration after sciatic nerve compression in rats. *J. Biol. Sci.* **2009**, *9*, 889–893. [CrossRef]

61. Noorafshan, A.; Omidi, A.; Karbalay-Doust, S. Curcumin protects the dorsal root ganglion and sciatic nerve after crush in rat. *Pathol. Res. Pract.* **2011**, *207*, 577–582. [CrossRef]

62. Noorafshan, A.; Omidi, A.; Karbalay-Doust, S.; Aliabadi, E.; Dehghani, F. Effects of curcumin on the dorsal root ganglion structure and functional recovery after sciatic nerve crush in rat. *Micron* **2011**, *42*, 449–455. [CrossRef]

63. Ma, J.; Liu, J.; Yu, H.; Wang, Q.; Chen, Y.; Xiang, L. Curcumin promotes nerve regeneration and functional recovery in rat model of nerve crush injury. *Neurosci. Lett.* **2013**, *547*, 26–31. [CrossRef]

64. Tello Velasquez, J.; Nazareth, L.; Quinn, R.J.; Ekberg, J.A.K.; St John, J.A. Stimulating the proliferation, migration and lamellipodia of Schwann cells using low-dose curcumin. *Neuroscience* **2016**, *324*, 140–150. [CrossRef]

65. Yüce, S.; Cemal Gökçe, E.; Işkdemir, A.; Koç, E.R.; Cemil, D.B.; Gökçe, A.; Sargon, M.F. An experimental comparison of the effects of propolis, curcumin, and methylprednisolone on crush injuries of the sciatic nerve. *Ann. Plast. Surg.* **2015**, *74*, 684–692. [CrossRef] [PubMed]

66. Lee, J.M.; Namgung, U.K.; Hong, K.E. Growth-promoting activity of sanyak (*Dioscoreae rhizoma*) extract on injured sciatic nerve in rats. *JAMS J. Acupunct. Meridian Stud.* **2009**, *2*, 228–235. [CrossRef]

67. Chen, B.; Niu, S.P.; Wang, Z.Y.; Wang, Z.W.; Deng, J.X.; Zhang, P.X.; Yin, X.F.; Han, N.; Kou, Y.H.; Jiang, B.G. Local administration of icariin contributes to peripheral nerve regeneration and functional recovery. *Neural Regen. Res.* **2015**, *10*, 84–89. [CrossRef]

68. Kou, Y.; Wang, Z.; Wu, Z.; Zhang, P.; Zhang, Y.; Yin, X.; Wong, X.; Qiu, G.; Jiang, B. Epimedium extract promotes peripheral nerve regeneration in rats. *Evid. Based Complement. Altern. Med.* **2013**, *2013*, 954798. [CrossRef]

69. Liu, B.S.; Yao, C.H.; Hsu, S.H.; Yeh, T.S.; Chen, Y.S.; Kao, S.T. A novel use of genipin-fixed gelatin as extracellular matrix for peripheral nerve regeneration. *J. Biomater. Appl.* **2004**, *19*, 21–34. [CrossRef]
70. Zuo, W.; Xu, F.; Zhang, K.; Zheng, L.; Zhao, J. Proliferation-enhancing effects of gastrodin on RSC96 Schwann cells by regulating ERK1/2 and PI3K signaling pathways. *Biomed. Pharmacother.* **2016**, *84*, 747–753. [CrossRef] [PubMed]
71. Zhu, Z.; Zhou, X.; He, B.; Dai, T.; Zheng, C.; Yang, C.; Zhu, S.; Zhu, J.; Zhu, Q.; Liu, X. *Ginkgo biloba* extract (EGb 761) promotes peripheral nerve regeneration and neovascularization after acellular nerve allografts in a rat model. *Cell. Mol. Neurobiol.* **2015**, *35*, 273–282. [CrossRef]
72. Hsu, S.H.; Chang, C.J.; Tang, C.M.; Lin, F.T. In vitro and in vivo effects of *Ginkgo biloba* extract EGb 761 on seeded Schwann cells within poly (DL-lactic acid-co-glycolic acid) conduits for peripheral nerve regeneration. *J. Biomater. Appl.* **2004**, *19*, 163–182. [CrossRef]
73. Lu, M.C.; Lai, T.Y.; Hwang, J.M.; Chen, H.T.; Chang, S.H.; Tsai, F.J.; Wang, H.L.; Lin, C.C.; Kuo, W.W.; Huang, C.Y. Proliferation- and migration-enhancing effects of ginseng and ginsenoside Rg1 through IGF-I- and FGF-2-signaling pathways on RSC96 Schwann cells. *Cell Biochem. Funct.* **2009**, *27*, 186–192. [CrossRef]
74. Ma, J.; Li, W.; Tian, R.; Lei, W. Ginsenoside Rg1 promotes peripheral nerve regeneration in rat model of nerve crush injury. *Neurosci. Lett.* **2010**, *478*, 66–71. [CrossRef]
75. Wang, L.; Yuan, D.; Zhang, D.; Zhang, W.; Liu, C.; Cheng, H.; Song, Y.; Tan, Q. Ginsenoside Re promotes nerve regeneration by facilitating the proliferation, differentiation and migration of Schwann cells via the ERK- and JNK-dependent pathway in rat model of sciatic nerve crush injury. *Cell. Mol. Neurobiol.* **2015**, *35*, 827–840. [CrossRef]
76. Renno, W.M.; Al-Maghrebi, M.; Al-Banaw, A. (-)-Epigallocatechin-3-gallate (EGCG) attenuates functional deficits and morphological alterations by diminishing apoptotic gene overexpression in skeletal muscles after sciatic nerve crush injury. *Naunyn. Schmiedebergs. Arch. Pharmacol.* **2012**, *385*, 807–822. [CrossRef]
77. Renno, W.M.; Al-Maghrebi, M.; Alshammari, A.; George, P. (-)-Epigallocatechin-3-gallate (EGCG) attenuates peripheral nerve degeneration in rat sciatic nerve crush injury. *Neurochem. Int.* **2013**, *62*, 221–231. [CrossRef] [PubMed]
78. Lin, Y.; Jiang, X.; Yin, G.; Lin, H. Syringic acid promotes proliferation and migration of Schwann cells via down-regulating MiR-451-5p. *Acta Biochim. Biophys. Sin. (Shanghai)* **2019**, *51*, 1198–1207. [CrossRef] [PubMed]
79. Liu, B.; Liu, Y.; Yang, G.; Xu, Z.; Chen, J. Ursolic acid induces neural regeneration after sciatic nerve injury. *Neural Regen. Res.* **2013**, *8*, 2510–2519. [CrossRef] [PubMed]
80. Wang, J.; Tian, L.; He, L.; Chen, N.; Ramakrishna, S.; So, K.F.; Mo, X. *Lycium barbarum* polysaccharide encapsulated poly lactic-co-glycolic acid nanofibers: Cost effective herbal medicine for potential application in peripheral nerve tissue engineering. *Sci. Rep.* **2018**, *8*, 8669. [CrossRef] [PubMed]
81. Wang, W.; Huang, C.Y.; Tsai, F.J.; Tsai, C.C.; Yao, C.H.; Chen, Y.S. Growth-promoting effects of quercetin on peripheral nerves in rats. *Int. J. Artif. Organs* **2011**, *34*, 1095–1105. [CrossRef]
82. Lu, M.; Yi, T.; Xiong, Y.; Wang, Q.; Yin, N. Cortex Mori Radicis extract promotes neurite outgrowth in diabetic rats by activating PI3K/AKT signaling and inhibiting Ca2+ influx associated with the upregulation of transient receptor potential canonical channel 1. *Mol. Med. Rep.* **2020**, *21*, 320–328. [CrossRef]
83. Hsiang, S.W.; Lee, H.C.; Tsai, F.J.; Tsai, C.C.; Yao, C.H.; Chen, Y.S. Puerarin accelerates peripheral nerve regeneration. *Am. J. Chin. Med.* **2011**, *39*, 1207–1217. [CrossRef]
84. Chen, H.-T.; Yao, C.-H.; Chao, P.-D.L.; Hou, Y.-C.; Chiang, H.-M.; Hsieh, C.-C.; Ke, C.-J.; Chen, Y.-S. Effect of serum metabolites of *Pueraria lobata* in rats on peripheral nerve regeneration: In vitro and in vivo studies. *J. Biomed. Mater. Res. Part B Appl. Biomater.* **2007**, *84B*, 256–262. [CrossRef]
85. Wang, Z.; Zhang, P.; Kou, Y.; Yin, X.; Han, N.; Jiang, B. Hedysari extract improves regeneration after peripheral nerve injury by enhancing the amplification effect. *PLoS ONE* **2013**, *8*, e67921. [CrossRef]
86. Wei, S.Y.; Zhang, P.X.; Han, N.; Dang, Y.; Zhang, H.B.; Zhang, D.Y.; Fu, Z.G.; Jiang, B.G. Effects of Hedysari polysaccharides on regeneration and function recovery following peripheral nerve injury in rats. *Am. J. Chin. Med.* **2009**, *37*, 57–67. [CrossRef]
87. Sheng, Q.S.; Wang, Z.J.; Zhang, J.; Zhang, Y.G. Salidroside promotes peripheral nerve regeneration following crush injury to the sciatic nerve in rats. *Neuroreport* **2013**, *24*, 217–223. [CrossRef]
88. Zuo, W.; Wu, H.; Zhang, K.; Lv, P.; Xu, F.; Jiang, W.; Zheng, L.; Zhao, J. Baicalin promotes the viability of Schwann cells in vitro by regulating neurotrophic factors. *Exp. Ther. Med.* **2017**, *14*, 507–514. [CrossRef] [PubMed]
89. Morani, A.S.; Bodhankar, S.L.; Mohan, V.; Thakurdesai, P.A. Ameliorative effects of standardized extract from *Trigonella foenum-graecum* L. seeds on painful peripheral neuropathy in rats. *Asian Pac. J. Trop. Med.* **2012**, *5*, 385–390. [CrossRef]
90. Zhang, Y.G.; Sheng, Q.S.; Wang, H.K.; Lv, L.; Zhang, J.; Chen, J.M.; Xu, H. Triptolide improves nerve regeneration and functional recovery following crush injury to rat sciatic nerve. *Neurosci. Lett.* **2014**, *561*, 198–202. [CrossRef] [PubMed]
91. Naik, A.K.; Latham, J.R.; Obradovic, A.; Jevtovic-Todorovic, V. Dorsal root ganglion application of muscimol prevents hyperalgesia and stimulates myelin protein expression after sciatic nerve injury in rats. *Anesth. Analg.* **2012**, *114*, 674–682. [CrossRef]
92. Wong, K.H.; Kanagasabapathy, G.; Naidu, M.; David, P.; Sabaratnam, V. *Hericium erinaceus* (Bull.: Fr.) Pers., a medicinal mushroom, activates peripheral nerve regeneration. *Chin. J. Integr. Med.* **2016**, *22*, 759–767. [CrossRef]
93. Wong, K.H.; Kanagasabapathy, G.; Bakar, R.; Phan, C.W.; Sabaratnam, V. Restoration of sensory dysfunction following peripheral nerve injury by the polysaccharide from culinary and medicinal mushroom, *Hericium erinaceus* (Bull.: Fr.) Pers. through its neuroregenerative action. *Food Sci. Technol.* **2015**, *35*, 712–721. [CrossRef]

94. Wong, K.H.; Naidu, M.; David, P.; Abdulla, M.A.; Abdullah, N.; Kuppusamy, U.R.; Sabaratnam, V. Peripheral nerve regeneration following crush injury to rat peroneal nerve by aqueous extract of medicinal mushroom *Hericium erinaceus* (Bull.: Fr) Pers. (Aphyllophoromycetideae). *Evid. Based Complement. Altern. Med.* **2011**, *2011*, 580752. [CrossRef] [PubMed]

95. Farha, M.; Parkianathan, L.; Amir, N.A.I.A.; Sabaratnam, V.; Wong, K.H. Functional recovery enhancement by tiger milk mushroom, *Lignosus rhinocerotis* in a sciatic nerve crush injury model and morphological study of its neurotoxicity. *J. Anim. Plant Sci.* **2019**, *29*, 930–942.

96. Kim, K.J.; Namgung, U.; Cho, C.S. Protective effects of Bogijetong decoction and its selected formula on neuropathic insults in streptozotocin-induced diabetic animals. *Evid. Based Complement. Altern. Med.* **2017**, *2017*, 4296318. [CrossRef] [PubMed]

97. Chen, Y.S.; Yao, C.H.; Chen, T.H.; Hsieh, C.L.; Lao, C.J.; Tsai, C.C. Effect of Buyang Huanwu decoction on peripheral nerve regeneration using silicone rubber chambers. *Am. J. Chin. Med.* **2001**, *29*, 423–432. [CrossRef] [PubMed]

98. Wang, Y.; Chen, Z.; Ye, R.; He, Y.; Li, Y.; Qiu, X. Protective effect of Jiaweibugan decoction against diabetic peripheral neuropathy. *Neural Regen. Res.* **2013**, *8*, 1113–1121. [CrossRef]

99. Wang, Z.Y.; Qin, L.H.; Zhang, W.G.; Zhang, P.X.; Jiang, B.G. Qian-Zheng-San promotes regeneration after sciatic nerve crush injury in rats. *Neural Regen. Res.* **2019**, *14*, 683–691. [CrossRef]

100. Nocera, G.; Jacob, C. Mechanisms of Schwann cell plasticity involved in peripheral nerve repair after injury. *Cell Mol. Life Sci.* **2020**, *77*, 3977–3989. [CrossRef]

101. Srinivasulu, C.; Ramgopal, M.; Ramanjaneyulu, G.; Anuradha, C.M.; Suresh Kumar, C. Syringic acid (SA) - A review of its occurrence, biosynthesis, pharmacological and industrial importance. *Biomed. Pharmacother.* **2018**, *108*, 547–557. [CrossRef]

102. Baliga, M.S.; Shivashankara, A.R.; Venkatesh, S.; Bhat, H.P.; Palatty, P.L.; Bhandari, G.; Rao, S. Phytochemicals in the prevention of ethanol-induced hepatotoxicity: A revisit. In *Dietary Interventions in Liver Disease: Foods, Nutrients, and Dietary Supplements*; Watson, R.R., Preedy, V.R., Eds.; Academic Press: Cambridge, MA, USA, 2019; pp. 79–89. ISBN 978-0-12-814466-4.

103. Li, Y.; Yao, J.; Han, C.; Yang, J.; Chaudhry, M.T.; Wang, S.; Liu, H.; Yin, Y. Quercetin, inflammation and immunity. *Nutrients* **2016**, *8*, 167. [CrossRef]

104. Desbarats, J.; Birge, R.B.; Mimouni-Rongy, M.; Weinstein, D.E.; Palerme, J.S.; Newell, M.K. Fas engagement induces neurite growth through ERK activation and P35 upregulation. *Nat. Cell Biol.* **2003**, *5*, 118–125. [CrossRef] [PubMed]

105. Schneider-Schaulies, J.; Kirchhoff, F.; Archelos, J.; Schachner, M. Down-regulation of myelin-associated glycoprotein on Schwann cells by interferon-γ and tumor necrosis factor-α affects neurite outgrowth. *Neuron* **1991**, *7*, 995–1005. [CrossRef]

106. Larsson, K.; Rydevik, B.; Olmarker, K. Disc related cytokines inhibit axonal outgrowth from dorsal root ganglion cells *in vitro*. *Spine* **2005**, *30*, 621–624. [CrossRef] [PubMed]

107. Kato, K.; Liu, H.; Kikuchi, S.I.; Myers, R.R.; Shubayev, V.I. Immediate anti-tumor necrosis factor-α (etanercept) therapy enhances axonal regeneration after sciatic nerve crush. *J. Neurosci. Res.* **2010**, *88*, 360–368. [CrossRef] [PubMed]

108. Newbern, J.M.; Li, X.; Shoemaker, S.E.; Zhou, J.; Zhong, J.; Wu, Y.; Bonder, D.; Hollenback, S.; Coppola, G.; Geschwind, D.H.; et al. Specific functions for ERK/MAPK signaling during PNS development. *Neuron* **2011**, *69*, 91–105. [CrossRef]

109. Ishii, A.; Furusho, M.; Bansal, R. Sustained activation of ERK1/2 MAPK in oligodendrocytes and Schwann cells enhances myelin growth and stimulates oligodendrocyte progenitor expansion. *J. Neurosci.* **2013**, *33*, 175–186. [CrossRef]

110. Harrisingh, M.C.; Perez-Nadales, E.; Parkinson, D.B.; Malcolm, D.S.; Mudge, A.W.; Lloyd, A.C. The Ras/Raf/ERK signalling pathway drives Schwann cell dedifferentiation. *EMBO J.* **2004**, *23*, 3061–3071. [CrossRef] [PubMed]

111. Agthong, S.; Kaewsema, A.; Tanomsridejchai, N.; Chentanez, V. Activation of MAPK ERK in peripheral nerve after injury. *BMC Neurosci.* **2006**, *7*, 45. [CrossRef] [PubMed]

112. Yamazaki, T.; Sabit, H.; Oya, T.; Ishii, Y.; Hamashima, T.; Tokunaga, A.; Ishizawa, S.; Jie, S.; Kurashige, Y.; Matsushima, T.; et al. Activation of MAP kinases, Akt and PDGF receptors in injured peripheral nerves. *J. Peripher. Nerv. Syst.* **2009**, *14*, 165–176. [CrossRef]

113. Huang, H.; Sun, Z.; Liu, H.; Ma, J.; Hu, M. ERK/MAPK and PI3K/AKT signal channels simultaneously activated in nerve cell and axon after facial nerve injury. *Saudi J. Biol. Sci.* **2017**, *24*, 1853–1858. [CrossRef] [PubMed]

114. Yang, D.P.; Kim, J.; Syed, N.; Tung, Y.J.; Bhaskaran, A.; Mindos, T.; Mirsky, R.; Jessen, K.R.; Maurel, P.; Parkinson, D.B.; et al. P38 MAPK activation promotes denervated Schwann cell phenotype and functions as a negative regulator of Schwann cell differentiation and myelination. *J. Neurosci.* **2012**, *32*, 7158–7168. [CrossRef]

115. Pines, J. Four-dimensional control of the cell cycle. *Nat. Cell Biol.* **1999**, *153*, E73. [CrossRef] [PubMed]

116. Lim, S.; Kaldis, P. Cdks, cyclins and CKIs: Roles beyond cell cycle regulation. *Development* **2013**, *140*, 3079–3093. [CrossRef]

117. Mairet-Coello, G.; Tury, A.; DiCicco-Bloom, E. Insulin-like growth factor-1 promotes G1/S cell cycle progression through bidirectional regulation of cyclins and cyclin-dependent kinase inhibitors via the phosphatidylinositol 3-kinase/Akt pathway in developing rat cerebral cortex. *J. Neurosci.* **2009**, *29*, 775–788. [CrossRef] [PubMed]

118. Syroid, D.E.; Zorick, T.S.; Arbet-Engels, C.; Kilpatrick, T.J.; Eckhart, W.; Lemke, G. A role for insulin-like growth factor-I in the regulation of Schwann cell survival. *J. Neurosci.* **1999**, *19*, 2059–2068. [CrossRef] [PubMed]

119. Hansson, H.A.; Dahlin, L.B.; Danielsen, N.; Fryklund, L.; Nachemson, A.K.; Polleryd, P.; Rozell, B.; Skottner, A.; Stemme, S.; Lundborg, G. Evidence indicating trophic importance of IGF-I in regenerating peripheral nerves. *Acta Physiol. Scand.* **1986**, *126*, 609–614. [CrossRef]

120. Cheng, H.L.; Randolph, A.; Yee, D.; Delafontaine, P.; Tennekoon, G.; Feldman, E.L. Characterization of insulin-like growth factor-I and its receptor and binding proteins in transected nerves and cultured Schwann cells. *J. Neurochem.* **1996**, *66*, 525–536. [CrossRef]

121. Apel, P.J.; Ma, J.; Callahan, M.; Northam, C.N.; Alton, T.B.; Sonntag, W.E.; Li, Z. Effect of locally delivered IGF-1 on nerve regeneration during aging: An experimental study in rats. *Muscle Nerve* **2010**, *41*, 335–341. [CrossRef]

122. Zhu, H.; Xue, C.; Yao, M.; Wang, H.; Zhang, P.; Qian, T.; Zhou, S.; Li, S.; Yu, B.; Wang, Y.; et al. MIR-129 controls axonal regeneration via regulating insulin-like growth factor-1 in peripheral nerve injury. *Cell Death Dis.* **2018**, *9*, 720. [CrossRef]

123. Sowndhararajan, K.; Deepa, P.; Kim, M.; Park, S.J.; Kim, S. Neuroprotective and cognitive enhancement potentials of baicalin: A review. *Brain Sci.* **2018**, *8*, 104. [CrossRef]

124. Stankoff, B.; Aigrot, M.-S.; Noël, F.; Wattilliaux, A.; Zalc, B.; Lubetzki, C. Ciliary neurotrophic factor (CNTF) enhances myelin formation: A novel role for CNTF and CNTF-related molecules. *J. Neurosci.* **2002**, *22*, 9221–9227. [CrossRef]

125. Xiao, J.; Hughes, R.A.; Lim, J.Y.; Wong, A.W.; Ivanusic, J.J.; Ferner, A.H.; Kilpatrick, T.J.; Murray, S.S. A small peptide mimetic of brain-derived neurotrophic factor promotes peripheral myelination. *J. Neurochem.* **2013**, *125*, 386–398. [CrossRef] [PubMed]

126. Wilhelm, J.C.; Xu, M.; Cucoranu, D.; Chmielewski, S.; Holmes, T.; Lau, K.S.; Bassell, G.J.; English, A.W. Cooperative roles of BDNF expression in neurons and Schwann cells are modulated by exercise to facilitate nerve regeneration. *J. Neurosci.* **2012**, *32*, 5002–5009. [CrossRef]

127. Tannemaat, M.R.; Eggers, R.; Hendriks, W.T.; De Ruiter, G.C.W.; Van Heerikhuize, J.J.; Pool, C.W.; Malessy, M.J.A.; Boer, G.J.; Verhaagen, J. Differential effects of lentiviral vector-mediated overexpression of nerve growth factor and glial cell line-derived neurotrophic factor on regenerating sensory and motor axons in the transected peripheral nerve. *Eur. J. Neurosci.* **2008**, *28*, 1467–1479. [CrossRef] [PubMed]

128. Yun, Y.R.; Won, J.E.; Jeon, E.; Lee, S.; Kang, W.; Jo, H.; Jang, J.H.; Shin, U.S.; Kim, H.W. Fibroblast growth factors: Biology, function, and application for tissue regeneration. *J. Tissue Eng.* **2010**, *1*, 218142. [CrossRef]

129. Holland, E.C.; Varmus, H.E. Basic fibroblast growth factor induces cell migration and proliferation after glia-specific gene transfer in mice. *Proc. Natl. Acad. Sci. USA* **1998**, *95*, 1218–1223. [CrossRef] [PubMed]

130. Hossain, W.A.; Morest, D.K. Fibroblast growth factors (FGF-1, FGF-2) promote migration and neurite growth of mouse cochlear ganglion cells *in vitro*: Immunohistochemistry and antibody perturbation. *J. Neurosci. Res.* **2000**, *62*, 40–55. [CrossRef]

131. Chen, B.; Hu, R.; Min, Q.; Li, Y.; Parkinson, D.B.; Dun, X. FGF5 regulates Schwann cell migration and adhesion. *Front. Cell. Neurosci.* **2020**, *14*, 1–12. [CrossRef]

132. Li, J.; Wuliji, O.; Li, W.; Jiang, Z.G.; Ghanbari, H.A. Oxidative stress and neurodegenerative disorders. *Int. J. Mol. Sci.* **2013**, *14*, 24438–24475. [CrossRef]

133. Areti, A.; Yerra, V.G.; Naidu, V.G.M.; Kumar, A. Oxidative stress and nerve damage: Role in chemotherapy induced peripheral neuropathy. *Redox Biol.* **2014**, *2*, 289–295. [CrossRef]

134. Li, Y.; Shi, Z.; Yu, X.; Feng, P.; Wang, X.-J. The effects of urotensin II on migration and invasion are mediated by NADPH oxidase-derived reactive oxygen species through the c-Jun N-terminal kinase pathway in human hepatoma cells. *Peptides* **2017**, *88*, 106–114. [CrossRef]

135. Ma, S.; Fu, A.; Lim, S.; Chiew, G.G.Y.; Luo, K.Q. MnSOD mediates shear stress-promoted tumor cell migration and adhesion. *Free Radic. Biol. Med.* **2018**, *129*, 46–58. [CrossRef]

136. Yu, J.S.L.; Cui, W. Proliferation, survival and metabolism: The role of PI3K/AKT/MTOR signalling in pluripotency and cell fate determination. *Development* **2016**, *143*, 3050–3060. [CrossRef]

137. Anitha, M.; Gondha, C.; Sutliff, R.; Parsadanian, A.; Mwangi, S.; Sitaraman, S.V.; Srinivasan, S. GDNF rescues hyperglycemia-induced diabetic enteric neuropathy through activation of the PI3K/Akt pathway. *J. Clin. Invest.* **2006**, *116*, 344–356. [CrossRef] [PubMed]

138. Chen, L.; Gong, H.Y.; Xu, L. PVT1 protects diabetic peripheral neuropathy via PI3K/AKT pathway. *Eur. Rev. Med. Pharmacol. Sci.* **2018**, *22*, 6905–6911. [CrossRef] [PubMed]

139. Shim, S.; Ming, G.L. Roles of channels and receptors in the growth cone during PNS axonal regeneration. *Exp. Neurol.* **2010**, *223*, 38–44. [CrossRef]

140. Wu, D.; Huang, W.; Richardson, P.M.; Priestley, J.V.; Liu, M. TRPC4 in rat dorsal root ganglion neurons is increased after nerve injury and is necessary for neurite outgrowth. *J. Biol. Chem.* **2008**, *283*, 416–426. [CrossRef]

141. Heo, D.K.; Chung, W.Y.; Park, H.W.; Yuan, J.P.; Lee, M.G.; Kim, J.Y. Opposite regulatory effects of TRPC1 and TRPC5 on neurite outgrowth in PC12 cells. *Cell. Signal.* **2012**, *24*, 899–906. [CrossRef] [PubMed]

142. Du, Y.; Tang, J.; Li, G.; Berti-Mattera, L.; Lee, C.A.; Bartkowski, D.; Gale, D.; Monahan, J.; Niesman, M.R.; Alton, G.; et al. Effects of P38 MAPK inhibition on early stages of diabetic retinopathy and sensory nerve function. *Investig. Ophthalmol. Vis. Sci.* **2010**, *51*, 2158–2164. [CrossRef] [PubMed]

143. Suryavanshi, S.V.; Kulkarni, Y.A. NF-Kβ: A potential target in the management of vascular complications of diabetes. *Front. Pharmacol.* **2017**, *8*, 798. [CrossRef] [PubMed]

144. Ruff, C.A.; Staak, N.; Patodia, S.; Kaswich, M.; Rocha-Ferreira, E.; Da Costa, C.; Brecht, S.; Makwana, M.; Fontana, X.; Hristova, M.; et al. Neuronal c-Jun is required for successful axonal regeneration, but the effects of phosphorylation of its N-terminus are moderate. *J. Neurochem.* **2012**, *121*, 607–618. [CrossRef]

145. Kato, N.; Matsumoto, M.; Kogawa, M.; Atkins, G.J.; Findlay, D.M.; Fujikawa, T.; Oda, H.; Ogata, M. Critical role of P38 MAPK for regeneration of the sciatic nerve following crush injury *in vivo*. *J. Neuroinflammation* **2013**, *10*, 757. [CrossRef]

146. Sang, Q.; Sun, D.; Chen, Z.; Zhao, W. NGF and PI3K/Akt signaling participate in the ventral motor neuronal protection of curcumin in sciatic nerve injury rat models. *Biomed. Pharmacother.* **2018**, *103*, 1146–1153. [CrossRef]

147. Holahan, M.R. A shift from a pivotal to supporting role for the growth-associated protein (GAP-43) in the coordination of axonal structural and functional plasticity. *Front. Cell. Neurosci.* **2017**, *11*, 266. [CrossRef] [PubMed]
148. Benowitz, L.I.; Routtenberg, A. GAP-43: An intrinsic determinant of neuronal development and plasticity. *Trends Neurosci.* **1997**, *20*, 84–91. [CrossRef]
149. Tang, X.; Chen, Y.; Gu, X.; Ding, F. *Achyranthes bidentata* Blume extract promotes neuronal growth in cultured embryonic rat hippocampal neurons. *Prog. Nat. Sci.* **2009**, *19*, 549–555. [CrossRef]
150. Fregnan, F.; Muratori, L.; Simões, A.R.; Giacobini-Robecchi, M.G.; Raimondo, S. Role of inflammatory cytokines in peripheral nerve injury. *Neural Regen. Res.* **2012**, *7*, 2259–2266. [CrossRef] [PubMed]
151. Li, J.; Wei, G.H.; Huang, H.; Lan, Y.P.; Liu, B.; Liu, H.; Zhang, W.; Zuo, Y.X. Nerve injury-related autoimmunity activation leads to chronic inflammation and chronic neuropathic pain. *Anesthesiology* **2013**, *118*, 416–429. [CrossRef] [PubMed]
152. Hobson, M.I.; Gree, C.J.; Terenghi, G. VEGF enhances intraneural angiogenesis and improves nerve regeneration after axotomy. *J. Anat.* **2000**, *197*, 591–605. [CrossRef]
153. Fang, Z.; Ge, X.; Chen, X.; Xu, Y.; Yuan, W.E.; Ouyang, Y. Enhancement of sciatic nerve regeneration with dual delivery of vascular endothelial growth factor and nerve growth factor genes. *J. Nanobiotechnol.* **2020**, *18*, 46. [CrossRef] [PubMed]
154. Lee, J.; Shin, J.E.; Lee, B.; Kim, H.; Jeon, Y.; Ahn, S.H.; Chi, S.W.; Cho, Y. The stem cell marker *Prom1* promotes axon regeneration by down-regulating cholesterol synthesis via Smad signaling. *Proc. Natl. Acad. Sci. USA* **2020**, *117*, 15955–15966. [CrossRef]
155. Jankowski, M.P.; McIlwrath, S.L.; Jing, X.; Cornuet, P.K.; Salerno, K.M.; Koerber, H.R.; Albers, K.M. Sox11 Transcription factor modulates peripheral nerve regeneration in adult mice. *Brain Res.* **2009**, *1256*, 43–54. [CrossRef]
156. Leibinger, M.; Müller, A.; Gobrecht, P.; Diekmann, H.; Andreadaki, A.; Fischer, D. Interleukin-6 contributes to CNS axon regeneration upon inflammatory stimulation. *Cell Death Dis.* **2013**, *4*, e609. [CrossRef] [PubMed]
157. Cox, A.A.; Sagot, Y.; Hedou, G.; Grek, C.; Wilkes, T.; Vinik, A.I.; Ghatnekar, G. Low-dose pulsatile interleukin-6 as a treatment option for diabetic peripheral neuropathy. *Front. Endocrinol. (Lausanne)* **2017**, *8*, 89. [CrossRef]
158. Snipes, G.J.; Suter, U.; Welcher, A.A.; Shooter, E.M. Characterization of a novel peripheral nervous system myelin protein (PMP-22/SR13). *J. Cell Biol.* **1992**, *117*, 225–238. [CrossRef]
159. Kuhn, G.; Lie, A.; Wilms, S.; Müller, H.W. Coexpression of PMP22 gene with MBP and P0 during *de novo* myelination and nerve repair. *Glia* **1993**, *8*, 256–264. [CrossRef]
160. Martin, L.J.; Kaiser, A.; Price, A.C. Motor neuron degeneration after sciatic nerve avulsion in adult rat evolves with oxidative stress and is apoptosis. *J. Neurobiol.* **1999**, *40*, 185–201. [CrossRef]
161. Liu, Y.; Wang, H. Peripheral nerve injury induced changes in the spinal cord and strategies to counteract/enhance the changes to promote nerve regeneration. *Neural Regen. Res.* **2020**, *15*, 189–198. [CrossRef]
162. Amiri, S.; Movahedin, M.; Mowla, S.J.; Hajebrahimi, Z.; Tavallaei, M. Differential gene expression and alternative splicing of survivin following mouse sciatic nerve injury. *Spinal Cord* **2009**, *47*, 739–744. [CrossRef] [PubMed]
163. Liu, Z.; Zhu, S.; Liu, L.; Ge, J.; Huang, L.; Sun, Z.; Zeng, W.; Huang, J.; Luo, Z. A magnetically responsive nanocomposite scaffold combined with Schwann cells promotes sciatic nerve regeneration upon exposure to magnetic field. *Int. J. Nanomedicine* **2017**, *12*, 7815–7832. [CrossRef] [PubMed]
164. Lundborg, G.; Dahlin, L.B.; Danielsen, N.; Gelberman, R.H.; Longo, F.M.; Powell, H.C.; Varon, S. Nerve regeneration in silicone chambers: Influence of gap length and of distal stump components. *Exp. Neurol.* **1982**, *76*, 361–375. [CrossRef]
165. Chang, Y.C.; Chen, M.H.; Liao, S.Y.; Wu, H.C.; Kuan, C.H.; Sun, J.S.; Wang, T.W. Multichanneled nerve guidance conduit with spatial gradients of neurotrophic factors and oriented nanotopography for repairing the peripheral nervous system. *ACS Appl. Mater. Interfaces* **2017**, *9*, 37623–37636. [CrossRef] [PubMed]
166. Vijayavenkataraman, S. Nerve guide conduits for peripheral nerve injury repair: A review on design, materials and fabrication methods. *Acta Biomater.* **2020**, *106*, 54–69. [CrossRef]
167. Fathi, S.S.; Zaminy, A. Stem cell therapy for nerve injury. *World J. Stem Cells* **2017**, *9*, 144–151. [CrossRef]
168. Willand, M.P.; Nguyen, M.A.; Borschel, G.H.; Gordon, T. Electrical stimulation to promote peripheral nerve regeneration. *Neurorehabil. Neural Repair* **2016**, *30*, 490–496. [CrossRef]
169. Turner, P.V.; Brabb, T.; Pekow, C.; Vasbinder, M.A. Administration of substances to laboratory animals: Routes of administration and factors to consider. *J. Am. Assoc. Lab. Anim. Sci.* **2011**, *50*, 600–613.
170. Katiyar, C.; Kanjilal, S.; Gupta, A.; Katiyar, S. Drug discovery from plant sources: An integrated approach. *AYU (An. Int. Q. J. Res. Ayurveda)* **2012**, *33*, 10–19. [CrossRef] [PubMed]
171. Atanasov, A.G.; Waltenberger, B.; Pferschy-Wenzig, E.M.; Linder, T.; Wawrosch, C.; Uhrin, P.; Temml, V.; Wang, L.; Schwaiger, S.; Heiss, E.H.; et al. Discovery and resupply of pharmacologically active plant-derived natural products: A review. *Biotechnol. Adv.* **2015**, *33*, 1582–1614. [CrossRef]
172. Singh, S.; Sharma, B.; Kanwar, S.S.; Kumar, A. Lead phytochemicals for anticancer drug development. *Front. Plant Sci.* **2016**, *7*, 1667. [CrossRef] [PubMed]
173. Lu, C.C.; Hsu, Y.J.; Chang, C.J.; Lin, C.S.; Martel, J.; Ojcius, D.M.; Ko, Y.F.; Lai, H.C.; Young, J.D. Immunomodulatory properties of medicinal mushrooms: Differential effects of water and ethanol extracts on NK cell-mediated cytotoxicity. *Innate Immun.* **2016**, *22*, 522–533. [CrossRef]
174. Seow, S.L.-S.; Naidu, M.; David, P.; Wong, K.H.; Sabaratnam, V. Potentiation of neuritogenic activity of medicinal mushrooms in rat pheochromocytoma cells. *BMC Complement. Altern. Med.* **2013**, *13*, 157. [CrossRef]

175. Wells, M.L.; Potin, P.; Craigie, J.S.; Raven, J.A.; Merchant, S.S.; Helliwell, K.E.; Smith, A.G.; Camire, M.E.; Brawley, S.H. Algae as nutritional and functional food sources: Revisiting our understanding. *J. Appl. Phycol.* **2017**, *29*, 949–982. [CrossRef]
176. Khan, M.I.; Shin, J.H.; Kim, J.D. The promising future of microalgae: Current status, challenges, and optimization of a sustainable and renewable industry for biofuels, feed, and other products. *Microb. Cell Fact.* **2018**, *17*, 1–21. [CrossRef]
177. Ariede, M.B.; Candido, T.M.; Jacome, A.L.M.; Velasco, M.V.R.; de Carvalho, J.C.M.; Baby, A.R. Cosmetic attributes of algae—A review. *Algal Res.* **2017**, *25*, 483–487. [CrossRef]
178. Borowitzka, M.A. Microalgae as sources of pharmaceuticals and other biologically active compounds. *J. Appl. Phycol.* **1995**, *7*, 3–15. [CrossRef]
179. Shannon, E.; Abu-Ghannam, N. Antibacterial derivatives of marine algae: An overview of pharmacological mechanisms and applications. *Mar. Drugs* **2016**, *14*, 81. [CrossRef]
180. Barbosa, M.; Valentão, P.; Andrade, P.B. Bioactive compounds from macroalgae in the new millennium: Implications for neurodegenerative diseases. *Mar. Drugs* **2014**, *12*, 4934–4972. [CrossRef] [PubMed]
181. Olasehinde, T.A.; Olaniran, A.O.; Okoh, A.I. Macroalgae as a valuable source of naturally occurring bioactive compounds for the treatment of Alzheimer's disease. *Mar. Drugs* **2019**, *17*, 609. [CrossRef] [PubMed]
182. Hannan, M.A.; Mohibbullah, M.; Hwang, S.Y.; Lee, K.; Kim, Y.C.; Hong, Y.K.; Moon, I.S. Differential neuritogenic activities of two edible brown macroalgae, *Undaria pinnatifida* and *Saccharina japonica*. *Am. J. Chin. Med.* **2014**, *42*, 1371–1384. [CrossRef] [PubMed]
183. Mohibbullah, M.; Bhuiyan, M.M.H.; Hannan, M.A.; Getachew, P.; Hong, Y.K.; Choi, J.S.; Choi, I.S.; Moon, I.S. The edible red alga *Porphyra yezoensis* promotes neuronal survival and cytoarchitecture in primary hippocampal neurons. *Cell. Mol. Neurobiol.* **2016**, *36*, 669–682. [CrossRef] [PubMed]
184. Tirtawijaya, G.; Mohibbullah, M.; Meinita, M.D.N.; Moon, I.S.; Hong, Y.K. The ethanol extract of the rhodophyte *Kappaphycus alvarezii* promotes neurite outgrowth in hippocampal neurons. *J. Appl. Phycol.* **2016**, *28*, 2515–2522. [CrossRef]
185. Pang, J.R.; Goh, V.M.J.; Tan, C.Y.; Phang, S.M.; Wong, K.H.; Yow, Y.Y. Neuritogenic and in vitro antioxidant activities of Malaysian *Gracilaria manilaensis* Yamamoto & Trono. *J. Appl. Phycol.* **2018**, *30*, 3253–3260. [CrossRef]

MDPI

St. Alban-Anlage 66

4052 Basel

Switzerland

Tel. +41 61 683 77 34

Fax +41 61 302 89 18

www.mdpi.com

Cells Editorial Office

E-mail: cells@mdpi.com

www.mdpi.com/journal/cells